2nd International Scientific and Practical Conference on Innovations in Engineering and Technology (ISPCIET 2019)

IOP Conference Series: Materials Science and Engineering
Volume 656

Veliky Novgorod, Russia
27 - 28 June 2019

Editor:

Sergey B. Sapozhkov

ISBN: 978-1-7138-0913-5
ISSN: 1757-8981

Printed from e-media with permission by:

Curran Associates, Inc.
57 Morehouse Lane
Red Hook, NY 12571

CURRAN ASSOCIATES INC.
proceedings
.com

Some format issues inherent in the e-media version may also appear in this print version.

Institute of Physics
Dirac House, Temple Back
Bristol BS1 6BE UK

Phone: 44 1 17 929 7481
Fax: 44 1 17 920 0979

techtracking@iop.org

Additional copies of this publication are available from:

Curran Associates, Inc.
57 Morehouse Lane
Red Hook, NY 12571 USA
Phone: 845-758-0400
Fax: 845-758-2633
Email: curran@proceedings.com
Web: www.proceedings.com

TABLE OF CONTENTS

IOP Conf. Series: Materials Science and Engineering **656** (2019) 011001 doi:10.1088/1757-899X/656/1/011001

Preface

The conference serves as a platform for discussion of advanced and innovative approaches to developing convergent solutions in engineering and technology and also in interdisciplinary scientific researches.

ISPCIET'2019 key topics rely on the principle of applicability, performance, and prospects for the development of presented innovative solutions for material science and mechanical engineering.

EDITOR

Sergey B. Sapozhkov

Polytechnic Institute of Novgorod State University - Federal State Budgetary Educational Institution of Higher Education "Yaroslav-the-Wise Novgorod State University" (NovSU)

E-mail: ietconf@novsu.ru

IOP Conf. Series: Materials Science and Engineering **656** (2019) 011002 doi:10.1088/1757-899X/656/1/011002

Peer review statement

All papers published in this volume of *IOP Conference Series: Materials Science and Engineering* have been peer reviewed through processes administered by the proceedings Editors. Reviews were conducted by expert referees to the professional and scientific standards expected of a proceedings journal published by IOP Publishing.

Selecting the design parameters of hydrostatic control system for a long-base truck train

A M Abramov[1]

[1]Yaroslav-the-Wise Novgorod State University, ul. B. St. Petersburgskaya, 41 173003 Veliky Novgorod, Russia

E-mail: Alexandr.Abramov@novsu.ru

Abstract. For the delivery of long and bulky goods today long trailer trucks are used. In order to comply with the regulated maneuverability indicators, the semi-trailer of the truck train must be equipped with swiveling axles and an appropriate control system. Analysis of control systems for long-base truck trains showed that the most common today are hydro-volume control systems. The article describes the hydro volume control system of the pivot axles of the long-base truck train. The kinematic schemes and the description of the method for calculating the design parameters of the hydrostatic control system of a long-base truck are given.

1. Introduction

At present, up to 68% of all freight traffic is carried out using long-base truck trains. But such vehicles make up only 12% of all trucks on secondary roads [1]. In the EU countries, there are legislative acts strictly regulating the external and internal radii of vehicle rotation, which limits the use of truck trains with uncontrolled axles, both in cities and on secondary roads.

In addition, at present there is a great need for the transportation of long, heavy and oversized cargo. For these purposes, multi-axle semi-trailers with a telescopic sliding drawbar, the length of which can reach 36 m, have become widely used. It is physically impossible for such vehicles to move on public roads without control systems of a trailing link [2].

It is also necessary to note the economic aspect, for example, according to calculations made by TRIDEC, the use of a control system with swiveling axles or wheels on a semi-trailer allows reducing tire wear by up to 66% [3].

For a long time in Russia, priority was given to mechanical control systems. Analysis of the control systems of the trailed links of long-base truck trains showed that the most common today are hydrostatic control systems (HCS) [4].

This article describes the methodology for selecting and calculating the design parameters of the hydrostatic control system of the pivot axes of a long-base truck train.

The use of the hydrostatic control system of the pivot axles of a long-base truck train significantly improves the controllability and maneuverability of a long-base truck train, as well as significantly improves traffic safety, especially in conditions of intensive traffic flows.

2. Rated parameters of truck train maneuverability

Today, the main regulatory document in force on the territory of the Russian Federation that regulates the requirements for the design of the vehicle is Technical Regulations of the Customs Union on the safety of wheeled vehicles, which in its turn is based on the requirements of the UNECE Regulations [5].

The regulated maneuverability indicators adopted in the UNECE Regulations are the dimensional outer turning radius R_a<12.5 m and the dimensional inside turning radius R_i>5.3 m, the turning width of the track wheels B_R<5.5 m (Germany) [6].

The Russian Federation regulates the minimum turning radius $R_{min} \leq 12$ m.

As a criterion for optimizing maneuverability, researchers often use B_T which is an overall lane, B_R is the swivel width of the track of the wheels, and C_d is the shift in the trajectory of the trailer link relative to the main trajectory of the tractor [7, 8].

Figure 1 presents the estimated indicators of maneuverability of the truck train.

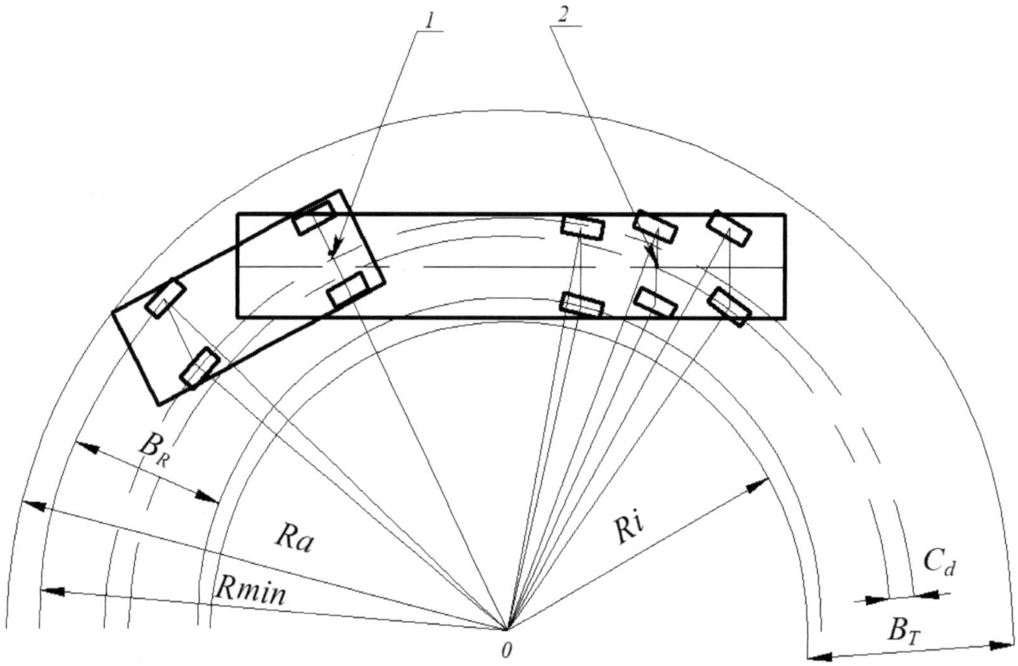

Figure 1. Estimated indicators of maneuverability of the truck train.
R_a – outer radius, R_i– internal radius; R_{min} – minimum turning radius; B_T – overall lane; B_R – the swivel width of the track of the wheels; C_d – trailer link trajectory shift; 1– base point of the tractor; 2 – base point of a semitrailer.

3. Selection of HCS design parameters

3.1. Selecting the type of axles to be driven

On modern semi-trailers, both individual wheels and axles of the semi-trailer can be controlled.

HCS are “passive”, i.e. without amplification. The force required to turn the steering wheels or axles is created by the master device (hydraulic cylinders), and the necessary forces or moments are realized due to the lateral reactions of the front steering wheels of the tractor.

The use of pivot axles allows using them at high axial loads, while at the same time less driving forces are required to rotate the axis. As a disadvantage, it is noted that the use of rotary axes requires large axle distances. An example of a control system with rotary axes is shown in figure 2.

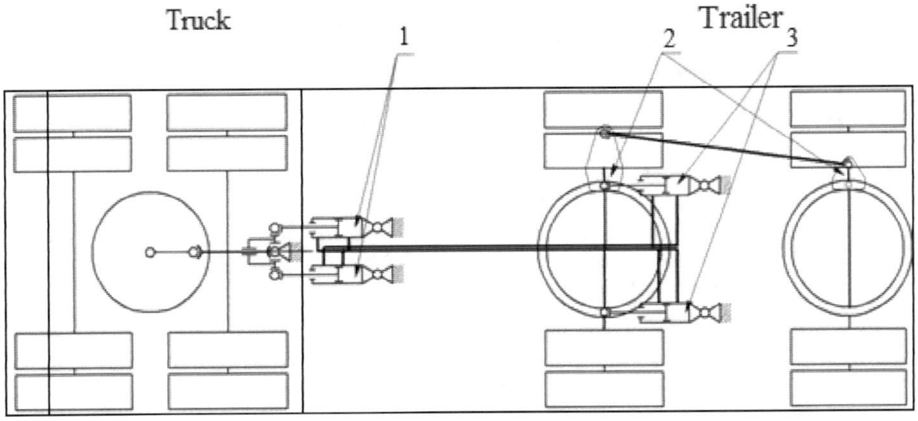

Figure 2. Kinematic scheme of a truck train with controlled axes of the company Tridec. 1 – master cylinders; 2 – pivot points; 3 – executive cylinders.

Driving wheels can be realized both with a dependent suspension in the form of a rigid axis and with an independent suspension. At the same time, their installation requires shorter center distances. Of the minuses worth noting is the axis price increase. An example of a control system with swivel wheels is shown in figure 3.

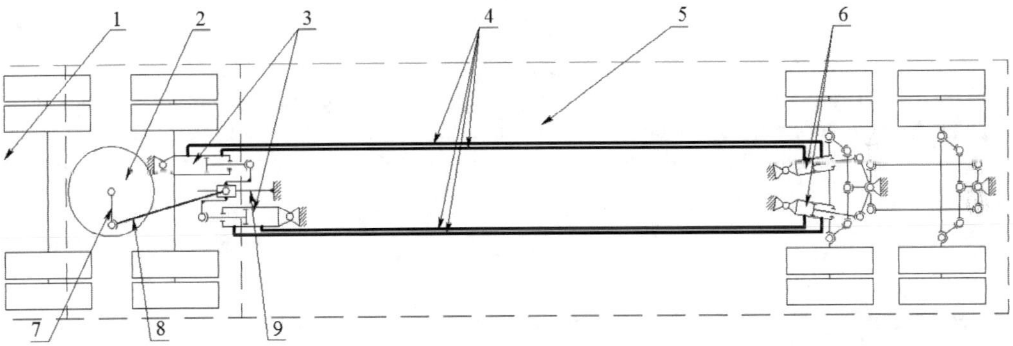

Figure 3. Kinematic diagram of a hydrostatic control system with rotating wheels of the company Tridec. 1 – tractor; 2 – tractor saddle; 3 – master cylinders; 4 – connecting hoses; 5 – semi-trailer; 6 – executive cylinders; 7 – crank; 8 – thrust; 9 – guide.

3.2. Kinematic analysis of indicators of maneuverability of a truck train and the choice of strategy for control system

When selecting the design parameters of the control system (CS), a kinematic analysis of the drive of the controlled axles or wheels is performed. This is based on the condition of rolling all the wheels of a train, with steady curvilinear motion, with a constant turning radius, in concentric circles with a single instantaneous turning center at point O.

To ensure the minimum displacement of the trajectory of the tractor and semi-trailer C_d or the minimum overall lane of traffic B_T (figure 4) all axles of the semi-trailer are required to be controlled. Naturally, this mode of movement requires turning all the wheels to different angles.

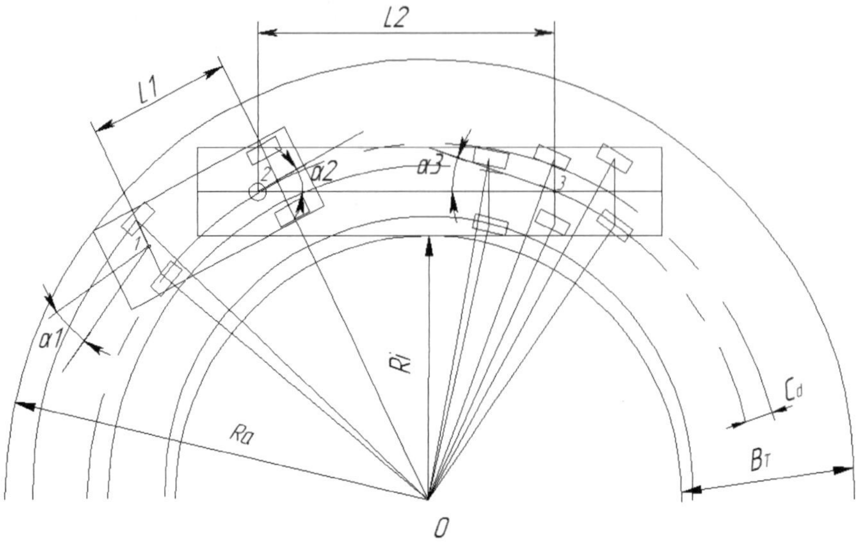

Figure 4. Criteria for optimizing maneuverability.

The control system is characterized by the gear ratio i_d, which is:

$$i_d = \frac{\propto_3}{\propto_2};$$

where $\propto_2$ – truck traing folding angle; $\propto_3$ – average steering angle of the semi-trailer steered wheels.

The choice of the gear ratio of the control system is determined by the design parameters of the semi-trailer: the base of the semi-trailer, the number of axles controlled and the strategy for optimizing the motion path.

The gear ratio i_d can have both a constant and a variable value.

If the gear ratio in the CS is constant, i.e., $i_d = const$, then during the movement of a truck train on a turn of 90^0 at the semi-trailer truck, point 3, there is a shift of the trajectory relative to point 2, to the outside when entering the turn and to the inside when leaving the turn. The shift of the trajectory negatively affects traffic safety, especially in conditions of intensive traffic flows.

Figure 5 shows a dependence graph of the angle of rotation of the steer wheels of the semi-trailer versus the folding angle of the truck train for systems with constant and variable gear ratio.

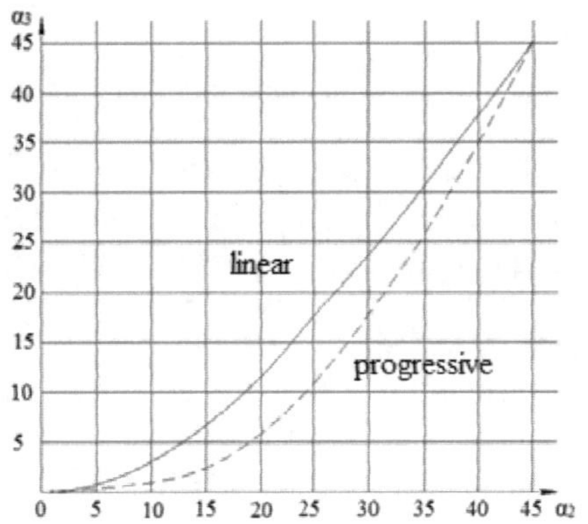

Figure 5. Dependence graph of the angle of rotation of the steer wheels of the semi-trailer versus the folding angle of the truck train for systems with constant and variable gear ratio.

Simulation of the movement of a truck train allows us to conclude that the control system with a progressive characteristic reduces the run-on of the wheels of the semi-trailer when turning to the outside. An example of such a control system is shown in figure 2.

4. Selection of hydraulic HCS

HCS is a closed hydraulic system, where a pair of cylinders is used as master cylinders connected with a chain link mechanism by a rocker mechanism, an actuator pair of hydraulic cylinders is connected with a mechanical drive of the pivot axles and a semi-trailer frame (figure 6). To increase the reliability of the HCS, the master and actuating hydraulic cylinders are connected by two independent hydraulic circuits.

To ensure the stable operation of the system, and in order to eliminate the self-oscillations of the steered axles (wheels), an overpressure of approximately 4.0 ... 6.0 MPa is created in the hydraulic drive. Pressure is created by including a pump station in the system. To maintain a constant pressure in each hydraulic circuit, pneumatic-hydraulic accumulators are used.

The pumping station is a separate unit containing: a pump, a direct-current driving electric motor with a voltage of 24V, a power of about 4 kW and a tank for the working fluid. Since the system is autonomous, a battery is used to power the electric motor, which could be charged by the tractor.

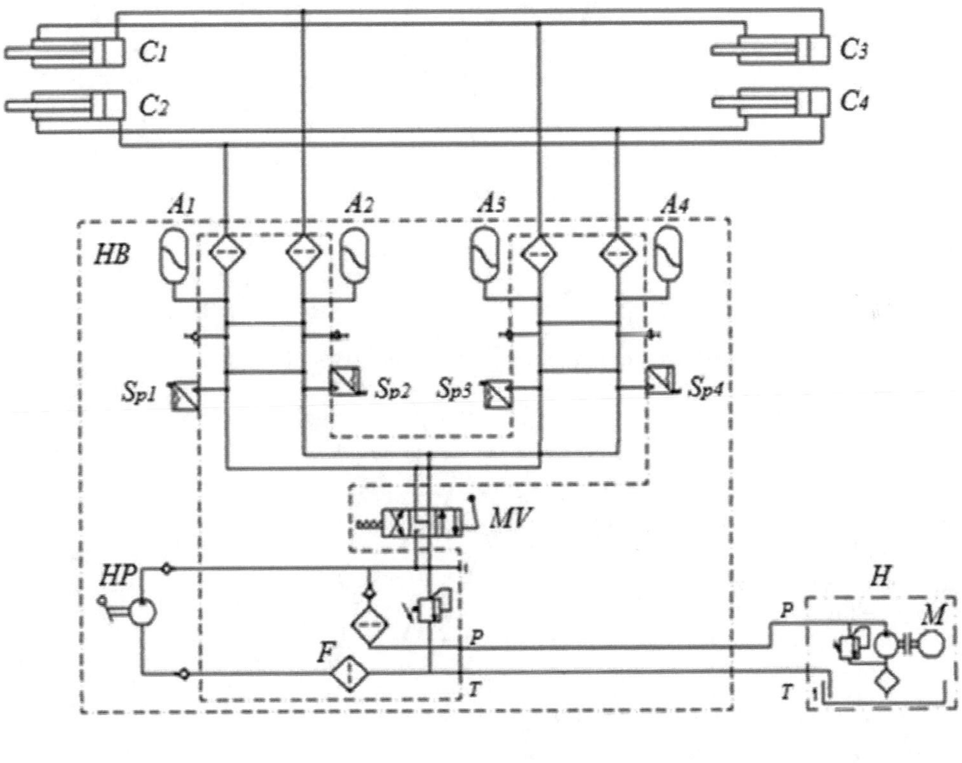

Figure 6. Kinematic diagram of a hydrostatic control system. C_1, C_2- master cylinders; C_3, C_4 - executive cylinders; A_1, A_2, A_3, A_4- pneumatic-hydraulic accumulators; $S_{P1}, S_{P2}, S_{P3}, S_{P4}$ – pressure sensor in the circuit; *HB* – hydraulic unit; *MV* - manual control valve; *H* - pumping unit; *HP* – hand pump; *F* - filter; *M* – electric motor.

A manual pressure boost is provided as an emergency circuit. Thus, it is possible by creating a pressure of 6.0 MPa to return the wheels of the semi-trailer to the initial state corresponding to a straight-line motion. In the event of a fault in the CS, it allows the train to continue moving with an unmanaged semi-trailer.

HCS is equipped with a self-diagnostic system, including axle position sensors, sensors and lamps for controlling the pressure of the working fluid.

5. Calculation of design parameters of HCS actuators

In HCS, there are, as a rule, two master and two actuating cylinders. The location of the hydraulic cylinders depends on the chosen layout of the mechanical drive (a system of rods and levers that provides the angles of rotation of the steered wheels or axles required according to the Ackermann law).

In HCS, at first the parameters of the actuating cylinders are determined, then the parameters of the actuating cylinders are selected. Figure 7 shows the design diagram of the actuating cylinders for the controlled axle of the semi-trailer.

The force created by the executive cylinders must exceed the moment of resistance to the rotation of the steered wheels at full load of the semi-trailer, which is required primarily in the manual control mode.

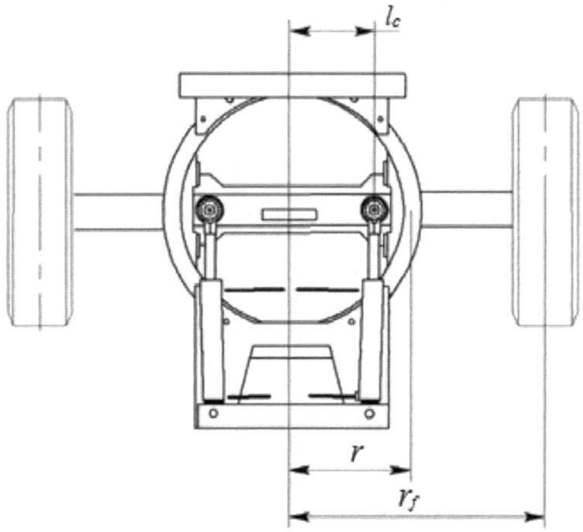

Figure 7. The design scheme for the controlled axis of the semi-trailer.

The total moment of resistance to the rotation of the axis is

$$M_{cz} = (M_{cf} + M_{c\varphi}) \cdot \eta / i_d \tag{1}$$

where M_{cf} – moment of resistance to rolling of the wheels when turning the axle;
$M_{c\varphi}$ – moment of resistance in the turntable;
η – efficiency of mechanical drive;
i_d – drive gear ratio.

$$M_{cf} = G_{atr} \cdot n \cdot f_k \cdot r_f \cdot g \tag{2}$$

where G_{atr} – load on a single axle of a semitrailer, kg;
n – number of axles driven;
f_k – wheel rolling resistance coefficient;
r_f – wheel running shoulder (half track of the axle bogie), m;
g – acceleration of gravity, m/s^2.

$$M_{c\varphi} = G_{tr} \cdot f \cdot r \cdot g, \tag{3}$$

where f – coefficient of friction-swing in the turntable;
r – turntable radius, m;
g – acceleration of gravity, m/s^2.
G_{tr} – set by semi-trailer axle manufacturer.

IOP Conf. Series: Materials Science and Engineering **656** (2019) 012001 doi:10.1088/1757-899X/656/1/012001

Knowing the total moment of resistance to the rotation of the axis, we determine the driving force on the cylinder rod:

$$P_c = M_c/l_c, \tag{4}$$

l_c – cylinder shoulder, m;

Knowing the force on the rod and asking the maximum value of fluid pressure, we determine the area of the cylinder

$$P_c = P_g \cdot F_c, \tag{5}$$

whrere P_g – fluid pressure in the system, MPa;

F_c – cylinder area, m^2;

The diameter of the cylinder is determined from its area, it is rounded up and selected on the basis of the standard number of hydraulic cylinders.

$$p_g \cdot \frac{\pi D^2}{4} \cdot 2 \cdot i_d = \frac{M_c}{l_c} \tag{6}$$

D – cylinder diameter, m

We express from the formula (6) the diameter of the cylinder:

$$D = \sqrt{\frac{2 \cdot \frac{M_c}{l_c}}{p_g \cdot \pi \cdot i_d}}, \tag{7}$$

The master cylinders in HCS are taken in the same dimensions as the actuators.

6. Selection of parameters of the pumping station and control unit

6.1. Pumping station

The pumping station in the HCS serves to discharge and maintain the necessary pressure in the drive of the control system, as well as to ensure the rotation of the controlled axes in the manual control mode.

Pump performance is determined by the ability to provide the required speed of rotation of the controlled axes in the manual control mode.

If the steering speed of the wheels is strictly regulated by the UNECE and EU directives for steering and is $\omega_{рк} = 10\ {}^{rad}/_{c}$, then, according to research literature data, for controlled axes of the semi-trailer it does not exceed 0.2 $^{rad}/c$ [2].

The pump capacity:

$$Q_p = \frac{V_c}{t_{min} \cdot \eta_p} \tag{8}$$

where V_c– cylinder volume;

taking into account the fact that there are two actuating cylinders, their total volume will be equal to

$$V_c = (F_c + F_{cst}) \cdot l_c \tag{9}$$

F_c– borehole cylinder area, m^2;

F_{cst} – rod cavity area of the cylinder, m^2;
l_c – cylinder stroke, m;
t_{min} – the minimum time for the axis to turn from one extreme to the other, s.

Axis rotation time from one extreme position to another

$$t = \frac{\alpha}{\omega}, \tag{10}$$

where α – angle of rotation of the axis from one extreme position to another, grad;
ω – axis rotation speed, grad/s.
A DC motor is used to drive the pump.
The choice of the electric motor is carried out taking into account the power required to drive the hydraulic pump and the maximum speed of the pump. The volumetric efficiency of the pump should be taken into account as well.
Motor power of hydraulic pump drive:

$$N_p = \frac{P_g \cdot Q_p}{1000 \cdot \eta_p}, \tag{11}$$

where P_g – fluid pressure, Pa;
Q_p – pump capacity, m^3/s;
η_p – volumetric efficiency of a pump.
After the capacity of the pump is determined, it is necessary to choose the volume of the tank of the pumping unit. It is selected with the calculation of 1 minute of the pump performance.

6.2. Control valve
To implement the manual control a valve is used. Its main parameters are the kinematic scheme of the distributor, the conditional flow area – the remote control, the supply voltage of the electromagnets. The control depends on the pump capacity and is selected from the table characteristics.

The hydraulic capacity of the valve is determined using the condition

$$Vp \geq V_{TP} \text{ if } P_H = P_{H\max},$$

$$V_{\text{TP}} \leq \frac{G_{max}}{F_{ns}} \sqrt{\frac{1}{2}\left(p_0 - \frac{P_H}{F_{ns}}\right)}, \text{ whence at } X = X_{max} \tag{12}$$

$$G_m \geq \frac{V_{TP} \cdot F_{ns}}{\sqrt{\frac{1}{2}\left(p_0 - \frac{P_H}{F_{ns}}\right)}}, \tag{13}$$

At the same time, the area of throttling windows

$$A_0 = \frac{G_m}{\mu\sqrt{\frac{2}{\rho}}} = \frac{V_{TP}}{\mu} \frac{F_{ns}}{\sqrt{\frac{1}{2}\left(p_0 - \frac{P_H}{F_{ns}}\right)}} \tag{14}$$

In practice, the choice of a hydraulic distributor can be carried out using the manufacturer's catalog sheets or their web-sites; one should take into account the hydraulic distributor circuit, its conditional flow area, inlet pressure, control type and nominal flow rate of the working fluid. When choosing a hydraulic distributor controlled by electromagnets, it is necessary to pay attention to the technical characteristics of the electromagnet: supply voltage, type of current and operating mode of the electromagnet.

6.3. Hydropneumatic battery
Hydro accumulator (HA) in HCS serves to maintain excess pressure in the hydraulic system.

The choice of HA basically comes down to determining the constructive (full) capacity V_k and the useful volume V_e of the liquid (the volume of liquid displaced by gas from the HA during its discharge, when the gas pressure decreases in a given range from P_{max} to P_{min}. P_{max} is the largest working pressure in the hydraulic system. P_{min} is the calculated operating pressure of the hydraulic system of the control system. To obtain the largest useful volume of liquid in HA, choose the initial air pressure P_a, equal to the lowest allowable pressure P_{min}.

In systems with automatic turning on of the pump, for maintenance of the set pressure range of working liquid the initial air pressure P_B should be slightly below P_{min}..

$$P_{min} - P_a = 0{,}2 \ldots 0{,}4 \text{ MPa}$$

The useful battery capacity V_e of HA in the isothermal process is determined from the following expressions:

$$V_e = V_k \left(1 - \frac{P_{min}}{P_{max}}\right) \quad (15)$$

where V_k – capacity of HA;

P_{min} – initial charging pressure of HA;

P_{max} – maximum working pressure.

7. Conclusion

HCS used today on long-base road trains provide acceptable indicators of maneuverability.

The present shift of the trajectories of the tractor and trailer link on transient modes of movement can be compensated by using HCS with a variable gear ratio to reduce the offset value on transient modes of movement up to 0.2 m, which will improve the controllability and stability of the long-base truck train and significantly improve the safety of the long-base truck train, especially in the conditions of intensive traffic flows.

References

[1] *Transport Statistics Great Britain 2004 30th edition* 2004 (London, UK: Department for Transport)

[2] Zakin Ia Kh 1986 *Manevrennost' Avtomobilia i Avtopoezda* [Manoeuverability of Vehicles and Truck-Trains] (Moscow: Transport) p 136

[3] *TRIDEC Product Flyer* 2019 Available at: http://www.tridec.nl/en/products/hydraulic-steering-systems.html (accessed: 18.04.2019)

[4] *Digital Catalogue* 2019 Available at: http://www.tridec.com/en/products.html (accessed: 18.04.2019)

[5] Technical Regulations of the Customs Union TR CU 018/2011 *"On the safety of wheeled vehicles"* **877** Available at: http://www.consultant.ru/document/cons_doc_LAW_125114/fb912286b5c44149bb594585163dbf84f712edb5/ (accessed: 18.04.2019) (In Russ.)

[6] Straßenverkehrs-Zulassungs-Ordnung (StVZO) § 32d Kurvenlaufeigenschaften 2019 Available at: http://www.gesetze-im-internet.de/stvzo_2012/ 32d.html (accessed: 18.04.2019)

[7] Tarasik V P 2006 *Teoriya Dvizheniya Avtomobilya* [Automobile Movement Theory] (St. Petersburg: BHV-Peterburg) p 478

[8] Litvinov A S and Farobin Y E 1989 *Avtomobil'. Teoriya Ekspluatatsionnykh Svoystv* [Automobile. Theory of Operational Properties] (Moscow: Mashinostroyeniye) p 237

IOP Conf. Series: Materials Science and Engineering **656** (2019) 012002 doi:10.1088/1757-899X/656/1/012002

Impact of the number of blades of the geokhod cutting body on the energy intensity of the rock destruction

V V Aksenov[1,6], A B Efremenkov[2,7], V Yu Sadovets[3], D A Pashkov[4] and V A Efremenkov[5,7]

[1] Research Center Siberian NPO Ltd, Sovetsky prospect, 56 650002, Kemerovo, Russia
[2]Yaroslav-the-Wise Novgorod State University, ul. B. St. Peterburgskaya, 41 173003, Veliky Novgorod, Russia
[3]T.F. Gorbachev Kuzbass State Technical University, ul. Vesennjaja, 28 650000, Kemerovo, Russia
[4] Institute of Coal, Federal Research Center for Coal and Coal Chemistry, Siberian Branch of the Russian Academy of Sciences, Leningradsky prospect, 10 650610, Kemerovo, Russia
[5]Peter the Great St.Petersburg Polytechnic University (SPbPU), Polytechnicheskaya, 29 195251, St.Petersburg, Russia

[6]E-mail: 55vva42@mail.ru
[7]E-mail: abe@novsu.ru

Abstract. The article presents the sequence of calculating the energy intensity of the rock destruction with the blade of the geokhod cutting body. The influence of the number of cutting elements of the geokhod body on the energy intensity of rock destruction is determined. To substantiate the relevance of the study, features of the new class of mining machines – geokhods and the main provisions of geokhod technology are presented.
The main research stages are formulated, based on the method for determining the energy intensity of the destruction of the rock face with the blade of the geokhod body. Geometrical features and power parameters of the geokhod blade cutting body that affect the energy intensity of rock destruction are identified.
To determine the influence of the number of blades of the geokhod body on the energy intensity of the destruction of the rock face, the geometrical parameters of the blade body and the parameters of the mining and technical conditions of the mine working are substantiated. For carrying out mathematical calculations, an algorithm has been developed for calculating the energy intensity of the destruction of a rock face with a blade body a geokhod. As a result of the study, it was revealed that the energy intensity of the destruction of rock with the blade cutting body increases unevenly with increasing number of blades.

1. Introduction

Existing mining systems and technologies of mine workings are poorly adapted, and in most cases are not able to solve the problem of cavity formation in the underground space at small depths [1–4].

A promising direction for the development of cavity formation technologies in the underground space is the use of geokhod technology, the basic element of which is the geokhod. This is an apparatus designed to carry out the excavation of underground workings for various purposes and location in space [6–9].

When developing technical and structural solutions for devices and elements of a geokhod, it is necessary to take into account the complex, helical movement of the machine to the bottom of the mine. In addition, the developed methods for determining the power parameters of devices and elements of geokhod, which interact with the external environment and between themselves, should take into account the complex nature of geokhod movement.

One of the main devices of the geokhod that has a direct impact on the power characteristics of the underground apparatus is a cutting body. In turn, the methodology for determining the power parameters of the body is influenced by its technical and design parameters.

The main indicator for evaluating the effectiveness of the geokhod cutting body is the energy intensity of rock destruction [10]. Therefore, the research aimed at improving the methodology for calculating the energy intensity of the rock face destruction, as well as determining the influence on it of the design parameters of the blade cutting body is relevant.

2. Research methodology

From the previous studies [11–16], the energy intensity of the destruction of the rock face with the blade cutting body of geokhod will be determined by the expression

$$p = \frac{P_{cd}}{F_{mid}}. \quad (1)$$

where R_{cv} – projection of the total force of the resistance of the soil to cutting, on a plane perpendicular to the axis of geokhod rotation, H;

F_{mid} – cross section of the slot area, mm^2.

For a geokhod blade cutting body, the total projection of the resistance of the soil to cutting on a plane perpendicular to the axis of rotation is equal to [14–19]

$$R_{cb} = n(R_{cb.cv} + R_{cb.side}), \quad (2)$$

where n– the number of blades of the geokhod body;

$R_{cb.cv}$– projection of the component of the force of soil resistance to cutting, depending on the width of the cut, on the plane perpendicular to the axis of rotation of the geokhod (figure 1), H;

$R_{cb.side}$– projection of the component of the force of soil resistance to cutting, not depending on the width of the cut, on the plane perpendicular to the axis of rotation of the geokhod (figure 1), H.

Projection of the component of the force of soil resistance to cutting, depending on the width of the cut, on the plane perpendicular to the axis of rotation of the geokhod [14-19]

$$R_{cb.cv} = \frac{\varphi m_{cv} h_B^2}{2\pi m \cos\gamma}\left[\frac{(\sin\beta_2 - \sin\beta_1)}{\sin\beta_2 \cdot \sin\beta_1} + ctg(\delta + \varphi_{fr}) ln\left|\frac{tg\frac{\beta_2}{2}}{tg\frac{\beta_1}{2}}\right|\right], \quad (3)$$

Projection of the component of the force of soil resistance to cutting, not depending on the width of the cut, on the plane perpendicular to the axis of rotation of the geokhod (figure 1), [14–19]

$$R_{cb.cv} = \frac{h_B}{n}\left(m_{side}\frac{h_B}{n} + m_{side.mid}\right)\left[\frac{\sin(\delta+\varphi_{fr}+\beta_1)+\sin(\delta+\varphi_{fr}+\beta_2)}{\sin(\delta+\varphi_{fr})}\right]. \quad (4)$$

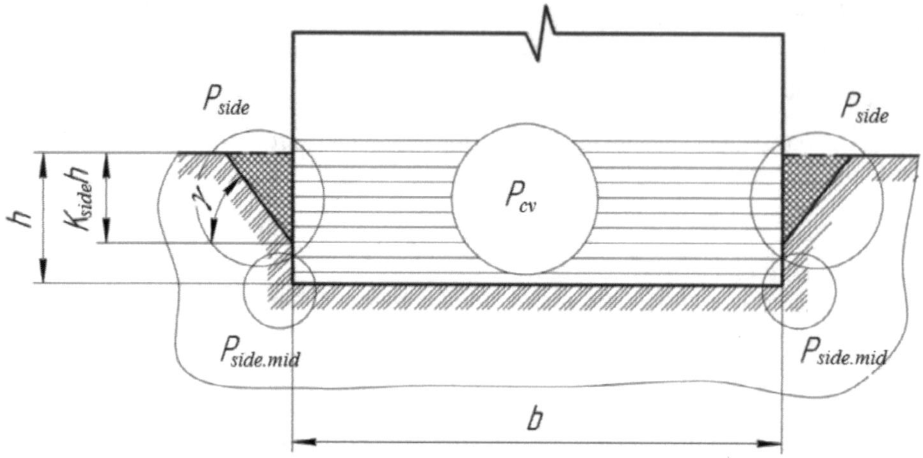

Figure 1. The areas of action of the components of the sharp blade cutting force.

From [14, 15, 16, 17, 18, 19] and Figure 1, the cross-sectional area of the slot will be determined

$$F_{mid} = bh + k_{side}^2 h^2 ctg\gamma, \tag{5}$$

where b– the blade width, m;
h– depth of cutting, m;
k_{side}– the depth ratio of the expansion part of the slot;
γ_{lim}– angle of inclination of the expansion part of the slot to the horizon, degree.

During one full turn geokhod moves a distance equal to the pitch of the helical blade h_B. Therefore, if we assume that there is only one blade on the cutting body, it should destroy the cut of thickness equal to h_B. When installing several knives on the geokhod cutting body, the cut thickness h will be equal to:

$$h = \frac{h_B}{n}, \tag{6}$$

Taking into account the expression (6) the cross-sectional area of the slot with one blade of the geokhod cutting body

$$F_{mid} = b\frac{h}{n} + k_{side}^2 \left(\frac{h}{n}\right)^2 ctg\gamma_{pr}, \tag{7}$$

Putting expressions (2) and (7) in expression (1) we get

$$p = \frac{n(R_{cb.cv} + R_{cb.side})}{n\left(b\frac{h}{n} + k_{side}^2\left(\frac{h}{n}\right)^2 ctg\gamma_{lim}\right)}, \tag{8}$$

We simplify expressions (8)

$$p = \frac{R_{cb.cv} + R_{cb.side}}{b\frac{h}{n} + k_{side}^2\left(\frac{h}{n}\right)^2 ctg\gamma_{lim}}, \tag{9}$$

From the expression (9) it can be concluded that the value of the energy intensity of the destruction of the rock face with the blade cutting body of the geokhod will be the same both for one blade and for the entire blade cutting body of the geokhod with several blades.

3. Results

To determine the influence of the number of blades of the geokhod blade cutting body on the energy intensity of the destruction of rock, we chose the parameters of the mining conditions of the mine workings and the geometrical parameters of the blade cutting body. The values of these parameters are presented in table 1.

Table 1. Mining conditions of working with geokhod and the geometrical parameters of the blade cutting body.

Name	Symbol	Unit of measurement	Value
Geokhod radius	R_g	m	0.3
Central nozzle radius	r_o	m	0.025
External propulsion step	h_B	m	0.3
Coefficient taking into account the influence of the cutting angle	φ		0.59
Specific cutting force in the frontal part of the slot at a cutting angle of 45^0	M_{cv}	H/m^2	97000
Strength of the destruction of the soil in the side parts of the slot	M_{side}	H/m^2	36000
Specific cutting force of one of the side edges of the blade	$M_{side.mid}$	H/m	8490
Angle of inclination of the radial blade to the plane perpendicular to the axis of rotation of the geokhod	γ	deg	0
Cutting angle	δ	deg	25
Friction angle	φ_{fr}	deg	31.4
Depth ratio of the expansion part of the slot	K_{side}		0.9
Angle of inclination of the expansion part of the slot to the horizon	γ_{lim}	deg	30

According to expression (9), we determined the dependence of the energy intensity of the destruction of the face rock with the blade cutting body of geokhod on the number of blades on the geokhod cutting body.

The procedure for calculating the geokhod blade cutting body is shown in figure 2. The developed algorithm was applied when designing a computer program for determining the energy intensity of the destruction of the face rock with the geokhod blade cutting body.

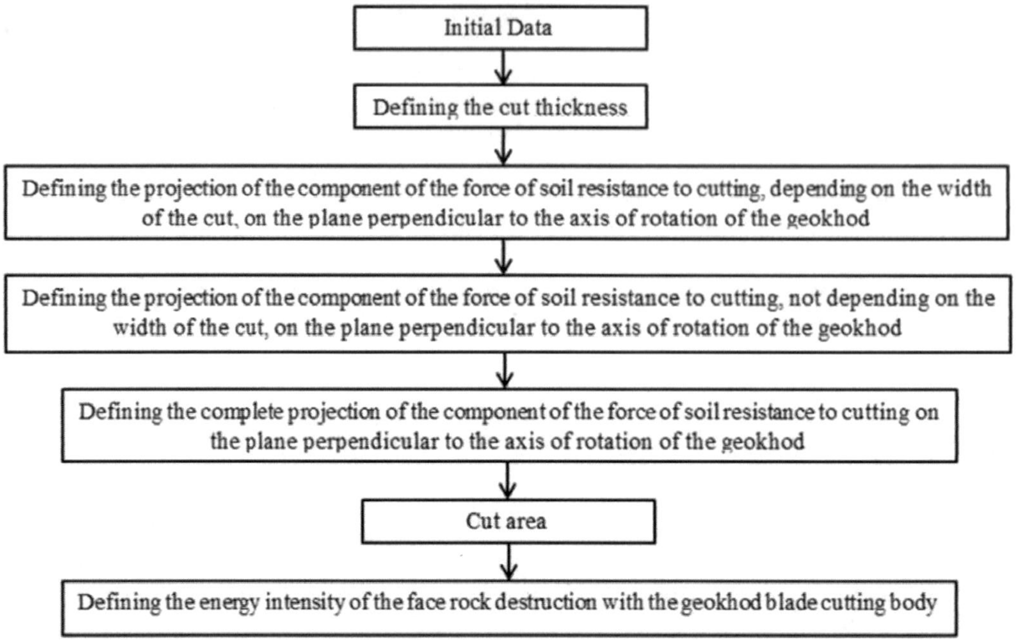

Figure 2. Algorithm for calculating the energy intensity of the face rock destruction with the geokhod blade cutting body.

A fragment of the calculation of the energy intensity of the face rock destruction with the geokhod blade cutting body with a varying number of blades using a computer program is shown in figure 3.

	A	B	C	D	E	F	G	H
1	**Symbol**	**Measurement unit**	**Value**		n	R_{cb}, H	p, MJ/m^2	
2	r_r	m	0.3		1	17808.55	0.08234541	
3	r_o	m	0.025		2	14334.54	0.09360826	
4	h_B	m	0.3		3	13176.53	0.09975507	
5	Ψ		0.59		4	12597.53	0.10362654	
6	m_{cv}	H/m^2	97000		5	12250.13	0.10628873	
7	m_{side}	H/m^2	36000		6	12018.53	0.1082317	
8	$m_{side.cv}$	H/m	8490		7	11853.1	0.10971222	
9	γ	degree	0		8	11729.02	0.11087783	
10	δ	degree	25		9	11632.52	0.11181935	
11	γ_{fr}	degree	31.4		10	11555.32	0.11259574	
12	κ_{side}		0.9					
13	γ_{fr}	degree	30					
14								

Figure 3. Calculation of of the energy intensity of the face rock destruction with the geokhod blade cutting body with a varying number of blades using the Microsoft Excel program.

On the basis of the obtained values, a graph of the dependence of the energy intensity of the rock destruction with the blade cutting body of the geokhod on the number of blades was plotted (figure 4).

The ordinate axis of the presented dependence shows the value of the energy intensity of the rock destruction with the blade cutting body of the geokhod (MJ/m^3), and the abscissa axis shows the change of the number of blades on the geokhod cutting body (pcs.).

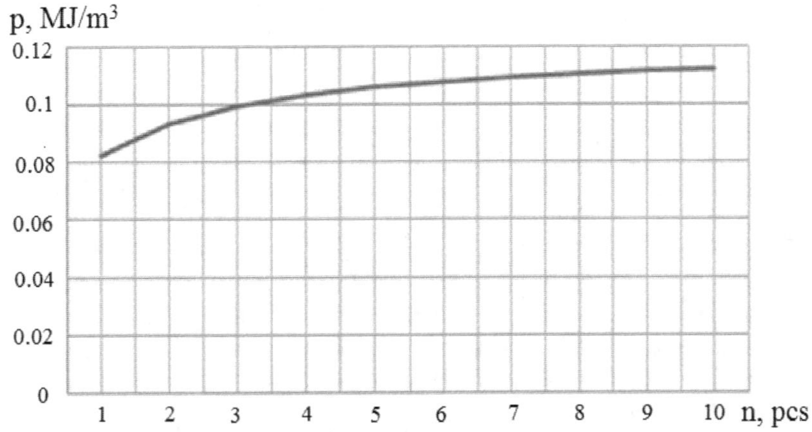

Figure 4. Dependence of the energy intensity of the face rock destruction with the blade cutting body of the geokhod on the number of blades.

4. Findings

From the graph presented in figure 4, it follows that:
- the value of the energy intensity of the rock destruction with the blade cutting body of the geokhod changes nonlinearly upwards with an increase in the number of blades on the cutting body of geokhod;
- with an increase in the number of blades n> 3 pcs., the energy intensity of the destruction of rock with blade cutting body continues to increase, but less intensively;
- the energy intensity of the destruction of the rock increases as a result of a relative decrease in the side expansions of the slot, within which the resistance of the rock is less than that in front of the frontal edge of the blade.

References

[1] Aksenov V, Sadovets V and Pashkov D 2017 *E3SWeb of Conferences* **21** 03008
[2] Aksenov V, Sadovets V, Rezanova E and Pashkov D 2017 *E3SWebConf* **15** 03015
[3] Aksenov V, Sadovets V and Pashkov D 2018 *E3SWebConf* **41** 03002
[4] Aksenov V V, Efremenkov A B, Sadovets V Yu and Pashkov D A 2018 *IOP Conference Series: Materials Science and Engineering* **441** 012002
[5] Nishi S and Seiki T 1997 Planning and design of underground space use *Mem Sch Eng Nagoya Univ* **1**
[6] Maid B, Schmid L, Ritz W and Herrenknecht M 2008 *Hardrock Tunnel Boring Machines* (Berlin: Ernst&Sohn) 343 p
[7] Wighman T 1998 *ENR: News* **Rec 4**
[8] Brierley G 1998 World Tunnel *Andsubsurface Excav* **9**
[9] Efremenkov A B 2011 *6th International Forum on Strategic Technology (IFOST-2011)* **2:1** 348
[10] Aksenov V V, Sadovets V Yu and Pashkov D A 2017 Determining the number of blades of the cutting body of geokhod for destruction of rocks of low hardness *Vestnik of Kuzbass State Technical University* **3 121** 116-126
[11] Sadovets V Yu, Beglyakov V Yu and Efremenkov A B 2015 Simulation of geokhod movement with blade actuator *Applied Mechanics and Materials* **770** 384–390

[12] Aksyonov V V, Sadovets V Yu and Beglyakov V Yu 2011 Impact of the bench on the stress-strain state of near face part of mine workings *Innovative Technologies and Economics in Mechanical Engineering* 575–580
[13] Walter A V and Aksenov V V 2015 Determining deviations in geometry of the geokhod shells *Applied Mechanics and Materials* **770** 439–444
[14] Aksenov V V, Efremenkov A B, Sadovets V Yu and Rezanova E V 2010 Creating innovative tools for the formation of underground space *Vestnik of Kuzbass State Technical University* **1 77** 42–47
[15] Sadovets V Yu and Pashkov D A 2016 The influence of the parameters of the helicoid generatrix on the shape of the geokhod blade cutting body *Natural and Intelligent Resources of Siberia Sibresurs Collection of materials of the XVI International Scientific and Practical Conference*
[16] Sadovets V Yu and Pashkov D A 2016 Determination of the function of the line of contact of the blade of the geokhod cutting body with the face surface *Technologies and Materials* **4** 9–14
[17] Aksenov V V, Sadovets V Yu, Pashkov D A and Rezanova E V 2018 Boundary conditions for determining the characteristic points of a geokhod blade cutting body *Vestnik of Kuzbass State Technical University* **2 126** 166–173
[18] Aksenov V V, Sadovets V Yu, Preis E B and Pashkov D A 2018 Improving the mathematical model for determining the power parameters of the geokhod blade cutting body *Mining Equipment and Electrical Engineering* **5 139** 16–22
[19] Carmody J and Sterling R 1993 *Undergroundspace Design* (New York: Reinhold)

IOP Conf. Series: Materials Science and Engineering **656** (2019) 012003 doi:10.1088/1757-899X/656/1/012003

Impact of the inclination angle of a blade of the geokhod cutting body on the energy intensity of rock destruction

V V Aksenov[1], A B Efremenkov[2,6], V Yu Sadovets[3], D A Pashkov[4] and V A Efremenkov[5,6]

[1] Research Center Siberian NPO Ltd, Sovetsky prospect, 56 650002, Kemerovo, Russia
[2]Yaroslav-the-Wise Novgorod State University, ul. B. St. Peterburgskaya, 41 173003, Veliky Novgorod, Russia
[3]T.F. Gorbachev Kuzbass State Technical University, ul.Vesennjaja, 28 650000, Kemerovo, Russia
[4] Institute of Coal, Federal Research Center for Coal and Coal Chemistry, Siberian Branch of the Russian Academy of Sciences, Leningradsky prospect, 10 650610, Kemerovo, Russia
[5]Peter the Great St.Petersburg Polytechnic University (SPbPU), Polytechnicheskaya, 29 195251, St.Petersburg, Russia

[6]E-mail: 55vva42@mail.ru
[7]E-mail: abe@novsu.ru

Abstract. The article presents the dependence of the energy intensity of rock destruction on the inclination angle of blade of geokhod cutting body for the destruction of low hardness rocks. To formulate the research tasks, the main provisions of the geokhod technology for making workings are presented; the features of the new class of mining equipment, geokhod, are given. The research relevance is formulated. To determine the energy intensity of the rocks destruction, the method of calculating the power and geometric parameters of the geokhod blade cutting body was chosen. An algorithm has been developed for calculating the energy intensity of rock destruction, on the basis of which a computer program has been developed for determining the power parameters of the geokhod blade cutting body. In addition, the article presents well-founded forms of face, which appear as a result of destruction with the blade cutting body of the geokhod. To determine the influence of the inclination angle of the geokhod blade cutting body on the energy intensity of destruction, the geometrical parameters of the blade cutting body and the parameters of the mining and technical conditions of the mine working were justified. As a result of the study, it was revealed that the energy intensity of rock destruction with the blade cutting body decreases unevenly as the inclination angle of the blade towards the reverse cone increases.

1. Introduction

The formation of a cavity in an underground space at shallow depths is a difficult and time-consuming task. Existing mining systems and technologies of mine workings are poorly adapted, and in most cases are not able to solve problems arising during the development of underground space at small depths [1, 2, 3, 4].

One of the promising approaches to the technology of making workings in small depths is the use of geokhod technology. The main device of the developed technology is a new class of equipment – geokhod. This is a technological apparatus intended for carrying out the excavation of underground workings for various purposes and location in space [4, 5, 6, 7, 8]. One of the variants of the geokhod design is presented in figure 1 [9, 10, 11].

Figure 1. A prototype of geokhod with a drum cutting body.

The use of geokhods with drum-type cutting bodies for the destruction of low hardness rocks [12, 13] is not effective from the point of view of safety and complexity of the design of the cutting body itself. A more promising constructive solution of the geokhod cutting body for the destruction of low hardness rocks is the use of blade of the earth-moving machine cutting body [11]. One of the constructive solutions for the geokhod blade cutting body with a diaphragm overlapping the working face is shown in figure 2.

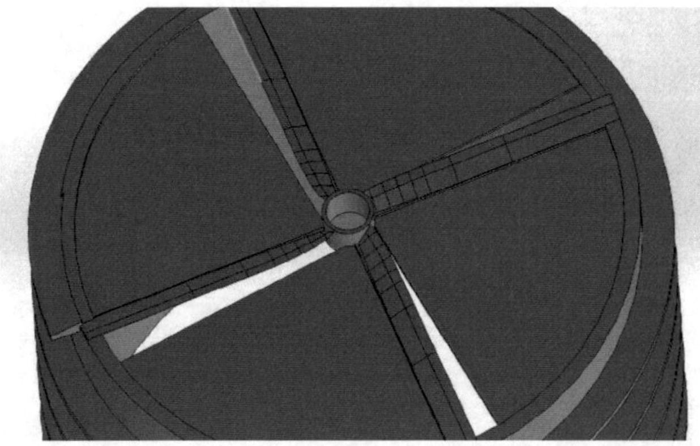

Figure 2. A constructive solution for the geokhod blade cutting body.

In the process of developing constructive solutions for devices and elements of the cutting body of a geokhod, it is necessary to take into account the complex movement of the machine towards the working face. Accordingly, when justifying the power parameters of devices and elements of the cutting body of a geokhod, it is necessary to take into account not only the peculiarities of the interaction of geokhod devices with the external environment, but also the interaction of elements among themselves.

To assess the performance of the executive body of the mining machine, the energy intensity of rock destruction is used [14, 15, 16]. In geokhod technology of cavity formation in underground space, the energy intensity of the destruction of low hardness rocks will be influenced by the geometrical parameters of the geokhod cutting body. Therefore, the research aimed at determining the influence of the geometrical parameters of the geokhod blade cutting body on the energy intensity of the rock destruction is relevant.

2. Research methodology

From the developed methodology for determining the energy intensity of rock destruction by earth-moving machines (p, MJ/m^3) proposed by Yu. A. Vetrov, the energy intensity is determined by the expression

$$p = \frac{P}{F_{mid}}, \tag{1}$$

where P– blade cutting force, H;

F_{mid} – slot cross section area, mm^2.

One of the parameters that affect the energy intensity of destruction is the slot cross section area. The impact of forces of one blade of geokhod cutting body for the destruction of low hardness rocks is presented in figure 3.

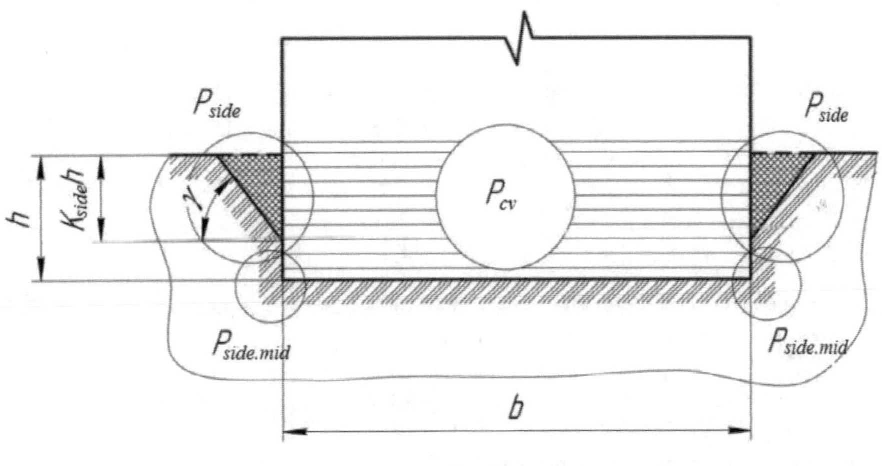

Figure 3. The areas of action of the components of the cutting force with a sharp blade.

For one blade of the geokhod cutting body, the cross-sectional area of the slot will be determined by the expression

$$F_{mid} = bh + k_{side}^2 h^2 ctg\gamma, \tag{2}$$

where b – the blade width, m;

h – depth of cutting, m;

k_{side}– the depth ratio of the expansion part of the slot;

γ– angle of inclination of the expansion part of the slot to the horizon, degree.

The second parameter affecting the energy intensity of rock destruction is the power factor. For the geokhod blade cutting body, the main power parameters will be the projection of the total force of the resistance of the soil to cutting on the axis of rotation of the geokhod P_0 and on the plane perpendicular to the axis of rotation of the geokhod R_{cb}, as well as the moment of resistance to cutting from this component of M_{cb} [14, 15, 16 , 17]. The scheme of action of power factors is presented in figure 4.

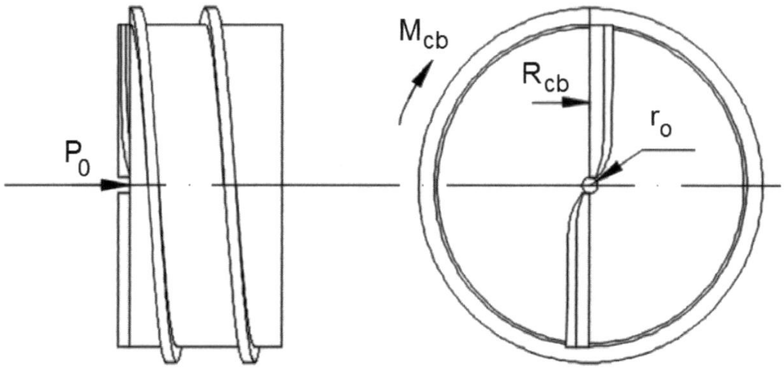

Figure 4. Calculation scheme for determining the total force of soil resistance to cutting with a blade cutting body of a geokhod.

To determine the energy intensity of rock destruction with a blade cutting body of a geokhod, the cutting force P in expression (1) will be equal to the projection of the total soil resistance to cutting on the plane perpendicular to the axis of rotation of the geokhod R_{cb}. Therefore, expression (1) takes the form

$$p = \frac{R_{cb}}{F_{mid}}, . \quad (3)$$

For a geokhod blade cutting body, the total projection of the resistance of the soil to cutting on a plane perpendicular to the axis of rotation is equal to [14–19]

$$R_{cb} = n(R_{cb.cv} + R_{cb.side}), \quad (4)$$

where n– the number of blades of the geokhod body;

$R_{cb.cv}$– projection of the component of the force of soil resistance to cutting, depending on the width of the cut, on the plane perpendicular to the axis of rotation of the geokhod, H;

$R_{cb.side}$ – projection of the component of the force of soil resistance to cutting, not depending on the width of the cut, on the plane perpendicular to the axis of rotation of the geokhod, H.

Projection of the component of the force of soil resistance to cutting, depending on the width of the cut, on the plane perpendicular to the axis of rotation of the geokhod [14–19]

$$R_{cb.cv} = \frac{\varphi m_{cv} h_B^2}{2\pi n \cos\gamma}\left[\frac{(\sin\beta_2 - \sin\beta_1)}{\sin\beta_2 \cdot \sin\beta_1} + ctg(\delta + \varphi_{fr}) ln\left|\frac{tg\frac{\beta_2}{2}}{tg\frac{\beta_1}{2}}\right|\right], \quad (5)$$

Projection of the component of the force of soil resistance to cutting, not depending on the width of the cut, on the plane perpendicular to the axis of rotation of the geokhod [18–23]

$$R_{cb.cv} = \frac{h_B}{n}\left(m_{side}\frac{h_B}{n} + m_{side.mid}\right)\left[\frac{\sin(\delta+\varphi_{fr}+\beta_1)+\sin(\delta+\varphi_{fr}+\beta_2)}{\sin(\delta+\varphi_{fr})}\right]. \quad (6)$$

Figure 5 shows the scheme of the geokhod blade cutting body, on which its main geometrical parameters are indicated. One of them is the inclination angle of the blade to the plane perpendicular to the axis of rotation of the geokhod – γ, the change of which affects the shape of the face. At $\gamma = 0°$, a flat face shape is formed (figure 6, a), and at $\gamma > 0°$ a conic face shape appears (figure 6, b).

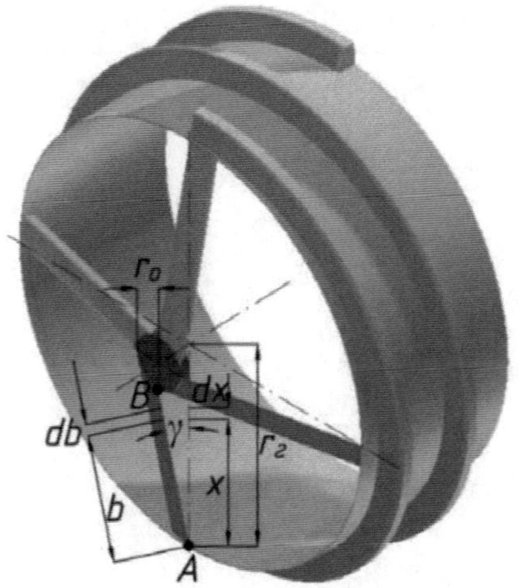

Figure 5. Scheme of a blade cutting body.

Taking into account the inclination angle of blade to the plane perpendicular to the axis of rotation of the geokhod (γ), the expression (2) takes the form

$$F_{mid} = \frac{bh}{\cos\gamma} + k_{side}^2 h^2 ctg\gamma_{pr}, \quad (7)$$

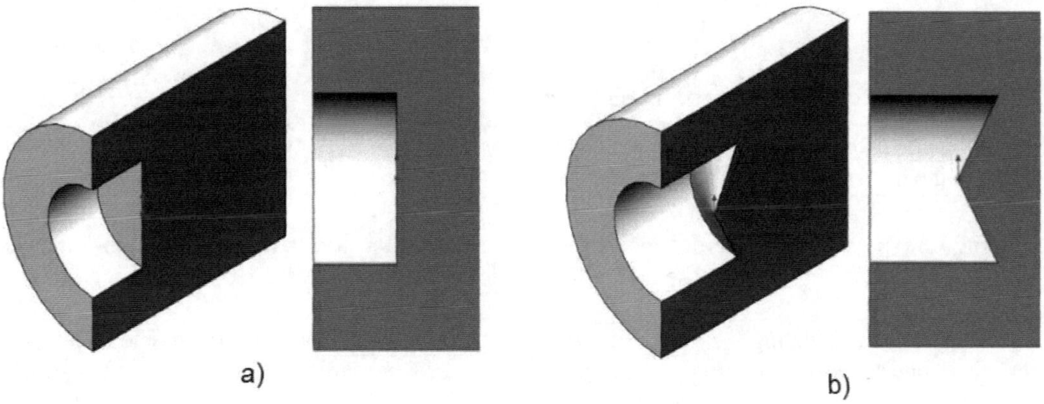

Figure 6. Scheme of a blade cutting body.

IOP Conf. Series: Materials Science and Engineering **656** (2019) 012003 doi:10.1088/1757-899X/656/1/012003

3. Results

To determine the influence of the inclination angle of the blade to the plane perpendicular to the axis of rotation of the geokhod on the energy intensity of rock destruction, we justified the parameters of mining technical conditions of making workings with geokhods, as well as the geometric parameters of the blade cutting body. The values of these parameters are presented in table 1.

Table 1. Mining technical conditions of making a working with a geokhod and the geometrical parameters of the geokhod blade cutting body.

Name	Symbol	Unit of measurement	Value
Geokhod radius	R_g	m	0.3
Central nozzle radius	r_o	m	0.025
External propulsion step	h_B	m	0.3
Coefficient taking into account the influence of the cutting angle	φ		0.59
Specific cutting force in the frontal part of the slot at a cutting angle of 45^0	M_{cv}	H/m^2	97000
Strength of the destruction of the soil in the side parts of the slot	M_{side}	H/m^2	36000
Specific cutting force of one of the side edges of the blade	$M_{side.mid}$	H/m	8490
Number of blades on a cutting body	n	pcs	1
Angle of inclination of the radial blade to the plane perpendicular to the axis of rotation of the geokhod	γ	deg	0
Cutting angle	δ	deg	25
Friction angle	φ_{fr}	deg	31.4
Depth ratio of the expansion part of the slot	K_{side}		0.9
Angle of inclination of the expansion part of the slot to the horizon	G_{pr}	deg	30

According to expression (3), the dependence of the energy intensity of rock destruction with the blade of the geokhod cutting body on the inclination angle of the blades of the geokhod cutting body is determined.

The procedure for calculating the geokhod blade cutting body is shown in figure 7. The developed algorithm was applied when designing a computer program for determining the energy intensity of the rock destruction with the geokhod blade cutting body.

IOP Conf. Series: Materials Science and Engineering **656** (2019) 012003 doi:10.1088/1757-899X/656/1/012003

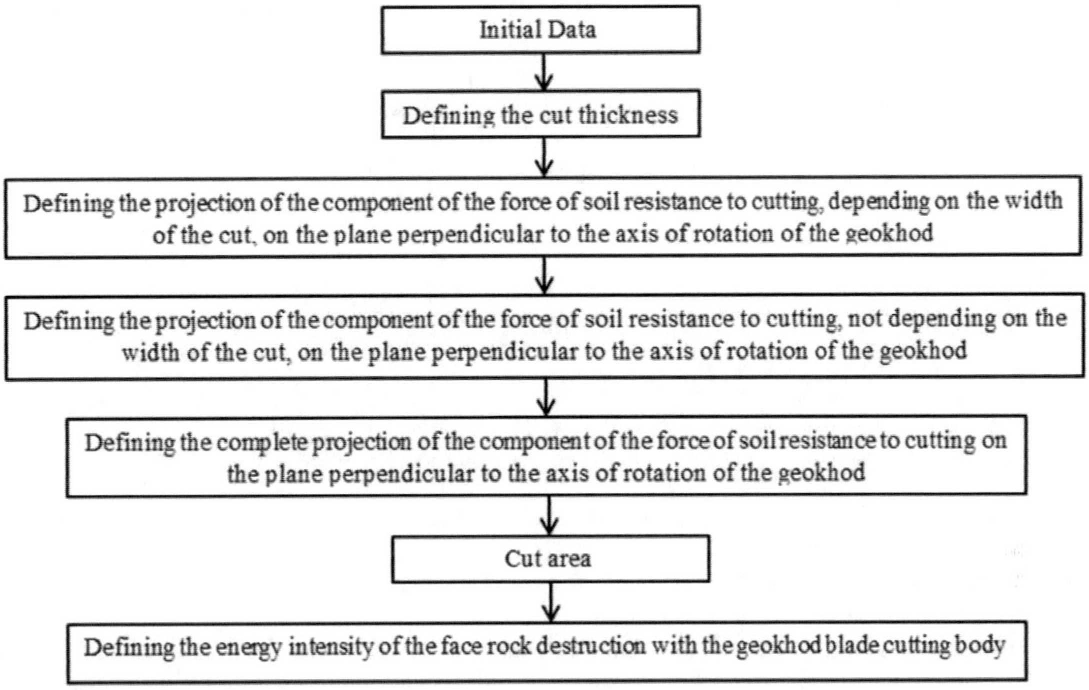

Figure 7. Algorithm for calculating the energy intensity of the rock destruction with the geokhod blade cutting body.

On the basis of the obtained values, we plotted a graph (figure 8) of the changes in the energy intensity of the rock destruction with the blade cutting body of the geokhod if the inclination angle of blades is changed.

The ordinate axis of the presented dependence shows the value of the energy intensity of the rock destruction with the blade cutting body of the geokhod (MJ/m^3), and the abscissa axis shows the change of inclination angle of blades of the geokhod cutting body (deg.).

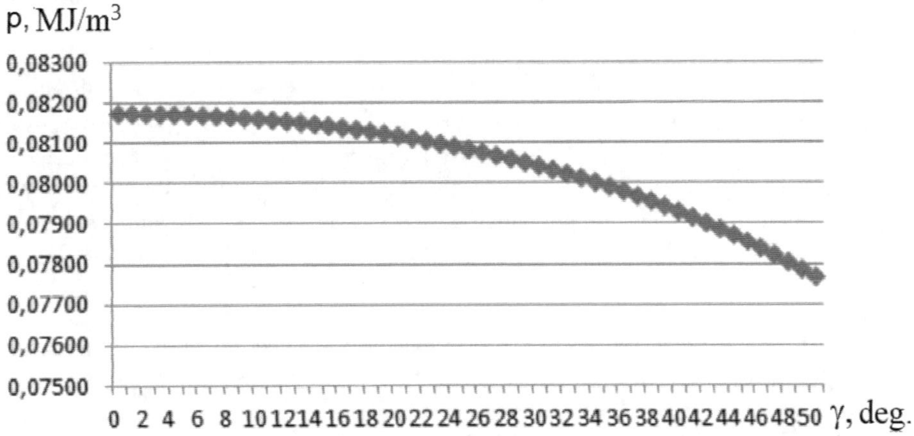

Figure 8. Dependence of the energy intensity of the rock destruction with the geokhod blade cutting body on the inclination angle of blade.

From the graph presented in figure 8, it follows that:

– the value of the energy intensity of the rock destruction with the geokhod blade cutting body changes nonlinearly downwards with an increase in the inclination angle of blades of geokhod cutting body;

– the energy intensity of the rock destruction decreases as a result of the relative increase in the area of the cut;

– from the point of view of the energy intensity of the rock destruction with the geokhod blade cutting body, the conic shape of the face is preferable, as shown in figure 6, b.

References

[1] Aksenov V, Sadovets V and Pashkov D 2017 *E3SWeb of Conferences* **21** 03008
[2] Aksenov V, Sadovets V, Rezanova E and Pashkov D 2017 *E3SWebConf* **15** 03015
[3] Aksenov V, Sadovets V and Pashkov D 2018 *E3SWebConf* **41** 03002
[4] Aksenov V V, Efremenkov A B, Sadovets V Y and Pashkov D A 2018 *IOP Conference Series: Materials Science and Engineering* **441** 012002
[5] Nishi S and Seiki T 1997 Planning and design of underground space use *Mem Sch Eng Nagoya Univ* **1**
[6] Maid B, Schmid L, Ritz W and Herrenknecht M 2008 *Hardrock Tunnel Boring Machines* (Berlin: Ernst&Sohn) p 343
[7] Wighman T 1998 *ENR: News* **Rec 4**
[8] Brierley G 1998 World Tunnel *Andsubsurface Excav* **9**
[9] Efremenkov A B 2011 *6th International Forum on Strategic Technology (IFOST-2011)* **2:1** 348
[10] Aksenov V V, Sadovets V Yu and Pashkov D A 2017 Determining the number of blades of the cutting body of geokhod for destruction of rocks of low hardness *Vestnik of Kuzbass State Technical University* **3 121** 116–126
[11] Sadovets V Yu, Beglyakov V Yu and Efremenkov A B 2015 Simulation of geokhod movement with blade actuator *Applied Mechanics and Materials* **770** 384–390
[12] Aksyonov V V, Sadovets V Yu and Beglyakov V Yu 2011 Impact of the bench on the stress-strain state of near face part of mine workings *Innovative Technologies and Economics in Mechanical Engineering* 575–580
[13] Walter A V and Aksenov V V 2015 Determining deviations in geometry of the geokhod shells *Applied Mechanics and Materials* **770** 439–444
[14] Aksenov V V, Efremenkov A B, Sadovets V Yu and Rezanova E V 2010 Creating innovative tools for the formation of underground space *Vestnik of Kuzbass State Technical University* **1 77** 42–47
[15] Sadovets V Yu and Pashkov D A 2016 The influence of the parameters of the helicoid generatrix on the shape of the geokhod blade cutting body *Natural and Intelligent Resources of Siberia Sibresurs Collection of materials of the XVI International Scientific and Practical Conference*
[16] Sadovets V Yu and Pashkov D A 2016 Determination of the function of the line of contact of the blade of the geokhod cutting body with the face surface *Technologies and Materials* **4** 9–14
[17] Aksenov V V, Sadovets V Yu, Pashkov D A and Rezanova E V 2018 Boundary conditions for determining the characteristic points of a geokhod blade cutting body *Vestnik of Kuzbass State Technical University* **2 126** 166–173
[18] Aksenov V V, Sadovets V Yu, Preis E B and Pashkov D A 2018 Improving the mathematical model for determining the power parameters of the geokhod blade cutting body *Mining Equipment and Electrical Engineering* **5 139** 16–22
[19] Carmody J and Sterling R 1993 *Undergroundspace Design* (New York: Reinhold)

IOP Conf. Series: Materials Science and Engineering **656** (2019) 012004 doi:10.1088/1757-899X/656/1/012004

The concept of creating perspective technological paradigm of formation (development) of the underground space on the basis of the leading development of new approaches in construction geotechnology and geotechnics. Premises and basic provisions (part 1)

V V Aksenov[1,2,6], A A Khoreshok[3] and V U Beglyakov[4,7] and A B Efremenkov[5,8]

[1]Scientific and research centre LLC "Siberian Research and Production Association", 650099, Sovetsky Av. 56, Kemerovo, Russia
[2]Institute of Coal of the Federal Research Center of Coal and Coal Chemistry, Siberian Branch of the RAS, 650610, 10 Leningradsky av, Kemerovo, Russia
[3]T.F. Gorbachev Kuzbass State Technical University, 650000, 28 Vesennyaya St., Kemerovo, Russia
[4]Yurga Technological Institute Branch of Tomsk Polytechnic University, 652052 Lenin-gradskaya Str. 26, Yurga, Russia
[5]Yaroslav-the-Wise Novgorod State University, ul. B. St. Petersburgskaya, 41 173003 Veliky Novgorod, Russia

[6]E-mail: 55vva42@mail.ru
[7]E-mail: begljakov@rambler.ru
[8]E-mail: abe@novsu.ru

Abstract. The article presents some reasons for the lagging of domestic mining engineering from foreign competitors, substantiates the need to form a new promising technological paradigm in the field of formation and development of underground space. A new systematic approach to the development of geotechnics and geotechnologies is proposed, which allows for creation of prerequisites for formation of a new technological structure and Russia's breakthrough into leading positions in the field of geotechnology and mining engineering.

The development of mining machines has long been on an extensive path, i.e. the increase in performance is achieved mainly due to the increase in power and weight and size characteristics [1, 2], with no significant breakthrough in the technical level. There is a noticeable lag from advanced industries, for example, from aircraft manufacturing. And this is typical not only for the Russian industry, but is a worldwide trend.

Technical characteristics of the aircraft are dramatically increased if weight and size characteristics are maintained or even if they are reduced. Aircraft designed with an interval of 25 years can not be confused during the arrangement by age (figure 1).

The growth of the technical level of mining machines, of course, is also there, but, alas, it cannot be compared to that of airplanes. Often, only a specialist can determine the age of a tunneling machine (figure 2).

Figure 1. Development of aircraft: Su-7 1955 on the left, Su-27 1980 on the right.

Figure 2. Development of tunneling machines: PKG-3 1956 on the left, KRP-3-660/1140 2000 on the right.

There was a need to create a new technological paradigm in the mining engineering [3]. And while creating a new paradigm one should take into account the experience of other, more dynamically developing industries.

Let's consider the approaches and solutions used, in particular, in aircraft construction and shipbuilding. As is known, in these areas, the results of those studies are used, which are based on the process of studying the motion of a solid body in air and water, respectively.

Tables 1 and 2 present the existing infrastructure for tracking the movement of a solid body in various environments.

Table 1. Infrastructure of tracking the movement of a solid in air and liquid environments.

Environment	Apparatus in the environment	Propeller	Mechanical engineering industry	Science of support	Test center
			Science of creation	Main purposes of science	
Aero	Airplane, helicopter, etc.	Propeller carrying screw (interaction with the environment)	Aircraft, helicopter engineering, rocket science	Aerodynamics of aircraft	Flight test center
				The study of forces arising on the surface of a solid moving in air environment	

<table>
<tr><td rowspan="2">Hydro</td><td rowspan="2">submarine, ship, etc.</td><td rowspan="2">Propeller (interaction with environment)</td><td rowspan="2">Shipbuilding</td><td>Hydrodynamics of underwater vehicles</td><td rowspan="2">Underwater test center</td></tr>
<tr><td>The study of forces arising on the surface of a solid moving in liquid environment</td></tr>
</table>

Table 2. Infrastructure of tracking the movement of a solid in a solid environment (geoenvironment).

<table>
<tr><td rowspan="2">Condition</td><td rowspan="2">Apparatus in the environment</td><td rowspan="2">Propeller</td><td>Mechanical engineering industry</td><td>Science of support</td><td rowspan="2">Test center</td></tr>
<tr><td>Science of creation</td><td>Main purposes of science</td></tr>
<tr><td rowspan="2">The status quo</td><td rowspan="2">Construction machines adapted to work in underground conditions</td><td rowspan="2">Wheel throw, caterpillar throw (interaction only at the contact of environments)</td><td rowspan="2">Mining machine engineering</td><td>absent</td><td rowspan="2">absent</td></tr>
<tr><td>absent</td></tr>
<tr><td rowspan="2">Proposition</td><td rowspan="2">Geokhod</td><td rowspan="2">Helical blade (interaction with the environment)</td><td rowspan="2">Geokhog engineering</td><td>Geodynamics of underground vehicles</td><td rowspan="2">Underground test center</td></tr>
<tr><td>The study of the forces arising on the surface of a solid moving in a solid environment</td></tr>
</table>

The aircraft industry is the most demonstrative. Originating at the turn of the last century and the past centuries, the industry in just over a hundred years has gone a long way from the first airplanes to modern airliners, helicopters and flying robots. Figure 3 conventionally shows the technological structure in the aircraft industry, and this structure was formed in the USSR at the turn of the 20s, 30s of the last century under difficult conditions of industry collapse, when the country was under technological and economic blockade, or in our time – under sanctions.

Technological structure in the aircraft industry has stood the test of time: made a great contribution to the creation of the "red" air fleet, winning the war, creating jet aircraft and, despite the difficult economic situation in the country in the 90s, kept it at the forefront in the world in the field of aircraft manufacturing.

The key elements (structures) of the technological paradigm in the aircraft industry are:

– Aerodynamics of aircraft [4, 5]. Fundamental science, studying the forces arising on the surface of a solid body moving in air environment.

The main objective of aerodynamics [5, 6] is the choice of a rational form of aircraft in order to obtain specified flight performance, as well as determine the aerodynamic loads and heat fluxes acting on the surface of the aircraft, for strength calculations.

– Aircraft [7, 8].

IOP Conf. Series: Materials Science and Engineering **656** (2019) 012004 doi:10.1088/1757-899X/656/1/012004

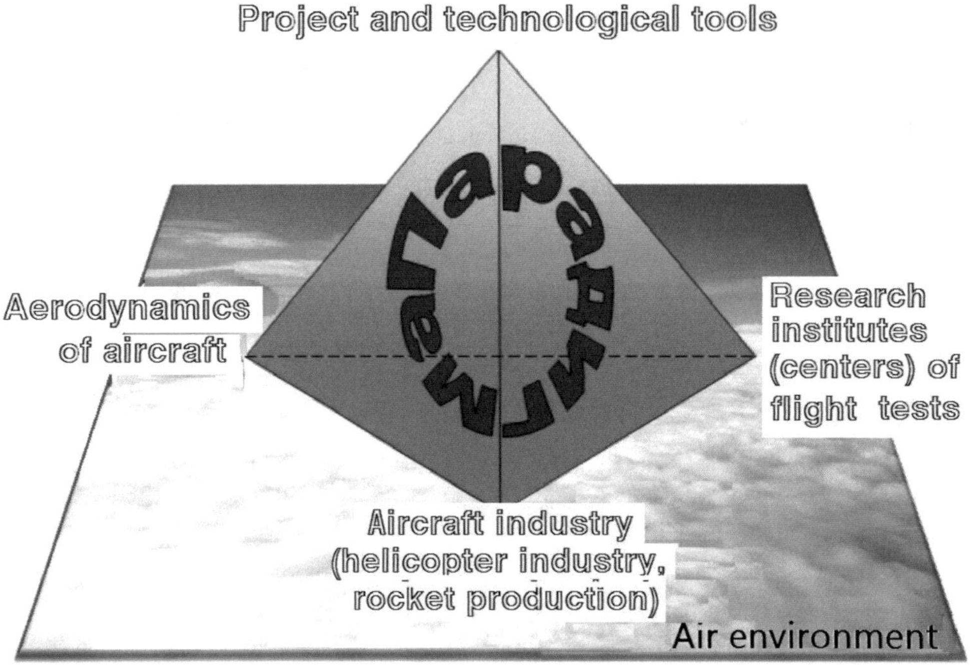

Figure 3. Scheme (paradigm) of technological structure in the aircraft industry.

1) Scientific discipline related to the design and construction of aircraft.

2) An aircraft manufacturing industry.

Aircraft (aircraft industry) at one time initiated the need for the industrialization of the country, and also became one of the main consumers of the created components of mechanical engineering: materials; bearings, mechanisms, etc.

– Research institutes (centers) of flight tests [9, 10]. The emergence of new professions: test pilot, test engineer, new methods and techniques

1) Conducting full-scale ground and flight tests of aircraft and their systems.

2) Participation in the development and upgrading of aircraft at all stages of its life cycle.

3) The introduction of tested aircraft and their systems in civil and military aviation.

Project and technological tools

1) Designing airports, etc.

2) Navigation provision, etc.

3) Staff training.

3) Creating and improving the regulatory framework.

4) Ensuring the safety of flights and operation of aircraft.

In the aggregate, the elements (structures) of the technological structure in the aircraft industry form a certain amount of knowledge – a paradigm of knowledge of the technological structure. In figure 3, the paradigm is conditionally represented as a volume of a tetrahedron.

With any change (increase) of knowledge in any element (structure) of the technological structure in the aircraft industry, the volume of the tetrahedron increases, that is, the general paradigm of knowledge of the technological structure increases.

The main advantage of the end-to-end technological structure in the aircraft industry is the reduction of the time period from the formation of the concept, the development and manufacture of new aircrafts and their systems, the development of products to their introduction into civil and military aviation.

Air environment
Figure 4 presents the Earth cutaway [12]. At present, the lower layers of the atmosphere are of practical importance for civil aviation: the troposphere and the lower part of the stratosphere (up to an altitude of 20 km) [6, 13]. When creating and operating the aircraft, the structure and parameters of the atmosphere are taken into account: pressure, density, temperature. In these layers, the pressure, density and temperature of the medium decrease when the altitude increases.

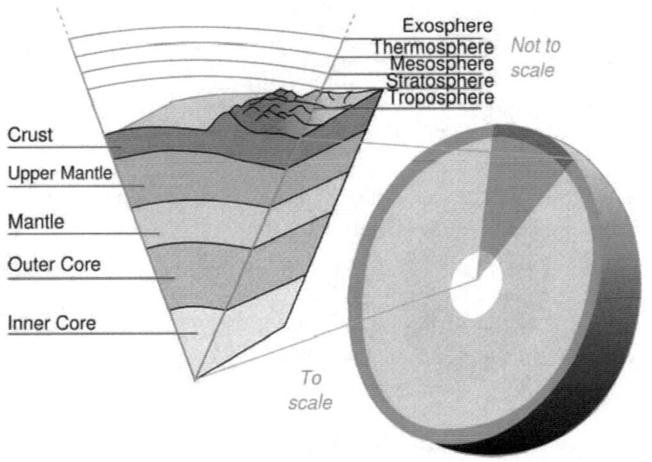

Figure 4. Earth, cutaway.

Geoenvironment
Geoenvironment is much more diverse and multifaceted in its properties and manifestations in comparison to the air environment.

Indicators of physical and mechanical properties of rocky and non-rocky soils (rocks), and there are more than 20 of them only by name [13], differ considerably among themselves.

When creating a promising technological mode of development (formation) of underground space on the basis of the advanced development of new approaches in building geotechnology and geotechnics, it seems necessary to fully take into account the experience and merits of the end-to-end technological structure created in the aircraft industry.

Purposes of the concept
Strategic purposes:

– Creation of a promising technological mode of development (formation) of underground space on the basis of the advanced development of new approaches in building geotechnology and geotechnics, providing a reduction in the period of time from the formation of design, development, production of new UV, fine-tuning of products and their systems to introduction into industrial and civil construction, units MES, MO.

– Creation of a new high-tech and knowledge-intensive engineering industry – geokhod engineering [14,15].

– Formation of a new segment of the mining engineering market and its capture.

Intermediate (tactical) purposes:

– Creation of experimental models of new types of products: geokhods, underground robots, mining structures and support [16].

– Creation and production of pilot batches of new types of products (geokhods of various standard sizes and functional purposes, resource-saving constructions of mine workings) [17–19].

– Creation of new geotechnologies for the formation and development of underground space, the development of underground transport infrastructure [20, 21].

– Creating a highly efficient and competitive mining technology of a new generation.

– Formation of the backbone infrastructure to speed up the process of creating and selling scientific and technical products in regions and Russia as a whole, bringing them to industrial use.

– Creation of production facilities for the manufacturing a new type of mining equipment, support of mine workings and lining of underground structures.

– Creating new jobs in depressed areas.

The way to achieve the goal

– Consolidation of efforts of the state, investors, scientific, design, production and educational organizations.

– Concentration of efforts on the chosen strategic and basic directions of activity.

– Creating competition in selected areas of activity.

The concept assumes the use of the program approach in solving the problem of formation of a new through technological structure.

Figure 5 conditionally shows the structure of a promising technological mode of development (formation) of underground space based on the advanced development of new approaches in building geotechnology and geotechnics.

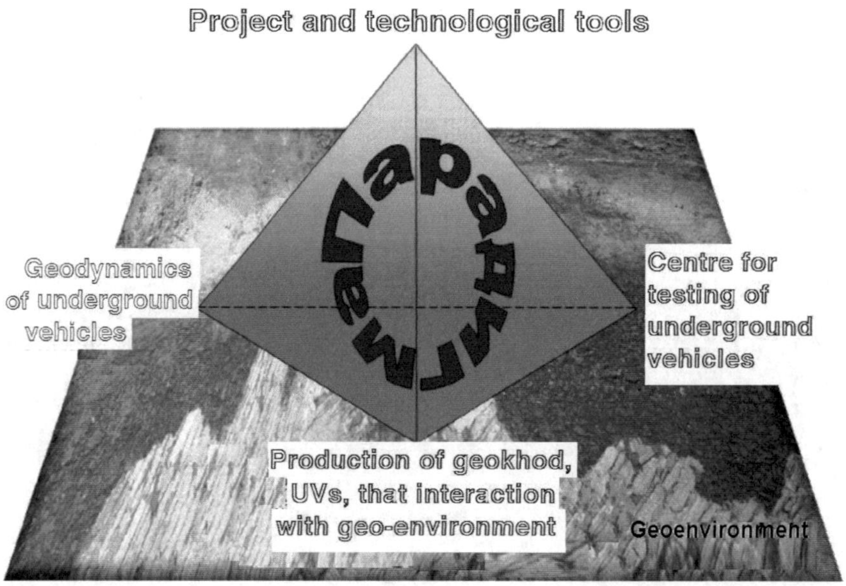

Figure 5. The paradigm of the promising technological structure of development (formation) of underground space based on the advanced development of new approaches in building geotechnology and geotechnics.

The key elements (structures) of the technological mode of development (formation) of underground space are:

IOP Conf. Series: Materials Science and Engineering **656** (2019) 012004 doi:10.1088/1757-899X/656/1/012004

– Geodynamics of underground vehicles. Fundamental science, studying the forces arising on the surface of a solid body moving in a solid environment.

The main task of geodynamics of an UV is to select a rational form of an UV for the purpose of obtaining specified technical characteristics, as well as determining geodynamic loads and heat fluxes acting on the surface of an UV, for strength calculations.

Geokhod engineering

1) The scientific discipline related to the design and construction of geokhods, UVs, moving in geo-environment using the environment itself, underground robots [22, 23].

2) The industry engaged in the production of geokhods, Uvs, moving in the geoenvironment using the environment itself, underground robots [24].

Geokhod engineering along with other industries should be the initiator of the new industrialization of the country, and also become one of the main consumers of products created in the framework of the implementation of the Concept of component engineering.

Center(s) of testing underground equipment. The emergence of new professions: test miner, test engineer of mining equipment, new methods and techniques.

1) Conducting full-scale ground and underground testing of UV moving in geoenvironment using the environment itself, underground robots and their systems.

2) Participation in the completion and modernization of UV moving in a geoenvironment using the environment itself, underground robots at all stages of their life cycle.

3) The introduction of tested UV, moving in the geoenvironment using the environment itself, underground robots and their systems in industrial and civil construction, units of the Ministry of Emergency Situations, MO.

Project and technological tools

1) Design of underground structures for various purposes, etc.

2) Navigation provision, etc.

3) Staff training.

4) Creating and improving the regulatory framework.

5) Ensuring the safety of flights and operation of work and operation of geotechnics.

In the aggregate, the elements (structures) of the prospective technological mode of development (formation) of underground space form a certain amount of knowledge – the Paradigm of knowledge of the technological mode. In figure 5, the paradigm is conditionally represented as the volume of a tetrahedron.

Main activity directions

– Development of scientific principles for designing geokhods and UVs moving in geoenvironment using the environment itself, underground multipurpose robots and their systems.

– Development of scientific bases of geodynamics of underground devices.

– Development of structures of geokhods, UVs, moving in the geoenvironment using the environment itself, underground robots, their elements and systems (geokhod engineering).

– Manufacturing of prototype models of geokhods, UVs, moving in geoenvironment using the environment itself, underground robots for various purposes.

– Testing experimental models of geokhods, UVs, moving in the geoenvironment using the environment itself, underground robots.

– Development on the basis of geokhods of mobile complexes for special purposes.

– Justification of the principles and the creation of new geotechnologies and the holding of underground workings.

– Development of technical solutions and methods for calculating support of mine workings and lining of underground structures for various purposes.

– Development of experimental models of support of mine workings and underground structures for various purposes.

– Manufacturing of experimental models of support.

– Testing experimental models of support.

– Creation, technical and scientifically-methodical equipment of the research center of underground tests.

– Creation of data acquisition and processing systems for the analysis of unique objects.

– Training of specialists for the design, manufacture, testing and operation of geokhods.

– Creation of scientific and technical products, software products.

– Design of underground structures, objects of underground transport infrastructure.

– Formation of a backbone scientific and project infrastructure, creation of manufactures for the production of a new type of mining equipment, support and lining of underground workings.

Brief description of products (strategic direction)

Table 3. A brief description of products.

Main products	1. Underground vehicles moving in the geoenvironment using the environment itself, underground robots of various sizes and purposes. 2. Lining of underground facilities of various sizes. 3. Single-purpose mobile complexes.
Type of products	1. A new type of mining equipment (expert opinion). 2. A new type of support of mine workings and lining of underground facilities.
Product analogue	Has no analogues
Application area	1. Mine workings of various location in space. 2. Construction of underground structures for various purposes: – city collectors; – magistral metro tunnels; – underground warehouses, storage facilities; – underground passages, garages; 3. Erection of fortification facilities. 4. Conducting rescue operations in the rubble.
Sales market	1. Enterprises of urban and municipal construction. 2. Transport construction enterprises. 3. Metro construction organizations, tunnel units. 4. Divisions of MO, MES, PMRU. 5. Underground coal mining enterprises.

Socio-economic results

The implementation of the concept and the creation of a promising technological structure for the development (formation) of underground space based on the advanced development of new approaches in building geotechnology and geotechnics will allow:

– to reduce the costs of building underground facilities for various purposes (transport tunnels, subway, mines, factories, urban collection facilities, nuclear power facilities, warehouses, etc.);

– to increase the country's defense;

– to create mountain rescue equipment for conducting rescue operations in the rubble;

– to create a highly efficient and competitive mining technology and a new generation technology;

– to export engineering equipment and technology;
– to create new jobs in depressed areas;
– to develop new Research and Development jobs;
– to create industrial enterprises for the production and marketing of new competitive products.

References

[1] Aksenov V V, Khoreshok A A, Efremenkov A B, Kazantsev A A, Beglyakov V Y and Valter A V 2015 Sozdanie novogo instrumentariya dlya formirovaniya podzemnogo prostranstva [Creating a new tool for the formation of underground space] *Gornaya technika* **1** 24–26

[2] Aksenov V V, Efremenkov A B, Sadovets V Y and Rezanova E V 2010 *Sozdanie innovacionnogo instrumentariya dlya formirovaniya podzemnogo prostranstva* [Creating innovative tools for the formation of underground space] *Vestnik KuzSTU* **1** 42–47

[3] *The concept of creating a promising industrial structure based on the advanced development of key engineering components* Available at: http://pro-kompas.info/strategy/ (accessed: 21.07.2019) [In Russ.]

[4] Petrov K P 1998 *Aerodinamika tel prostejshih form* [Aerodynamics of bodies of the simplest forms] (Factorial Publ)

[5] Mhitaryan A M 1976 *Aerodinamika* [Aerodynamics] (M.: Mechanical engineering) pp 217–232

[6] Pyshnov V 2013 *Aerodinamika samolyota* [Aircraft aerodynamics] (Ripol Classic Publ)

[7] Krivov G A at al 2008 Grazhdanskoe samoletostroenie v nachale XXI stoletiya [Civil aircraft in the beginning of the XXI century] *Activities of leading global manufacturers*

[8] Yakovlev A S 1980 Samolyotostroenie [Aircraft engeneering] *The development of aviation science and technology in the USSR*

[9] Vasilchenko K K, Leonov V A and Pashkovskij I M 1994 *Letnye ispytaniya samoletov* [Flight tests of aircraft] (Moscow: Mashinostroenie Publ) p 407

[10] Akimov A I, Brestov L M and Mikheev R A 1980 *Letnye ispytaniya vertolyotov* [Flight tests of helicopters] (Moscow: Mashinostroenie Publ) p 399

[11] *Paradigm* (Wikipedia) Available at: https://ru.wikipedia.org/wiki/Парадигма (accessed: 21.07.2019) [In Russ.]

[12] *Structure of Earth* (Wikipedia) Available at: https://ru.wikipedia.org/wiki/Структура_Земли (accessed: 21.07.2019) [In Russ.]

[13] Rzhevskij V V and Novik G Y 1964 *Osnovyj fiziki gornyh porod* [Fundamentals of rock physics] (Science Publ)

[14] Aksenov V V, Beglyakov V Y, Valter A V, Efremenkov A B and Kazantsev A A 2016 Opyt uchastiya yurginskogo tekhnologicheskogo instituta (filiala) NI TPU v proekte po organizacii vysokotekhnologichnogo proizvodstva [Experience of participation of the Yurginsky Technological Institute (branch) of TPU in the project on the organization of high-tech production] *Technologies and materials* **2** 10–17

[15] Aksenov V V, Beglyakov V Y, Valter A V and Efremenkov A B 2016 Opyt uchastiya v proekte po organizacii vysokotekhnologichnogo proizvodstva [Experience of participation in the project on the organization of high-tech production] *Mining equipment and electrical engineering* **8(126)** 8–15

[16] Aksenov V V, Beglyakov V Y, Blashchuk M Yu, Efremenkov A B, Kazantsev A A, Khoreshok A A and Valter A V 2016 Geokhod: zadachi, harakteristiki, perspektivy [Geokhod: objectives, characteristics, prospects] *Mining equipment and electrical engineering* **8 (126)** 3–8

[17] Aksenov V V and Kazantsev A A 2013 Armiruyushchaya zakonturnaya krep' gornyh vyrabotok – novyj podhod k stroitel'stvu podzemnyh sooruzhenij [Reinforcing marginal support of mine workings – a new approach to the construction of underground structures] *Mining Information and Analytical Bulletin (scientific and technical journal)* **S6** 411–418

[18] Aksenov V V, Kazantsev A A and Dortman A A 2012 Obosnovanie neobhodimosti sozdaniya sistem krepi gornyh vyrabotok pri prohodke po geovinchesternoj tekhnologii [Justification of

the need to create support systems for mine workings when driving on geowinchester technology] *Mining Information and Analytical Bulletin (scientific and technical journal)* **S3** 138–143

[19] Aksenov V V, Kazantsev A A and Dortman A A 2013 Obosnovanie neobhodimosti sozdaniya profilya dlya vintovoj ramnoj krepi v geovinchesternoj tekhnologii [Justification of the need to create a profile for helical frame support in geowinchester technology] *Mining engineer* **1** 280–287

[20] Aksenov V V 2004 *Nauchnye osnovy geovinchesternoj tekhnologii provedeniya gornyh vyrabotok i sozdaniya vintopovorotnyh agregatov* [The scientific basis of geovinchesternoy technology for mine workings and the creation of screw-turn units] (Kemerovo, Institute of Coal and Coal Chemistry, Siberian Branch of the Russian Academy of Sciences)

[21] Aksenov V V and Efremenkov A B 2008 Geovinchesternaya tekhnologiya i geohody – innovacionnyj podhod k osvoeniyu podzemnogo prostranstva [Geovinchester technology and geokhods – an innovative approach to development of underground space] *Expert Technique* **1** 54–58

[22] Beglyakov V U, Aksenov V V, Kazantsev A A, Kostinets I K 2017 Razrabotka zakonturnoj oporno-dvizhitel'noj sistemy geohodov [Development of outside the contour support-propulsion system of geokhod] *Vestnik KuzSTU* **6(124)** 175–181

[23] Beglyakov V U, Aksenov V V, Kostinets I K and Khoreshok A A 2017 Determining the forces of interaction of main geokhods systems with geo-environment and with each other *Mining science and technology* **4** 23–28 [In Russ.] doi.org/10.17073/2500-0632-2017-4-23-28

[24] Aksenov V V and Valter A V 2013 Specifika geohodov kak ob'ektov proizvodstva i problemy sozdaniya tekhnologij ih izgotovleniya [The specifics of geokhods as objects of production and the problems of creating technologies for their manufacture] *Mining engineer* **1** 222–223

IOP Conf. Series: Materials Science and Engineering **656** (2019) 012005 doi:10.1088/1757-899X/656/1/012005

The concept of creating perspective technological paradigm of formation (development) of the underground space on the basis of the leading development of new approaches in construction geotechnology and geotechnics. Premises and basic provisions (part 2)

V V Aksenov[1,2,6], A A Khoreshok[3], V U Beglyakov[4,7] and A B Efremenkov[5,8]

[1]Scientific and research centre LLC "Siberian Research and Production Association", 650099, Sovetsky Av. 56, Kemerovo, Russia
[2]Institute of Coal of the Federal Research Center of Coal and Coal Chemistry, Siberian Branch of the RAS, 650610, 10 Leningradsky av, Kemerovo, Russia
[3]T.F. Gorbachev Kuzbass State Technical University, 650000, 28 Vesennyaya St., Kemerovo, Russia
[4]Yurga Technological Institute Branch of Tomsk Polytechnic University, 652052 Lenin-gradskaya Str. 26, Yurga, Russia
[5]Yaroslav-the-Wise Novgorod State University, ul. B. St. Petersburgskaya, 41 173003 Veliky Novgorod, Russia

[6]E-mail: 55vva42@mail.ru
[7]E-mail: begljakov@rambler.ru
[8]E-mail: abe@novsu.ru

Abstract. The article presents some reasons for the lagging of domestic mining engineering from foreign competitors, substantiates the need to form a new promising technological paradigm in the field of formation and development of underground space. A new systematic approach to the development of geotechnics and geotechnologies is proposed, which allows for creation of prerequisites for formation of a new technological structure and Russia's breakthrough into leading positions in the field of geotechnology and mining engineering.

The main areas of development of underground space include [1, 2]:

– adaptation of natural cavities (caves, karst cavities);

– reuse of existing man-made cavities (mine workings, mothballed objects of civil defense, etc.) in a new capacity;

– construction of new underground facilities for civil and special purposes.

The development of civilization is accompanied by the expansion of infrastructure and population growth, which inevitably leads to reduction of the amount of the undeveloped territories on the surface of the planet as well as search for new places to accommodate buildings, industrial and other objects of human activity.

In fact, there are three of such places [3, 4]: space, water bodies and the Earth's interior, and subsequently the surface and the interior of other planets. At the present stage of development of our society, underground space is of the greatest interest.

Worldwide, the construction of underground facilities for various purposes has become one of the priorities [5]. The number of them in developed countries doubles every 10 years [3–5], and in the future we should expect a further increase in the rate of exploration of underground space.

The market for the construction of underground workings in Russia is growing dynamically. Growth rates in natural terms range from 9 to 27%.

By function, underground facilities can be divided into four main groups [3, 4]:

1. Industrial and infrastructure facilities for technical purposes (energy, mining and industrial enterprises, transport complexes, warehouses, etc.).
2. Social facilities (libraries, hospitals, commodity bases and storages, gyms, shops, etc.).
3. Objects for environmental purposes (repositories for the disposal of radioactive and industrial waste, hazardous substances, hazardous manufactures).
4. Objects of defense purposes.

The modern state of technology in the field of underground space formation by a mechanized method is represented by two classes of mining machines [7, 8]: tunneling machines and tunneling shields. At the same time, if the production of tunneling machines in Russia is somehow represented by Ltd Yurga Machine-Building Plant and JSC Kopeysk Machine-Building Plant, then the production of panel-tunnel mining aggregates is much more complicated, it simply does not exist.

In the USSR, tunneling shields were designed and manufactured by domestic enterprises. Import of such important equipment was almost absent [5, 7, 8]. In the late 90s, production was almost completely stopped due to the crisis in the economy of our country, scientific and design institutes were closed. Since the 2000s, tunneling shields have been purchased from foreign manufacturers. Nowadays in Russia, most of the equipment (especially shields of large diameters from 6 m) is of foreign manufacture.

Russia has fallen into a situation in which it is extremely dependent on imports.

It should be noted that the world's first mining combines and sewage treatment plants for mining reservoir deposits were created and introduced in the USSR in the 40s and 50s. Further development and introduction of new equipment (not equipment upgrades) was a very long process.

The main types of products offered by foreign manufacturers for tunneling are: tunneling machines of all diameters for all rocks and environmental requirements, machines for driving in hard rocks, open tunneling machines with a system of thrust (Gripper) against the walls of the tunnel, tunneling machines for oblique tunneling, expansion tunneling machines , tunneling shields, tunneling machines, etc.

Major competing companies [5, 8]**:**

– ***Herrenknecht AG (Germany).*** It is the only company in the world that manufactures machines for the construction of tunnels with diameters ranging from 0.10 to 19 meters in any engineering and geological conditions;

– ***The Robbins Company (USA).*** An international organization with four main production enterprises, has an extensive network of representative offices throughout the world (at the moment there is no official representative office in Russia). Over 60 years on the market;

– ***Mitsubishi heavy industries, Ltd (Japan).*** Large diversified holding with a century and a half history (the company was founded in 1868);

– ***Lovat Tunnel Equipment (Canada).*** Engaged in the design, maintenance, production and repair of shield tunnel complexes, specializes in shield microtunnel complexes of diameter from 0.75 to 15 m;

– ***WIRTH GmbH (Germany).*** Produces equipment for laying and maintenance of tunnels, which is used throughout the world.

Currently, the introduction of fundamentally new domestic technology in coal and mining companies, tunneling organizations in Russia is almost impossible, since a prerequisite for its use is the customers' requirement to provide the results of using the machine in other enterprises. This is caused by the discrepancy between the technical levels of the consumer and the manufacturer of technological machines, i.e. the destruction of the technological paradigm. (Paradigm – technological structure or Techno-economic paradigm is a set of related productions that have a single technical level and develop synchronously).

This makes impossible the revival and development of the mining engineering industry without restoring or creating a new promising industrial and technological paradigm.

At present, with the active participation of Ltd "Siberian NPO", the Concept of creating a promising industrial structure based on the priority development of key engineering components has been formed and is being implemented. One of the approaches of the Concept is the thesis [9]:

"It is difficult to enter the industrial structure formed by global corporations with standard products, but one can set a goal to replace this structure to the maximum extent. If we propose more economical technologies, we can significantly expand our participation in the international division of labor, as well as ensure our priority in creating more advanced mechanisms and machines."

A constraining factor in creating a promising industrial paradigm in the part relating to mining engineering, both in Russia and abroad, is the current approach to the creation of mining machines.

The traditional representation of excavation as the process of cavity formation in the rock massif, has always defined and still defines the directions of:

– scientific research;

– improvement of geotechnologies for the construction of underground structuresand, accordingly

– creation of tunneling equipment for the formation of underground space.

At present, in underground conditions, the following external propulsive devices are used to move the tunneling apparatus: tracked, wheeled, wheel-rail or distance-walking. The propellers that have performed well when working on the earth's surface (at the contact of solid and air environments) are not adapted for movement in geoenvironment.

In the course of work of the tunneling machine or the shield, in order to create the force of thrust and pressure force on the executive body, the external geoenvironment itself is not involved in any way, what involved is only the solid surface of the excavation at the interface of the geo-and air environments, or with the shield method of penetration – a strong permanent support.

This circumstance leads to the main problems of modern technologies for the construction of underground workings:

– the impossibility of creating a universal mining apparatus capable of moving in any direction of the Earth's interior;

– the dependence of the maximum force on the gravity vector, and as a result, the impossibility of creating large pressure forces on the executive body to destroy hard rock.

As a result, in order to create sufficient pressure forces, designers are forced to increase the mass of mining combines, the mass of which already exceeds 130 tons, and the task of reducing the overall metal consumption of the machine is not even considered.

Further development of work in the field of geotechnologies and geotechnics can go in two directions [2, 6]:

1) modernization of the existing mining equipment and its improvement by creating systems of a new technical level;

2) the search and creation of a fundamentally new, alternative geotechnological tools (technologies, geotechnics) for exploration of the subsoil and formation of underground space.

Known technologies for mining, developing along the path of increasing the power and metal consumption of equipment, have practically exhausted their capabilities in increasing productivity, ensuring safety of work and expanding the scope.

In search of ways to create a fundamentally new tools for mine workings, the team of authors reviewed alternative approaches and solutions used, in particular, in aircraft construction and shipbuilding. As is known, in these areas results of the research based on the process of studying the motion of a solid body in air and water, respectively, are used.

Aircraft designers and shipbuilders in the design and construction of vehicles intended for movement in airspace (aircraft, helicopter, etc.) and in the aquatic environment (ship, submarine, etc.) found the possibility of using the environment itself to create thrust when the corresponding device moves. For this purpose, propellers operating inside the environment (propeller, propeller screw, etc.), or "into coverage" of the environment (water jet and other jet propellers), performance of which does not directly depend on gravity, were created. As a result, the possibility of moving a solid body (apparatus) in any direction of the air or water space was created.

Over the years, a team of scientists and engineers has been working to create a fundamentally new type of mining equipment – geokhodes [10–39].

A scientific and practical groundwork has been created in the field of developing new technology for mine workings [10, 11], and a new type of mining technology [12–15], the patent purity of the developed technical [15–24] and technological [11, 25-29] solutions has been confirmed.

With the participation of employees of UTI TPU, OJSC "UMP", KuzSTU, OJSC Siberian NPO and OJSC "KORMZ", the Comprehensive Project "Creation and Production of a Multi-Purpose Screen Tunneling Units of Geological Multipurpose Objects" was implemented - the winner of the Ministry of Education and Science (2013-218-04) in the selection of organizations for the right to receive subsidies for the implementation of complex projects for the creation of high-tech production [12, 13].

The accumulated scientific background allowed the team of authors to implement a complex of research and developmental works, which resulted in a prototype model of geokhod "model 401" (figure 1) [14].

a)

b)

Figure 1. A prototype of a geokhod model "401" with a diameter of 3.2 m. a) in the manufacture of the OJSC "KORMZ"; b) on the test bench.

IOP Conf. Series: Materials Science and Engineering **656** (2019) 012005 doi:10.1088/1757-899X/656/1/012005

The results obtained not only opened up new areas of research in the field of mining and determined the need for the emergence of a new branch of engineering – geo-engineering, but also created the prerequisites for Russia to become a leader in the field of geotechnology and mining engineering.

Patent search and analysis of scientific publications showed that the research team is a leader in this direction. There are no analogues of the research conducted in mining and the results obtained both in Russia and in the world ***yet***.

During the designing and manufacturing of the prototype, the absence of a domestic component base for many components was revealed, which, along with the imposed sanctions, will be a deterrent when creating new samples of geokhods.

In addition, there are no specialized centers for testing developed new geotechnics, its fine-tuning and implementation at existing enterprises.

The implementation of the project gave as a result not only a prototype of the machine, but also revealed new ways to improve the design of the geokhod, problems of testing and experimental studies of this type of machines, problems of creating a working environment for mastering the production of new types of products. The results obtained form the basis of ***innovative geotechnical tools for the formation of underground space –*** a complex that includes:

– a new approach to the conduct of underground mining and the formation of underground space;
– new technologies for underground mining;
– new class of mining technology;
– a new type of support for mine workings and lining of underground structures;
– new scientific and methodological support.

A new approach to the underground mining and the formation of underground space – the excavation of underground mining is considered as a process of movement of a solid body (equipment) in the environment of surrounding rocks (geological environment).

The marginal array (geoenvironment) is used here:

– as a supporting element involved in formation of the driving force of the underground vehicle – geokhod, incl. for the formation of a pressure force on the executive body;
– for the perception of reactive forces during the movement of the tunneling unit (underground apparatus);
– to perform basic technological operations, including fixing the production of a permanent support.

New technologies for underground mining – geo-technology – the process of mechanized mining with the formation and use of a system of marginal screw and longitudinal channels in which operations for the development of the face, cleaning the rock mass, fixing the open space, and moving the entire tunneling system to the face are being done in combined mode.The involvement of the marginal rock array is achieved by the introduction of an additional technological operation – the formation of a system of marginal channels [30-36].

New class of mining equipment, systems (GPS) – geokhods, underground robots.

Geokhod is a machine moving in the rock array using geological environment. The basic element of geokhod technology. Representing a new class of mining machines, geokhods are designed for driving underground workings for various purposes and location in space.

Underground robots. In the line of robots for various purposes, there are aircraft, ground, floating and underwater robots. There is an urgent need to create underground robots, especially for performing specialized tasks of the Ministry of Emergency Situations and other in dangerous and hazardous conditions. The principle of geokhod operation assumes fully automatic control of the entire system. According to experts of the Ministry of Emergency Situations geokhod is the closest to the appearance of an underground robot.

A new type of lining of mine workings and lining of underground structures is a geo-built (marginal) support [37-39].

Geo-built (marginal support is based upon the use of screw and longitudinal channels behind the contour of the current generation for formation of a spatial support system "support – marginal array of rocks".

New scientific and methodological support – new scientific disciplines, educational programs, methods of substantiating the parameters of the main systems of geokhod and geo-built support, etc.

The solution of problems of designing a new type of mining equipment – geokhods– requires the creation of new scientific directions:

– geodynamics of underground devices is a science that studies the forces arising on the surface of a solid (underground device) moving in a solid environment (geological environment);

– geokhod engineering is a scientific discipline related to the design and construction of geokhods (underground robots).

One of the defining directions for the development of mining (development of the subsoil and formation of underground space) is the improvement of its ***design and technological tools***, which in turn is based on specialists, scientific and engineering personnel and scientific and methodological support.

Findings:

– The introduction of a fundamentally new domestic technology in coal and mining companies is almost impossible in Russia, since a prerequisite for its use in the enterprise is the requirement to provide the results of using the machine in other enterprises.

– The current state of technology in the field of underground space formation in a mechanized way is represented by two classes of mining machines: tunneling machines and tunneling shields. At the same time, if the production of tunneling machines in Russia is somehow represented by Ltd Yurga Machine-Building Plant and JSC Kopeysk Machine-Building Plant, then the production of panel-tunnel mining aggregates is much more complicated, it simply does not exist.

– In Russia, with the active participation of OJSC Siberian NPO, the Concept of creating a promising industrial paradigm based on the priority development of key engineering components has been formed and is being implemented.

– Known technologies for mining, developing along the path of increasing the power and metal consumption of equipment, have practically exhausted their capabilities in increasing productivity, ensuring safety of work and expanding the scope.

– A new approach to the process of construction of underground workings and the formation of underground space has been formulated – the excavation of underground workings was initially considered as a process of movement of a solid body (equipment) in the environment of surrounding rocks (geo-environment).

– A scientific and practical groundwork has been created in the field of developing a new technology for mine workings, and a new type of mining equipment – geokhod.

– Many geokhod systems have no analogues in mining engineering, they are fundamentally different in purpose and principle of operation from all existing systems of mining tunnels and tunneling shields. In addition, the manufacture of geokhods and their systems are very specific in production.

– The results obtained not only opened up new areas of research in the field of mining and determined the need for the emergence of a new branch of engineering – geo-engineering, but also created the prerequisites for Russia to become a leader in the field of geotechnology and mining engineering.

– The absence of a domestic component base for many components has been revealed, which, along with the imposed sanctions, will be a deterrent when creating new samples of geokhods.

– There are no specialized centers for testing the developed geotechnics, its fine-tuning and implementation at existing enterprises.

– One of the defining directions for the development of mining (development of the subsoil and formation of underground space) is the improvement of its ***design and technological tools.***

– There is an urgent need to create a promising technological paradigm for the development (formation) of underground space based on the rapid development of new approaches in building geotechnology and geotechnics.

References

[1] Belyaev V L and Belyaev V B 2012 Osvoenie podzemnogo prostranstva gorodov v aspekte ih ustojchivogo razvitiya [The development of the underground space of cities in the aspect of their sustainable development] *MSSU* 198 p

[2] Aksenov V V, Khoreshok A A, Efremenkov A B, Kazantsev A A, Beglyakov V Y and Valter A V 2015 Sozdanie novogo instrumentariya dlya formirovaniya podzemnogo prostranstva [Creating a new tool for the formation of underground space] *Gornaya technika* **1** 24–26

[3] Kartoziya B A 2015 Osvoenie podzemnogo prostranstva krupnyh gorodov. Novye tendencii [The development of the underground space of large cities. New tendencies] *Mining Information and Analytical Bulletin (scientific and technical journal)* **1** 615–629

[4] Kartoziya B A *at al* 2001 Mine and underground construction. Vol 1: Textbook for universities *Academy of Mining Sciences* p 607

[5] *Marketing research of the tunneling market and estimation of the volume of works on the construction of underground workings (structures) in Russia 2012–2015* Available at: https://www.megaresearch.ru/issledovaniya/syre/mineraly-gornye-porody/22904 (accessed: 21.07.2019) (In Russ.)

[6] Aksenov V V, Efremenkov A B, Sadovets V Y and Rezanova E V 2010 *Sozdanie innovacionnogo instrumentariya dlya formirovaniya podzemnogo prostranstva* [Creating innovative tools for the formation of underground space] **1**

[7] Safokhin M S 1995 Mining machines and equipment: a textbook for universities *Nedra* p 463

[8] Brenner V A, Zhabin A B, Schegolevsky M M, Polyakov A V and Polyakov A V 2009 Shield tunneling complexes: Textbook *Gornaya kniga MSU* p 447

[9] *The concept of creating a promising industrial structure based on the advanced development of key engineering components* Available at: http://pro-kompas.info/strategy/ (accessed: 21.07.2019) (In Russ.)

[10] Aksenov V V 2004 Nauchnye osnovy geovinchesternoj tekhnologii provedeniya gornyh vyrabotok i sozda-niya vintopovorotnyh agregatov [Scientific basis of geovinchesternoy technology for mine workings and the creation of screw-rotating units] *ICC SU RAS*

[11] Aksenov V V, Izakson V Y, Cherdantsev N V, Anferov B A and Kuznetsova L V Method for underground mining of circular cross section *Patent for invention* RUS 2369742 26.03.2008

[12] Aksenov V V, Beglyakov V Y, Valter A V, Efremenkov A B and Kazantsev A A 2016 Opyt uchastiya yurginskogo tekhnologicheskogo instituta (filiala) NI TPU v proekte po organizacii vysokotekhnologichnogo proizvodstva [Experience of participation of the Yurginsky Technological Institute (branch) of TPU in the project on the organization of high-tech production] *Technologies and materials* **2** 10–17

[13] Aksenov V V, Beglyakov V Y, Kazantsev A A, Valter A V and Efremenkov A B 2016 Opyt uchastiya v proekte po organizacii vysokotekhnologichnogo proizvodstva [Experience of participation in the project on the organization of high-tech production] *Mining equipment and electrical engineering* **8(126)** p 8–15

[14] Aksenov V V, Beglyakov V Y, Blashchuk M Yu, Efremenkov A B, Kazantsev A A, Khoreshok A A and Valter A V 2016 Geokhod: zadachi, harakteristiki, perspektivy [Geokhod: objectives, characteristics, prospects] *Mining equipment and electrical engineering* **8 (126)** 3–8

[15] Lelukh B F, Aksenov V V, Efremenkov A B, Khoreshok A A, Masalitin B G, Beglyakov V Y and Timofeev V Y Birotative tunneling shield unit *Patent for invention* RUS 2412354 31.07.2009

[16] Aksenov V V, Slyadnev A V, Efremenkov A B, Khoreshok A A, Masalitin B G, Beglyakov V Y and Timofeev V Y Geokhod (tunnel shield unit) *Patent for invention* RUS 2552539

04.12.2013

[17] Aksenov V V, Efremenkov A B, Lelukh B F and Sadovets V Y Duplex geokhod *Patent for invention* RUS 2469192 24.05.2011

[18] Gorbunov V F, Eller A F, Aksenov V V, Nagorny V D and Skomorokhov V M Tunneling shield unit *Patent for invention* RUS 1229354 24.04.1984

[19] Eller A F, Aksenov V V, Nagorny V D and Gorbunov V F Tunneling shield unit *Patent for invention* RUS 1328531 23.10.1985

[20] Eller A F, Gorbunov V F, Aksenov V V, Pushkina N B, Saruev L A and Krauin'sh P Y Tunneling shield unit *Patent for invention* RUS 1668678 02.08.1989

[21] Eller A F, Gorbunov V F and Aksenov V V Tunneling shield unit *Patent for invention* RUS 1719642 04.05.1987

[22] Aksenov V V, Efremenkov A B, Blashchuk M U, Lelukh B F and Beglyakov V Y Tunneling shield unit *Utility patent* RUS 112269 13.12.2010

[23] Eller A F, Aksenov V V and Pushkina N B Tunneling shield unit *Patent for invention* RUS 2066762 **4**

[24] Aksenov V V, Efremenkov A B, Timofeev V Y, Beglyakov V U, Blashchuk M U Tunneling shield unit (geokhod) *Patent for invention* RUS 2418950 05.10.2009

[25] Aksenov V V, Valter A V and Beglyakov V U 2014 Obespechenie geometricheskoj tochnosti obolochki pri sborke sekcij geohoda [Ensuring the geometric accuracy of the shell when assembling sections of geohod] *Metal processing (technology, equipment, tools)* **4 (65)** 19–28

[26] Valter A V, Aksenov V V, Beglyakov V U and Chazov P A 2015 Opredelenie pogreshnosti raspolozheniya sektorov stabiliziruyushchej sekcii geohoda na osnove dannyh koordinatnogo kontrolya [Determination of the error of the location of the sectors of the stabilizing section of the geohod based on the coordinate control] *Metal processing (technology, equipment, tools)* **4 (69)** 31–42

[27] Valter A V and Aksenov V V 2015 Varianty obespecheniya tochnosti obolochek i sobiraemosti korpusov geohoda [Options to ensure the accuracy of the shells and the collection of buildings geohod] *Mechanics of XXI century* **14** 89–92

[28] Valter A V and Aksenov V V 2015 Opredelenie otklonenij geometricheskoj formy obolochek korpusnyh izdelij geohoda [Determination of deviations of the geometric shape of geokhod hulls] *Collection of works of the International Scientific and Practical Conferenceof Yurga Institute of Technology* 165–170

[29] Aksenov V V and Valter A V 2013 Specifika geohodov kak ob'ektov proizvodstva i problemy sozdaniya tekhnologij ih izgotovleniya [The specifics of geokhods as objects of production and the problems of creating technologies for their manufacture] *Mining engineer* **1** 222–223

[30] Ermakov A N, Aksenov V V, Khoreshok A A and Ananiev K A 2014 Obosnovanie trebovanij k ispolnitel'nym organam formirovaniya zakonturnyh kanalov geohoda [Justification of the requirements for the executive bodies of the formation of marginal channels of geokhod] *Vestnik KuzSTU* **2** 5–7

[31] Aksenov V V, Kostinets I K and Beglyakov V U 2013 Osobennosti raboty vneshnego dvizhitelya geohoda [Features of the external propeller of a geokhod] *Mining Information and Analytical Bulletin (scientific and technical journal)* **S6** 419–425

[32] Aksenov V V, Beglyakov V U, Kazantsev A A, Kostinets I K and Koperchuk A V 2016 Klassifikaciya geometricheskih parametrov vneshnego dvizhitelya geohoda [Classification of the geometric parameters of the external propeller of geokhod] *Mining equipment and electrical engineering* **8** 33–39

[33] Beglyakov V U, Aksenov V V, Kazantsev A A and Kostinets I K 2017 Razrabotka zakonturnoj oporno-dvizhitel'noj sistemy geohodov [Development of the boundary propeller support system of geokhods] *Vestnik KuzSTU* **6 (124)** 175–181

[34] Aksenov V V, Beglyakov V U and Kostinets I K 2013 Razrabotka trebovanij k vneshnemu dvizhitelyu geohoda [Development of requirements for external propulsion of a geokhod]

Mining engineer **1** 182–188

[35] Beglyakov V U, Aksenov V V, Kostinets I K and Khoreshok A A 2017 Skhemy nagruzheniya pri modelirovanii processa vzaimodejstviya vneshnego dvizhitelya geohoda prikonturnym massivom porod [Loading schemes for modeling the interaction of an external propulsion geokhod marginal rock mass] *Mining science and technology* **3** 3–8

[36] Beglyakov V U, Aksenov V V, Kostinets I K and Khoreshok A A 2017 Opredelenie sil vzaimodejstviya osnovnyh sistem geohoda s geosredoj i mezhdu soboj [Determination of the forces of interaction of the main geokhod systems with the geoenvironment and between each other] **4** 23–28

[37] Aksenov V V and Kazantsev A A 2013 Reinforcing edge support of mine workings - a new approach to the construction of underground structures [Armiruyushchaya zakonturnaya krep' gornyh vyrabotok – novyj podhod k stroitel'stvu podzemnyh sooruzhenij] *Mining Information and Analytical Bulletin (scientific and technical journal)* **S6** 411–418

[38] Aksenov V V, Kazantsev A A and Dortman A A 2012 Obosnovanie neobhodimosti sozdaniya sistem krepi gornyh vyrabotok pri prohodke po geovinchesternoj tekhnologii [Justification of the need to create support systems for mine workings when driving on geowinchester technology] *Mining Information and Analytical Bulletin (scientific and technical journal)* **S3** 138–143

[39] Aksenov V V, Kazantsev A A and Dortman A A 2013 Obosnovanie neobhodimosti sozdaniya profilya dlya vintovoj ramnoj krepi v geovinchesternoj tekhnologii [Justification of the need to create a profile for helical frame support in geowinchester technology] *Mining engineer* **1** 280–287

Mathematical modelling of the dosing process of bulk materials on the example of the sowing apparatus of a seeder

D N Algazin[1]

[1]Yaroslav-the-Wise Novgorod State University, ul. B. St. Petersburgskaya, 41 173003 Veliky Novgorod, Russia

E-mail: dna@novsu.ru

Abstract. A mathematical model of the dosing process of bulk materials has been developed, it allows to determine the optimal technological and design parameters of the metering device of the sowing apparatus of a seeder. A new type of a sowing apparatus of a seeder is proposed, allowing to seed small-seeded crops.

1. Introduction

Uniform and specified in certain values dosing of bulk materials [1], including the dosing of seeds during sowing, is a matter of current interest [2]. As a result of the analysis of technical solutions for sowing small-seeded crops, such as rapeseed, alfalfa, and tributary, it was found that the main disadvantages of the existing dosing systems are the uneven distribution of seeds (seeding pulsation) and increased seed injury. To eliminate these shortcomings, a metering device (sowing device) of bobbin-screw type [3] was developed, as is the synthesis of a bobbin sowing device and a screw dosing device.

2. Model

We consider a mathematical model of the process of dosing bulk materials on the example of a bobbin-screw type sowing apparatus [4]. Scheme of the bobbin-screw type sowing apparatus is presented in figure 1.

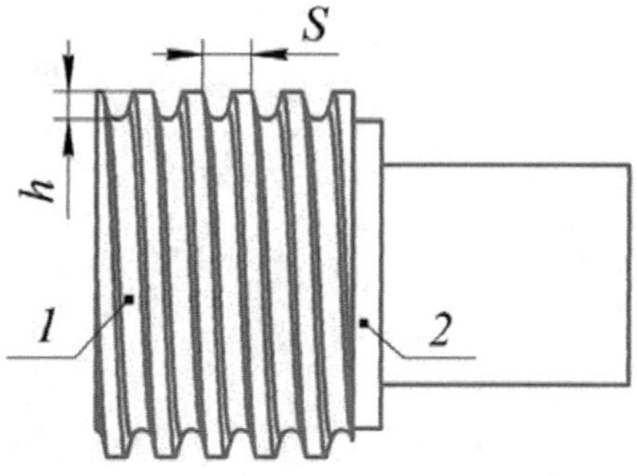

Figure 1. Bobbin-screw type sowing apparatus
1 – flute; 2 – groove; S – helix pitch, mm; h – depth of the rabbet, mm.

The volume of seeds V_0 m^3 ejected by the bobbin in one turn consists of the volume of seeds trapped in the flutes V_F and the volume of seeds ejected from the active layer V_A, i.e.:

$$V_0 = V_F + V_A, \tag{1}$$

where V_F is the seed volume sown by the flutes in one turn of the bobbin, m^3;

V_A is the seed volume sown by the active layer one turn of the bobbin, m^3.

The determination of the volume of seeds sown by the active layer V_A is complicated by a number of circumstances and primarily by the fact that the speed of the seeds movement is not constant in the thickness of layer C (figure 2), and besides the type of function can only be determined by experiment.

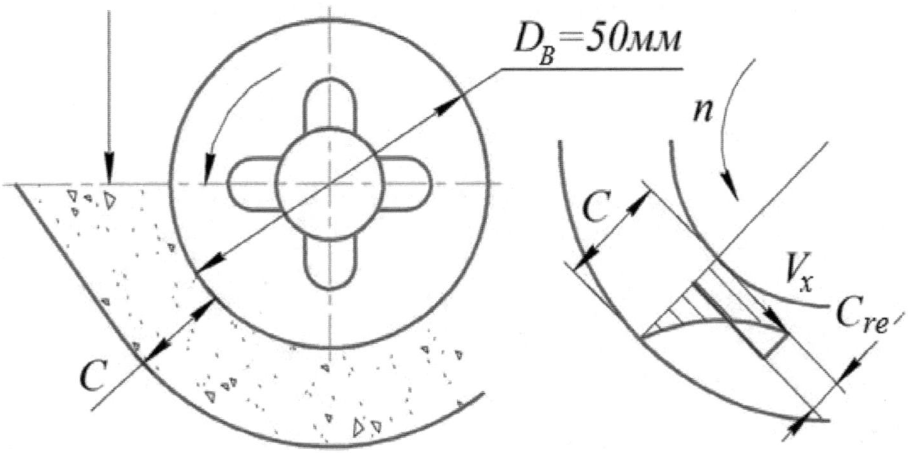

Figure 2. Scheme of the bobbing type sowing apparatus operartion C is the thickness of the layer; C_{re} is the reduced thickness of the active layer; n is the sowing machine bobbin rotation frequency.

Experimental study of the operation of the bobbin-screw type sowing apparatus showed that the pattern of change in the velocity V_x of seed movement in the active layer can (as a first approximation) be expressed by the following relationship [5]:

$$V_x = \pi \cdot D_B \cdot n\left(1 - \frac{x}{C}\right)^m, \tag{2}$$

where m is an exponent, determined experimentally;

n is the sowing machine bobbin rotation frequency, c^{-1}.

We will conduct a simulation of the working process of the sowing apparatus, focusing on the reduced thickness of the center of the reduced layer C_{re}, which is easily determined experimentally. The given thickness of the active layer is the thickness C_{re} of a certain layer of seeds moving with a constant velocity V_{sa} (Figure 2).

If we use the expression 2, then the relationship between C and C_{re} can be determined from the following equality:

$$V_{sa} = \int_0^C \left(1 - \frac{x}{C}\right)^m dm = C_{re} \cdot V_{sa} \tag{3}$$

Whence:

$$C = C_{re}(m+1). \tag{4}$$

IOP Conf. Series: Materials Science and Engineering **656** (2019) 012006 doi:10.1088/1757-899X/656/1/012006

The values of C_{re} vary slightly with the change in the length of the working length of the bobbin I_B and the frequency of its rotation n, figure 2.

Thus, the volume of seeds sown by the active layer is determined by the formula:

$$V_A = \pi \cdot D_B \cdot C_{re} \cdot l_B \,, \tag{5}$$

where D_B is the diameter of the bobbin of the sowing apparatus, m;

C_{re} is the reduced thickness of the active layer, m;

I_B is the working length of the bobbin of the sowing apparatus, m;

The amount of seeds sown by the bobbin flutes can be determined by calculation, because the flute of the bobbin screw sowing apparatus is a helix with only one flute, and there is a single-thread screw, or with several flutes if there is a multiple-thread screw. Therefore, the volume of the flute is determined:

$$V_F = F_F \cdot L, \tag{6}$$

where F_F is the area of the section of the bobbin's flutes

L is the length of the flute of the bobbin, m.

The flute in cross section is a parabola, which can be represented as consisting of two elements: an inverted trapezium and a segment. Geometrical parameters of the flute are presented in figure 3.

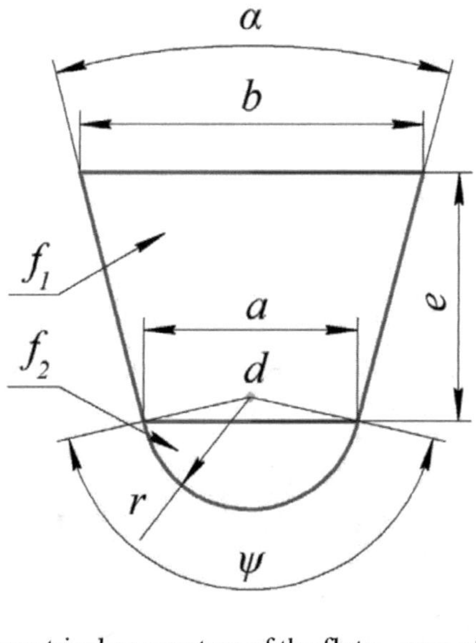

Figure 3. Geometrical parameters of the flute – area of a trapezium; – area of a segment; a is the length of the smaller base of the trapezium; b – the length of the bigger base, m; e – the height of the trapezium; r – radius of the segment; α – tilt angle of the verge; ψ – central angle.

The sectional area of the flute is determined as the sum of the trapezoid and segment areas (figure 3):

$$F_F = \left(b - h \cdot tg\frac{\alpha}{2}\right)h + \frac{r^2}{2}\left(\frac{\pi}{180}\Psi - sin\Psi\right). \tag{7}$$

We will find L – the length of the flute of the screw bobbin of the sowing apparatus:

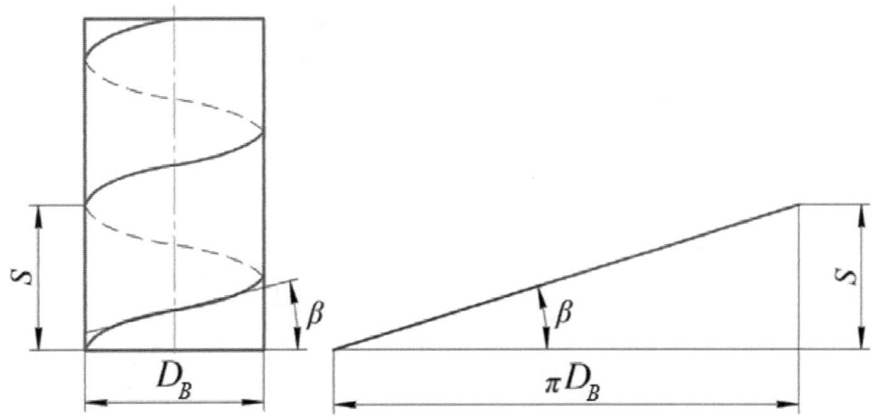

Figure 4. The main parameters of the helix D_B is the diameter of the bobbin, m; β is the elevation angle of the helix, degrees; L is the length of the flute of the sowing apparatus' screw bobbin, m; S is the helix pitch, m.

As can be seen from figure 4, the arc length of one pitch l, is determined by:

$$l = \frac{\pi D_B}{cos\beta}, \tag{8}$$

where D_B is the diameter of the bobbin, m;
β is the elevation angle of the helix, degrees.
The helix pitch is determined by:

$$S = \pi \cdot D_B \cdot tg\beta \tag{9}$$

At an angle of $0^o < \beta < 90^o$, the flute is located along the helix with the parameters shown in figure 4, the number of turns is determined by:

$$z = \frac{l_B}{p \cdot cos\beta}, \tag{10}$$

Therefore, the length of the helix is equal to:

$$L_F = l \cdot z. \tag{11}$$

To determine the length of the flute of the bobbin, we substitute in equation 11 equations 8, 9 and 10:

$$L_F = \frac{2l_K}{\sin 2\beta}. \tag{12}$$

Due to a number of factors: the shape and size of the seeds, the profile of the flute of the bobbin, etc., it is necessary to enter the fill factor of the flute ξ, which is determined by the ratio of the volume of the flute filled with seeds to the total volume of the flute, and this is how we get the volume of seeds sown by the flute for one turn of the bobbin:

$$L_F = \xi\left(b - h \cdot tg\frac{\alpha}{2}\right)h + \frac{r^2}{2}\left(\frac{\pi}{180}\Psi - sin\Psi\right)\frac{2l_B}{\sin 2\beta}. \tag{13}$$

Then the volume of seeds sown by the bobbin per turn (the working volume of the bobbin or the sowing capacity), based on formulas 1, 12 and 20, expresses in:

$$V_0 = \xi\left(b - h \cdot tg\frac{\alpha}{2}\right)h + \frac{r^2}{2}\left(\frac{\pi}{180}\Psi - sin\Psi\right)\frac{2l_K}{\sin 2\beta} + \pi \cdot D_B \cdot C_{re} \cdot l_B. \quad (14)$$

Knowing that the number of seeds is determined by: where γ is the volumetric weight of seeds, kg/m^3, and δ is the weight of 1000 seeds, kg, we determine the number of seeds sown per one turn of the bobbin, while introducing the correction speed coefficient k_n, which shows how much the number of sown seeds varies from the frequency of rotation of the bobbin. Formula 14 takes the final look:

$$N = \left[\left(\left(b - htg\frac{\alpha}{2}\right)h + \frac{r^2}{2}\left(\frac{\pi}{180}\psi - sin\psi\right)\right)\frac{2l_K\xi}{sin2\beta} + \pi \cdot D_B \cdot C_{re} \cdot l_B\right] \cdot \delta^{-1} \cdot \gamma - k_n n \quad (15)$$

This expression shows how many seeds are sown in one turn of the bobbin of the sowing apparatus.

3. Results and discussion

Using the method of substituting the constant values of the parameters of the most common seeder into expression 15, we model the seed dosing process, i.e. the dependence of the metering capacity of the metering device on the technological and constructive parameters of the bobbin screw sowing apparatus. Using the data obtained, we construct two-dimensional sections of response surfaces (figure 5) and analyze them.

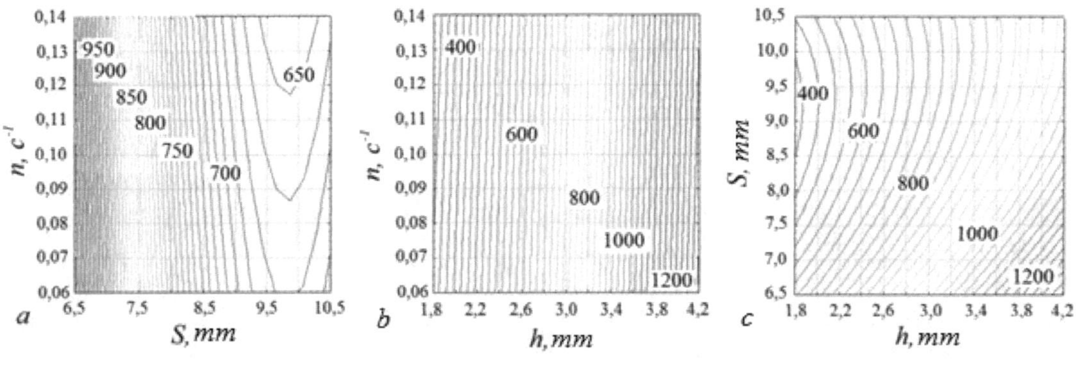

Figure 5. Two-dimensional section of response surface: $a - N = f(n, S)$, %, if h = 3 mm; $b - N = f(n, h)$, %, if S = 8 mm; c – $N = f(S, h)$, %, if n = 0,10 c^{-1}.

From figure 5 (*a*) it can be seen that the dependence of the seeding capacity *N* on the frequency of rotation of the bobbin *n* is inverse linear, that is, if there is an increase in the frequency of rotation of the bobbin *n*, the seeding capacity *N* decreases. The dependence of the seeding capacity *N* on the helix pitch *S* is a power-law.

Figure 5 (*b*) shows that the dependence of the seeding capacity *N* on the depth of the flute *h* is straight linear; the deeper the flute, the greater the number of seeds sown.

Figure 5 (*c*) shows the combined effect of the parameters: the depth of the flute *h* and the helix pitch *S* on the seeding capacity *N*.

We will conduct a comparative analysis of the sowing capacity of the designed sowing apparatus. For one turn of the sowing apparatus of a seeder, it is necessary to sow 830 pieces of rapeseed at a seeding rate of 3.0 million pieces/ha of seeds, consequently this condition is met by the parameters: flute depth $h = 4$ mm, helix pitch $S = 8$ mm with the frequency of rotation of the bobbin $n = 0{,}07 - 0{,}13$ c^{-1}. With these parameters, 800 – 1200 pieces of seeds are sown, which allows both to increase and decrease the seeding rate.

4. Conclusion

As a result of mathematical modeling of the seed dosing process, namely the dependence of the metering capacity of the dosing device on the technological and design parameters of the bobbin screw sowing apparatus, the technological and constructive parameters of the dosing device of the sowing apparatus that are optimal for the sowing of small seed crops are determined. Constructive parameters: depth of the flute $h = 4$ mm, helix pitch $S = 8$ mm. Technological parameters: bobbin rotation frequency $n = 0{,}07\ldots0{,}13$ c^{-1}.

The developed model is recommended to be used for development of new types of dosing devices for dosing of various bulk materials.

References

[1] Isaev Y M, Kryuchin N P, Semashkin N M and Kryuchin A N 2018 Theoretical studies of movement of loose material in a dosing device [Ulyanovsk State Agrarian University Samara state agricultural academy] *International Journal of Mechanical Engineering and Technology* Vol 9 **5** pp 834–840

[2] Aufderheide D and Di Matteo L 2017 A framework and taxonomy for the classification of bulk material metering and dosing equipment within the cement industry *Di Matteo Group Beckum ZKG International* Vol 69 **3** p 47–53

[3] Kem A A and Algazin D N 2009 Ravnomerny vysev melkosemennykh kultur [Uniform sowing of small seed crops] *Selsky mekhanizator* **10** p 12–13

[4] Algazin D N and Kem A A 2009 *Screw-bobbin sowing apparatus. Patent № 85789 Russian Federation, MPK A 01 P 7/12* (Siberian Research Institute of Agriculture and Peat, publ. 20.08.2009) Bull **23**

[5] Domrachev V A, Korobkin I O, Algazin D N, Kem A A and Shevchenko A P 2014 Vyseivauschie sistemy posevnykh mashin [Sowing systems of sowing machines] *FEIHE* (Omsk State Agrarian University named after P.A. Stolypin) p 190

IOP Conf. Series: Materials Science and Engineering **656** (2019) 012007 doi:10.1088/1757-899X/656/1/012007

Improving the reliability of technological equipment in the production of thermoelectric elements

V V Alisin[1,3] and V V Shishova[2]

[1]A A Blagonravov Mechanical Engineering Research Institute of the Russian Academy of Sciences (IMASH RAN), M. Kharitonyevskiy Pereulok 4, 101990 Moscow, Russia
[2]OOO «Vamiva», Moscow, Russia

[3]E-mail: vva-imash@yandex.ru

Abstract. The article is devoted to increasing the service life of the bearing of the mixing assembly in the installation for the synthesis of bismuth telluride. Comparative tribological experiments were carried out. It is shown that the implementation of a bearing bush from a nanostructured crystal of partially stabilized zirconium dioxide can increase by 4 times the reliability and service life of the installation.

1. Introduction

The quality of high-performance thermoelectric modules based on nanostructured highly stable crystals of bismuth telluride is largely determined by the reliability of the synthesis plants designed to obtain non-oriented thermoelectric material from the initial components. Blocks of semiconductor material in the installation are created from the melt by a given thermal cycle of heating and holding the melt in a neutral medium (helium) in the installation shown in figure 1.

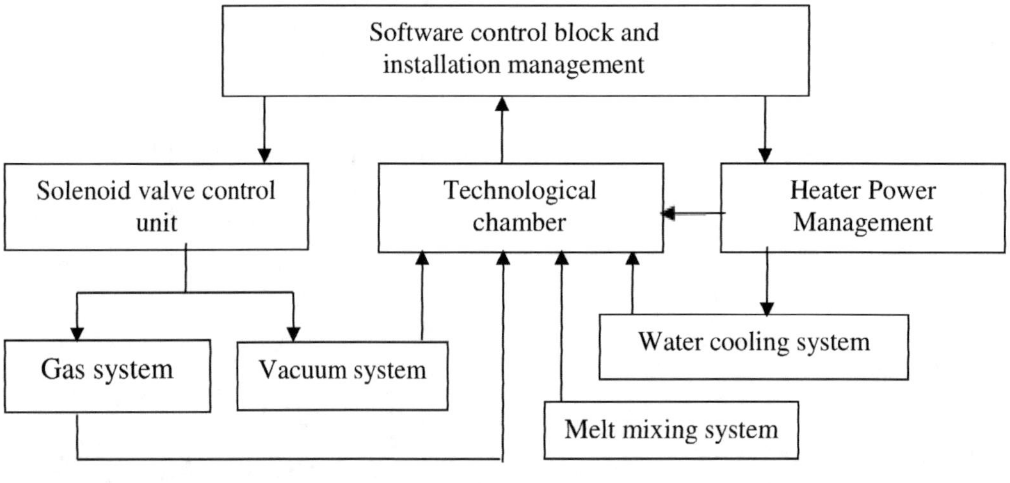

Figure 1. Structural functional diagram of the installation for the synthesis of bismuth telluride.

The least reliable and durable unit of synthesis plants is a mixing device, which is an original unit, the service life of which is limited by the wear of supporting elements made of graphite and operating under conditions of friction without lubrication, high temperatures (500 °C) and in helium medium.

The production of nanostructured crystals is carried out in a continuous technological cycle. According to technical requirements, it is necessary to ensure, within 6 months (4320 hours), trouble-free operation of technological equipment with the reliability function P(t) = 0.95. The existing design of the mixing unit does not satisfy these requirements. An extremely aggressive environment forces the entire installation to be made of graphite. The limiting unit limiting the service life of the plant are the sliding bearings of the stirring unit, the wear resistance of which is low.

The appearance of new ceramic composite ceramics [1–4] and nanostructured crystals of partially stabilized zirconium dioxide (PSZ crystals) [5–8] opens up the possibility of increasing the service life of the mixing device.

The purpose of the work is to increase the resource and reliability of the sliding supports in the melt mixing device by installing bushings made of materials based on compounds of refractory elements (ceramics or crystals), working in tandem with tool steel У10А. The choice of this pair is due to the requirements of ensuring high wear resistance, viscosity, temperature resistance.

2. Materials and equipment

The objects of study were: samples of nanostructured ZrO_2 +3.7 mol.% Y_2O_3 crystal, the amount of stabilizing additive was substantiated in [9]; sintered zirconia ceramic materials with additions of cerium oxides and lanthanum corundum hexaaluminate nanoceramic composite obtained by the sol-gel technology. Samples were made in the form of fingers with a square cross section of $5*5.10^{-3}$m and a length of $7.5*10^{-3}$ m. The end working surfaces of finger samples are flat with a roughness of Ra = 0.8. The counterbody is made in the form of a disk with a diameter of 12×10^{-2} m, made of hardened tool steel У10А, HRC 49–52.

Tribological studies were carried out on the universal friction machine УМТ-1 according to the method [9] with a change in average contact pressures in the range of 2.5<p<25 MPa and sliding speed of 0.2 <v <2 m/sec. The model friction unit was made according to the "disk-finger" scheme. Finger ceramic samples were made in the form of rods with a diameter of 8 mm, the counterbody – disks with a diameter of 120 mm were made of tool steel У10А.

3. Results and discussion

The choice of materials for sleeve bearings is due to the high corrosion resistance of materials and strength properties. Table 1 shows a comparison of the mechanical characteristics of ceramic materials that are promising and have found application in friction units. The absence of grain boundaries in PSZ crystals leads to higher mechanical characteristics as compared with ceramics.

Table 1. Mechanical characteristics of ceramic materials.

Material	Elasticity module (GPa)	Hardness (GPa)	Fracture toughness ($MPa*m^{1/2}$)	Compression strength (MPa)	Bending strength (MPa)
ZrO_2+ +3mol.Y_2O_3	353	15	10	2700	924
SiC	430	29	4	–	–
Si_3N_4	300–320	15	5–8	3500	800
Al_2O_3	380–390	22	2.3–3.5	4000	450–700

To test the corrosion resistance of the PSZ crystals, the samples were kept in a chamber in an aggressive environment under working conditions for 1 month. As a result of the experiment, no traces of corrosion were found. PSZ crystals retain their performance in the temperature range from – 140°C to +1400°C (melting point ~ 2800°C), have the highest crack resistance compared with other ceramics, and hence a higher wear resistance and antifriction. These properties determine the competitive advantages of nanostructured PSZ crystals over the tested samples of technical ceramics due to the unique combination of strength and tribological properties. The sizes of the PSZ crystal domains were estimated on the basis of measurements and analysis of the broadening of diffraction reflexes caused by violations of the periodicity of the three-dimensional crystal lattice. The main types of violations are: limited size of the region of the crystal lattice that scatters radiation coherently (the size of the region of coherent scattering – RCS, corresponding to the characteristic size of the domains), and disturbance of interplanar distances in the lattice – microstrain. The data for calculating the regions of coherent scattering (D) and microstrains (ε) in synthesized PSZ crystals that are promising for tribological applications are given in table 2. A standard (not broadened) profile was obtained by recording the standard of polycrystalline silicon with the same shooting parameters.

Table 2. Parameters of the domain structure of the PSZ crystals.

No.	Structure	D, nm	ε, 10^{-3}
1	ZrO_2–2.5 mol.% Y_2O_3	38	4.7
2	ZrO_2 –3 mol.% Y_2O_3	115	2.7
3	ZrO_2–3.5 mol.% Y_2O_3	64	1.8
4	ZrO_2-4 mol.% Y_2O_3	63	1.5

The wear rate and friction coefficients of PSZ crystals + 3 %Y_2O_3 annealed in vacuum are close to the most wear-resistant samples, have a high stability during reproduction, namely, the standard deviation is 0.030, which is 36% less than that of non-annealed samples. A distinctive feature of these crystals is the practical absence of chips during testing, which makes them attractive for use in friction units.

The sliding speed of the samples in the conducted tests for friction and wear was selected from the condition of the onset of luminescence of the samples, which corresponded to a temperature of about 500°C. With an increase in the sliding speed, the friction coefficient decreased (figure 2): at P = 500 H – by 1.53 times from f_{yc} = 0.52 m/sec to f_{yc} = 0.34 m/sec.

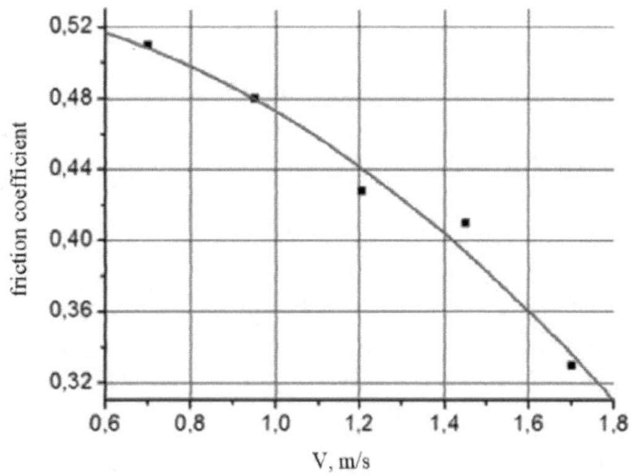

Figure 2. The dependence of the friction coefficient of the PSZ crystal on the sliding speed.

In the course of the experiment, the friction heating of the friction unit occurred and at v = v_{cr} the crystal sample began to glow. The critical velocity v_{cr} = 1.41 m/sec corresponded to the load P = 500 H (p = 10MPa); load P = 936 H (p = 18.6 MPa) – v_{cr} = 0.23 m/sec. For friction regimes, the densities of heat fluxes are as follows:

P = 500 H (p = 10 MPa), f = 0.52, v = 1.41 m/sec → q = fpv = 7.33 $\cdot 10^6$ W/m^2
P = 936 H (p = 18,6 MPa), f = 0.63, v = 0.23 m/sec→ q = fpv = 2.69 $\cdot 10^6$ W/m^2

The falling character of the dependence v = φ(v) can be explained by the influence of the temperature factor. As the sliding speed in the contact of conjugated bodies increases, the frictional heating of the samples occurs, the mechanical strength of the surface layers decreases and, as a consequence, the strength of adhesion bonds decreases in the areas of actual contact of the conjugated bodies.

The tribotechnical characteristics of the crystals depend on the contact pressure. Test conditions: contact pressure p varies in the range of p = 5–80 MPa; counterbody – steel У10A, friction without lubrication. The sliding speed at each pressure p was chosen from the condition that the bulk temperature of the ceramic sample did not exceed 400°C, which was controlled by the glow of the sample. The test results are presented in figure 3.

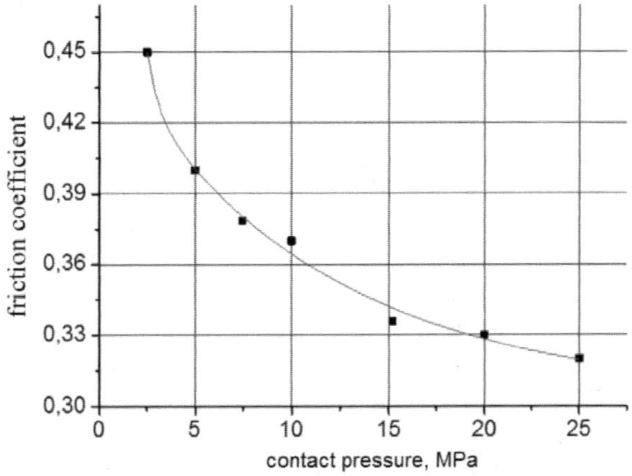

Figure 3. The dependence of the friction coefficient of the PSZ crystal on the load.

Comparative wear resistance tests were carried out under friction mode without lubrication at a specific load of 6 MPa and a sliding speed of v = 0.2 m/sec. The choice of load and sliding speed is based on the results of the strength and tribological calculation of the node mixing node according to the methods of [10]. The crystals of partially stabilized zirconium dioxide (PSZ crystals) are distinguished by better antifriction and wear resistance (table 3). To a greater extent, these benefits relate to crystals annealed after growing.

Table 3. The results of comparative tribological tests.

No.	Material	Basic structure, atom. %	f	I*10
1	Original crystal ZrO_2 + 3 mol.% Y_2O_3	Zr-83.1 Y-10.1	0.35	2.5
2	Same crystal annealed in vacuum	Zr-36.5 Ce-50	0.30	1.8
3	Ceramics ZrO_2 + 17mol.% CeO_2.	Zr-70.1 Y-17.3 Mg- non-determ.	0.52	2.9
4	Nanocomposite based on corundum	Al - base La-91.3 Y-6.9	0.56	3.2

The wear rate and friction coefficients of PSZ crystals + 3 %Y_2O_3 annealed in vacuum are close to the most wear-resistant samples, have a high stability during reproduction, namely, the standard deviation is 0.03, which is 36% less than that of non-annealed samples. A distinctive feature of these crystals is the practical absence of chips during testing, which makes them attractive for use in friction units.

Let us look at a dry friction bearing. The case, as well as the shaft with working blades, is made of graphite and constructively form two slide bearings. The shaft is driven by a flexible coupling from the electric motor. Sliding bearings are loaded with a radial force from the tension of a flexible connection (belt). Workload fluctuations are minor. The shaft rotates, the sleeve is not movable. Technical requirements for materials of a friction pair are formulated on the basis of the requirements for reliability and service life of process equipment. The mixing device in the unit for the synthesis of the blocks works for 1 hour during the day, therefore during the calculation period, with a duration of 4320 hours, 180 hours are in operation. Based on the required level of equipment reliability P(t) = 0.95, we determine the required safety factor (m) based on the wear rate of materials of a friction pair (m = 1.7).

Let us determine the contact parameters of the sliding bearing by calculation methods [11, 12]: the average pressure at the contact area p_c = 1.31 MPa (with a load factor β=0.016 half the contact angle φ_0=14°, sliding speed V = 0.256 (m/sec). Wear rate [4] for graphite material according to the criterion of Archard is equal to I/p=2*10^{-9}, the limit gap Δ_{lim} = 100 microns.

The service life of the bearing was determined as the number of revolutions of the shaft and its adequate operation time, at which the total wear of the bodies in frictionlessness will reach the value of the maximum permissible wear. The accuracy of the calculations is determined by the accuracy of the determination of the elemental law of wear J(p), which includes the main parameters and the degree of their influence on the wear of materials. The method is universal, can be applied to various typical friction units. The service life of the support device mixing in graphite performance is equal to

$$T = \frac{\Delta_{lim}}{m(^{I}/_{p})p_c V} = 24.3\ (hrs) \qquad (1)$$

The result of the calculation and the practice of operating the synthesis plants indicate that the mixing unit is not very wear-resistant and that it must be replaced regularly.

Tribotechnical characteristics of materials are defined in [7, 9, 13] and the life of the support calculated by the formula (1) is given in table 4. The wear rate and friction coefficients of PSZ + 3 %Y_2O_3 crystals annealed in vacuum are close to the most wear-resistant samples, have high stability during reproduction, namely the value of the standard deviation is equal to 0.030, which is 36% less than that of non-annealed samples. A distinctive feature of these crystals is the practical absence of chips during testing, which makes them attractive for use in friction units.

Table 4. Tribological properties of the slewing with sleeves of promising structural ceramics and crystals.

No.	Base material	Additives, %	Friction coeff. f	Wear coeff. I/p $(MPa)^{-1}$	Estimated life of the support (hrs.) at P(t) = 0.95
			Zirconium dioxide crystals		
1	ZrO_2 ++3% Y_2O_3	original	0.30	5.0*10^{-10}	100
2	orient. (001)	annealed	0.27	4.5*10^{-10}	111
3	ZrO_2 ++3% Y_2O_3	CeO_2+Er_2O_3	0.12	2.4*10^{-10}	208
			Ceramics		
4	Si_3N_4 (NC-132)	–	0.48	1.0*10^{-9}	86
5	Nanocompos.	+ La, + Y	0.56	6.4*10^{-10}	50
6	Al_2O_3 +ZrO_2+ +m mol%	17% CeO_2	0.52	5.8*10^{-10}	92

The study of surface morphology showed that there is a partial change from the initial state associated with the appearance of films of secondary structures. Figure 4a and figure 4b present a general view of the secondary structures of the friction surfaces under study in samples of PSZ crystals. Films (light areas) are present on rubbing surfaces, and the overall pattern of film distribution over the surface, the level of film continuity, and the area of individual film sections are different for samples with different concentrations of stabilizing Y_20_3.

Figure 4. Friction surface at magnification: 2a) – (x100); 2b) – (x300).

Analysis of the morphology of the surface secondary structures allows us to conclude that the destruction of the surface layers of the samples occurs mainly by local destruction and peeling of the surface films. In this case, the process of destruction is localized in the layer of secondary structures.

4. Conclusion

On the basis of the obtained data on the wear resistance of the samples, the expected life of the slide bearings of the mixing device was calculated with the probability of failure-free operation (reliability function) equal to P(t) = 0.95. The calculation data show that the only way to ensure a given level of reliability is the installation of micro-doped cerium nanostructured PSZ in sliding bearings. The manufacture of bushings from any other material among those tested will lead to a decrease in the level of reliability.

References

[1] Su Yu, Zhang Ya, Song J *at al* 2017 Tribological behavior and lubrication mechanism of self-labricating ceramic/metal composites. The effect of matrix type on the friction and wear properties *Wear* **V 372-373** 130–138

[2] Su Yu, Zhang Ya, Song J *at al* 2017 Tribological behavior and lubrication mechanism of self-labricating ceramic/metal composites. The effect of matrix type on the friction and wear properties *Wear* **V 372-373** 130–138

[3] Dudkin B N, Bugayeva A Yu, Zainullin G G and Filippov V N 2004 Properties of the ceramic composite of lanthanum corundum hexaaluminate *Refractories and technical ceramics* **12** 14–18

[4] Horng-W W, Yang-Y Ch and Jeng-Haur Horng J-H 2015 Contact temperature under three-body dry frictions *Wear* **V 339-331** 85–92

[5] Alisin V V, Boric M A and Lomonova E E 2005 Zirconia-based nanocrystalline synthesized by directional crystallization from the melt *Mater Sci &Eng* **C25** 577–583

[6] Alisin V V 2007 Nanostructured technical crystals and ceramics for friction units *Friction and lubrication in machines and mechanisms* **9** 21–279

[7] Alisin V V, Lukyanov A I and Morozov A A 2008 Methodology for estimating the parameters of friction-contact fatigue of technical ceramics and crystals *Friction and wear* **29(3)** 225–230

[8] Alisin V V, Borik M A, Vishnyakova M A *et al* 2007 Yigh-strength and wear-resistant materials based on nanostructured zirconia crystals (PSZ) *Perspective materials* **1** 194

[9] Frolov K V, Osiko V V, Alisin V V *et al* 2006 Study of the mechanical and tribological properties of nanocrystalline material of the new generation based on zirconium dioxide *Problems of mechanical engineering and machine reliability* **4** 3–8

[10] Reshetov D N 1974 *Detali Mashin* [Machine Parts] (Moscow: Mashinostroyeniye) p 655
[11] Pronikov A S 2002 *Parametricheskaya nadezhnost' mashin* [Parametric reliability of machines] (Moscow: Izdatel'stvo MGTU imeni N E Baumana) p 560
[12] *Tribology-Lubrication Friction and Wear* 2001 ed I V Kragelsky and V V Alisin (London and Bury St Edmunds UK: Professional Engineering Publishing Limited) p 948
[13] Borik M A, Vishnyakova M A and Kulebyak A V 2012 Content and annealing on the structural transformations of ZrO2 *Inorganic Materials* **48(2)** 156–160

IOP Conf. Series: Materials Science and Engineering **656** (2019) 012008 doi:10.1088/1757-899X/656/1/012008

Reserves for increasing electricity production on operating nuclear power plants

V N Belyakov[1], P Kondurov[2,3], P A Kruglikov[2,3] and O V Nikulenkov[1,4]

[1]Yaroslav-the-Wise Novgorod State University, ul. B. St. Petersburgskaya, 41, 173003 Veliky Novgorod, Russia
[2] I. I. Polzunov Scientific and Development Association on the Research and Design of Power Equipment, 3/6 Atamanskaya str., St. Petersburg, 191167, Russian Federation

[3]E-mail: nov-tm@mail.ru
[4]E-mail: Oleg.Nikulenkov@novsu.ru

Abstract. Ways to increase electricity production on operation nuclear power plants by means of improving circuit equipment performance are considered. The paper gives specific quantitative examples of underproduction of electric power by individual nodes of the thermal circuit of a nuclear power unit. The authors emphasize the importance of the development of a thermal-hydraulic model of an NPP unit at the stage of data processing of thermal tests with a view to agreeing the energy and material balances of the thermal scheme. This approach allows to increase the accuracy of the analysis of thermal efficiency of energy equipment.

1. Introduction

Currently, the task of increasing power generation at operating nuclear power plants (NPP) is extremely relevant. This is primarily due to global trends in the growth of electricity consumption and energy saving. Around the world, operating organizations are working to increase electricity production at operating NPPs. According to [1] in the United States in the period 1977-2005 due to upgrades of different levels, the electric capacity of operating NPPs was increased by 4,417 MW, which is comparable with the commissioning of 4 new power units.

Work on increasing power generation is carried out in different ways: by increasing the reactor's thermal capacity allowed for operation; modernization of the most important equipment in terms of power generation; reducing the measurement error of the thermal power of the reactor and, thereby, its maximum approaching the licensed capacity. Examples of such events include:

– increase in reactor thermal power up to 104% at nuclear power plants with ВВЭР-1000 AO (JSC) “Kontsern Rosenergoatom”;

– increasing the electric power of the RBMK -1000 units at NPP AO “Kontsern Rosenergoatom” by 32 MW by increasing the length of the blades of the last stage of the K-500-65/3000 turbines [2];

– increase in thermal power of the reactor due to a decrease in the error in determining its thermal power at power unit No. 1 of NPP “Limerick” LCC Exelon from 2% to 0.35% [3].

2. Results and discussion

As practice shows, at present the quality of the work of the thermal-mechanical equipment of NPPs of the Russian Federation is not adequately controlled. The main reasons for this phenomenon are: firstly, the lack of detailed reporting regarding thermal efficiency; secondly, the use of operational control devices for the calculation of technical and economic indicators (TEI). The current reporting of NPPs in the form of 3-ТЭК actually only states the electrical power and electricity generation by each power unit, and the analysis of the quality of the power unit is practically absent: only a comparison of some characteristics with the standard for this power unit is given. According to the 3-TEK form, it is impossible to compare power units with each other and identify the causes of differences in generation and, thus, identify the reserves for increasing electrical power. It is necessary to pay attention to the fact that, despite the identical reporting form of 3-TEK, the performance indicators of different NPPs are not quite comparable: for example, the thermal losses of the reactor at each NPP are different and lie within 2 ÷ 24 MW. Analysis of operational data shows that the actual electrical power of the power units may vary significantly. This especially applies to power units of different NPPs. The reasons may be:

- various heat output to the working body of the turbine unit;
- different temperature and flow rate of cooling water;
- features of the thermal scheme of a specific power unit;
- conducted equipment upgrades;
- technical condition of the equipment;
- insufficient quality of repair work and level of operation.

As for the use of operational control devices for the calculation of TEI, they have errors that do not meet the requirements relating to the calculation of TEI. In essence, the operational control devices are connected with the system of protection, interlocks and signaling equipment of the unit (mainly turbines). In [4] it is said that the value of the analysis largely depends on the accuracy of the source data. Measurement errors can lead to incorrect conclusions that can be costly.

This article presents the main results of thermal tests at power units of single-circuit NPP with demonstration of the potential for energy saving.

Figure 1 shows the general approach applied in conducting thermal tests and analyzing their results.

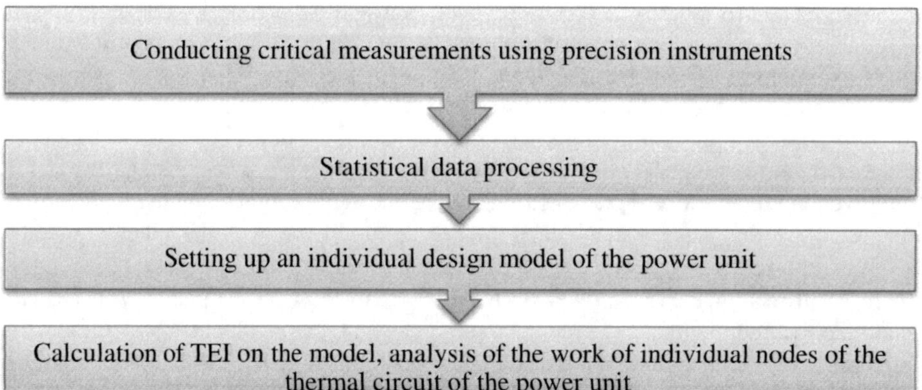

Figure 1. The general approach when testing a power unit.

Separately, it is worth noting the importance of the adjustment phase of the thermal-hydraulic model, which is, in fact, the final stage of the data processing algorithm and characterizes the integral verification of the laws of thermodynamics throughout the scheme. At this stage, the energy and material balances are consistent according to the scheme and unreliable readings that, for example, lead to unattainable heat transfer coefficients in heat exchange equipment, are identified.

Measurement errors can be both up and down from the true value, so when setting up the model, they partially compensate each other. Figure 2 shows a histogram of the probability density distribution of errors in the verification of the thermal-hydraulic model of a power unit of one of the NPPs (the setting was made for 4 modes of operation).

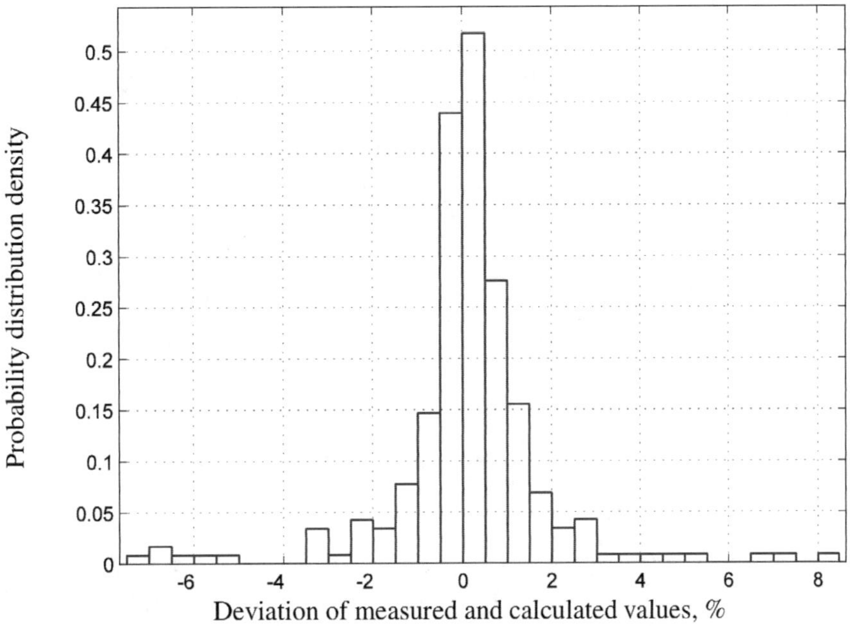

Figure 2. Histogram of the distribution of verification errors of the thermal-hydraulic model of the power unit.

Inconsistency in the energy and material balances of the thermal scheme leads to incorrect estimates of capacity under-utilization, which essentially reduces the reporting efficiency to zero. The table, for example, presents a comparison of the results of the evaluation of the power unit under-production using a model set up according to test data and station reporting.
Based on the experience of testing, the following main measures can be distinguished, bearing the potential for energy saving:

1. Improving the accuracy of measuring the flow of feed water;
2. Improvement of the work of heat exchange equipment of the regeneration system and heating installations.

Table 1. Comparison of power shortage by the power unit.

	Thermal circuit node under-production, MW				
	separators-steam superheaters	low pressure compressor	low pressure heater	Deaerator	Total
Station reporting	0.36	0.25	0.47	0.09	1.17
Calculation by CKTI	0.65	0.12	1.08	2.44	4.29

The feedwater flow rate determines the thermal power supplied to the working fluid of the turbines. The change in feed water consumption by 1% leads to approximately the same change in electrical power. The maximum flow rate of feed water, in addition to the limitations on the turbine capacity, determines the thermal capacity of the reactor, which should not exceed the value specified in the operating license. In Europe and the USA, this problem is solved by installing high-precision ultrasonic flow meters (UFM), which allow to measure the feedwater flow with an accuracy of ± 0.32% [3]. UFMs are used in parallel with regular narrowing devices (ND), such as diaphragms or nozzles, and are used for online calibration of NDs [1]. Moreover, if the discrepancy between the readings of the control system and the remote control device exceeds the permissible value, the thermal power of the reactor is immediately reduced in order to establish the reasons that caused the indicated deviation. Currently, at most NPPs of Russia, mainly diaphragms are used as flow meters; the measurement error of feedwater flow is within 2%. Using the experience of American NPPs, the additional generation of electricity for a 1000 MW power unit will be ~ 16 MW.

The quality of work of heat exchange equipment can be characterized by a heat transfer coefficient. As practice shows and tests carried out, the best regenerative heaters at nuclear power plants have a heat transfer coefficient of ~ 3500 ÷ 3600 W/(m2 · K); turbine exhaust steam condensers ~ 3200 ÷ 3500 W/(m2 · K). Table 2 provides data on the comparison of the operational performance of heat exchange equipment obtained during the thermal tests of the NPP power unit.

Table 2. Underproduction of electricity by the unit due to the deterioration of the performance of heat exchange equipment.

Thermal circuit element	ΔN_e, MW		Comments
	turbine No.1	turbine No.2	
low pressure heater-1	0.3	0.4	For turbine No.1: heat transfer coefficient 2335 W/(m^2*K) vs. standard 3264 W/(m^2*K); for turbine No.2: heat transfer coefficient 2222 W/(m^2*K) vs. standard 3273 W/(m^2*K).
low pressure heater-2	0.7	0.8	For turbine No.1: heat transfer coefficient 1423 W/(m^2*K) vs. standard 3316 W/(m^2*K); For turbine No.2: heat transfer coefficient 1319 W/(m^2*K) vs. standard 3326 W/(m^2*K).
low pressure heater-3	0.7	0.3	For turbine No.1: heat transfer coefficient 1679 W/(m^2*K) vs. standard 3411 W/(m^2*K); For turbine No.2: heat transfer coefficient 2401 W/(m^2*K) vs. standard 3421 W/(m^2*K).
low pressure heater-4	0.1	0.2	For turbine No.1: heat transfer coefficient 3042 W/(m^2*K) vs. standard 3518 W/(m^2*K); For turbine No.2 heat transfer coefficient 2846 W/(m^2*K) vs. standard 3529 W/(m^2*K).
low pressure compressor	4.8	0.1	For turbine No.2: heat transfer coefficient 2730 W/(m^2*K) vs. standard 3221 W/(m^2*K), also accounting for muffled pipes; For turbine No.1: accounting for muffled pipes
Total:	6.5	1.7	Taking into account the influence in the circuit

Non-condensable gases, which impair the heat transfer from the steam side, have a strong influence on the operation of steam-water heaters. Often, it is poorly organized withdrawal of non-condensable gases and leads to deteriorated performance of heat exchange equipment. In turbine condensers, in addition to the effective suction of the vapor-air mixture, there is also a strong influence of sediments from the side of cooling water. Currently, the problem of combating sediments in tube bundles is successfully solved by using efficient ball cleaning systems. In regenerative heaters, this problem is not so acute, since reactor-quality water is used, which then goes directly to the reactor, as in single-circuit NPPs or in a steam generator in the case of two or three loop stations.

When analyzing the operation of heat exchange equipment, consideration should be given to the issue of the operation of the heaters of the heat and power plant. Such plants, especially at single-circuit NPPs, were created with enormous reserves of heat performance, therefore, insufficient attention was paid to monitoring their thermal efficiency. Table 3, for example, presents the results of the analysis of the work of a group of boilers for heat supply of one of the nuclear power plants.

Table 3. Analysis of the work of a group of boilers for single-loop heating NPP.

Parameter	Meas. unit	Boiler No.1	Boiler No.2	Boiler No.3
Boiler thermal power	MW	6.3	7.4	4.9
Boiler underheating	°C	0.5	0.4	0.2
Boiler heat transfer coefficient	$W m^2 \cdot K$	120	802	1297
Standard heat transfer coefficient of the boiler	$W m^2 \cdot K$	1587	1696	1773

As the experience of the studies performed shows, the reduced heat engineering indicators of the operation of heat supply boilers are determined mainly by the deteriorated organization of the removal of non-condensable gases.

3. Conclusion

The example of specific units and equipment of the thermal circuit of a power unit shows the reserves for increasing electrical power. It should be noted that there is the need for regular rapid testing of power units, with the aim of timely determining the deterioration of the thermal performance of equipment, as well as the ability to determine the nature of their flow in time. Another approach, which is actively used in Europe and the United States, is the creation of an independent control system for TEI, which monitors equipment performance in real time and, in the event of overshooting, carries out diagnostics indicating possible causes of the decrease in thermal performance of equipment. It is important to note that such a system does not work with operational control devices, but with special laboratory precision devices.

References

[1] *The role of instrumentation and control systems in power uprating projects for nuclear power plants* 2008 (Vienna: IAEA) **NP-T1.3** [in Austria]

[2] Davydenko N N, Solomeev V A et al 2010 Modernizaciya protochnoj chasti CND parovoj turbiny K-500-6,4/3000 dlya povysheniya ee moshchnosti [Modernization of the flow section of the low-pressure cylinder of the K-500-6.4 / 3000 steam turbine to increase its capacity] *Thermal Engineering* **9** pp 21–25[in Russian]

[3] *Reactor Core Thermal Power Uncertainty Calculation Unit 1* 2009 LLC Calculation LE-0113 (Exelon Generation Company) Rev. 0

[4] Cotton K., Schofield P. 1971 Analiz izmeneniya rabochih harakteristik parovyh turbin [Analysis of changes in the performance of steam turbines] *Power Machines and Plants* **2** pp 62–72 [in Russian]

IOP Conf. Series: Materials Science and Engineering **656** (2019) 012009 doi:10.1088/1757-899X/656/1/012009

The effect of generating reactive stresses in the spring elements of titanium nickelide

V N Belyakov[1]

[1]Yaroslav-the-Wise Novgorod State University, ul. B. St. Peterburgskaya, 41 173003, Veliky Novgorod, Russia

E-mail: val-belyakov@mail.ru

Abstract. The article is devoted to the study of the effect of generation of reactive stresses in tension springs. The studies of the kinetics of generation and relaxation of these stresses in the TiNi alloy are presented. Regularities of the change in reactive stresses during long-term thermal cycling in the range of complete and incomplete MPs are considered. It was established that the level of reactive stresses depends on several factors, such as the degree of pre-deformation, the type of stress state, the stiffness of the opposing system, the strength properties of the alloy.

1. Introduction

The effect of generation of reactive stresses in materials with shape memory is becoming more widely used in various fields of technology and even medicine [1, 2]. Helical springs [3] are used as executive elements in many mechanisms, which allow creating large axial displacements and regulating the developed reactive forces by changing the spring's geometrical parameters. However, despite such wide possibilities of alloys with SME, their use is limited by the instability of the thermomechanical behavior under various heat shift conditions. Therefore, the study of the laws governing the change in reactive stresses during long-term thermal cycling in the range of complete and incomplete MP spring elements is relevant.

2. Objectives and research methods

In this work, we studied the effect of generation of reactive stresses in tension springs (figure 1) from titanium nickelide of equiatomic composition.

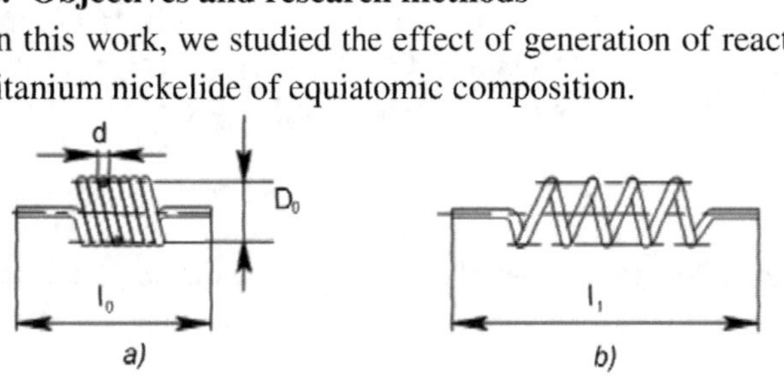

Figure 1. Extension spring: a) – initial condition; b) –deformed state.

Springs were coiled from nickel-titanium wire of equiatomic composition with various geometrical parameters d, D_0, n on special mandrels. The specified form was obtained by annealing in a pinched state at a temperature of 500 °C for one hour.

For the calculation of stresses and relative deformation, we used the formulas of the resistance of materials taking into account the elastoplastic behavior of EPP alloys. Reactive stresses were calculated by the formula (1)

$$\tau = \frac{6PD_0}{\pi d^3} \quad (1)$$

where: P – is the axial force, H; D_0 – is the average diameter of the spring, mm; d – is the diameter of the wire, mm;

Relative deformation was calculated by the formula (2)

$$\gamma = \frac{\Delta ld}{\pi D_0^2 n} \quad (2)$$

where: Δl – is the axial deformation of the spring, mm; n– the number of turns of the spring;

Similar formulas for calculations were used by the authors in studies of the thermomechanical behavior of springs [4, 5]. The magnitudes of reactive stresses and relative deformations determined on springs of different geometric shapes with the same loading conditions were the same, which indicates the validity of using these formulas.

The study was conducted on the installation, the scheme of which is shown in figure 2.

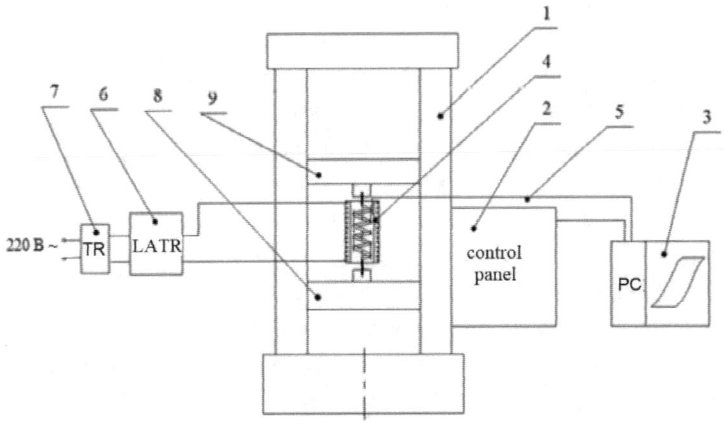

Figure 2. Installation diagram: 1– tensile machine; 2– control panel; 3 –computer; 4 – tube furnace; 5 – thermocouple; 6 – LATR; 7 – transformer; 8 – lower traverse; 9 – upper traverse.

When the spring was gradually heated by a laboratory transformer (6), reactive forces were generated in the material through an interval of martensitic transformations, which were recorded by the control panel of the bursting machine (2). The heating temperature was controlled by a thermocouple installed in the furnace and connected to the input of the computer. The signal from the control panel is also connected to the computer input and a hysteresis loop generating reactive forces during heating and their relaxation during cooling was recorded on the screen.

The kinetics of generation and relaxation of these stresses in the TiNi alloy is shown in figure 3. The sample was pre-stretched in the martensitic state at 270 K to a residual elongation of 2.7% and fixed in supports, the rigidity of which was infinite relative to the rigidity of the sample. Thermal cycling was carried out both through the entire interval of phase transformations and the incomplete interval ΔT^H.

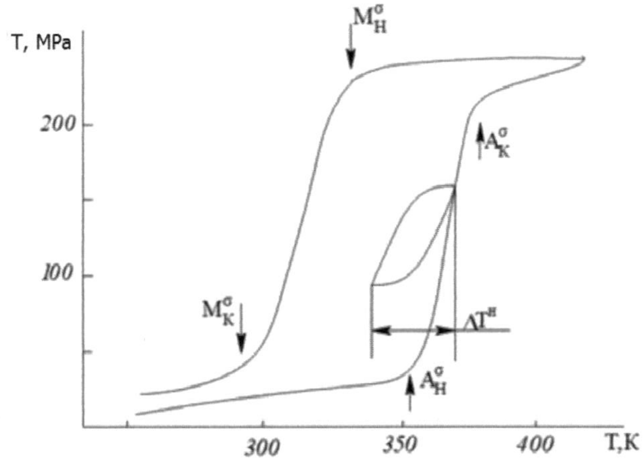

Figure 3. The effect of generation and relaxation of reactive stresses.

3. Results and discussion

The magnitude of the generated stresses closely correlates with the amount of the high-temperature phase that appeared during the warm-up process, and the temperatures A_b^σ and A_e^σ correspond to the beginning and end of the reverse martensitic reaction of the stressed metal. During cooling, relaxation of stresses is observed in accordance with the kinetics of the development of plasticity of direct transformation. The fall in stresses begins with a temperature M_b^σ and ends at a temperature lying in the interval M_b-M_e. The temperature M_b^σ, as well as the temperature A_e^σ, strongly depends on the magnitude of the generated stresses. For titanium nickelide, the elevation of these temperatures over M_b and A_e reaches one hundred and more degrees [6].

The generation of reactive stresses is observed after preliminary deformation. The maximum level of reactive stresses σ_r^{max} increases with increasing degree of pre-deformation (ε_{pr}) and stiffness of the reaction system (K_{pr}). Under conditions of torsion and bending, an increase in ε_{pr} or γ_{pr} is accompanied by a smooth increase in σ_r^{max} and τ_r^{max} (curves 1, 2 in figure 4a). Under the conditions of stretching, the $\sigma_r^{max} - \varepsilon_{pr}$ dependence (curve 3, figure 4b) has a sharp bend, due to the fact that the maximum stress level is determined by the yield strength of austenite σ_r. When reactive stresses reach σ_r, their further increase is limited by the plastic flow stress of austenite. At $K_{pr} = 0$, the molding restoration in a state free from stresses leads to the complete or partial return of a given deformation. With an absolutely rigid clamping of the $K_{pr} \rightarrow \infty$, the restoration of the form is prohibited, and the material generates efforts, the magnitude of which is determined by the degree of preliminary deformation and the mechanical properties of the material.

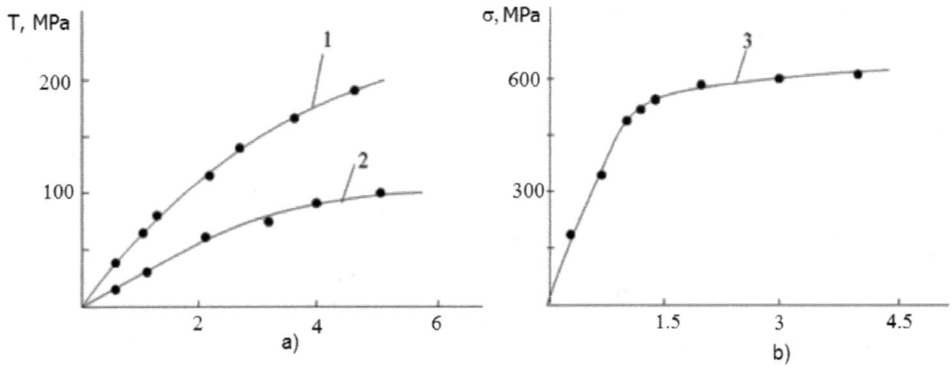

Figure 4. The effect of pre-strain on the level of reactive stresses in TiNi alloy.

With the final stiffness of the pinching, both the generation of stresses and the return of deformations occur simultaneously. The magnitude of the generated stresses increases with increasing stiffness of the reaction.

The level of maximum reactive stresses σ_r^{max} can be estimated from the loading curve of a material with an SME in the austenitic state (figure 5).

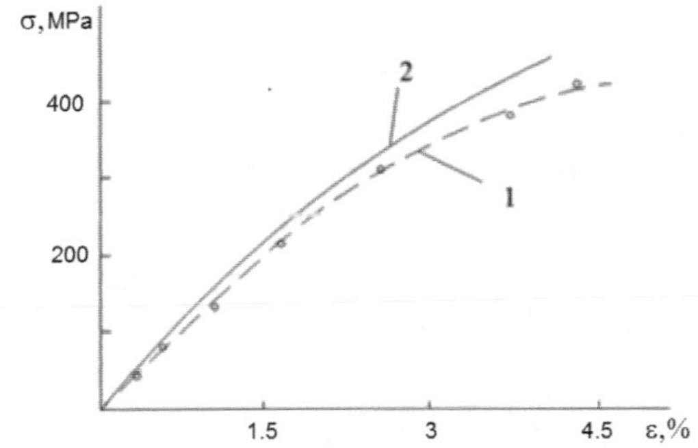

Figure 5. Dependence of reactive stresses σ_r on the value of preliminary deformations ε_{pr} (curve 1) and the tension diagram at a temperature of 150 °C (curve 2) for TiNi alloy.

From figure 5 it can be seen that reactive stresses differ little from tensile stresses of austenite. This result is of practical importance, since it is easy to calculate the magnitude of reactive stresses for arbitrary sizes of curves of rods with the required values of force using the well-known austenite stretching diagram.

A characteristic feature of the kinetics of reactive stresses is the ability to reproduce a virtually unlimited number of times the stress-temperature hysteresis during repeated heating and cooling [7]. If the material that has experienced complete or partial stress relaxation is reheated in a clamped state to the temperature A_e^{σ}, then the stresses will arise according to the original scheme. Repetitive heat shifts either slightly reduce the level of reactive stresses, or slightly increase it depending on the chemical composition of the alloy, its structural state, but after a certain number of thermal cycles, the hysteresis loop stabilizes.

A cyclical study of TiNi alloys in the form of springs, in the first cycles, a sufficiently effective drop in the level of reactive stresses is observed, and then after 10–20 cycles the process stabilizes, that is, the further nature of σ_r change for titanium nickelide is determined by a linear law. It was noted that after 1600 heat cycles even a certain increase in reactive stresses is observed, in accordance with figure 6, curve 1, due to phase hardening and an increase in the dislocation yield strength. In the range of 2000–4500 cycles, the PH practically does not change.

The character of the curve $\sigma_r=\sigma_r(N)$ in the case of a soft pinching (a spring with a SME – counter spring) is somewhat different. First, there is no marked anomaly in the thermocyclic behavior of titanium nickelide in the case of finite stiffness, in accordance with figure 6, curve 2, and, second, the full stabilization of the process of thermomechanical deformation occurs only after 6500 cycles.

The authors of [8] suggested that the nature of the thermal cyclic behavior of TiNi alloys is significantly affected by the pinching stiffness, the degree of preliminary deformation, the cycling interval, the composition of the alloy, etc.

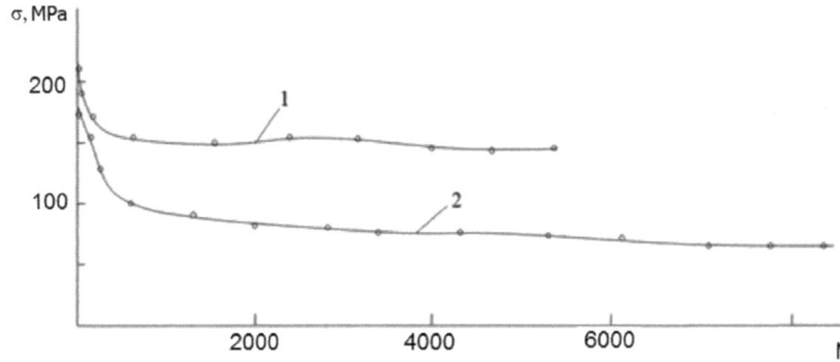

Figure 6. The change in reactive stresses during thermocycling of titanium nickelide:
1 – the case of severe pinching, γ_{pr} = 3.5%; 2 – mild opposition γ_{pr} = 3.0%.

In the mechanisms of two-sided action, the working spring – counter-spring material with SME works under conditions of finite stiffness of the reaction, generating reactive stresses $\tau^{н}_{r}$. when heated to a predetermined temperature of the $A^{н}_{к}$. During subsequent heat changes in a given temperature range $\Delta T^{н}=A^{н}_{к}-M^{н}_{к}=T_{max}-T_{min}$, cooling takes place during cooling, and during heating, the generation of stresses, the change of which is characterized by a voltage swing $\Delta\tau$. Consequently, the thermocyclic operability of interconnected elements under conditions of incomplete MPs will be determined by the stability of the reactive stresses generated by the alloy $\tau^{н}_{r}$ and the magnitude of the stresses $\Delta\tau$.

Thermocyclic performance studies were carried out on spring samples made of an alloy of Ti = 50.2% Ni with given geometrical sizes. The thermocycling interval $\Delta T^{н}=A^{н}_{к}-M^{н}_{к}$ was chosen in such a way as to ensure the condition of incomplete phase transformation. The change in reactive stresses during heat changes was studied with two values of the stiffness of the counteraction of the $K_{пр}$, determined by the stiffness of the counter spring. The value of the preliminary deformation was taken equal to $\gamma_{пр}$ = 3%.

The effect of preliminary thermal testing on the cyclic performance of the material was also investigated. To do this, before testing, the samples were cycled (20 and 40 cycles) through the full MP interval under a constantly applied stress τ = 50 MPa. The dependence of reactive stresses and stress range on the number of thermal cycles is presented in figure 7:

1, 6 – K_{pr}=41 GPa; 2, 7 – K_{pr}=15.2 GPa – without thermal testing;

3, 8 – K_{pr}=41 GPa; 4, 9 – K_{pr}=15.2 GPa – thermal testing 20 cycles;

5, 10 – K_{pr}=15.2 GPa – thermal testing 40 cycles.

Studies show that the nature of changes in reactive stresses is almost the same for different values of the stiffness of pinching and varies greatly depending on the preliminary thermocyclic treatment.

Thermal testing leads to stabilization of the stresses $\tau^{н}_{r}$ and $\Delta\tau$, which remain unchanged from the first thermal cycles. Only at the initial stage there is a slight decrease in $\tau^{н}_{r}$.

An increase in the duration of thermal testing from 20 to 40 cycles has practically no effect on the functional and mechanical properties, which indicates that twenty preliminary thermal cycles are sufficient to stabilize the reactive stresses during thermal cycles in a narrow temperature range.

In untrained samples with thermal changes, an insignificant increase in $\tau^{н}_{r}$ is first observed. This is associated with a decrease in the characteristic temperatures of martensitic transformations in the process of cycling, which, at a fixed $A^{н}_{к}$ temperature, leads to an increase in the degree of reverse phase transition and, accordingly, to an increase in $\tau^{н}_{r}$.

The subsequent decrease in stresses is determined by the processes of thermocyclic creep, which are most pronounced in samples that have not been subjected to thermal cycling.

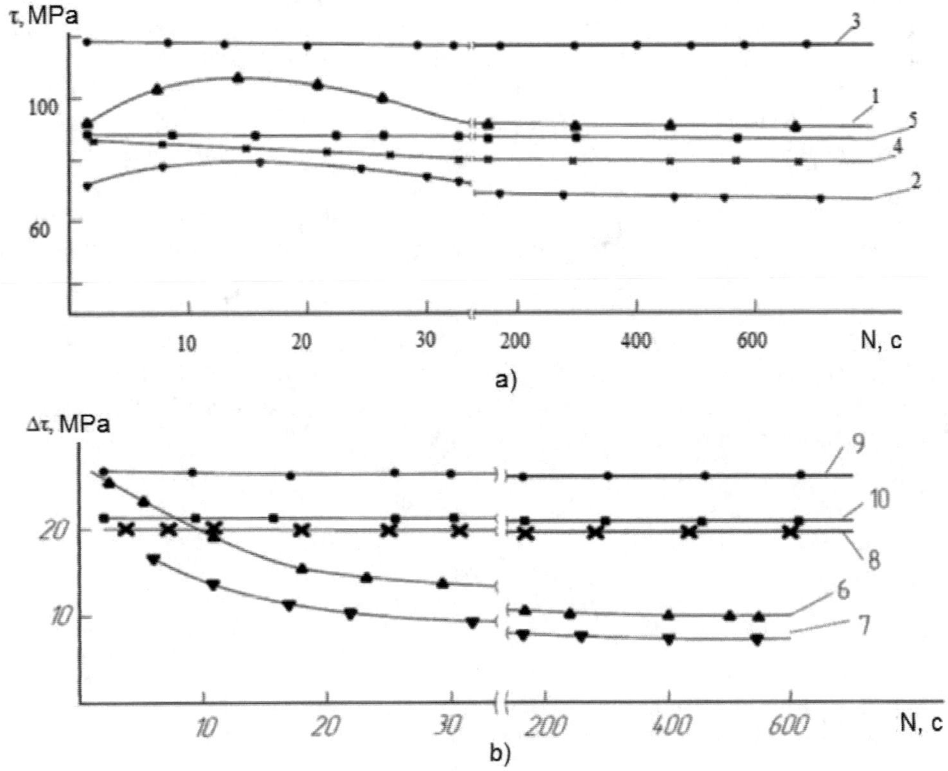

Figure 7. The change in the level of reactive stresses (a) and the range of stresses (b) in the Ti alloy is 50.2% at % Ni during thermal cycling in the interval of incomplete MP. ΔT^{H} = 15K in conditions of the finite stiffness of counteraction of the K_{pr} (γ_{pr}=3,0%).

4. Conclusion

The conducted studies show that the preliminary thermal training is necessary for the stabilization of the thermomechanical properties under conditions of incomplete MPs, which remain unchanged for a long term thermal cycling. Thus, we can conclude that the level of reactive stresses depends on several factors, such as the degree of pre-deformation, the type of stress state, the stiffness of the opposing system, and the strength properties of the alloy. The magnitude of the reactive forces during heating is proportional to the amount of austenite formed, and the nature of the relaxation of forces is determined by the growth of crystals of the new phase (martensitic) in the field of external stresses with decreasing temperature.

References

[1] Jiang P, Zheng Y, Tong Y, Chen B, Tian, L Li, Gunderov D and Valiev R 2014 Transformation hysteresis and shape memory effect of an ultrafine-grained TiNiNb shape memory alloy *Intermetallics* **54** 133–135

[2] Gusev D, Kollerov M and Sharonov A 2015 A study of reactive stresses generated by implants from titanium nickelide-based alloys *Titanium* **1** 42–49

[3] Khusainov M, Larionov A, Malukhina O and Tambulatov B 2001 *Thermal Valve* RF patent for invention No. 2171937 from 01/27/2001 Publ. 08/10/01 Bul **22**

[4] Andronov I, Demina M, Kormshchikova Z and Matveeva O 2017 The influence of incomplete thermal cycles on the performance of a helical coil spring made of titanium nickelide *Deformation and fracture of materials* **8** 51–58

[5] Abdrakhmanov S, Dotalieva Zh and Abdyzhapar A 2017 Analytical study of the deformation-force behavior of conical springs with shape memory effect *Deformation and fracture of materials* **9** 32–46

[6] Mashikhin A, Kazarina S and Movchan A 2016 Generation of reactive stresses in the investigation of oriented transformation of a shape memory alloy in a bent state *Bulletin of Tambov University* **3** 1152–55

[7] Tong Y, Chen F, Guo B, Tian B, Li L, Zheng Y, Gunderov D and Valiev R 2013 Superplasticity and its stability of an ultrafine-grained Ti49.2Ni50.8 shape memory alloy processed by equal channel angular presing *Materials Science and Engineering*: A Vol 587 61–64

[8] Belyaev S, Volkov A and Evard M 2017 Modeling irreversible deformation and destruction of titanium nickelide during thermal cycling *Deformation and fracture of materials* **4** 65–73

Technological parameters for the production of steel and titanium blanks by the method of helical rolling

V N Belyakov[1,2] and O S Naumova[1]

[1]Yaroslav-the-Wise Novgorod State University, ul. B. St. Peterburgskaya, 41, 173003, Veliky Novgorod, Russia

[2]E-mail: val-belyakov@mail.ru

Abstract. The article is devoted to the technology of producing steel and titanium blanks by the method of cross-helical rolling. The grades of the investigated alloys and technological processing modes are given. Studies of the effect of rolling parameters (deformation temperature, feed angle) on the production of round billets are presented. The general nature of the influence of technological factors of the process on the conditions of metal deformation is established. The change in the microstructure of titanium alloys after hot deformation was investigated. The expediency of using this method is substantiated.

1. Introduction

The desire to reduce the technological losses of expensive materials, to reduce the costs of redistribution has a stable relevance in the production of rods from alloyed metals and expensive alloys. Modern conditions of metal products turnover exacerbate the problem with new requirements. Of particular value is the possibility of prompt production and delivery of quality steel in limited quantities in an expanded brand size assortment.

Stationary helical rolling, naturally and inseparably connecting the discontinuity and continuity with its essence, occupies a uniquely preferable position in the indicated direction. Largely due to this feature, helical rolling is present on the widest range of industrial applications, including: pipe; detail rolling and rolling production.

The use of the process of helical rolling is widely used to produce round solid steel from various materials [1, 2]. The technology is based on the method of helical rolling of continuous billets in a three-roll stand with elevated values of partial reductions, which is achieved mainly by increasing the feed angles. This method allows to obtain during hot rolling in one pass the degree of deformation in the axial zone of the rod up to 130% and the drawing ratio μ=2,9 for steel billets [3].

2. Material and research methods

The paper presents studies of the effect of rolling parameters (deformation temperature, feed angle) on the production of round billets from O9Г2C (O9G2S), 40X (40KH), 60C2A (60S2A) steels and industrial titanium alloys, the chemical composition of which is given in table 1.

IOP Conf. Series: Materials Science and Engineering **656** (2019) 012010 doi:10.1088/1757-899X/656/1/012010

Table 1. The chemical composition of titanium alloys.

Alloy	Original structure	Chemical composition, %						
		Al	**Mn**	**Mo**	**V**	**Cr**	**Fe**	**Zr**
BT1 (VT1)	α							
BT5 (VT5)	α	5	-	-	-	-	-	-
OT4 (OT4)	Pseudo-α	3	1,5	-	-	-	-	-
BT20 (VT20)	Pseudo-α	5.5–6	-	0.8–1.5	1.2	-	-	2.2
BT22 (VT22)	α + β	5	-	4.5	4.5	1.2	1.0	-

The study was carried out on an industrial rolling mill IRM 30/10, in which the rolls are inclined to the rolling axis at an angle of 15 °. Rolling was subjected to billets with a diameter of 15 ... 25 mm with a pre-deformed structure obtained by the method of longitudinal rolling.

Studied the possibility of obtaining maximum compression per pass at a given rolling temperature and feed angle. The degree of deformation (drawing ratio) was varied by calibrating the rolls to a given diameter of the rolled steel, and the conditional speed of rotation of the workpiece – by changing the speed of rotation of the rolls. For helical rolling, working gauges are formed by converting the rolls to a predetermined amount. The degree of maximum compression at each temperature was limited to obtaining high-quality bars, which were evaluated by the state of the surface, the accuracy of geometric dimensions.

3. Results and discussion

According to the research results, the dependences of the maximum compression per pass on the rolling temperature were obtained (figure 1).

Research results show that with increasing rolling temperature, the maximum compression per pass increases and reaches a maximum at a certain temperature for each alloy grade (figure 1). This behavior is explained by an increase in the plasticity of the material with an increase in the deformation temperature in the single-phase region. The decrease in the maximum compression with increasing temperature above T_{opt} is explained by the fact that due to the high ductility, the metal is squeezed into the gaps between the rollers, the rolling process is disturbed, and ovality is formed in the cross section of the workpiece.

Conducting rolling with a lower degree of deformation per pass leads to the formation of a structure that is not uniform across the cross section. The structure of the workpiece across the cross section is unevenly worked out, the metal is quickly cooled down and, as a result, surface defects appear (flaws); the productivity of the rolling process is reduced. When setting a greater degree of deformation per pass during rolling, uneven heating of the rod occurs, which also causes structural heterogeneity of the metal and instability of the mechanical properties of the cross section of the rod.

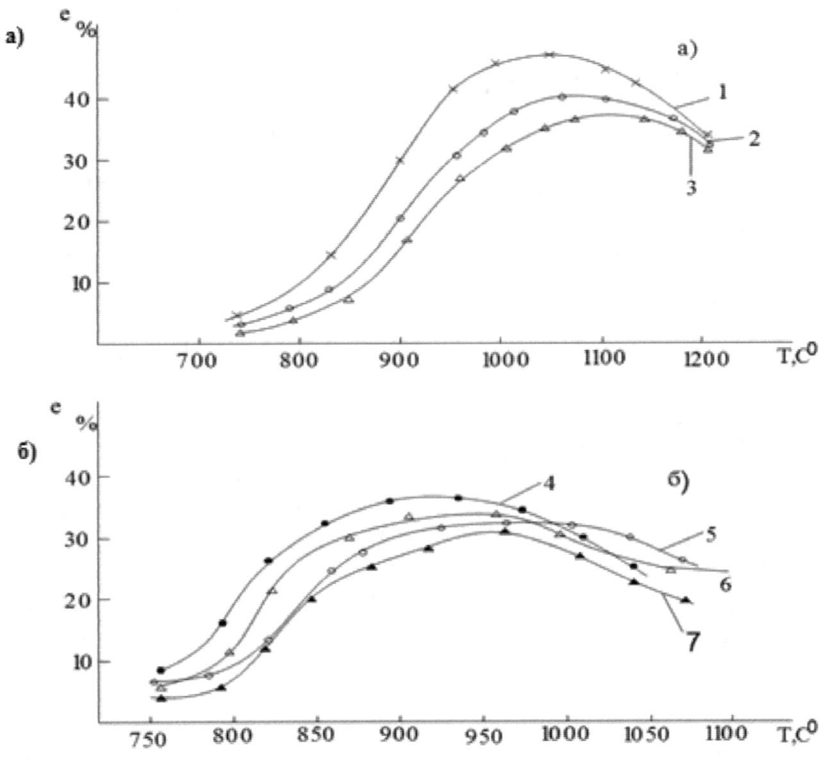

Figure 1. The dependence of the maximum compression per pass from rolling temperature for steels (a): 1 – 09Г2С (O9G2S); 2 – 40X (40KH); 3 – 60C2A (60S2A); for titanium alloys (b): 4 – BT1 (VT1); 5 – BT5 (VT5); 6 – OT4 (OT4); 7 –BT22 (VT22).

We should note the extreme nature of the temperature dependence of the limiting coefficient of drawing with a maximum at a rolling start temperature 80 ... 100° C below the temperature of the maximum plastic properties, determined by plastometric tests [4]. This indicates intense deformation heating and its noticeable effect on the deformation conditions of the metal.

It should be noted that there is a dependence of the maximum compression on the conditional speed of rotation of the workpiece for materials with low thermal conductivity (such as alloy BK22 (VK22)). An increase in the rotational speed leads to localization and intensification of the deformation heating in the zone of maximum velocities and a decrease in the degrees of deformation. Reducing the speed of rotation of the workpiece contributes to the weakening of the non-uniformity of the temperature field, reducing local maxima near the area of ring failure due to more complete completion of heat transfer processes and obtaining higher quality rolled products [5].

One of the most important technological parameters affecting the processing performance and the quality of the resulting blanks is the feed angle β, which is set by the angle of the rolls to the axis of the workpiece and the calibration of the working part of the roll.

In studies, the feed angle was changed by changing the calibration of the mushroom-shaped rolls within 10 ... 22°. It was established that with an increase in the feed angle, an increase in the axial movement of the workpiece in the rolling process, and, accordingly, an increase in the productivity of the process, occurs. But on the other hand, an increase in the feed angle impairs the process of metal capture by the rollers and does not allow to obtain maximum reduction values in a single pass. The positive effect of large feed angles on the rolling capacity of the workpieces reflects a significant increase in the uniformity of the distribution of strain-rate and temperature conditions in the deformation zone. This is facilitated by the growth of private reductions and reduction in the number of deformation cycles. The analysis of the parameters of shaping showed, in particular, that an increase in the feed angle from 12 to 20° when rolling a bar with a diameter of 18 mm from a billet with a diameter of 25 mm leads to an increase in the maximum partial reduction from 12.5 to 15.6%, to a reduction in the number of deformation cycles of 1, 67 times. A decrease in the depth of the axial extension of the end face of the workpiece by 13% also indirectly indicates a decrease in the non-uniformity of deformation. When rolling steel O9Г2C40X the optimum feed angle is in the range of 16–18°.

The process of helical rolling of all the studied steels and alloys on conic calibrations of rolls with a tilt angle of 10 ... 15 ° forming to the rolling axis is characterized by high stability during the seizure of blanks and in the steady state with a wide range of feed angles (12 ... 24°) and coefficients of extracts per pass (1.1 ... 4.5). Increasing the feed angles to 20 ... 25° worsens the gripping conditions, violates the stability of rotation and axial movement of the deformable workpiece.

The general nature of the established laws of the influence of technological factors of the process on the conditions of metal deformation was noted when studying the characteristics of helical rolling of other alloys [6, 7]. The individual characteristics of a particular material are manifested mainly in the choice of the temperature range of rolling and the level of deformability. Direct experiments have shown for the first time the possibility of defect-free deformation of billets in the process of helical rolling according to the modes presented in table 2.

Table 2. Modes of helical rolling of steels and titanium alloys.

Steel /alloy type	Rolling temperature,°C	Feed angle, degrees	Maximum compression per pass, %
09Г2C (09G2S)	1000	15	45
40X (40 KH)	1050	16	40
60C2A (60S2A)	1060	16	37
BT1 (VT1)	900	14	35
BT5 (VT5)	980	12	33
OT4 (OT4)	950	12	33
BT20 (VT20)	950	12	31
BT22 (VT22)	960	12	30

The limitation of the maximum rolling temperature is determined not only by the conditions for obtaining rolled products of the correct geometric shape, but also by obtaining the necessary structure of the material. With an increase in the heating temperature, the speed of the oxidation process increases, the depth of the decarburizing layer increases, which is especially important for medium and high carbon steels.

IOP Conf. Series: Materials Science and Engineering **656** (2019) 012010 doi:10.1088/1757-899X/656/1/012010

Experiments show that when heated to a temperature of 1100° C and held for an hour for samples of 30 mm in diameter from 60C2A (60S2A) steel, the size of the decarburizing layer is 0.6 mm. Therefore, when choosing the optimal rolling temperature, it is necessary to take into account both the plastic properties and the structural features of the steel.

Microstructure studies have shown that heating titanium alloys to temperatures below phase transformations and the subsequent rolling at these temperatures have virtually no effect on grain growth. After rolling BT1 and BT5 (VT5) alloys at a temperature of 900° C and a degree of reduction of 40%, the average grain size was 0.1–0.2 mm, which corresponds to 4.5 type of macrostructure in a ten-point system. The initial structure of the samples annealed at 850° C was 5 points of the macrostructure. An increase in the rolling temperature of the BT5 (VT5) alloy to 1000° C made it possible to increase the degree of reduction to 80%, but the average grain size accordingly increased to 0,25÷0,3 mm (6 point of the macrostructure).

It should be noted that in the temperature range of 1000-1100° C, the plastic properties of the metal deteriorate sharply in the entire speed range of deformation, which is associated with overheating of the metal. This is confirmed by a sharp increase in grain size. In addition, at temperatures above 900° C, a brown scale, which is titanium dioxide, is formed on the surface. The depth of the oxide layer at 1000° C and holding for 10–15 min reached 0.1–0.15 mm.

Rolling alloys OT4 (OT4), BT20 (VT20) with a pseudo-α structure showed that at a temperature of 850–900° C, the alloys have the least ductility, which increases dramatically when passing through the polymorphic transformation point. However, an increase in the rolling temperature to 1050° C led to an increase in the macrograin to 0.25 with an initial structure of 0.1 mm.

Based on the research obtained on the microstructure of titanium alloys, the following can be noted:

1. The grain size and the microstructure score after rolling substantially depends on the nature of the initial structure of the sample.
2. At temperatures of hot deformation located below the polymorphic $\alpha + \beta \Leftrightarrow \beta$ transformations, the macrostructure score practically does not increase, the macrograin is noticeably crushed.
3. At temperatures of hot deformation above the point of polymorphic $\alpha + \beta \Leftrightarrow \beta$ transformation, a significant growth of micrograin is observed and the macrostructure score increases. The deformation here significantly affects the grain size. The increase in deformation crushes the grain, but with deformations of 80-90% it is not possible to obtain a fine structure, which is easily achieved with small deformations in the temperature range below the polymorphic transformation limit.
4. Significant grinding of the initial coarse macrostructure is possible only at the deformation temperature below the polymorphic transformation point.

With correctly selected modes of helical rolling, it was established that in the macro- and microstructure of the metal rolled according to the new technology, there are no discontinuities of deformation origin.

The specific nature of the development of metal forming, accompanied by intense shear deformations on the background of general compaction, has a positive effect on almost all levels of the metal structure. In carbonaceous alloyed steels 40X (40KH), 60C2A (60S2A), high dispersion and uniformity of isolated particles of the carbide phase were recorded, practically in the absence of mesh and line formations [8].

Such a structure of carbides differs sharply from the characteristic carbide streakiness obtained by rolling in calibers, forging, and approaches the structure of carbides in steels obtained by quenching and subsequent tempering.

In the process of stationary helical rolling and subsequent standard heat treatment of alloyed titanium alloys BT20 (VT20), BT22 (VT22), a dense uniform macrostructure is formed with an almost complete absence of different grain sizes over the cross section and bar length [9].

It was established that, along with high surface quality, the deformation scheme for helical rolling allows to steadily obtain rods with a cross section ovality of less than 0.5% of the nominal diameter and a curvature of no more than 0.75 mm per meter. The obtained results provide the possibility of increasing the yield ratio and the efficiency of redistribution as a whole by reducing the volume or eliminating the repair of intermediate blanks and finished bars.

It is especially important to note that when solving all the above problems, the process of helical rolling, as a method of deformation, is applied in its rational form without significant corrections to the properties of the material. Technological recommendations for a particular material are mainly reduced to the definition of a rational heating temperature, taking into account the above-noted features of the temperature mode, allowable reductions per pass and the frequency of rotation of the rolls [10].

The properties of a particular steel or alloy are present in the process modes in the form of a temperature range of processing, a permissible degree of compression of the coefficient per pass, or in the number of passes for a given total deformation and frequency of rotation of the rolls. And the rational configuration of the deformation center, the roll calibration and the parameters of the internal trajectories are practically unrelated, both with compression per pass and the number of passes performed, i.e. have a high degree of invariance with respect to material properties. In addition, in rational form-changing modes, there are no restrictions on maximum crimps on the conditions of natural gripping and rotation of the workpiece, and minimum on filling the caliber and stability of the roll due to their concomitant implementation.

4. Conclusion

Thus, when determining the optimal rolling parameters, it is necessary to take into account the technological features of the helical rolling, roll calibration, the structure and properties of the rolled alloy. These circumstances create the basis for the development of the principle of stationary helical rolling into a new constructive-technological quality, specifically focused on the production of bars of small sections.

References

[1] Ivanov M, Penkin A, Kolobov Yu, Golosov E, Nechaenko D and Bozhko S 2010 Warm cross-helical rolling in tapered rolls as a method of intense plastic deformation *Deformation and fracture of materials* **9** 13–18

[2] Miyazaki T, Terada D, Miyajima Y, Suryanarayana C, Murao R, Yokoyama Y, Sugiyama K, Umemoto M, Todaka Y and Tsuji N 2011 Synthesis of non-equilibrium phases in immiscible metals mechanically mixed by high pressure torsion *Mater. Sci* V 46 42964301

[3] Galkin S, Kharitonov E and Mikhailov V 2003 Reverse radial-shear rolling. Essence, opportunities, advantages Research methods *Titanium* **1** 39–43

[4] Bogatov A, Panov E 2013 The effect of the stress-strain state during helical rolling on the structure and plasticity of metals and alloys *Metallurg* 63–69

[5] Naydenkin E, Mishin I and Ratochka I 2015 The effect of helical rolling and subsequent annealing on the structure and mechanical properties of titanium alloy BT22 *Deformation and fracture of materials* **4** 32–46

[6] Roh J-H, Seo J-J, Hong S-T, Kim M-J, Han H and Roth J 2014 The mechanical behaviour of 5052-H32 aluminum alloys under a pulsed electric current Int. *J. Plasticity* Vol 58 84–89

[7] Derevyagina L, Gordienko A and Pochivalov Yu 2018 The effect of helical rolling on the fracture characteristics of low carbon steel at negative temperatures. *Deformation and fracture of materials* **2** 32–46

[8] Naizabekov A, Arbuz A 2015 Effect of the helical rolling on the microstructure of steel 40X *Bulletin of KazNTU Engineering science* **5** 249–255

[9] Ivanov K, Naydenkin E, Lykova O, Ratochka I, Mishin I and Vinokurov V 2017 Evolution of the structure and mechanical properties of the alloy BT6 during the helical rolling and subsequent deformation and heat treatments *University Bulletin Physics* 60(**7**) 126–132

[10] Tekoglu C, Hutchinson J and Pardoen T 2015 On localization and void coalescence as a precursor to ductile fracture Phil *Trans R A* Vol 373 20140121

IOP Conf. Series: Materials Science and Engineering **656** (2019) 012011 doi:10.1088/1757-899X/656/1/012011

Study of the thermal stability of process lubricants for heavy-duty modes of contact friction

E G Berdichevsky

Yaroslav-the-Wise Novgorod State University, ul. B. St. Petersburgskaya, 41 173003 Veliky Novgorod, Russia

E-mail: bersev@mail.ru

Abstract. Results of research of thermal stability of the lubricants used in industrial technologies related to heavy-duty modes of contact friction are given. Certain mathematical models of thermal destruction of lubricants process allow to give approximate quality assessment. Considering complexity of process of thermal destruction experimental studies are necessary and orientation for them is offered. The two-stage process of thermal destruction of lubricants is shown; the rate of pyrolysis is studied. The conclusion is drawn on the influence of flying products of thermal decomposition of lubricants on a tribological situation in the zone of contact friction.

Thermal stability being a functional property of process lubricants (PL) plays the most important role in heavy-duty modes of boundary contact friction, which is characteristic of technological processes (forging, rolling, abrasive machining, etc.) Nevertheless, insufficient attention has been paid to the study of thermal stability in scientific and technical literature, and methods of quantitative and qualitative assessment of thermal stability of PL are not standardized. Therefore, the purpose of this study which is to consider theoretical and applied aspects of PL behavior in high-gradient temperature fields typical of heavy-duty conditions of contact friction, seems actually relevant.

Thermal decomposition of any PL, especially hydrocarbon ones, is a series of variable and sequential chemical reactions with the formation of a large number of volatile and residual products. Intense thermal effect on the boundary layer of the PL is produced by the heated metal and the heat of friction and by plastic deformation [1].

According to the generally accepted point of view, the reason for the thermal decomposition of PL is an increase in the energy of vibration motion of the parts of molecules relative to each other with the rise of temperature. The energy of the oscillatory motion is evenly distributed over all the bonds in the molecules. At a certain temperature, the average energy of vibrations becomes greater than the strength of the weakest bonds, which leads to their breaking. Further temperature rise may cause stronger bonds destruction. Breaking of the bonds in a molecule leads to formation of "fragments" with unsaturated valency - free radicals, with high reaction capacity. As a result of recombination and repeated interactions of radicals, a radical-chain reaction develops. Due to the continuous interaction of radicals with stable molecules, new decay products and new radicals are formed. The complexity of the mechanism of thermal decomposition explains the complexity of the resulting gaseous mixture composition.

Thermodynamic and kinetic calculations of modern PL pyrolysis are impossible due to the exceptional complexity of the accompanying processes.

Approximate and empirical dependencies are also of little value for thermal stability of PL assessment, since, one hand, they comprise many empirical coefficients, and, on the other hand, they describe pyrolysis for a specific situation. Thus, in [1], the dependency for calculating the burnout rate of oils is proposed:

$$U_v = \frac{0.001\alpha(T_r - T_{boil})M}{\rho_l(\Delta H^0_{vap} + C_p(T_{boil} - T_0))} , \qquad (1)$$

where: U_v is the burnout rate of the liquid;

α is the heat transfer coefficient from the heated body to the liquid;

ρ_l - density of the liquid at the initial temperature T_0;

C_p - average heat capacity of liquid in the temperature range from T_0 to T_{boil};

ΔH^0_{vap} - heat of evaporation of the liquid.

For a specific lubricant the values of α, C_p, T_{boil}, ΔH^0_{vap} are unknown and difficult to find. In this regard, the main means of identifying the thermal stability of PL is to conduct an experiment with special equipment in conditions as close as possible to real ones.

Experimentally, thermal stability of PL is estimated indirectly by the parameters of the gas mixture resulting from pyrolysis. Here two approaches are possible. On the one hand, it is possible to consider thermal stability as a significant physical and technical property of a lubricant. On the other hand, thermal stability can be considered as a result of the process of contact interaction of heated surfaces separated by a PL layer. In both cases, thermal stability is considered as a gas-making ability of PL subjected to high temperature exposure.

In the first approach, the specified volume of lubricant is loaded into metal capsule, after which the capsule is sealed and placed in a temperature-controlled heating device. The working cavity of the capsule is connected with a highly sensitive manometer through a capillary. The gaseous products resulting from the pyrolysis of PL cause an increase in pressure in the capsule cavity, which is recorded by a pressure gauge with accuracy of ± 0.4%. A capsule with a working volume of 0.048 m^3 was used for the experiment conducted with the use of this technique. The chamber of the capsule stored 5g of lubricant. The maximum heating temperature of the capsule was 1100°C.

To implement the second approach, a plant has been developed, the scheme of which is presented in figure 1.

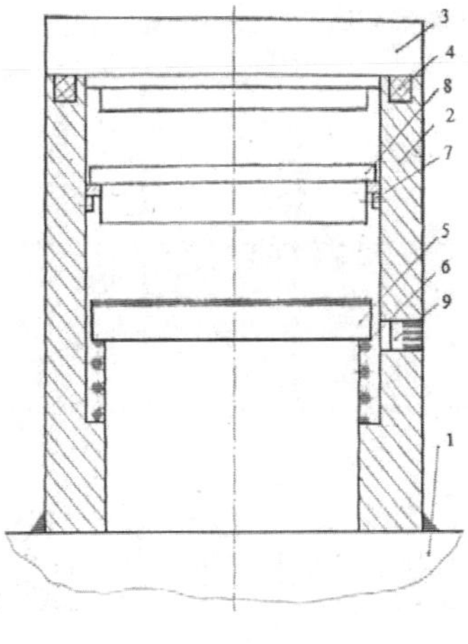

Figure 1. Plant diagram for the study of thermal stability.

The case of the plant 2 was placed on the magnetic plate 1. The joint between the case and the plate was carefully sealed. The top cover 3 is connected with the case by a clamping device. The sealing of the upper connector is provided by gaskets 4. The case is equipped with a bottom die 5, which is held in the lower non-working position by the attraction of a magnetic plate overcoming the force of the spring 6. Lubricant under test is applied to the surface of the bottom die. A ring-shaped holder 7, in which the heated sample 8

IOP Conf. Series: Materials Science and Engineering **656** (2019) 012011 doi:10.1088/1757-899X/656/1/012011

is placed, is fixed on small protrusions in the case. After installing the sample 8, the connector is sealed and the magnetic plate is turned off. The lubricated bottom die 5 is released and with the help of springs forces the heated specimen 8 to the device cover. The lubricant layer burns out (the process of pyrolysis). The pressure and temperature of the resulting gas mixture are registered by a miniature sensor 9. Due to the presence of gas leaks, the accuracy in defining thermal stability in the plant is ± 10% with confidence level of 0.95.

For a quantitative assessment of the gas-making ability of PL, it is advisable to use the U parameter, called the decomposition rate of PL and defined by the formula:

$$\boldsymbol{U} = \frac{P_f - P_s}{m(T_f - T_s)} \left(\frac{kPa}{g \cdot {}^{\circ}C}\right), \qquad (2)$$

where: P_s and P_f are the initial and final values of the pressure of gaseous products of thermal decomposition;

T_s and T_f - absolute temperatures of the beginning and end of thermal effects on the PL;

M – is the mass of lubricant in the tribo-active zone.

It has been established that lubricants for hot stamping of leading foreign firms are characterized by high gas-generating value [3]. The results presented in figure 2 indicate that gas pressure in the stamping zone can reach values comparable with deformation energy. Substantial nonlinearity of gas pressure dependence on temperature during punching and on the composition of the lubricant was also confirmed.

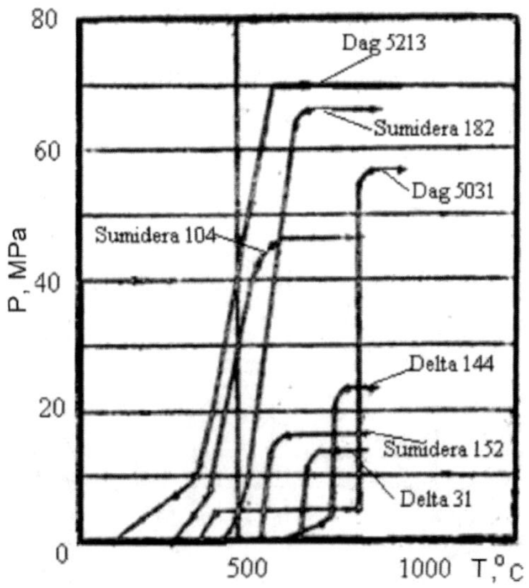

Figure 2. Dependence of gas pressure on temperature during punching.

Release of volatile components of the PL pyrolysis is of particular interest. There is every reason to believe that in some cases, the volatile components of thermal destruction of PL can have a radical impact on the tribological situation in the zone of loaded content [1].

Volatile components composition depends on many factors: on the initial chemical composition of the lubricating film, on the kinetics of the pyrolysis process, on temperature gradients, on the catalytic action of juvenile metal surfaces.

Typical yields of pyrolysis products of organic compounds under various process conditions are presented in table 1. Sometimes hydrogen and CO are found in volatile products. As can be seen, even insignificant heat exposure time (0.1 s) causes formation of considerable amount of volatile components. Relative volatile yield can be calculated by the formula [4]:

$$\xi = 1 - exp\left[-\tau k_0 exp\left(-\frac{E}{Rt_4}\right)\right], \quad (3)$$

where: ξ is relative yield of volatiles;

k_0 is the rate constant for the reaction of thermal decomposition. We believe that it is subject to the dependence of Arrhenius;

τ is the time of heat exposure;

R is the universal gas constant;

E – activation energy, KJ / mol;

T_4 – maximum process temperature, K.

This dependence can be used for the qualitative analysis of the process of pyrolysis of PL. Given the complexity and multicomponent structure of real lubricants, experimental identification of k_0 and E constants is extremely difficult.

Table 1. Typical yields of organic pyrolysis products.

	Product Yield (%) under various conditions			
	785÷800°C	815÷840°C	840÷870°C	870÷925°C
Product	1,2 c	0,05c	0,35c	0,1c
CH_4	15,6	16,6	16,8	16,7
C_2H_4	23,0	29,9	29,3	33,3
C_3H_6	13,6	12,7	12,2	11,7
C_4H_6	2,2	3,8	4,2	4,8
C_5	32,8	29,7	27,8	23,9

The methodology for the study of materials pyrolysis (more precisely, their gas-making ability) is as follows. 5 g of the test PL was loaded into the working chamber of the plant for the study of thermal stability. After sealing the chamber was fixed inside the heating furnace. The temperature inside the chamber was indicated with a thermocouple. Inside the chamber the volatile products of pyrolysis created pressure, which was registered with a high-precision model manometer. To find out the actual gas-generating value of the PL, the reading of the air pressure at the given temperature was subtracted from the pressure gauge indication.

Gas-generating value of some lubricants are presented in figure 3.

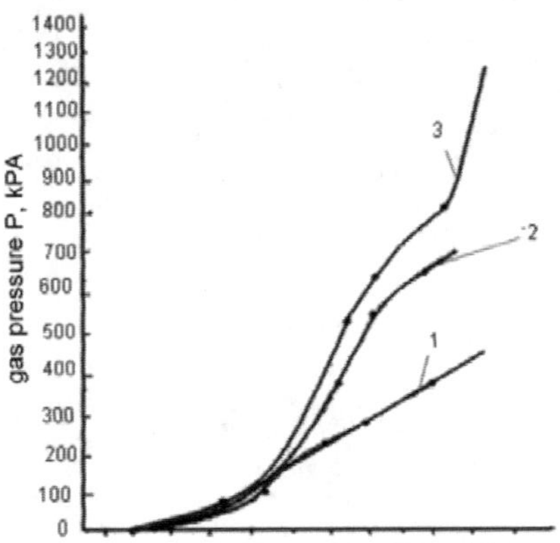

Figure 3. Experimental studies of gas-generated value of PL: 1 – Aluminol; 2 – Pressol-KF; 3 – Graphitol E-21 (concentrate).

From table 2 it can be seen that the composition of the medium is of critical importance for the thermal decomposition of the PL and thermal stability caused by it. Graphite enhances its gas-generating value sharply. Gas pressure reaches values comparable with the energy of deformation and can significantly distort the stress state pattern in the zones of friction and plastic deformation. Silicone oil has little gas-generating value. In some cases, the air dissolved in the lubricating film may influence the gas pressure.

Table 2. The rate of pyrolysis of PL.

The area of pyrolysis	PL	Rate of pyrolysis, kPa / g • °C
I	Aluminol E (concentrate)	$16,4\cdot 10^2$
	IHO	$19,6\cdot 10^2$
	Graphitol E-21 (concentrate)	$24,5\cdot 10^2$
II	Aluminol E (concentrate)	$55\cdot 10^2$
	IHO	$98,8\cdot 10^2$
	Graphitol E-21 (concentrate)	$165,2\cdot 10^2$

Summarizing the above, we can conclude that the thermal decomposition of PL is accompanied by complex physicochemical processes. Experimental and theoretical study of the products of PL thermal decomposition helps deepen the understanding of the mechanism of contact boundary friction and affect the tribological situation radically.

References

[1] Berdichevskij E G 2011 *Friction and lubricants in plastic processing of metals*: monograph (Veliky Novgorod: NovSU Publ) p 183
[2] Zhorov Yu I 2019 *Pyrolysis of hydrocarbons* (Moscow: Himiya Publ) p 292
[3] Nakamura D N 2016 Global ethylene capacity increases slightly in 06 *Gil and Jurnal* **105** 92 – 114
[4] Adamyan V A 2018 *Theory of combustion and explosion* (St-Peterburg: Lan' Publishing House) p 116
[5] Lemonidov F F and Vasalos I A 2016 Preparation and evaluation of catalysts for the production of ethylene via steam cracking *Applied Catalysts* **54** 119–196

IOP Conf. Series: Materials Science and Engineering **656** (2019) 012012 doi:10.1088/1757-899X/656/1/012012

Energetic activation of the processes of abrasive diamond metal processing

E G Berdichevsky

Yaroslav-the-Wise Novgorod State University, ul. B. St. Petersburgskaya, 41 173003 Veliky Novgorod, Russia

E-mail: bersev@vmail.ru

Abstract. The possibility of energetic activation of the processes of abrasive diamond metal processing is considered. Emphasis is placed on thermal activation of coolants and lubricants and on electrical activation of the grinding zone. It is shown that under certain conditions, the external energy supply can significantly improve the performance of abrasive-diamond processing. The role of the electrode potential during electrical activation is disclosed. Physical aspects of energetic activation are discussed. The prospects of other schemes of electric activation are demonstrated.

1. Introduction

Abrasive diamond processing of metals is one of the leading technologies in machine building and instrument making. It is this processing method that often provides the final required dimensional accuracy and surface quality of the product. Therefore, much attention is paid to the improvement of abrasive diamond processing in the scientific and engineering sphere. Traditional methods of improving the efficiency of abrasive diamond processing processes, such as improving the tool, optimizing the processing modes, and improving the quality of liquid coolants, are so deeply developed that they give a slight increase in efficiency. Innovative and promising is the energetic activation of the abrasive diamond processing processes, consisting in the directional influence of external energy flows on a detail being processed, on an abrasive diamond tool, on a coolant or, in a complex, on a treatment area.

Of the many known methods of energetic activation for the study of prospects in the abrasive diamond processing processes, two were selected: thermal and electric.

2. Results and Discussion

The mechanism of energetic activation of species and processes is complex and poorly studied [1]. It is assumed that as a result of the action of the energy supplied by external currents, the potential and kinetic energy of the atoms and molecules of the object being activated increases, and the activation energy of the processes in which the object participates accordingly decreases. The activation energy is an energy barrier that must be overcome in order for the desired interaction between objects to occur.

For most physicochemical processes associated with abrasive diamond processing, the activation energy values under normal conditions are quite large, so the necessary reactions do not take place at all or are very sluggish.

The relationship between the velocity of processes and their activation energy is described by the Arrhenius law [2], from which it follows that even a slight decrease in the activation energy is accompanied by a strong acceleration of the interaction:

IOP Conf. Series: Materials Science and Engineering **656** (2019) 012012 doi:10.1088/1757-899X/656/1/012012

$$k = e^{\frac{-E}{RT}}, \qquad (1)$$

where: k is the interaction velocity;

A is a constant;

E is the activation energy, kcal/mol;

R is the universal gas constant;

T is the temperature.

For processes between valent-saturated molecules, which are characteristic for traditional abrasive diamond processing, E~1-10 kcal/mkm.

When additional external energy is activated by activated objects, the potential and kinetic energy of atoms and molecules of the object may increase so much that they will decay with the formation of new reactive particles (ions, electrons, free radicals). As a result, not only ordinary reactions are significantly accelerated, but also new unpredictable types of interaction can occur.

Thermal activation, the most accessible for an applied implementation, lies in temperature rise of any of the interacting objects by supplying thermal energy. From the aforecited Arrhenius equation it follows that a change in temperature in an arithmetic progression causes a change in the reaction rate constant in a geometric progression. So, if it is impossible to synthesize water from hydrogen and oxygen at 20 °C (it would take 54 billion years), then at 500 °C, this would take only 50 minutes, and at 700 °C, the reaction proceeds immediately [2].

The most accessible scheme was studied - thermal activation of coolant liquids used in the processes of abrasive diamond processing.

When heated, widely used water-based coolant lubricants (emulsions, chemical solutions of salts and polymers) significantly improve its many physical and technical properties that positively affect the performance of the process. So, when coolants heat up, their viscosity decreases significantly, the wetting ability with respect to metals and ceramics improves, the surface tension decreases, and the washing potential increases. In some situations, thermal activation of coolant liquids can lead to negative results due to a decrease in the cooling effect, destruction of hydrocarbon components, a decrease in anti-wear properties of coolant liquids based on doped oils. In each case, experimental testing is required.

The study of the effect of the temperature of a standard 5% water-oil emulsion was carried out on a flat grinding operation of hardened 40X steel circles 24A95CM1K5 at a cutting speed of 35 m/s, a reciprocating speed of 8 m/min, and a vertical feed of 0.02 mm/double move.

Experiments have shown (figure 1) that the temperature of the emulsion significantly affects the specific productivity g, the dimensional wear ΔR_y and durability of the circle T. The effect of temperature on the indicator R_a (arithmetic average deviation of the profile) of the treated surface is insignificant. The dependence of specific productivity and dimensional wear on temperature is extreme. The optimum is observed at 45 °C. Circle strength decreases linearly when the emulsion temperature increases. The value of the optimal temperature depends on the composition of the coolant liquid and processing conditions and is in the range of 40–60 °C.

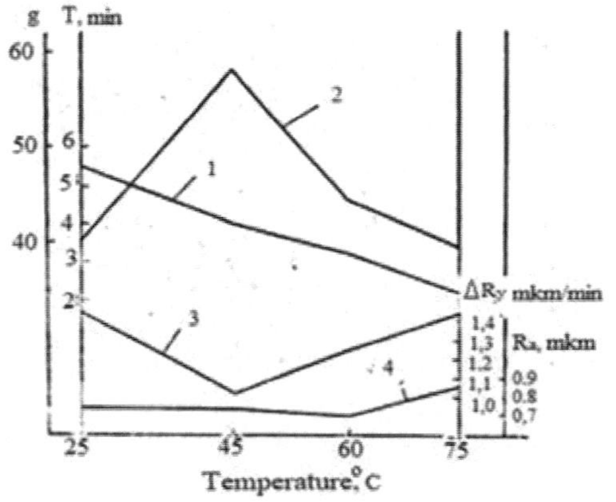

Figure 1. The effect of heating temperature and emulsions on the performance of the process of flat grinding of steels. 1 — circle strength; 2 — specific productivity; 3 — specific dimensional wear of the wheel; 4 — indicator.

In the well-known works on thermal activation of coolant liquids, static isothermal heating of the entire mass of liquid was used. Apparently, large reserves are laid in the dynamic high-gradient heating mode, as well as in the effects of a non-uniform thermal field. Model experiments in this sphere have shown a number of interesting effects. Thus, it has been established that when a liquid droplet spreads over a solid surface in the non-isothermal mode, thermocapillary and thermoosmotic effects, which can act in opposite directions, play a significant role. Therefore, when the liquid flows in the direction of decreasing temperature, the speed of spreading increases, while when it flows in the opposite direction, the movement slows down. As the temperature gradient increases along the sample, the flow rate of the liquid increases from the heated end of the sample to the cold one. With this flow, the process' kinetics is described not by a cubic parabola, as in case of isothermal spreading, but by a quadratic one [3]. Thus, by creating the required temperature gradient on a detail being processed or a tool, it is possible to intensify the flow of the coolant into the treatment zone [4].

Electric activation of the processes of abrasive-diamond processing can be implemented according to various schemes. The analysis showed that the preferred scheme is the transmission of electric current through the circuit "coolant — the detail being processed." The physical model of such activation in aquatic environments is as follows. When a metal surface comes into contact with an aqueous medium, a double electric layer and an electrode potential arise spontaneously at the interface. As a result, all the physicochemical processes occurring on the wetted metal surface are influenced by the field of the electrical double layer and depend on the magnitude of the electrode potential. Such processes include wetting, adsorption, the formation of phase films, etc.

Studies have shown that such properties of the metal being processed as microhardness, surface energy, dispersibility, friction coefficient, deformability, etc., depend on the magnitude of the electrode potential [5]. The electrode potential reacts sensitively to structural and phase transformations in the surface layer of the metal being processed and is associated with the magnitude of the residual stresses [6].

The substantial dependence of the strength of the metal, as well as the nature and kinetics of the physicochemical processes occurring on the surface of the detail, on the magnitude of the electrode potential makes it possible to activate the processing by displacing the electrode potential in one direction or another. The electrode potential can be shifted by polarizing the detail being processed, i.e. imposing external potential on it. Depending on the sign of the imposed potential, the polarization can be cathodic and anodic.

The effects of electrical polarization on the diamond honing process showed that during cathodic polarization, a decrease in the electrode potential below the potential of a zero iron charge is accompanied by a significant decrease in the surface roughness of the machined surface and an increase in the efficiency of the process. Polarization was carried out from an external power source assembled according to the galvanostat scheme. As a current source, ordinary batteries were used. The current flowed through the "detail being honed – coolant liquid" circuit. Electrolytes and emulsions were used as coolant. In a wide range of changes in the electrode potential, the processing productivity varies along an extreme curve. The results obtained are explained by the manifestation of the electrocapillary effect [3]. Hydrogen absorption of the metal surface, which is inevitable during cathodic polarization, is insignificant.

The use of polarization of carbide plates made of T15K6 alloy with AChK circles 125×10×3 ACP 50/40 K1 100% being grinded in the electrolyte medium made it possible to reduce diamond consumption by 20–60%. In this case, the polarizing current was 10 mA, the voltage in the circuit "detail – coolant" was 15 V, the current density was 2 mA/cm^2.

Of considerable practical interest is the possibility of activating the processes of abrasive diamond processing in hydrocarbon media [7]. In experiments, the process of round grinding with corundum abrasives was activated by a constant electric field. Mineral oil was applied to the surface to be processed by contact using a porous gasket. A metal electrode was placed under the gasket, to which, like the detail, an electric charge was applied. In fact, the part and the electrode performed the function of the capacitor plates. The detail was isolated from the walls.

Activation was accompanied by an improvement of many indicators of the processing process. In particular, cutting forces have decreased, the abrasive wheel resistance has almost doubled, the surface geometry has improved. The voltage of the permanent electric field varied in the range of 20–100 V/cm. A positive result of activation can be explained by an increase in the strength of lubricating adsorption films on the metal, as well as by better oil wetting of hard-to-reach areas of the surface micro-profile.

The electrostatic field can activate the processes taking place at the "macro level", i.e., to increase the heat transfer from the detail to the coolant, to influence the hydrodynamics of the coolant. In order to improve pumping and cooling action of the coolant during grinding, in the USA it was proposed to supply sufficiently large dissimilar charges to the jet of liquid and the detail being processed. This method was modified by the introduction of a second electrostatic field, which allowed for infinite adjustment of the power and direction of the coolant flow (figure 2).

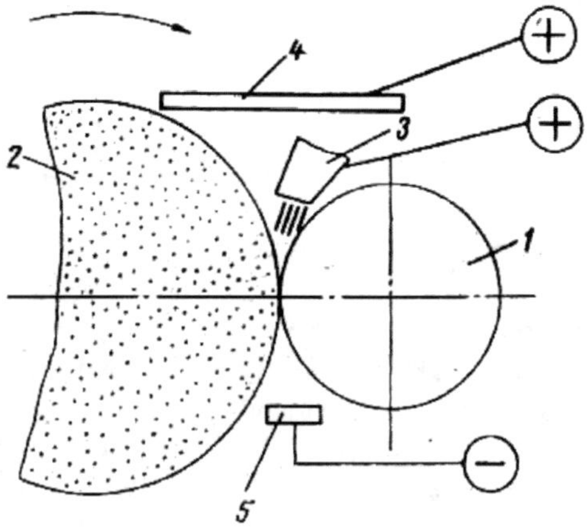

Figure 2. Scheme of regulation of hydrodynamic processes during grinding by imposing electrostatic fields.

A negative charge from an external source is supplied not only to the detail 1 being grounded by grinding 2, but also to the plate 5 installed under the processing area and having the opportunity to occupy an arbitrary position in space. The coolant jet, which is fed to the cutting zone through the nozzle 3, imparts a positive charge, with results in a potential difference between the jet and the detail 1 being grounded through the machine components. Above the zone of ejection of the coolant from the nozzle the plate 4 is installed, it is charged with the same charge as the jet of coolant and has the ability to occupy arbitrary positions in space. An additional electrostatic field (between the plates 4 and 5) is turned relative to the zone of contact between the circle and the detail so that the resultant between it and the main field corresponds to the required hydrodynamic parameters.

It can be assumed that important applied results will be obtained when activating the processing process with variable and asymmetric electric fields, as well as when leading to the processing zone of other types of energy (ultrasonic, magnetic, radiation, etc.)

3. Conclusion

Energetic activation of the processes of abrasive diamond processing is both an innovative direction in the scientific foundations of engineering technology, and a means of improving efficiency in real production. The obtained positive results on the consideration of thermal and electrical activation is an incentive to study other schemes of activation effect on the processes of abrasive diamond processing.

References

[1] Shevelya V V and Oleksandrenko V P 2016 *Tribochimiya i reologia iznosostoykosti* [Tribo chemistry and rheology of wear resistance] (Moscow: Mir) p 278
[2] Karpetiyants M H 2013 *Vvedenie v teoriu khimicheskikh processov* [Introduction to the theory of chemical processes] (Vyschaya schkola Publ) p 316
[3] Inoce K 2014 *Kapillyarnaya khimiya* [Capillary Chemistry] (Moscow: Mir) p 272
[4] Berdichevskij E G 2011 *Trenie i smazka pri plasticheskoj deformacii* [Friction and lubrication during plastic deformation] (Veliky Novgorod: NovSU Publ) p 184
[5] Khudobin L V and Zhdanov V F 2016 Primenenie elektricheskogo polya dlya povysheniya effektivnosti SOZH pri shlifovanii [The use of an electric field to increase the efficiency of coolant during grinding] *Chistovaya obrabotka detalej mashin* **8** p 40–46
[6] Komatsu W 2017 Kinetics of solid–state reactions *Pros. W. Intern. Simp. Reativity of Solids* p 182–186
[7] Myshkin N K and Petrovec M I 2014 *Trenie, smazka, iznos* [Friction, lubrication, wear] (Phizmatlin Publ) p 36

IOP Conf. Series: Materials Science and Engineering **656** (2019) 012013 doi:10.1088/1757-899X/656/1/012013

The effect of nanoceramic materials on the tribological properties of technological lubricants

E G Berdichevsky

Yaroslav-the-Wise Novgorod State University, ul. B. St. Petersburgskaya, 41 173003 Veliky Novgorod, Russia

E-mail: berse@mail.ru

Abstract. The effect of some nanotechnological materials such as boehmite and β-sialon on the tribological characteristics of technological lubricants has been studied. It is shown that even a low percentage of nanoceramics used for doping mineral oils can significantly improve the tribological situation in the zone of boundary contact friction. Technical and economic characteristics of the considered nanoceramics suggest them promising components in the development of innovative technological lubricants for heavy duty friction.

1. Introduction

The possibility of improving the tribological properties of lubricants by doping them with nanocomponents has been shown in many studies. So in [1] it was established that such nanoproducts as fullerene black, fullerene soot and other nanocarbons increase the functional – tribological properties of lubricants operating under conditions of heavy friction contact. There is a lot of information about the successful use of nanocarbons in motor and transmission oils [2], in plastic deformation of metals [3], in boundary sliding friction [4]. However, carbon nanopowders have high cost, instability in lubricants and low heat resistance. As a result, carbon friction nano-modifiers are not widely used in practice.

2. Nanoceramics

It is interesting and promising to consider nanoceramic materials as alloying components of technological lubricants. Nanoceramics are much cheaper and thermostable than nanocarbons. Many nanomaterials have a layered disulfide and molecular structure, like graphite and molybdenum disulfide, and in this case they can actively exhibit antifriction properties [5]. Nano- and finely dispersed powders of layered silicates added to various oil compositions significantly improved the tribological situation in many friction sites. The achieved tribo effect was explained by the interaction on friction surfaces of broken atomic bonds of particles of compositions with active bonds. Initially, the composition consists of solid-dispersed basic particles of layered silicate with a dimension of $0.5 \div 20$ microns with thin from 10 to 500 nanometers marshallite inclusions.

Silicate contains siloxane (bridged) Si-O-Si bonds. Due to high-tech fine grinding, bonds are broken so that the number of uncompensated broken bonds is as large as possible. Broken siloxane bonds are active hydrogen acceptors. When they interact, compensated Si-OH silanol groups are crteated. Thus, by adding finely divided layered silicate into the lubricant, which has a large specific surface and a large number of broken siloxane bonds, we create at their expense the conditions for binding active hydrogen in the friction zone, i.e. prevent its interaction with the metal and prevent hydrogen wear of the latter.

The presence of marshalls inclusions stimulates, according to the opinion of developers, the adhesive interaction of geosilicate friction modifier with the metal surface of the friction unit, contributing to the formation of anti-friction sliding surfaces film. In addition, marshallites also adsorb bound hydrogen.

The authors of the composition note that the use of layered silicate powders as an additive to lubricants is especially effective when high temperatures and contact pressures are present in the friction zone. In this case, the thermal decomposition of silicate occurs with the formation of new chemical products that actively interact with each other and the metal of the rubbing pair. As a result of numerous transformations, disintegrations, transformations, repeated interactions on the friction surfaces a mineral film of high density, hardness and adhesion to the substrate is formed. Mineral film provides a significant reduction in wear, prevents surface chips.

3. Doping with nanoceramic boehmite

Of great scientific and applied interest is the study of the tribological properties of lubricants doped with nanoceramic boehmite.

Boehmite belongs to the hydroxide subclass and has the form γ-AlOOH. The mineral represents an aluminum hydroxide and is one of the main components of bauxite. The composition of boehmite is similar to the diaspora, isostructural. The basis of the layered crystalline structure of boehmite is the alternation of packs of two internal oxygen and two external hydroxyl layers with aluminum atoms located in voids.

Boehmite is insoluble in acids, does not melt in front of the blowpipe, splits cleavage. The density of boehmite is 3.02–3.06 g/cm^3, the hardness on the Mohs scale is 3.5. The mineral contains impurities of various oxides: Fe_2O_3, SiO_2, TiO_2, and others.

Nanocrystalline boehmite is produced according to the technology of hydrothermal synthesis from quartzite. During the synthesis, aluminum particles of up to 10 microns in size are transformed into nanocrystalline boehmite. The stable properties of the initial aluminum powders and developed technological process make it possible to synthesize boehmite powders with constant properties. Unlike additives based on natural raw materials, this allows for greater stability of the properties of artificial tribomodifiers [6].

Tribological effectiveness of lubricants doped with boehmite was evaluated on the SMT-1 friction machine according to the roller-to-block scheme. The list of tested compounds is given in table 1.

Table 1. List of tested compounds.

Name of tested additive	Additive concentration in oil M-10DM
Boehmite from ZAO "Concern Nanoindustry" with surfactant after ultrasonic treatment	Boehmite 0.08 g/l, surfactant 0.3–0.5 %
Boehmite produced by GOSNITI with surfactant after ultrasonic treatment	Bemite 0,08 g/l, surfactant 0.3–0.5 %
Boehmite produced by GOSNITI with silica powder and surfactant after ultrasonic treatment	Boehmite and silicon oxide at 0.04 g/l, Surfactant 0.3–0.5%
Potassium polititanat and surfactant after ultrasonic treatment	Potassium polititanat 0.08 g/l, surfactant 0.3–0.5 %

SAS (surface-active substance) – surfactant – based on fat soluble acids. The predominant particle size of boehmite is 2–3 microns, the maximum – up to 10 microns. Aggregates of particles were crushed by ultrasound.

Figure 1 shows the dependence of the friction coefficient on the load for the tested compositions and for pure M-10 DM oil.

Tests have shown that the minimum friction and temperature optimal concentration of boehmite is 0.5%, at that the coefficient of friction is reduced by one third.

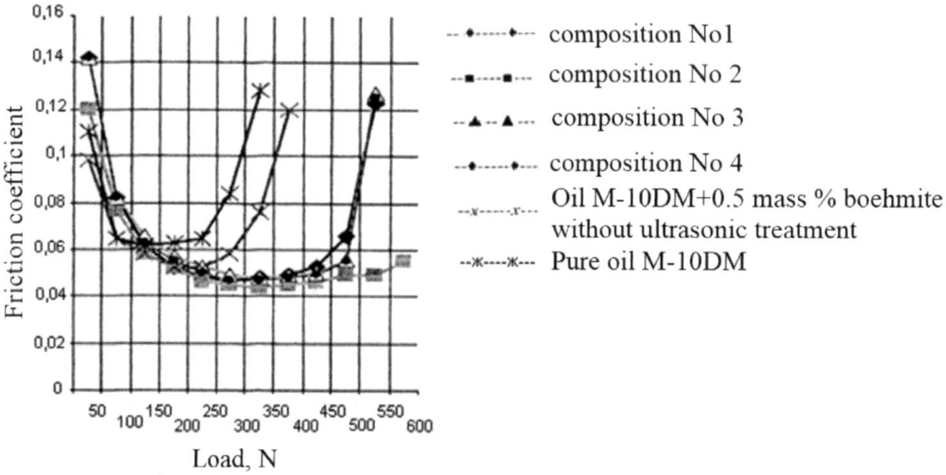

Figure 1. Dependence of friction coefficient on the load and lubricating composition.

The best results showed lubricants with surfactants and ultrasonic treatment. Each group of lubricant compositions has its own ranges of optimum concentration in oils and loads in friction units, at which their greatest efficiency is ensured.

Tribological properties of boehmite were studied as an doping component of the emulsion. 3% emulsion of standard emulsol was accepted as a model composition. 5%, 10% and 15% boehmite was introduced into the emulsol with mechanical mixing. Dispersion of emulsol in water was carried out first with a propeller stirrer, then by ultrasound. The lubricating properties of the emulsions were evaluated on a four-ball friction machine. Coefficient of plastic friction was determined on the stand by plane-symmetric sink of aluminum rings by the method described in [3]. The test results are shown in table 2.

Table 2. Tribological properties of 3% emulsions from emulsols boehmite doped.

Boehmite concentration in emulsion, %	Bedding load, R_k, N	Welding load, R_s, N	Wear spot diameter, mm	Coefficient of plastic friction (bench tests)
0	210	280	5.6	0.38
5	240	320	4.8	0.32
10	270	360	4.1	0.28
15	300	410	3.7	0.24

From the results it follows that an increase in boehmite concentration in emulsion and in emulsion leads to a systematic improvement of all tribological properties considered. The load of breaking lubricating film at 15% of the content of boehmite increased by almost 50%, the load of welding balls – by 40%. Nanocrystalline boehmite modifier made it possible to reduce friction coefficient by almost 80%. It was established that emulsions containing boehmite are more dispersed, stable, and less prone to foaming. It is likely that boehmite particles, being adsorbed on the boundary between the oil and water phases, impede the coalescence of oil phase droplets, i.e. have a separating-stabilizing effect.

4. Nanosized β-sialon powder

In addition to boehmite, nanosized β-sialon powder is promising. Nanoparticles have an average size of 50 microns, show rounded shape and represent Al_2O_3 and AiN solid solution in β-Si_3N_3 with the general formula $Si_{6x}Al_xO_xN_{8x}$ (where x=0.8–4.2) obtained by the plasma-chemical method. Medium-viscous industrial oil I-40 thickened with cerisin was used as a lubricant base.

The results of testing lubricant compositions doped with nanopowder β-sialon are presented in table 3. The tests were carried out on four-ball friction machine by standard method [3].

Table 3. Tribological characteristics of lubricating compositions doped with β-sialon.

Lubricating composition	Bedding load, R_K, N	Welding load, R_S, N	Chips index, C	Friction coefficient
Oil I-40 + 1% cerasin	65	122	26	0.17
Oil I-40 + 1% cerasin + 1% β- sialon	110	204	54	0.15
Oil I-40 + 1% cerasin + 2% β-sialon	188	276	69	0.14
Oil I -40 + 1% cerasin + 3% β-sialon	290	414	90	0.14
Oil I -40 + 1% cerasin + 3% β-sialon + 2% colloidal graphite	306	420	90	0.12

Analysis of test results showed that introduction of nano-sized doping ceramic component made it possible to increase bearing capacity of friction pair by 4.5 times, significantly increase the load of lubricant film breakthrough (3.5 times), and slightly reduce friction coefficient. The addition of the classic antifriction additive – graphite – insignificantly improved the tribological indicators on the background of nanoceramic modifier. The obtained positive results can be explained by complex transformations of β-sialon on friction surfaces, which, under the influence of nanoceramics, change not only their microgeometry and morphology, but also elastoplastic properties [7].

"Concern Nanoindustry" (Russia) has produced preparation for doping transmission and engine oils based on nanoceramic of supersonic grinding. The advantages of lubricating oils with nanopowders compared with oils doped with traditional additives are shown in table 4.

Table 4. Comparison of advantages of oils doped with nanoceramics and traditional additives.

Characteristics	Traditional additives	Nanoceramic modifier "Stribile"
Friction surface hardening	Not happening	Happening
Microrelief alignment	Not happening	Optimum clearance is being added
Compatibility with used oils	Requires careful selection	Compatible with all oils
Friction coefficient	Not less than 0.015	0.008–0.03
Vibration	Declining	Declining
Anticorrosion resistance	Low	High

Note that the discussed nanoceramic product showed significant efficiency in friction pairs with relatively soft and spare loads (transmission, internal combustion engines, gearboxes).

5. Conclusion

Nanoceramic powder materials used for doping technological lubricants have positive effect on tribological situation in contact friction zone. Due to the low cost and thermal stability usual for all ceramics, nanoceramics can serve as the basis for creating innovative tribological lubricants for heavy regimes of boundary friction. Physical-chemistry of behavior of the products such as boehmite and β-sialons in lubricating compositions needs further study.

References

[1] Berdichevsky E G 2018 Friction Process Control in Forging and Hot Stamping Technologies *IOP Conf. Ser.: Mater. Sci. Eng.* **441** 012010
[2] Pogodaev L I 2006 *Structural and energy models of materials reliability and machine parts* (SPb: ATPF Publ) p 608
[3] Berdichevsky E G 2011 *Friction and lubrication during plastic deformation* (Veliky Novgorod: NovSUPubl) p 184
[4] Ginsburg B M *at al* 2000 Effect of fullerene C60, fullerene soot and other carbon materials on the boundary slipping friction of metals *Technical Physics* **70 (12)** p 87–97
[5] Davydov N A and Slitsky B U 2009 Practical implementation of effectiveness of new nanotechnological materials for rolling equipment tribological systems of railway transport *Collection of scien. papers of the IX intern. conf. "Tribology and Reliability"* (SPb) p 96–101
[6] Almyasheva O V, Korytkova E N at al 2005 Synthesis of aluminum oxide nanocrystals under hydrothermal conditions *Inorganic Materials* **41 (5)** p 80–91
[7] Kostetsky B 2014 The structural-energetic coctpt in the theory of friction avd wear (synergism and self-organization *Wear* **159** p 1–15

Modeling of process of regulating production, processing and marketing of milk and milk products

M K Chernyakov[1,3] and M M Chernyakova[2]

[1]Novosibirsk State Technical University, Karl Marx Ave., 20, Novosibirsk, 630073, Russia
[2]Siberian Institute of Management of the Russian Academy of National Economy and Public Administration

[3]E-mail: mkacadem@mail.ru

Abstract. The article is devoted to the investigation of problems of regulating processes in dairy industry. One of such problems is the absence of effective informative technological technology of simulating these processes. To achieve the goal set, it has been supposed to apply the simulation based on the theory of Petri nets. Methodological base of the investigation includes the formal theory of Petri nets as well as the theory of its practical application. The work has demonstrated main stages of evolution of Petri nets from simple to innodiversification ones, the latter being suggested for simulating technological processes and systems of organizations of dairy industry. Specific feature of innodiversification Peri nets is the conjunction of combinations of all known kinds of Petri nets and the application of their advantages to solve particular tasks. It has been found that the instruments of innodiversification Petri nets may be effectively applied to simulate organizations of dairy industry of any level.

1. Introduction

Dairy industry develops under digitalization influence as well as due to development of economic and mathematical theories, and their impact on the national economy by means of formulating recommendations and suggestions in the field of public administration. Today, leading economists agree that state economic policy has been formed under the influence of developing economic and mathematical theories. However, there are disagreements concerning scales, mechanisms and goals of such influence. In this connection, there occurs a task to consider characteristics of forming main theoretical approaches to the problems of public administration, taking the dairy industry as an example.

State regulation of economy in terms of digitalization leads to modification of the system of measurements of legislative, executive and controlling character which must be implemented by authorized government agencies and community organizations in order to adapt economic and social systems to the conditions of digital economy. Today, the main driving forces of digital transformation are new products and services, new information and management technologies, innovative business-models.The key drivers of digital transformation are industry digital platforms, including the dairy industry.

2. Research Methods and Approaches

Dairy industry is characterized by net structure of its organizations. Mathematical apparatus for the systems of net structure has been theoretically developed and can be implemented for this field organizations as a simulation model. According to the authors, such structure may be implemented based on the Petri net theory.

According to the authors, it is the further development of Petri nets that allows to apply them to simulate the work of organizations taking impacting decisions of the subsystem of regulating the dairy industry. In the research field, a subregulator may be considered indirectly, by means of a model. By manipulating the subregulator model of the dairy industry, it will be possible to get new information about organizations, avoid dangers, expensiveness, or discomfort of analyzing the real system itself. As a rule, models are based mathematically. Today, further development of the Petri net theory has the following main directions:

– development of main means, methods, and notions which are necessary in order to further develop the formal Petri net theory;

– application research of the Petri net theory in the field of modeling systems, analyzing them, and achieving results of deep penetration into processes and systems being modeled.

Innovative approach to technological process of organizations of the dairy industry is characterized by complex dynamic system, including various units, such as equipment, control and management means, auxiliary and transport devices, processing instruments or media which constantly move, interact and change, production objects, operators (people, robots, manipulators) performing processes and managing them. Analyzing complex technological processes suggests decomposition of the production system into subsystems and subsubsystems of different level of depth. Due to the decomposition of the system into subsystems, hierarchy of the production system structure may be made, this allows to consider it at different levels of detailing.

Due to high complexity and labor intensity of processes of designing technological processes and systems of the dairy industry, it is advisable to use digital technologies.

In order to forecast results of projecting technological processes and systems, to use information and communication technologies (ICT) is advisable with the help of simulation modeling methods.

The goal function of conception of innovative simulation modeling of processes and systems in the dairy industry is to make project calculations of productivity and efficiency characteristics, considering the analysis of different structural variants of organizations, degrees of risks of digital economy [2], as well as the impact of disturbing factors of surrounding and inner spheres. To achieve the goal set, it is necessary to solve the following tasks of designing a model:

– forecasting main characteristics of production systems and technological processes;

– obtaining statistic data and other characteristics of technical and economic efficacy;

– using the obtained results of projecting a model to search for the best variant of production systems, or technological processes;

– studying optimal variants of structure of technological process operations with the help of the instruments of digital economy model.

Thanks to the instruments of projecting a model, it is possible, for example, to describe a workstation (WKS), a dairy farm, agricultural holding, or the whole industry. The model projecting of production systems or technological processes allows to reproduce parallel, successive, or parallel and successive schemes of functioning, considering stochastic events and their impact on systems and processes. The model projecting allows to make the detailed analysis of projected variants of operative and route structures, and of influence of various factors on productivity, loading factor, and other economic parameters which are necessary to manage, considering technological risks of digital economy [2].

3. Research Results

One of the directions of digitalization of dairy organizations is to make workstations. In this connection, due to control response (leasing, subsidies, loans, investments, allowances, etc.) business process changes and transition from manual labor to automated one are stimulated. For example, to automate the universal technological process of milk production (figure 1a), milking machines are installed. As a result of such control response, the following transformation of workplace will occur (see figure 1b).

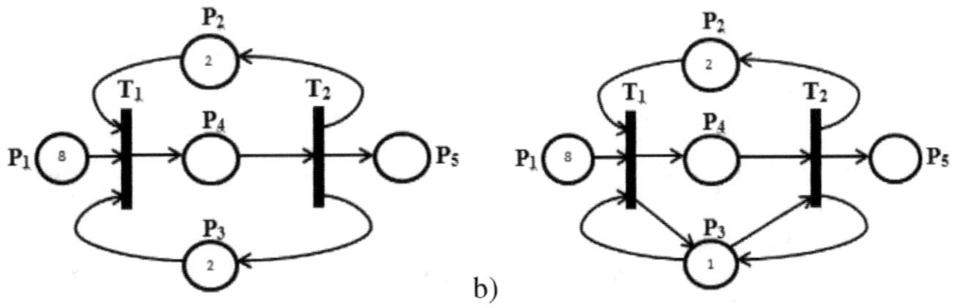

Figure 1. Petri net graph of universal (a) and automated (b) workstation.

Figure 1 presents the technological operations of starting (T_1) and completion (T_2) of milk production processes in manual (a) and automated (b) work regimes. To start the operation (T_1), cows (P_1), implements (P_2), and operators (P_3) must be present, i.e. there must be tags in these positions ($P_1>0$, $P_2>0$, $P_3>0$). For the operation (T_2) it is required to complete the milk production (P_4), and for the automated regime there must be free operators in the position (P_3). After completing the operation (T_2), ready products are placed into the accumulator (P_5).

Due to automation (figure 1*b*), the operator does not follow the technological process of milk production, but just installs a milking unit before milking and removes it after milking, this allows to reduce significantly the labor effort and halve the personnel of the given workplace (the number of tags in the position P_3). Besides, machine milking is more productive than manual one. The disadvantages include increased social risks of digital economy [2], particularly, unemployment which may influence negatively and requires to be regulated socially. First of all, there are social and psychological risks related to labor income and especially unemployment. And besides, risks related to labor income in digital economy tend to change rapidly and dynamically [3].

However, the world practice has shown that automation and robotization contribute to labor productivity growth and effective work of enterprises, including dairy industry ones. Particularly, during machine milking the operator can perform other operations, or, together with the automated operation, perform the technological operation of milk production in manual regime (figure 2a).

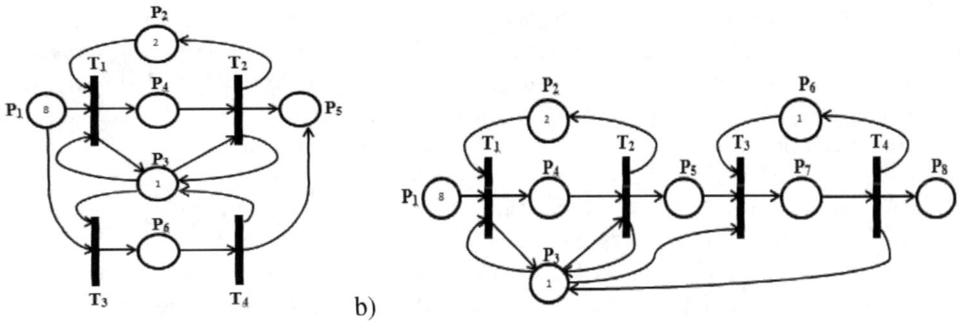

Figure 2. Graph of Petri net of combined workplace for parallel (a) and sequential (b) scheme.

Figure 2a presents the combination of operations: P_4 – automated cow milking and P_6 – manual cow milking, as well as starting (T_1) and completion (T_2) of technological processes of automated milk production, starting (T_3) and completion (T_4) of manual milking processes.

An example of other kind of work is to place the product (milk) in transporting means (figure 2b).

IOP Conf. Series: Materials Science and Engineering **656** (2019) 012014 doi:10.1088/1757-899X/656/1/012014

Figure 2a also includes positions: P_6 – capacity for products to transport in the standby mode, P_7 – transportation capacities in the filling mode, P_9 – capacities ready for transportation, as well as starting (T_3) and completion (T_4) of the process of preparing capacities for transportation.

The analysis of figure 2 has shown that a situation may arise when the process does not proceed rationally, as it is possible to start the transition of manual milking T_3 (figure 2) earlier than that of automated one T_1. In this case, following the manual milking, the operator may not start the process of automated milk production. To exclude such development of events, it is necessary to correct the initial Petri net and add priorities of transitions. Priority Petri Net is a standard Petri net, with each transition T being matched by a certain number P_{rt} called the priority of the given transition. The priority is used to solve conflicts. For example, of two transitions T_1 and T_3 which clash due to a certain common resource in position P_3, advantage is given to that of them which has higher priority. Particularly, if all transitions of the given net have different priorities, then there will not be any conflicts in such a net. It is possible to determine priorities for potentially conflict transitions only. Let us determine the following priority rules:

- automated operations must have higher priority than universal (manual) ones;
- unloading (final) operations must have higher priority than loading (initial) ones.

Considering these rules, it is possible to avoid conflicts in the net by the way of setting transition priorities (numbers in round brackets) (figure 3).

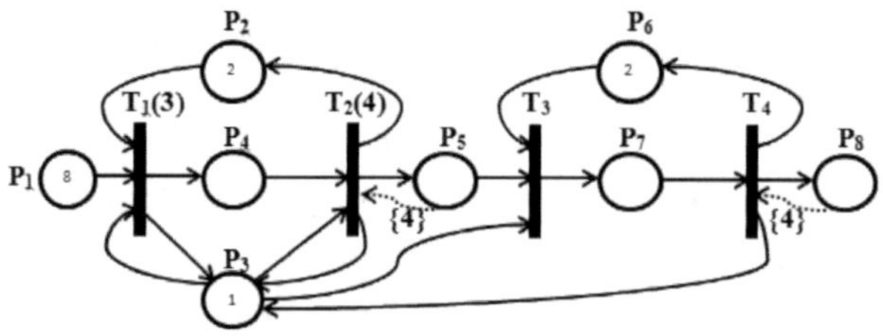

Figure 3. Petri net graph with priorities and delay arrows.

Figure 3 shows that, of all possible transitions, the transition T_2 will be made in the first turn at the moment of starting transitions, since it is of the highest priority. Priority may not be assigned to the transition T_4, as it will be always performed in order to free the operator, without the latter it is impossible to run other transitions.

Besides, it should be noted that in real organizations there are some limitations for certain processes. For example, product storages have a certain capacity size. To introduce such limitations in Petri nets, delay (inhibiting) arrows may be applied (figure 3). In inhibitory Petri nets, a special kind of inhibitory (delaying) arrows is added to standard relations leading from positions to transitions. Such an arrow hinders activation of the corresponding transition, provided there are tags in the corresponding position more than the given in the limitation. So, such transition will be activated only when in the given position the number of tags is fewer than it is given in the limitation. Thus, delay arrows allow to check evidently the number of tags in the given position.

The analysis of figures 1–3 has also demonstrated that the storage of ready products (the capacity of milk obtained) P_5 is not limitless, and it is suggested to be limited by the volume of the milk obtained from 4 cows. As it is necessary to empty the storage for subsequent milk parties, or install the next one, it is required to limit the transition T_2 with a delay arrow from the position P_5, as it is shown in figure 3.

In figure 3 the delay arrows are shown by point vectors with the limitation values in braces.

Organizations function dynamically, so it is advisable to introduce temporary delays in the apparatus of Petri nets (figures 1–3), «which may be made by means of considering temporary delays of the markers in the positions and time of deploying the transitions permitted. Such delays allow to represent the dynamic character of changing conditions and procedural aspect of using resources in the functioning of systems being simulated» [1]. Duration of operations (delays) is of stochastic character (figures 1–3), i.e. delays are random variables. As all random variables, they are evaluated by mathematical expectation and mean square deviation.

Besides, like the duration, the events themselves are of stochastic character of indistinctness (figures 1–3), i.e. they are random variables which may or may not occur. For example, there is a probability for machines or operators to fail, disturbing the normal functioning of processes as well as of the whole organization. It is possible to introduce probabilities of failing implements and operators in stochastic Petri nets. Delays for restoring the system's performance and its processes are of random character as well.

In Petri nets, delays may be defined as some-variable functions. The number of tags in positions, the state of transitions, limitations, priorities, turns, etc. are used as arguments. In simulating technological processes in dairy industry organizations, many dynamic objects of different types are used, and it is necessary to use original algorithms of behaving in the net for each of them. To solve this problem, it is advisable for each tag to have at least one parameter designating the type of a marker. Such parameters are usually distinguished by colors. Colors of tags can be used as an argument in functional nets. Such functional net is assumed to be called a colored one [4].

The analysis of figure 3 has shown that in the positions P_1 (the number of cows), P_5 (amount of ready products – milk), P_8 (packed products – milk poured in cans) there are different objects with different units of measuring production which should be purposefully marked by different colors. Colored Petri nets (figure 3) allow to assign individual meanings (colors) to tags, this making possible to distinguish them. Tag meanings may be simple, or arbitrary complex. Transition in colored Petri nets (figure 4) defines correlations between the values of input tags and those of output tags. For example, yields from 4 cows will be enough to fill a milk can with capacity 25 liters. Using the property of graph multiplicity, this can be presented as 4 input connections from the position P_5 and one leaving for the position P_7 to transit T_3 (figure 4).

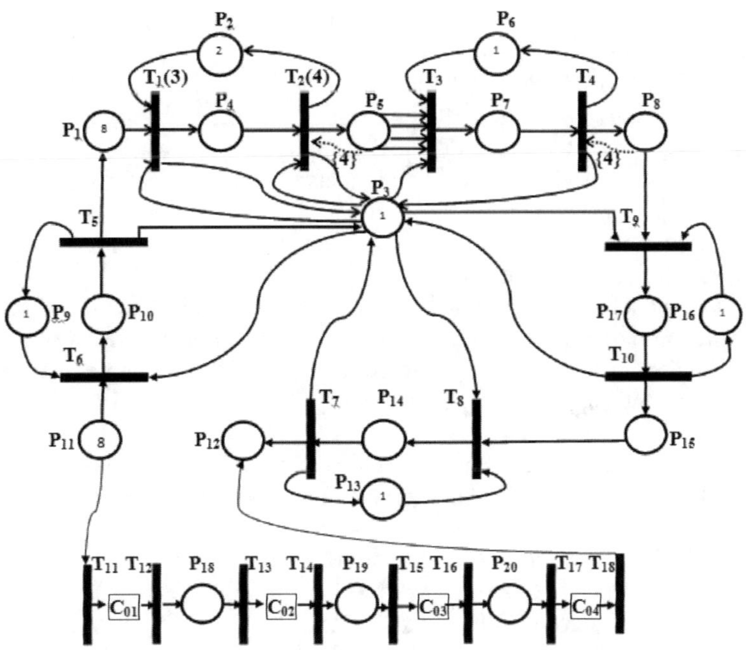

Figure 4. Hierarchical 3-level simulation model of the organization for the production, processing and sale of milk.

As figure 4 shows, the precondition, which defines how many tags are absorbed by the given transition and what are the meanings of the tags, is assigned to the input connections of the transition T_3. Output connection of the transition T_3 defines (with the help of expressions) the number of tags which will be placed in corresponding positions. With the help of colored tags, the graph in figure 4 allows to simulate the process of transforming provisions (cows) into semi-finished products (milk) as a result of the first technological operation (milking), and then, as a result of the second technological operation (filling cans with milk), into products (cans filled with milk).

The process is not completed with these operations. Preliminary maintenance operation (care for cows and placement) and sales of finished products (the milk poured in cans) are still necessary. Preliminary maintenance operation suggests the time expenses of the operator to maintain the placement according to sanitary norms, as well as feeding of animals. Sales of finished products may have some different options: sales of milk to suppliers, converters, delivery to retail outlets, or independent sales in the market or other places. Besides, some farms can independently process milk into dairy products (sour cream, cottage cheese, butter, yogurt and others) with subsequent sales of more technological and, as a result, more expensive products. These procedures increase a variety of tag colors in the Petri net. Considering the above mentioned, the graph can be transformed into a simulation model of personal subsidiary farm (PSF) for production and sale of dairy products (figure 4).

Figure 4 shows the implements necessary to maintain the placement according to sanitary norms, as well as to feed animals, in the positions P_9 – a standby mode, P_{10} – an operating mode. Position P_{11} is a capacity (storage, placement) with feed for cows. Transitions characterize the operations of starting (T_6) and completion (T_5) of the technological operation for maintaining the placement according to sanitary norms and feeding animals.

Sales of finished products is implemented in outlets (suppliers, trade organizations, markets, etc.) in the positions P_{13} – a standby mode, P_{14} – an operating mode. Transitions characterize the operations of starting (T_8) and completion (T_7) of the technological operations for sales of finished products. Results in the form of incomes (cash and non-cash) from the operation for sales of products are placed in the accumulator (account, wallet), position P_{12}.

Independent processing of milk into dairy products is implemented with the help of special equipment and instruments in positions P_{16} – a standby mode, P_{17} – an operating mode. Transitions characterize the operations of starting (T_9) and completion (T_{10}) of the technological operation for processing milk into dairy products. Finished products (sour cream, cottage cheese, butter, yogurt and others) are placed into the accumulator (product packaging), position P_{15}. If a farm does not process milk into dairy products, then the positions P_{16}, P_{17}, transitions T_9 and completions T_{10} shown in figure 4 are excluded from the simulation model.

The simulation models given in figures 1–4 present modules (bricks) which may be used to construct a simulation model of dairy industry organizations of any complexity level, and which may be composed of just one module (figures 1–3). However, if a farm has more than one operator (people, robots, manipulators) and each of them has its own specialization, then there may be several modules, e.g. 4, as it is shown in figure 4.

Figure 5 shows a 2-level simulation model as the component of 3-level one for milk production and sales. Its upper level includes 4 modules of the lower level, which implement the technological operations:

C_{01} – for maintaining the placement according to sanitary norms and feeding animals, serviced by the operator P_{16}.

C_{02} – of cow milking, serviced by the operator P_3.

C_{03} – of preparing capacities for transportation P_{18}.

C_{04} – of selling products P_{17}.

Figure 5 shows that the two-level structure including 4 nets of the lower level is reflected in the hierarchical net. In this case, the lower level nets can be considered as predetermined processes

IOP Conf. Series: Materials Science and Engineering **656** (2019) 012014 doi:10.1088/1757-899X/656/1/012014

(subsystems), and then the process of simulating of different multi-level (hierarchical) systems of any depth of nesting can be implemented.

The main characteristic of hierarchical Petri nets that differs them from other ones is that they are characterized by two transition types: simple and composite. Simple transition is similar to those examined earlier, and composite transition includes Petri nets of lower levels.

4. Results Discussion

Synthesis of Petri nets of higher level is implemented due to including modules and nets of lower level. Agricultural cooperation (AGC) combining several personal subsidiary farms (PSF) on a voluntary basis may be taken as an example. The example of such cooperation of PSF composed of a module and PSF as an hierarchical net including 4 modules is shown in figure 4.

In most cases, cooperatives are composed of a great number of members (n), so their simulation models are advisable to construct, using the property of nesting. Figure 5 shows the hierarchical simulation model AGC composed of parallel functioning n of modules PSF, each of them can present a simple or composite hierarchical Petri net of any nesting degree.

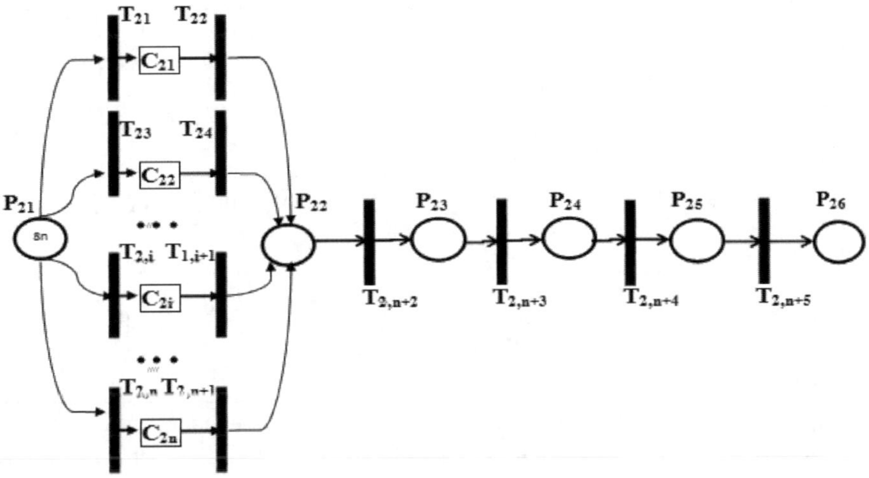

Figure 5. Hierarchical simulation model of AGC with the help of WF-networks of milk production, processing and sales organization.

Figure 5 shows «that the transition process to hierarchical Petri nets is based on the rule that any positions and transitions can include Petri nets» [4] simulating the routes of technological processes functioning at enterprises, including the simplest models located in the corresponding net top. The common kind of activities for PSFs forming an AGC can be, for example, launching a production line for bottling milk in containers, as this is more technological product with greater value. Simulating of operating flows in WorkFlow-systems with the help of WF-Petri nets, called also the nets of operating flows, is given in Figure 5 (the positions from P_{22} to P_{26}).

Main properties of WF-Petri nets given in Figure 6 are applied to check the operating flow nets for zero defects. Zero defects mean the absence of structural conflicts such as deadlocks and lacks of synchronization [5].

Classification of frequently applied combined nets which are dependent on preceding nets they inherit are given in figure 7.

Limitation	• There is only one initial position $\mathbf{P_{22}}$ of WF –nets in which there are no transitions entering $\mathbf{P_{22}}$ • There is only one final position 0, in which there are no transitions leaving $\mathbf{P_{26}}$ • Each node of the given net is located on the way from $\mathbf{P_{22}}$ to $\mathbf{P_{26}}$.

Zero Defects	• For any marking reached from initial condition there is succession of operations which converts the marking into final condition. • Condition is the only one that is reachable from the condition and contains at least one tag in the position. • WF-net does not have any deadlock transitions, i.e. those that never operate.

Figure 6. A list of main properties of WF-nets.

Temporary colored	• when it is necessary to simulate the change in time of condition of the object which possesses a number of characteristics that are not discrete variables.
Hierarchical colored	• when the simulated system is composed of a large number of parts subordinated to each other.
Hierarchical temporary colored	• allow to simulate any arbitrary object or system of any complexity, this being their main advantage.
Indistinct hierarchical temporary colored	• to simulate any arbitrary object but they are much more effective than the previous kind, when operating with objects which do not succumb to strict formalization.
Inno-diversification	• to simulate any arbitrary object in any sphere including: • temporary stochastic indistinct colored functional hierarchical nets; • with additions of priority inhibitory WF-nets.

Figure 7. The purpose of combined networks.

Extension of interpreted indistinct hierarchical temporary colored Petri nets [6] combined with stochastic functional and of some addition priority inhibitory WF-nets, which we call inno-diversification nets, is especially interesting for simulation modeling of functioning dairy industry organizations.

Inno-diversification nets possess the following advantages:

– Simplicity of describing the functioning of upper level organizations by means of Petri nets.

– Simplicity of construction of modeling technological processes which proceed in lower level organizations, or their structural divisions, by means of Petri nets.

– Possibilities of increasing the efficacy of functioning inno-diversification nets due to applying properties of multi-level organizations and their structural divisions.

– Tags of inno-diversification nets include transferred and transformed data of different types.

– Movements of tags over the inno-diversification net simulate data transferring over technological processes proceeding in organizations, and their transformation. Assignment of information about technological processes proceeding in organizations to tags allows to make multi-level nets capable to simulate the work of any organization regardless to its level and with reference to various kinds of limitations, making them closer to real conditions of functioning.

Technique of projecting by system-innodiversification approach is implemented by assignment of higher level of nested Petri nets of lower levels which describe actions implemented with the help of matching elements to positions and transitions [7]. Figure 5 shows simulation models of PSF as hierarchical Petri nets composed of three levels. It should be especially emphasized that the number of levels of hierarchical Petri nets may increase in transition to organizations of higher level. Growth of the number of levels of Petri nets increases adequacy of modeling and accuracy of models.

Application of innodiversification Petri nets allow to overcome the disadvantages of other kinds of Petri nets [7].

5. Conclusions

Based on the forgoing, it should be noted that the instruments of innodiversification Petri nets may be effectively applied to simulate organizations of dairy industries of any level. The task of defining parameters of dairy industry organizations, considering various priorities, can be obtained, using simulation modeling based on innodiversification Petri nets. The possibility of making simulation models for higher levels of hierarchy should be considered.

Acknowledges
This work was financially supported by the Novosibirsk State Technical University (Project C19-12).

References

[1] Chernyakov M K, Chernyakova M M and Akberov K C 2018 Simulation design of manufacturing processes and production systems *Advances in Engineering Research* B **157** pp 124–128

[2] Chernyakov M K and Chernyakova M M 2018 Technological Risks of the Digital Economy *Journal of Corporate Finance Research* 2018 B 1 **(4)** pp 99–109

[3] Chernyakov M K, Chernyakova M M and Akberov K C 2019 Dynamic model of social risks in the digital economy *Advances in Economics, Business and Management Research* B **81** (MTDE 2019) pp 373–378

[4] Wang Rui, Zheng Wei, Liang, Ci and et al. 2016 An integrated hazard identification method based on the hierarchical Colored Net *Safety Science* B **88** pp 166–179

[5] Boubeta-Puig Juan *et al* 2019 MEdit4CEP-CPN: An approach for complex event processing modeling by prioritized colored Petri nets *Information Systems* B **81** pp 267–289

[6] Zaw Win 2017 Modeling and Simulation of ISO 8583 *Transaction Using Timed Colored Petri Net Int. Conf. on Advanced Computing and Applications* (ACOMP) pp 133–136

[7] Chernyakov M K, Chernyakova M M and Akberov K C 2018 Innodiversification model of the digital economy of the agricultural sector *Advances in Social Science, Education and Humanities Research* B **240** SICNI 2018 pp 262–267 DOI: 10.2991/sicni-18.2019.114.

Determination of external impacts on the bearing unit of the geokhod

A A Dronov[1,5], V Yu Beglyakov[1,6], V V Aksenov[1,2,7], A B Efremenkov[3,8] and V A Efremenkov[4,8]

[1]Yurga Institute of Technology, TPU Affiliate, 652050, 26 Leningradskaya Str., Yurga, Russia
[2] Scientific and research center LLC «Siberian Research and Production Association», 650002, 1 Sosnoviy bul., Kemerovo, Russia
[3] Yaroslav-the-Wise Novgorod State University, ul. B. St. Peterburgskaya, 41 173003, Veliky Novgorod, Russia
[4]Peter the Great St.Petersburg Polytechnic University (SPbPU), 195251, Russia, St.Petersburg, Polytechnicheskaya, 29

[5]E-mail: aa-dronov@mail.ru
[6]E-mail: begljakov@rambler.ru
[7]E-mail: 55vva42@mail.ru
[8]E-mail: abe@novsu.ru

Abstract. The mathematical model of interaction of the bearing unit of the geokhod with geo-environment and related systems was developed. The model takes into account the variability of the bearing unit of the geokhod and opens up the possibilities for further determining the inter-actions of the elements of the bearing unit of the geokhod with each other.

1. Introduction
Designing new mining machines is an urgent task [1–6]. When designing new mining machines it is necessary to create mathematical models [7–12]. The geokhod is a tunneling shield that implements movement in the rock mass due to interaction with the system of spiral contour channels formed by it [13]. The geokhod is a new class of mining equipment, as a result a number of difficulties arise in its design and manufacture. Creation of mathematical models of interaction of systems of the geokhod with each other is an actual problem [14–20]. To justify the design parameters of the bearing unit of the geokhod, it is necessary to determine the nature of external influences and internal interactions of the bearing unit of the geokhod.

2. Materials and methods
Considers two variants for the layout of the bearing unit. The shell of the bearing unit can be paired with the head section of the geokhod (figure 1, a) and carry out rotational and translational motion with it. Or it can be paired with a stabilizing section of the geokhod (figure 1, b) and carry with it only the translational motion.

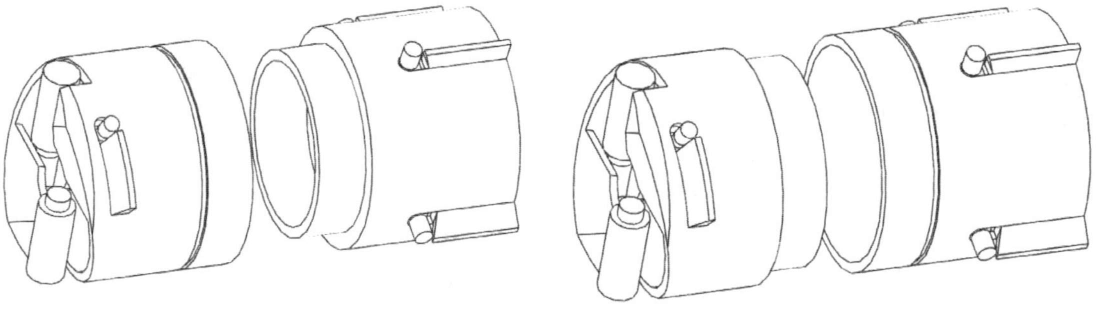

a) b)

Figure 1. Variants for the layout of the bearing unit of the geokhod: a) the shell of the bearing unit paired with the head section b) the shell of the bearing unit paired with the stabilizing section.

The calculation scheme of external forces acting on the shell of the geokhod was compiled (figure 2). The symbols used in the scheme are shown in table 1.

In the first variant, the friction force of the bearing unit of the geokhod against rock is directed along the line of location of the helical blade of the geokhod. It is named as T_{FB} (var a). In the second variant, the friction force of the bearing unit of the geokhod against rock is directed along a path of movement of geokhod. It is named as T_{FB} (var b).

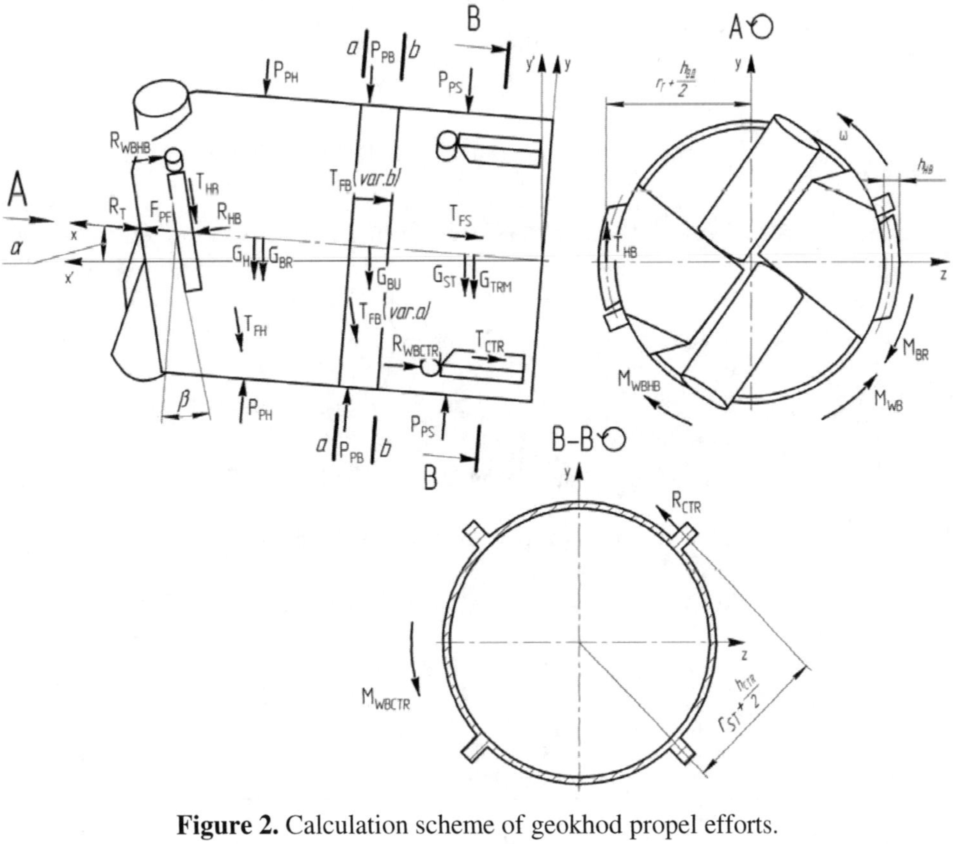

Figure 2. Calculation scheme of geokhod propel efforts.

Table1. Legend to calculation scheme (figure 2).

Symbol	Unit of measure	Name
F_{PF}	[N]	Screw propeller Pulling force
P_{PH}	[N]	Normal component of rock pressure on the head section
P_{PS}	[N]	Normal component of rock pressure on the stabilizing section
P_{PB}	[N]	Normal component of rock pressure on the bearing unit of the geokhod
R_T	[N]	Total cutting resistance projection on sections axis of rotation
R_{HB}	[N]	Reaction of contour rocks on the helical blade
R_{WBHB}	[N]	Reaction of contour rocks on the helical blade working body
R_{WBCTR}	[N]	Reaction of contour rocks on the counter-rotation elements working body
T_{HB}	[N]	Helical blade friction against rock
T_{CTR}	[N]	Total friction force of stabilizing section counter-rotation elements against rock
T_{FH}	[N]	Total force of the head section shell friction against rock
T_{FB}	[N]	Total force of the bearing unit of the geokhod shell friction against rock
T_{FS}	[N]	Total force of the stabilizing section shell friction against rock
M_{WB}	[N ·m]	Moment of cutting resistance on the main working body
M_{WBHB}	[N ·m]	Moment of cutting resistance on the helical blade working body
M_{WBCTR}	[N ·m]	Moment of cutting resistance on the counter-rotation elements working body
M_{BR}	[N ·m]	Moment of broken rocks shift up from the geokhod bottom
G_H	[N]	Weight of the head section with executive bodies, loading devices and other equipment
G_{BU}	[N]	Weight of the bearing unit of the geokhod with loading devices and other equipment
G_{ST}	[N]	Weight of the stabilizing section with loading devices and other equipment
G_{BR}	[N]	Weight of broken rock mass inside the unit
G_{TRM}	[N]	Weight of transported rock mass inside the unit
$r_H = r_{ST}$	[m]	Head and stabilizing sections radiuses
h_{HB}	[m]	Helical blade height
h_{CTR}	[m]	Counter-rotation elements height
α	[degrees]	Mine working gradient angle
β	[degrees]	Helical blade angle
ω	[sec^{-1}]	Angular rotating velocity of geokhod head section

The geokhod was considered as a solid body, which under the influence of external forces is in equilibrium. The section method was applied to the compiled calculation scheme. Cross-section of the geokhod on planes *a-a* and *b-b* was made. Received three parts of the calculation scheme. They were named: the head part of the calculation scheme (figure 3); the tail part of the calculation scheme (figure 4); the middle part of the calculation scheme (figure 5).

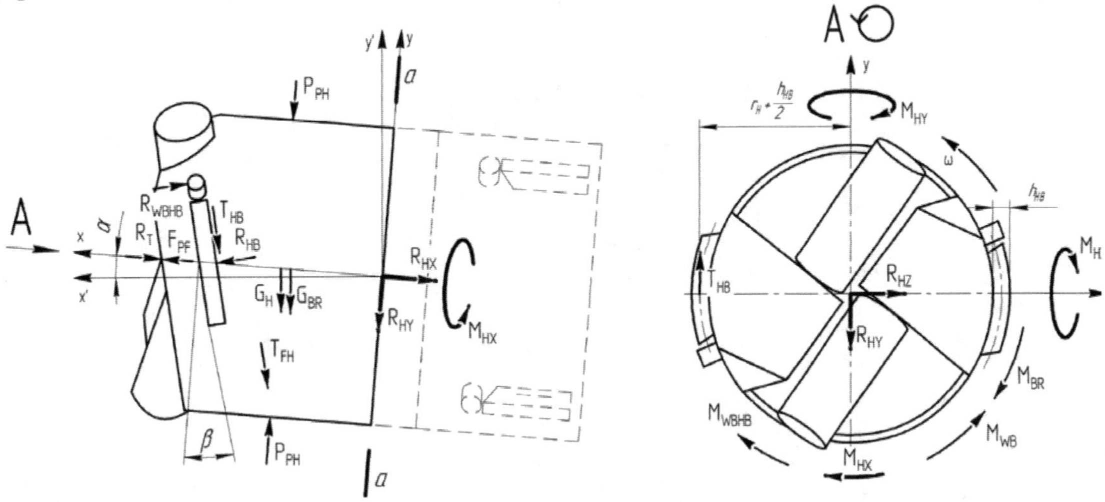

Figure 3. The head part of the calculation scheme.

3. Results and discussion

On the head part of the calculation scheme (figure 3) introduced a countervailing forces R_{HX}, R_{HY}, R_{HZ} and countervailing moments of forces M_{HX}, M_{HY}, M_{HZ}.

The following assumptions and limitations were made:

– the force manifestation of the rock pressure is uniformly distributed over the shell of the geokhod;

– shell of geokhod has absolute rigidity;

– geokhod carries out absolutely rectilinear movement with an arbitrary mine working gradient angle. Therefore $R_{HZ}=0$, $M_{HY}=0$, $M_{HZ}=0$.

Was drawn up the system of equations:

$$\begin{cases} F_{PF} - R_T - R_{WBHB} \cdot \cos\beta - (G_H + G_{BR}) \cdot \sin\alpha - T_{HB} \cdot \sin\beta - T_{FH} \cdot \sin\beta - R_{HX} = 0 \\ M_{WB} - M_{WBHB} \cdot \cos\beta - M_{BR} - T_{HB} \cdot \left(r_H + \dfrac{h_{HB}}{2} \right) \cdot \cos\beta - R_{ВД} \cdot \left(r_H + \dfrac{h_{CTR}}{2} \right) \cdot \sin\beta - \\ - T_{FH} \cdot r_H \cdot \cos\beta + M_{HX} = 0 \\ R_{WBHB} \cdot \sin\beta - T_{HB} \cdot \cos\beta - (G_H + G_{BR}) \cdot \cos\alpha - T_{FH} \cdot \cos\beta - R_{HB} \cdot \sin\beta - R_{HY} = 0 \end{cases} \quad (1)$$

The values T_{HB} и F_T are equal:

$$T_{HB} = |R_{HB}| \cdot f_{FR} \quad (2)$$

$$F_{PF} = R_{HB} \cdot \cos\beta, \quad (3)$$

where f_{FR} – coefficient of friction of steel against rock.

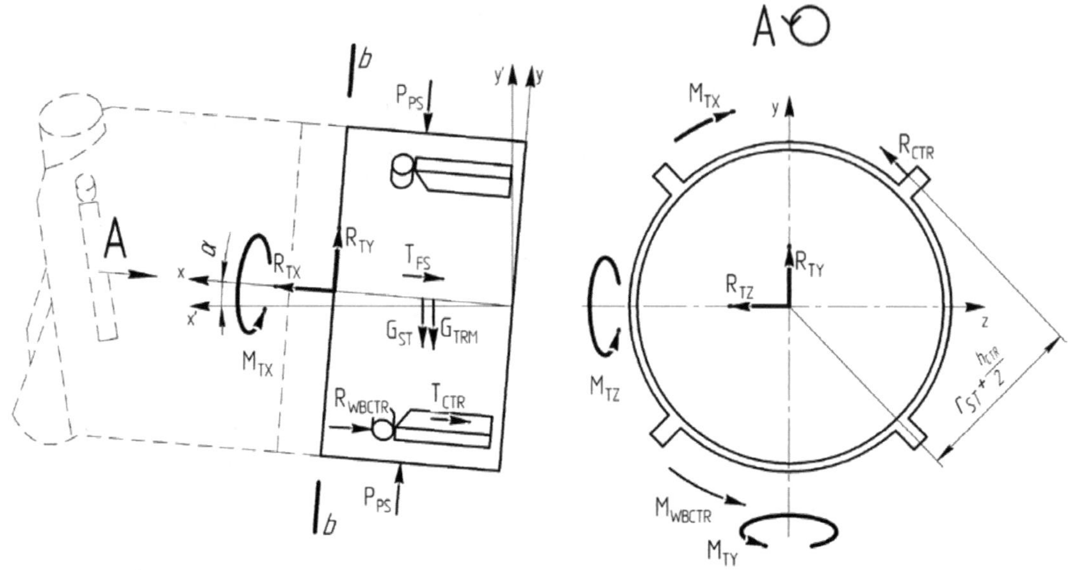

Figure 4. The tail part of the calculation scheme.

Then the system of equations (1) took the form:

$$\begin{cases} R_{HB}\cdot\cos\beta\cdot(1-f_{FR}\cdot tg\beta)-R_T-R_{WBHB}\cdot\cos\beta-(G_H+G_{BR})\cdot\sin\alpha-T_{FH}\cdot\sin\beta-R_{HX}=0 \\ M_{WB}-M_{WBHB}\cdot\cos\beta-M_{BR}-R_{HB}\cdot\left(r_H+\dfrac{h_{HB}}{2}\right)\cdot\cos\beta\cdot(f_{FR}+tg\beta)- \\ -T_{FH}\cdot r_H\cdot\cos\beta+M_{HX}=0 \\ R_{WBHB}\cdot\sin\beta-R_{HB}\cdot f_{FR}\cdot\cos\beta-(G_H+G_{BR})\cdot\cos\alpha-T_{FH}\cdot\cos\beta-R_{HB}\cdot\sin\beta-R_{HY}=0 \end{cases} \quad (4)$$

Values R_{HX}, R_{HY}, M_{HX} were obtained from equations of system (4):

$$R_{HX}=R_{HB}\cdot\cos\beta\cdot(f_{FR}\cdot tg\beta-1)+R_T+R_{WBHB}\cdot\cos\beta+(G_H+G_{BR})\cdot\sin\alpha+T_{FH}\cdot\sin\beta \quad (5)$$

$$R_{HY}=R_{WBHB}\cdot\sin\beta-R_{HB}\cdot f_{FR}\cdot\cos\beta-(G_H+G_{BR})\cdot\cos\alpha-T_{FH}\cdot\cos\beta-R_{HB}\cdot\sin\beta \quad (6)$$

$$M_{HX}=-M_{WB}+M_{WBHB}\cdot\cos\beta+M_{BR}+R_{HB}\cdot\left(r_H+\frac{h_{HB}}{2}\right)\cdot\cos\beta\cdot(f_{FR}+tg\beta)+T_{FH}\cdot r_H\cdot\cos\beta \quad (7)$$

In expressions (5–7) unknown is R_{HB}. We introduce constants independent of the unknown R_{HB}

$$R_1=R_T+R_{WBHB}\cdot\cos\beta+(G_H+G_{BR})\cdot\sin\alpha+T_{FH}\cdot\sin\beta \quad (8)$$

$$M_1=-M_{WB}+M_{WBHB}\cdot\cos\beta+M_{BR}+T_{FH}\cdot r_H\cdot\cos\beta \quad (9)$$

$$R_2=R_{WBHB}\cdot\sin\beta-(G_H+G_{BR})\cdot\cos\alpha-T_{FH}\cdot\cos\beta \quad (10)$$

Then the expression (5-7) took the form:

$$R_{HX}=R_{HB}\cdot\cos\beta\cdot(f_{FR}\cdot tg\beta-1)+R_1 \quad (11)$$

$$M_{HX} = R_{HB} \cdot \left(r_H + \frac{h_{HB}}{2} \right) \cdot \cos\beta \cdot \left(f_{FR} + tg\beta \right) + M_1 \tag{12}$$

$$R_{HY} = R_2 - R_{HB} \cdot \cos\beta \cdot \left(f_{FR} - tg\beta \right) \tag{13}$$

On the tail part of the calculation scheme (figure 4) introduced a countervailing forces R_{TX}, R_{TY}, R_{TZ} and countervailing moments of forces M_{TX}, M_{TY}, M_{TZ}.

Was drawn up the system of equations:

$$\begin{cases} R_{TX} - T_{FS} - (G_{ST} + G_{TRM}) \cdot \sin\alpha - T_{CTR} - R_{WBCTR} = 0 \\ M_{WBCTR} + R_{CTR} \cdot \left(r_H + \frac{h_{CTR}}{2} \right) - M_{TX} = 0 \\ R_{TY} - \left(G_H + G_{TRM} \right) \cdot \cos\alpha = 0 \end{cases} \tag{14}$$

The value T_{CTR} is equal:

$$T_{CTR} = \frac{M_{ROT}}{\left(r_H + \frac{h_{CTR}}{2} \right)} \cdot f_{FR}, \tag{15}$$

where M_{ROT} – desired torque of the geokhod transmission, [N·m].
Then the system of equations (14) took the form:

$$\begin{cases} R_{TX} - T_{FS} - (G_{ST} + G_{TRM}) \cdot \sin\alpha - \frac{M_{ROT}}{\left(r_H + \frac{h_{CTR}}{2} \right)} \cdot f_{FR} - R_{WBCTR} = 0 \\ M_{WBCTR} + R_{CTR} \cdot \left(r_H + \frac{h_{CTR}}{2} \right) - M_{TX} = 0 \\ R_{TY} - \left(G_H + G_{TRM} \right) \cdot \cos\alpha = 0 \end{cases} \tag{16}$$

Values R_{TX}, R_{TY}, M_{TX} were obtained from equations of system (16):

$$R_{TX} = T_{FS} + (G_{ST} + G_{TRM}) \cdot \sin\alpha + \frac{M_{ROT}}{\left(r_H + \frac{h_{CTR}}{2} \right)} \cdot f_{FR} + R_{WBCTR} \tag{17}$$

$$R_{TY} = \left(G_H + G_{TRM} \right) \cdot \cos\alpha \tag{18}$$

$$M_{TX} = M_{WBCTR} + R_{CTR} \cdot \left(r_H + \frac{h_{CTR}}{2} \right) \tag{19}$$

In expressions (17–19) unknown are R_{CTR} and M_{ROT}. We introduce constants independent of the unknown R_{CTR} and M_{ROT}.

$$R_3 = T_{FS} + (G_{ST} + G_{TRM}) \cdot \sin\alpha + R_{WBCTR} \tag{20}$$

$$R_4 = r_H + \frac{h_{CTR}}{2} \tag{21}$$

Then the expression (17, 19) took the form:

$$R_{TX} = R_3 + \frac{f_{FR}}{R_4} \cdot M_{ROT} \tag{22}$$

$$M_{TX} = M_{WBCTR} + R_4 \cdot R_{CTR} \tag{23}$$

Consider the middle part of the calculation scheme (figure 5).

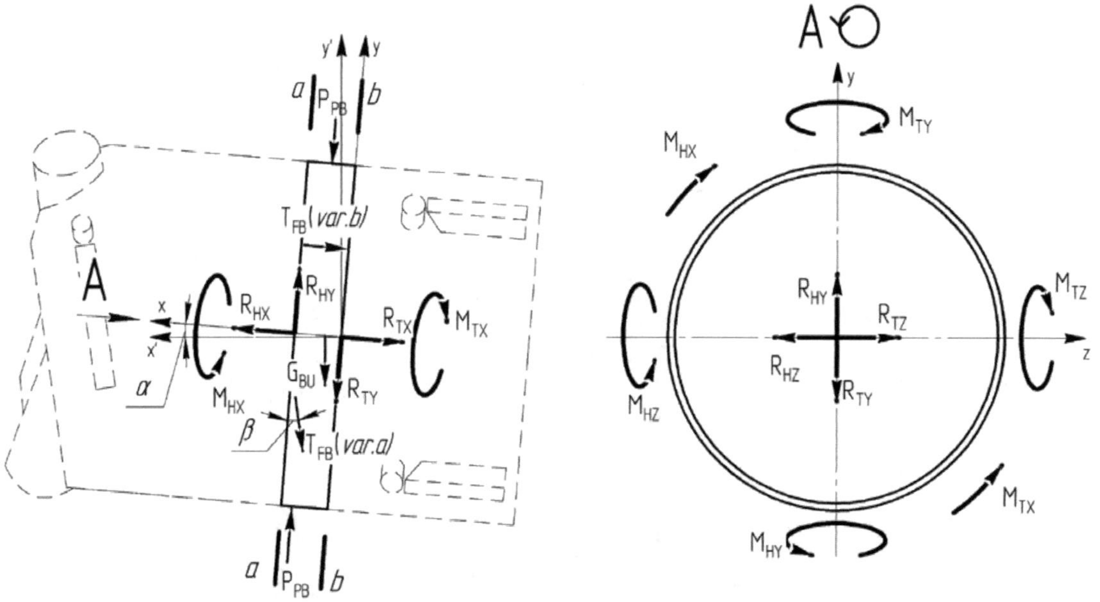

Figure 5. The middle part of the calculation scheme.

As mentioned earlier, there are two variants of the configuration of the bearing unit of the geokhod. Consider the first variant. Was drawn up the system of equations:

$$\begin{cases} R_{HX} - R_{TX} - G_{BU} \cdot \sin\alpha - T_{FB} \cdot \sin\beta = 0 \\ M_{TX} - M_{HX} - T_{FB} \cdot r_H \cdot \cos\beta = 0 \\ R_{HY} - R_{TY} - G_{BU} \cdot \cos\alpha - T_{FB} \cdot \cos\beta = 0 \end{cases} \tag{24}$$

Taking into account the expressions (11–13, 18, 22, 23), the system of equations (24) took the form:

$$\begin{cases} R_{HB} \cdot \cos\beta \cdot (f_{FR} \cdot tg\beta - 1) + R_1 - R_3 - \dfrac{f_{FR}}{R_4} \cdot M_{ROT} - G_{BU} \cdot \sin\alpha - T_{FB} \cdot \sin\beta = 0 \\ M_{WBCTR} + R_4 \cdot R_{CTR} - R_{HB} \cdot \left(r_H + \dfrac{h_{HB}}{2} \right) \cdot \cos\beta \cdot (f_{FR} + tg\beta) - M_1 - T_{FB} \cdot r_H \cdot \cos\beta = 0 \\ R_2 - R_{HB} \cdot \cos\beta \cdot (f_{FR} - tg\beta) - (G_H + G_{TRM}) \cdot \cos\alpha - G_{BU} \cdot \cos\alpha - T_{FB} \cdot \cos\beta = 0 \end{cases} \quad (25)$$

We introduce constants independent of the unknown R_{HB}, R_{CTR} and M_{ROT}.

$$R_5 = R_1 - R_3 - G_{BU} \cdot \sin\alpha - T_{FB} \cdot \sin\beta \quad (26)$$

$$R_6 = R_2 - (G_H + G_{TRM}) \cdot \cos\alpha - G_{BU} \cdot \cos\alpha - T_{FB} \cdot \cos\beta \quad (27)$$

$$M_2 = M_{WBCTR} - M_1 - T_{FB} \cdot r_H \cdot \cos\beta \quad (28)$$

Find the unknown R_{HB}, R_{CTR} and M_{ROT}:

$$R_{HB} = \frac{R_6}{\cos\beta \cdot (f_{FR} - tg\beta)} \quad (29)$$

$$R_{CTR} = \left[\frac{R_6 \cdot \left(r_H + \dfrac{h_{HB}}{2} \right) \cdot (f_{FR} + tg\beta)}{f_{FR} - tg\beta} - M_2 \right] \cdot \frac{1}{R_4} \quad (30)$$

$$M_{ROT} = \left[\frac{R_6 \cdot (f_{FR} \cdot tg\beta - 1)}{f_{FR} - tg\beta} + R_5 \right] \cdot \frac{R_4}{f_{FR}} \quad (31)$$

Consider the second variant of the configuration of the bearing unit of the geokhod. Was drawn up the system of equations:

$$\begin{cases} R_{HX} - R_{TX} - G_{BU} \cdot \sin\alpha - T_{FB} = 0 \\ M_{TX} - M_{HX} = 0 \\ R_{HY} - R_{TY} - G_{BU} \cdot \cos\alpha = 0 \end{cases} \quad (32)$$

The system of equations (32) was solved similarly to the solution of the system (24). Unknown were named R'_{HB}, R'_{CTR} and M'_{ROT}.

$$R'_{HB} = \frac{R_6 + T_{FB} \cdot \cos\beta}{\cos\beta \cdot (f_{FR} - tg\beta)} \quad (33)$$

$$R'_{CTR} = \left[\frac{(R_6 + T_{FB} \cdot \cos\beta) \cdot \left(r_H + \frac{h_{HB}}{2} \right) \cdot (f_{FR} + tg\beta)}{f_{FR} - tg\beta} - M_2 - T_{FB} \cdot r_H \cdot \cos\beta \right] \cdot \frac{1}{R_4} \quad (34)$$

$$M'_{ROT} = \left[\frac{(R_6 + T_{FB} \cdot \cos\beta) \cdot (f_{FR} \cdot tg\beta - 1)}{f_{FR} - tg\beta} + R_5 + T_{FB} \cdot (\sin\beta - 1) \right] \cdot \frac{R_4}{f_{FR}} \quad (35)$$

4. Conclusion

The expressions obtained are the mathematical model of the interaction of the bearing unit with the rock mass and another systems of geokhod.

The results obtained will allow:

– to carry out a comparative analysis of different layouts of the bearing unit of the geokhod;

– define internal interactions of the bearing unit of the geokhod.

References

[1] Maidl B, Schmid L, Ritz W, Herrenknecht M, Wehrmeyer G and Derbort M 2008 *Hardrock Tunnel Boring Machines* (Berlin: Ernst & Sohn Verlag für Architektur und technische Wissenschaften GmbH und Co.KG)

[2] Maidl B, Thewes M and Maidl U 2013 *Handbook of Tunnel Engineering Wilhelm Ernst & Sohn* (Berlin, Germany: Verlag für Architektur und technische Wissenschaften GmbH & Co KG)

[3] Maidl B, Thewes M and Maidl U 2014 *Handbook of Tunnel Engineering II Basics and Additional Services for Design and Construction Wilhelm Ernst & Sohn* (Berlin Germany: Verlag für Architektur und technische Wissenschaften GmbH & Co KG)

[4] Maidl B, Herrenknecht M, Maidl U and Wehrmeyer G 2012 *Mechanised Shield Tunnelling Second Edition Ernst & Sohn* (Berlin Germany: Verlag für Architektur und technische Wissenschaften GmbH & Co KG)

[5] Robbins R 1987 Machine tunnelling in the twenty-first century *Tunnelling and Underground Space Technology incorporating Trenchless* **2 (2)** 147–154

[6] Nishi J and Seiki T 1997 Planning and design of underground space use *Memoirs of the School of Engineering Nagoya University* **49 (1)** 48–93

[7] Zhang Q, Kang Y, Qu C, Wang Y, Huang T and Cai Z 2010 Mechanical model for operational loads prediction on shield cutter head during excavation *IEEE/ASME International Conference on Advanced Intelligent Mechatronics* AIM **5695778** 1252–1256

[8] Deng K, Huang J and Wang H 2009 Layout optimization of non-equidistant arrangement for thrust systems in shield machines Automation in Construction **49 PA** 135–141

[9] Shen J, Jin X, Li Y and Wang J 2009 *Numerical simulation of cutterhead and soil interaction in slurry shield tunneling Engineering Computations* (Swansea, Wales) **26 (8)** 985–1005

[10] Shi H, Yang H, Gong G and Wang L 2011 Determination of the cutterhead torque for EPB shield tunneling machine *Automation in Construction* **20 (8)** 1087–1095

[11] Beloglazov I I and Ikonnikov D A 2016 Computer Simulation Methods for Crushing Process in an Jaw Crusher *IOP Conf. Ser.: Mater. Sci. Eng.* **142** 012074

[12] Bazhin V Y, Beloglazov I I and Feshchenko R Y 2016 Deep conversion and metal content of Russian coals *Eurasian Mining* **2** 28–32

[13] Aksenov V V, Walter A V, Gordeyev A A and KosovetsA V 2015 Classification of geokhod units and systems based on product cost analysis and estimation for a prototype model production *IOP Conf. Ser.: Mater. Sci. Eng.* **91** 012088

[14] Aksenov V V, Beglyakov V Y, Kazantsev A A and Doroshenko I V 2016 Development of Requirements for a Basic Standardized Mathematical Model of Geokhod *IOP Conf. Ser.:*

Mater. Sci. Eng. **127** 012031

[15] Blaschuk M, Dronov A, Koperchuk A, Chernukhin R and Litvinenko V 2017 Kinematic Parameters of Rotary Transmission with Hydraulic Cylinders *E3S Web of Conferences* **15** 03003

[16] Blaschuk M, Dronov A and Ganovichev S 2016 Calculation of Free Interior Dimensions in Geokhod Transmission with Hydraulic Cylindres *IOP Conf. Ser.: Mater. Sci. Eng.* **127** 012033

[17] Aksenov V, Chicherin I, Kostinez I, Kazantsev A and Efremenkov A 2017 Substantiation of the Necessity for Design of Geohod Control System *E3S Web of Conferences* **21** 03001

[18] Beglyakov V, Aksenov V, Kostinets I, Efremenkov A and Khoreshok A 2018 Influence of the Supporting Surface Inclination Angle of the External Geokhod Propulsor on the Deflected mode of Boundary Rock Massif *IOP Conf. Ser.: Mater. Sci. Eng.* **441** 012008

[19] Aksenov V, Efremenkov A, Sadovets V and Pashkov D 2018 Substantiation of Characteristic Bending Points of the Blade Operating Body of the Geokhod *IOP Conf. Ser.: Mater. Sci. Eng.* **441** 012005

[20] Sadovets V, Beglyakov V and Aksenov V 2015 Development of Math Model of Geokhod Bladed Working Body Interaction With Geo-environment *IOP Conf. Ser.: Mater. Sci. Eng.* **91** 012085

IOP Conf. Series: Materials Science and Engineering **656** (2019) 012016 doi:10.1088/1757-899X/656/1/012016

Pre-investment quantitative assessment of the geological risk of the development of coal deposits

A B Efremenkov[1,6], T B Rogova[2] and S V Shaklein[3,4] and V A Efremenkov[5,6]

[1]Yaroslav-the-Wise Novgorod State University, ul. B. St. Peterburgskaya, 41 173003, Veliky Novgorod, Russia
[2]Kuzbass State Technical University, ul. Vesennjaja, 28 650000, Kemerovo, Russia
[3]Federal Research Center for Coal and Coal Chemistry, Siberian Branch of the Russian Academy of Sciences, Leningradsky prospect, 10 650610, Kemerovo, Russia
[4]Institute of Computational Technologies of SB RAS, Kemerovo Branch, ul. Rukavishnikova, 21 650025, Kemerovo, Russia
[5]Peter the Great St.Petersburg Polytechnic University (SPbPU), Polytechnicheskaya, 29 195251, St.Petersburg, Russia

[6]E-mail: abe@novsu.ru

Abstract. The article analyzes the main goals, objectives, and directions of modeling and optimization of technological processes and construction industry. The basic tasks and factors of organizational and technological modeling of the construction industry, as well as their influence on the change in production volumes, labor productivity growth and the efficiency of construction enterprises have been determined. The main directions of optimization of technological processes and construction are formulated on the basis of modernization and development of production and technological capabilities of construction enterprises.

1. Introduction

The consequence of the geological risk manifestation is the increase in production costs, and sometimes the impossibility of fulfilling the contractual conditions both in terms of volume and quality of the supplied raw materials. Therefore, a reliable assessment of the level of geological risk is significant not only for mining companies, but also for banks interacting with them, traders, investment, leasing, and insurance structures.

The practice of the Kuzbass coal extraction enterprises shows that gross errors in assessing the quality of geological information lead to serious economic consequences (up to liquidation of the business), and ordinary ones lead to a decrease in the expected technical and economic indicators of the enterprise by 5–25%.

One of the tools to reduce the consequences of the manifestation of all types of risks inherent in the mining business is the created reserve of the enterprise mine capacity. According to the experience of the coal industry in Russia, the volume of mining by open-cast coal mining enterprises is on average 15% less than their production capacity, and underground mining enterprises (mines) – by 40%. This is due to the fact that the efficiency of underground mining of coal to a much greater degree than open mining depends on the geological conditions and, consequently, is more sensitive to non-confirmation of geological ideas about them which are underlying the design decisions of enterprises. These

IOP Conf. Series: Materials Science and Engineering **656** (2019) 012016 doi:10.1088/1757-899X/656/1/012016

differences in the level of capacity utilization are mostly explained by the need to counter the manifestations of geological risk.

The importance of taking into account the confidence of geological information has been widely acknowledged and long recognized by the world business community. Its objectives are reports on solid minerals resources, prepared according to public reporting codes, the main ones being the "CRIRSCO" family codes. The main task of such codes is to uphold the interests of investors and companies consuming resources by preventing the conscious or erroneous underestimation of the level of reliability of geological data at mining enterprises.

The degree of geological risk is determined by the level of confidence of geological data about the field, which is the object of the mining business. The assessment of confidence is carried out primarily by an expert, i.e. in a very subjective way, which does not provide a high level of estimation reliability and does not exclude the possibility of dishonesty and corruption. It is generally accepted that the higher the inherent risk, the greater the amount of evidence is required for the formation and expression of expert opinion. Due to the high significance of geological risk and taking into account the subjective nature of expert assessments, the use of quantitative estimates of the error results of geological exploration of deposits is becoming increasingly important in the world. The use of such methods is encouraged by present-day international evaluation codes [4, 5, etc.], and the Russian state resource classification [6] provides for the mandatory use of them.

2. Quantitative methods for assessing the confidence of the results of geological exploration of deposits

In most mining industries, geostatistical methods are used to estimate the confidence of mineral reserves (estimates of kriging dispersion, relative kriging errors, distances from variogram, etc.). However, for coal deposits, this approach is ineffective. This is due to the nature of the most significant characteristics assessed (hypsometry of the seam, including disjunctive dislocation, continuity of the thickness of the formation), sampling and uneven placement of drill holes. The need to take into account these features is recognized by the world expert community. The Australian manual [7] indicates that the variogram can only help in determining the continuity distances between observation points. Sole use of the variogram is risky. In isolation this is not considered appropriate as it fails to consider all the other necessary factors contributing to the confidence in the estimate, such as sample geometry and local geological features.

In this regard, the assessment methods that have a geometric basis are more acceptable for coal deposits. They are based on the theory of the geochemical field of Prof. P. K. Sobolevsky formulated at the beginning of the XX-th century. The main postulates of this theory are extremely close to the provisions of geostatistics that appeared much later. In accordance with the theory of P.K. Sobolevsky, the field of any geological indicator corresponds to four basic properties, one of which is the property of unambiguity: the value of the indicator at each point of the geological space can have only one single value.

It follows from the above postulate that a model that ideally describes the considered characteristic of a field should also respond to it. And, therefore, the emergence of ambiguity of constructions in the process of modeling is the evidence of the inadequacy of the model being created and the real object. Moreover, the greater the degree of this inadequacy is, the larger the ambiguity of the model is. Thus, the assessment of the geological materials confidence should be based on an assessment of the degree of ambiguity of the model.

Actually, the ambiguity of constructions can be quantitatively evaluated only in the presence of redundant measurements, i.e. in conditions where there are at least two independent measurements of the feature at the same point in the underground space. However, in exploration practice such measurements are extremely rare and undesirable. Therefore, the approach to creating a method for measuring the degree of ambiguity of a model can only be based on the idea of artificial creation of indirect redundant definitions. The implementation of such an idea within a quadrilateral cell of the exploration grid is as follows [8].

We consider a convex quadrilateral with vertices – drill holes (figure 1). In such a quadrilateral one can draw two diagonals (1–3 and 2–4) intersecting at point A. Using an interpolation method, one can determine the value of the feature at point A from each diagonal ($P_{1\text{-}3}$ and $P_{2\text{-}4}$). It is clear that theoretically $P_{1\text{-}3} = P_{2\text{-}4}$. However, due to measurement errors and interpolation, the values $P_{1\text{-}3}$ and $P_{2\text{-}4}$ will not coincide with each other. Therefore, their difference, being the difference of two independent indirect definitions, can be considered as a numerical measure of the ambiguity of constructions.

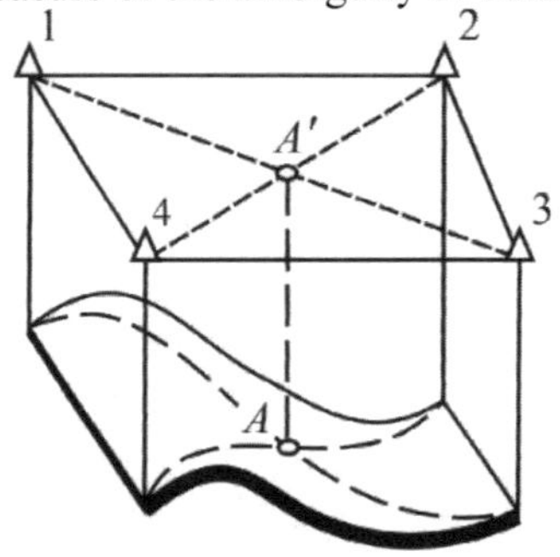

Figure 1. Creating indirect redundant definitions in the contour of the cell of the exploration grid.

Thus, the criterion of the reliability of the study of the indicator in the contour of the quadrilateral of the exploration grid can be defined as $K = | P_{1\text{-}3} - P_{2\text{-}4}|$.

Cubic spline interpolation is chosen as an interpolation method used in assessing the reliability of the study of seam hypsometry. In this case, the criterion for the exploration of hypsometry, called the lambda criterion, is calculated by the formula (according to the scheme for designating conventional numbers of holes in figure 1)

$$\lambda = \sec\delta_T \Big| \, [F_1(x_1)T(\delta_1) - F_2(x_1)T(\delta_3)]R_1 + (Z_3 - Z_1)F_3(x_1) + Z_1 - [F_1(x_2)T(\delta_2) - F_2(x_2)T(\delta_4)]R_2 - (Z_4 - Z_2)F_3(x_2) - Z_2 \, \Big|, \quad (1)$$

where R_1, R_2 –projections lengths of diagonals 1–3 and 2–4 of a quadrilateral; x_1, x_2 – relations, respectively, of the length of the projection of the distance from hole 1 to point A to length R_1 and from hole 2 to point A to length R_2; Z_i – the elevation of the seam at the point of the i-th seam subsection; δ_i – dip angle of a seam in the i-th hole in the direction of the diagonal; δ_T – dip angle of a seam at point A; $T(\delta_i)$ – tangent of a dip angle δ_i, which is marked «–», if the directions of the diagonal (from hole 1 to hole 3 and from hole 2 to hole 4 – figure 1) are directed to one side, or otherwise it is marked «+».

The functions F_j are determined by the values of the parameters x_1 or x_2 by the formulas:

$$F_1(x) = (1-x)^2 x; \qquad F_2(x) = (1-x)x^2; \qquad F_3(x) = (3-2x)x^2$$

The criterion calculated by the formula (1) characterizes the ambiguity of the hypsometric constructions in the normal direction to the seam.

For evaluation of capacity models and coal quality indicators, linear interpolation turned out to be quite satisfactory, the use of which allows calculating a criterion, called the delta criterion:

$$\Delta = | \, (P_3 - P_1)x_1 + P_1 - (P_4 - P_2)x_2 - P_2 \, |, \quad (2)$$

where P_i – the value of the estimated indicator in the hole with the i-th vertex of the quadrilateral (evaluation unit).

For practical purposes, the relative delta criterion δ is usually used expressing the value of Δ in percentage with respect to the average value of the indicator at point A:

$$\Delta = 200 \frac{\Delta}{(P_3-P_1)x_1+P_1+(P_4-P_2)x_2-P_2}, \% \quad (3)$$

IOP Conf. Series: Materials Science and Engineering **656** (2019) 012016 doi:10.1088/1757-899X/656/1/012016

As a result of the research conducted, it has been established that the criteria for exploration are representative if the geometric shape of the quadrangular grid cell meets the following conditions:

– parameters x_1 and x_2 are in the range from 0.3 to 0.7;

– the ratio of the length of the longer diagonal to the length of the less extended one does not exceed 2.6;

– the inside corners of the quad are within the range from 25 to 155°;

– the average length of the two longest sides of the quadrilateral is less than four times the average length of the two short sides.

The calculation of the criteria for exploration is permissible to carry out only if the achieved density of the exploration holes grid ensures the correctness of interpolating the values of the indicator in the interhole space.

To assess the validity of the use of lambda exploration criteria, the validity of interpolation of seam elevations between holes is estimated. The assessment is carried out on the basis of building a relationship between the average values of the lambda-criteria for exploration and the average areas of quadrilaterals (the exploration curve). Its construction is carried out at least by three "points", obtained according to the initial and rarefied approximately two and four times grid of holes. A uniform rarefaction of observations network and the subsequent averaging of the obtained values ensures the equal complexity of the object for each of the points of the exploration curve, i.e., neutralizes the effect of the complexity factor.

Since, as the areas of quadrilaterals grow (i.e. as the density of the exploration grid decreases), the reliability of the simulation results should objectively decrease (the criteria should increase). Under the conditions of the interpolation validity, this dependence should have the character of a monotonically increasing function. Otherwise, interpolation is not valid. Thus, the criterion for the validity of the interpolation of seam elevations is the shape of the exploration curve. When analyzing its form, if the adjacent points of the exploration curve have slightly (less than 20%) different values of the lambda criteria, the hypothesis about the statistical equality of their values is checked.

The resulting view of the exploration curve is analyzed using a decision table (figure 2), on the basis of which conclusions are drawn about the validity of building a hypsometric plan.

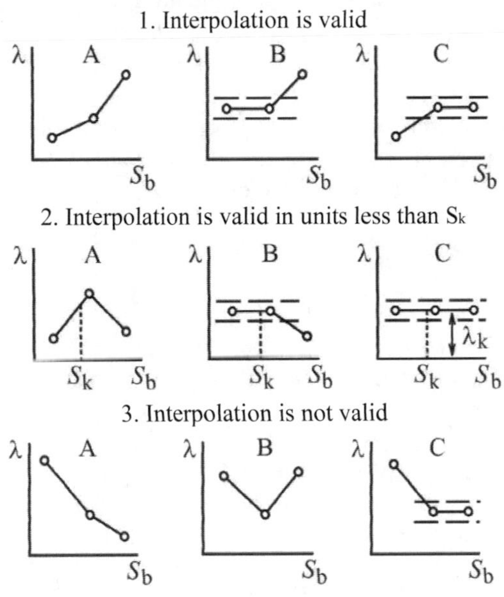

Figure 2. Decision table for assessing the validity of interpolation of seam elevations (statistically equal criteria values are highlighted by framing them with dotted lines).

The decision table singles out three main types of exploration curves.

The first type (Types 1A, 1B, and 1C in Figure 2) includes the exploration curves that have the theoretically expected form.

The second type of exploration curves (types 3A, 3B and 3C in Figure 2) has a contradictory to theoretical form. With this form of the exploration curve, the lambda criteria are non-informative and cannot be interpreted.

The third, intermediate, curve type (types 2A, 2B, and 2C in Figure 2) combines both of the preceding types. At the beginning, as the areas of evaluation blocks grow, the ambiguity of constructions grows (the theoretically expected trend is the first type exploration curve), and then its decrease occurs (the second type exploration curve). For curves of this type, a conclusion is made about the limited legitimacy of interpolation of elevations in the interhole space: for evaluation units with an area of less critical (S_k) it exists, and for the larger one it does not. The value of S_k is taken as the average area of the blocks of the first rarefaction grid (the second point of the curve), calculated without taking into account the blocks whose area is equal to or greater than the average area of all the blocks of the first rarefaction grid. Subtype 2C can also appear in the case of substantial object over-exploration, i.e. under the conditions of ubiquitous interpolation of marks. Therefore, if the criteria for exploration are less than 4 m, two additional dilutions of the holes grid are additionally performed and the obtained curve is analyzed.

The validity of the use of the delta criteria of the seam capacity exploration and coal quality indicators is checked on the basis of the ratio of the total and "non-geometrized" variability measured by the formula

$$\gamma = \frac{\Delta_s}{\sigma}, \tag{4}$$

where Δ_s – the average value of the delta-test of exploration, characterizing the "non-geometrized" variability; σ – standard deviation characterizing total variability.

Interpolation of indicators in the interhole space is considered valid if $\gamma \leq 0{,}8$.

3. The use of geometric criteria of exploration to address the risk assessment issues of an investment project

The assessment of the geological risk of an investment project of subsurface development is made on the basis of determining the expected errors of the mining and geometric models of the field. As a result of studies [8], it was found out that the errors of mining and geometric models are directly proportional to their ambiguity (exploration criteria). Therefore, the calculation of the expected error in the geometrization of the hypsometry, the thickness of the seam and the coal quality indicators (hypsometric plans, plans of thickness and ash content, etc.) is performed by multiplying the obtained values of the exploration criteria by the proportionality coefficients K, the value of which is determined by the given probability that the actual errors will not exceed the expected

$$R_g = \lambda K_g, \qquad R_p = \Delta K_p, \tag{5}$$

where R_g and R_p – expected errors in geometrization of seam hypsometry and linearly interpolated parameters (thickness, ash distribution, etc.); K_g – coefficient of proportionality, taken equal to 0.6 for the required probability estimates of 0.68 and 1.2 – with probability of 0.8; K_p – coefficient of proportionality, taken equal to 1.3 when the required probability estimates 0.68 and 2.0 – with probability 0.8.

For already operated fields, the proportionality ratios can be adjusted considering the accumulated experience of mining operations based on the use of a special technology – monitoring the confidence of resources. The information basis for monitoring is the data obtained on the already developed part of the field. To do this, a quantitative assessment is made of the confidence of the exploration data only, and then the assessment of the confidence of the same data by comparison with the results of mining operations. As a result, a method of quantitative assessment of the confidence of geological information adapted to the subsurface area is formed.

According to the results of calculation of exploration criteria for numerous quadrangular cells of drill holes grid that overlap each other and recalculation into expected errors for each stratum according to the considered indicators, exploration cartograms are built on which areas with different degrees of confidence are indicated by conventional signs. In addition, a generalized cartogram of exploration is being constructed, providing for the cumulative accounting of the confidence of all indicators. The obtained estimates can be interpreted in the categories of mineral resources (figure 3).

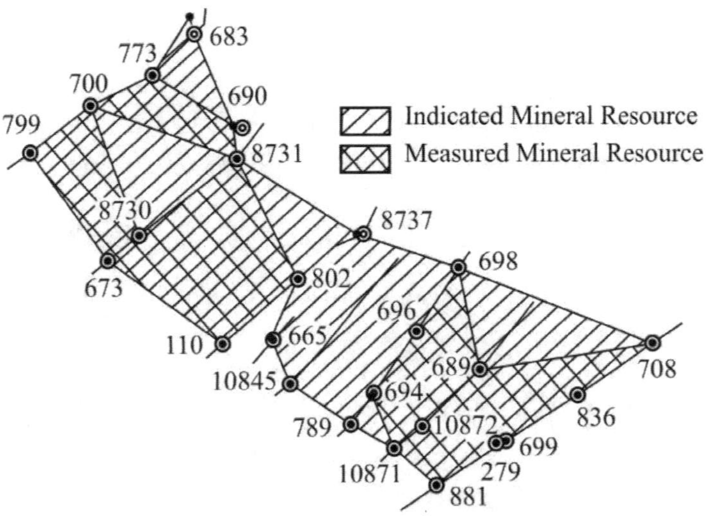

Figure 3. Seam exploration cartogram.

The recommended boundary separating measured and indicated resources by the hypsometry factor is an error of 8 m, and by factors of the seam thickness and ash content – a relative error of up to 20–30%.

Extensive practice of applying the criteria of exploration has shown that their use improves the quality of the assessment of the confidence of exploration information by three to four times compared with traditional expert approaches. At the same time, it does not matter at all exactly how the field models were built and the resources were calculated: polygonal or geostatistical way.

Exploration cartograms are an informational basis for assessing the geological risk of relatively short periods and are used to determine the discount rate that includes the risk premium.

When assessing the risk of developing a field as a whole or a considerable part of it, a simulation method is actively used. When it is applied, a change in the project's performance indicators is investigated, depending on the degree of "swaying" of the values of the initial parameters, which typically include prices, production costs, production, resources, investments, etc. At the same time, taking into account the geological risk is mainly associated with an estimate of the expected non-confirmation of resources. In coal practice, when determining the expected amount of resources, it is necessary to take into account not only the error of their calculation, but also the error caused by the non-confirmation of the mining and geological conditions of their occurrence in the measure excluding technical possibility of extraction.

Statistical studies conducted on materials of existing enterprises established the dependence of the share D of such unconfirmed resources in the total calculated resources of the enterprise from the values of the criteria for exploration.

The type of this dependence depends on the method of mining:

– with long-pillar method

$$D = 3{,}0 + 2{,}8\lambda_U + 0{,}10\bar{\delta}\text{, \%;} \tag{6}$$

– with heading-and-stall method

$$D = 2{,}1 + 2{,}0\lambda_U + 0{,}07\bar{\delta}\text{, \%;} \tag{7}$$

– with open-pit method

$$D = 1{,}5 + 1{,}0\lambda_U + 0{,}04\bar{\delta}\text{, \%,} \tag{8}$$

where $\bar{\delta}$ – the average value of the relative delta criteria of thickness exploration in the estimated contour, %; λ_U – the average value of the specific values of the lambda-criteria of the exploration of seam hypsometry, determined by the formula:

$$\lambda_U = \frac{\lambda}{S} \text{ with } S \geq 100 \text{ thousand m}^2,$$
$$\lambda_U = \lambda \text{ with } S < 100 \text{ thousand m}^2, \tag{9}$$

where S – the average area of projections of quadrangular evaluation units that are within the evaluation contour in hundreds of thousands m^2.

Practice has shown that the use of formulas (7) – (9) is correct in cases where the estimated contour includes at least 20 quadrangular evaluation blocks.

4. Conclusion

Quantitative geometric methods for assessing the confidence of resources have a high degree of reproducibility, which allows for their use by specialists of a non-geological and non-mining profile. These methods provide the possibility of their application to the materials of coal deposits, characterized by high irregularity of exploration networks. Their use allows increasing the confidence of determining the degree of investment geological risk of field development, as well as to determine the boundaries of the contours within which the risk will have the greatest values, that determines the possibility of reducing the risk by conducting additional geological exploration.

References

[1] Dowd P 2018 Quantifying the impacts of uncertainty *Handbook of Mathematical Geosciences Springer Cham* 349–373

[2] Kulak V, Voloshin V and Frianov V 2015 Evaluation of influence exerted by coal exploration quality on mine planning efficiency *Mining informational and analytical bulletin* **5** 13–22

[3] Kopytov A, Rogova T and Shaklein S 2018 Investment Risks Reduction of Developing Coal Deposits on the Basis of Quantitative Assessment Method of the Geological Information Confidence *The 9th Russian-Chinese Symposium. Coal in the 21st Century: Mining, Intelligent Equipment and Environmental Protection. Advances in Engineering Research* Vol 176 322–326

[4] *Australasian Code for Reporting of Exploration Results, Mineral Resources and Ore Reserves. The JORC Code* 2012 (Melbourne: The Joint Ore Reserves Committee of The Australasian Institute of Mining and Metallurgy, Australian Institute of Geoscientists and Minerals Council of Australia) p 44

[5] *Russian Code for the Public Reporting of Exploration Results, Mineral Resources and Mineral Reserves (NAEN Code)* 2013 (Moscow: National Association for Subsoil Examination (NAEN)) p 64

[6] *Classification of Reserves and Prognostic Resources of Solid Minerals* 2006 (Order of the Ministry of Natural Resources of the Russian Federation No 278 of December 11, 2006) p 9
[7] *Australian Guidelines for the Estimation and Classification of Coal Resources* 2014 (Melbourne: Guidelines Review Committee on behalf of the Coalfields Geology Council of New South Wales and the Queensland Resources Council) p 47
[8] Rogova T and Shaklein S 2011 The confidence of coal reserves. Quantitative assessment and monitoring (Saarbrücken: LAP LAMBERT Academic Publishing GmbH & Co) p 508

IOP Conf. Series: Materials Science and Engineering **656** (2019) 012017 doi:10.1088/1757-899X/656/1/012017

Research of the process of hogging of crankshafts with a single crank neck when stamping fillets

V N Emelyanov[1,2] and O M Mirzakhamdamov[1]

[1]Yaroslav-the-Wise Novgorod State University, ul. B. St. Petersburgskaya, 41 173003 Veliky Novgorod, Russia

[2]E-mail: valem1219@mail.ru

Abstract. The article describes the results of the research of the process of hogging of crankshafts (CS) with a single crank neck when stamping fillets and a computer program with the help of which it is possible to calculate the expected magnitude and direction of hogging of CS and also to select a combination of stamping modes at which the amount of hogging that needs to be done is minimal.

1. Introduction

Crankshafts (CS) are widely used in modern engineering. However, despite the extensive experience gained by experts in the design, manufacture and operation of CS, they are often destroyed in operation due to insufficient fatigue resistance. According to prof. V. P. Uskov, such destruction is observed in 2–8% of engines entering repair [1]. Destruction occurs mainly through the fillets. The most effective and easily feasible from the point of view of increasing the resistance to fatigue CS is the operation of hardening their fillets by surface plastic deformation (SPD): by rolling rollers, stamping, etc. However, as practice and numerous experiments show, the hardening of fillets CS SPD leads to hogging of CS [2]. Moreover, the more effective is the method from the point of view of increasing the resistance to fatigue, the more it leads to a greater distortion of the crankshaft. With the intensification of hardening modes, the fatigue resistance of the crankshaft increases, but at the same time the magnitude of its distortion increases too.

At the Department of Mechanical Engineering Technology of NovSU the main principles of the theory of warping of CS during the hardening of their SPD fillets have been developed [2]. In particular, theoretical, experimental and computer studies of the process of distortion of spatial crankshafts with six crankpins (CP), which are arranged in pairs at angles of 120° [2], and with five CP located at an angle of 72° to each other were carried out [3]. The purpose of this work is to study the process of hogging with one CP during the stamping of fillets and the development of a computer program that adequately describes this process.

2. Results of the grapho-analytical research

As the object of study a CS was chosen, its design scheme is presented in figure 1. The behavior of the "ideal" CS was studied. The cheeks of "ideal" CS are of the same thickness – 20 mm, the diameters of the main crank necks (CN) and CP are the same – 48 mm, the radii of the CN and CP fillets are equal to each other – 2.2 mm, the crank radius r = 40 mm, the distance from the left support to the section AA b = 100 mm, the distance between the supports L = 95 mm.

IOP Conf. Series: Materials Science and Engineering **656** (2019) 012017 doi:10.1088/1757-899X/656/1/012017

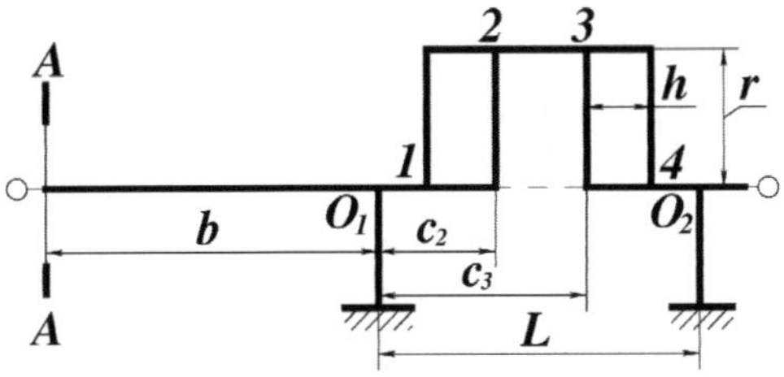

Figure 1. A scheme of an "ideal" CS.

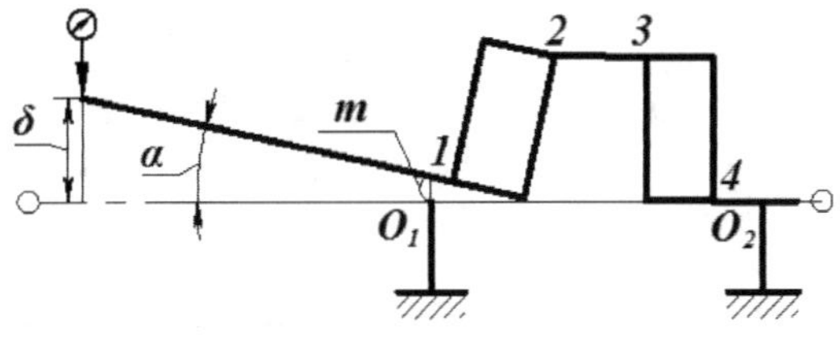

Figure 2. Hogging of a CS when stamping a fillet 2.

When stamping the fillet 2, the left cheek of the CS will bend in the area of the fillet 2 belonging to the lower generatrix CP CS (Figure 2), and CS will assume the position shown in Figure 2, which shows the full value of hogging and the corresponding angle α between the left CN and the horizontal. Since the CS has mass, it will turn around the $\boldsymbol{O_2}$ point counterclockwise until the left CN comes into contact with the $\boldsymbol{O_1}$ support (Figure 3). At the same time, the left CN forms an angle $\boldsymbol{\alpha_1}$ with the horizontal. Using the indicator, you can trace (measure) the measured hogging value $\boldsymbol{\delta_{ms}}$ in section A–A and calculate the corresponding angle $\boldsymbol{\alpha_1}$ using the formula: $\boldsymbol{tg\ \alpha_1 = \delta_{ms} / b}$. However, the total hogging value $\boldsymbol{\delta}$ and the corresponding angle $\boldsymbol{\alpha}$ remain unknown.

There are several questions:

1. What are the total hogging value $\boldsymbol{\delta}$ and the corresponding full hogging angle $\boldsymbol{\alpha}$?
2. What is the share of the measured hogging value $\boldsymbol{\delta_{ms}}$ in the total hogging value $\boldsymbol{\delta}$ (accordingly, what is the proportion of the measured angle $\boldsymbol{\alpha_1}$ in the full hogging angle $\boldsymbol{\alpha}$)?
3. Is the ratio between the measured hogging value $\boldsymbol{\delta_{ms}}$ and the total hogging value $\boldsymbol{\delta}$ (accordingly, the ratio between the measured hogging angle $\boldsymbol{\alpha_1}$ and the total angle $\boldsymbol{\alpha}$) constant, or does it change with changes in the stamping conditions?

To study these issues, as well as for the grapho-analytical study of the behavior of the "ideal" CS, T-FLEX CAD computer-aided design tools were used.

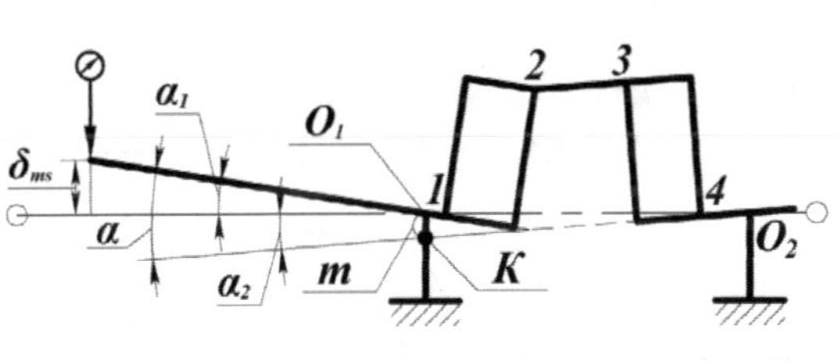

Figure 3. Full hogging angle $\boldsymbol{\alpha}$ and its elements.

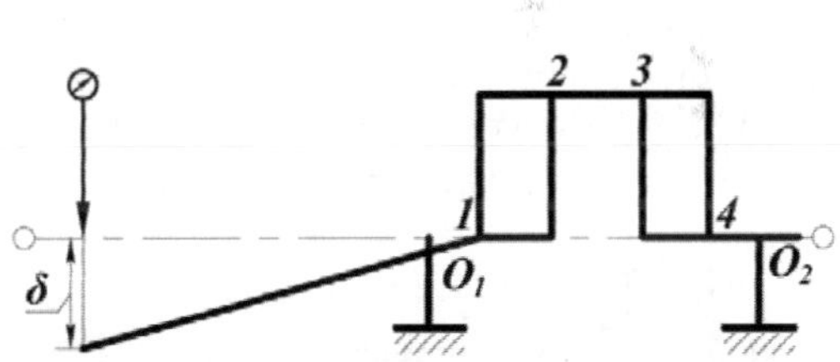

Figure 4. Hogging of CS during stamping its fillet 1.

When turning CS around point $\boldsymbol{O_2}$, the left CN will fall on support $\boldsymbol{O_1}$ by the value $\boldsymbol{m}$ (figures 2, 3). Put the value $\boldsymbol{m}$ from the point $\boldsymbol{O_1}$ down and draw a straight line through the points $\boldsymbol{O_2}$ and $\boldsymbol{K}$. This forms an angle $\boldsymbol{\alpha_2}$, the angle to which the CS turned when stamping fillet 2, it can be calculated using the formula: $\boldsymbol{tg\ \alpha_2 = m / L}$.

From Figure 3 it can be seen that the angles $\boldsymbol{\alpha_1}$ and $\boldsymbol{\alpha_2}$ together form an angle $\boldsymbol{\alpha}$. Knowing the value of the angle $\boldsymbol{\alpha}$, one can calculate the hogging value $\boldsymbol{\delta}$, using the formula: $\boldsymbol{\delta = (b + c_2)\ tg\ \alpha}$.

(The total hogging value $\boldsymbol{\delta}$ must also be known when calculating the stress state of CS elements after stamping its fillets).

When stamping fillets, 3 CS behaves similarly, i.e. the left CN rises and CS will turn around point $\boldsymbol{O_2}$ counterclockwise.

When stamping fillet 1, the situation will be opposite: the left CN will go down (figure 4), this will be prevented by the $\boldsymbol{O_1}$ support therefore, the CS will turn around point $\boldsymbol{O_2}$ clockwise (figure 4). When stamping fillet 4 CS behaves similarly. It is clear that when stamping the "ideal" CS fillets with unchanged chasing modes, the total hogging value δ and the corresponding angle α remain constant when processing any fillet.

(To calculate the total hogging value of CS, it is necessary to algebraically sum up the hogging values of the CS from stamping of all the fillets).

Thus, to find the total hogging value δ from the stamping of any fillet (for example, fillet 2), you must do the following:

1. Process the fillet 2 and set the value of δ_{ms}.
2. Calculate the angle α_1 using the formula: $tg\ \alpha_1 = \delta_{ms} / b$.
3. Calculate the angle α_2 using the formula: $tg\ \alpha_2 = m / L$.
4. Determine the angle α, corresponding to the total hogging value δ, as the sum: $\alpha = \alpha_1 + \alpha_2$.
5. Determine the total hogging value δ by the formula: $\delta = (b + c_2)\ tg\ \alpha$.

However, it is impossible to measure the value m directly during the operation of stamping fillet 2 or after it. It is possible to determine the value m, knowing the value δ_{ms}, using the computer-aided design system T-FLEX CAD, but this requires time-consuming geometric constructions in each particular case, which increases the time and labor-intensiveness of work. The situation would be greatly simplified if the fraction of the angle α_1 in the angle α had a constant value under any circumstances. Then, using the value of δ_{ms}, one could calculate the angle α_1, and from its value calculate the angle α as $\alpha = f(\alpha_1)$. Then, from the angle α, determine the total hogging δ. These actions can be carried out without the involvement of T-FLEX CAD. Therefore, the problem of conducting a grapho-analytical study of the behavior of an "ideal" CS under various conditions arises.

With an increase in impact energy, both the total magnitude of hogging δ and its measured value δ_{ms} increase. (The total angle α and the measured angle α_1 increase accordingly). In the work [2], the formula (1) for the calculation of the hogging value when stamping fillets is given:

$$\delta = K \frac{HB^{0.75}(1540-HB)(1-\nu)}{E} \cdot \frac{r}{h^2} \cdot \frac{c}{2} \cdot \sqrt[4]{E_Y D_1 \Psi}, \quad (1)$$

where K is a coefficient determined experimentally;

HB is the Brinell hardness;

ν is the Poisson's ratio;

E is the modulus of elasticity of the I-kind, MPa;

r is the radius of the crank CS, mm;

h is the thickness of the CS cheek, mm;

c is distance from the left support to the processed fillet;

E_Y is the impact energy of the peen strike, J;

D_1 is the diameter of spherical head of a peen, mm;

Ψ is the coefficient taking into account the incomplete processing of a SPD cheek.

The effect of increasing the total angle α on the measured angle α_1 is shown in table 1, the results of which are used to plot the graph in figure 5. From these data it follows that with an increase in the angle α from 0.1° to 30°, the fraction of the angle α_1 increases from 63.38% to 73.02%. accordingly, with δ increasing from 0.24 mm to 71.73 mm, the share of δ_{ms} increases from 46.99% to 56.06%.

Table 1. The influence of the total angle α on the measured angle α_1, the effect of the total warping δ on the measured hogging value δ_{ms}.

α	0.1°	0.2°	0.4°	0.6°	0.8°	1°	10°	20°	30°
α_1, % from α	63.38	63.42	63.5	63.57	63.64	63.72	66.84	70.03	73.02
δ, mm	0.24	0.47	0.94	1.41	1.88	2.35	23.18	46.56	71.73
δ_{ms}, mm	0.11	0.22	0.44	0.67	0.89	1.11	11.72	24.93	40.21
δ_{ms}, % from δ	46.99	47.01	47.08	47.15	47.25	47.31	50.55	53.54	56.06

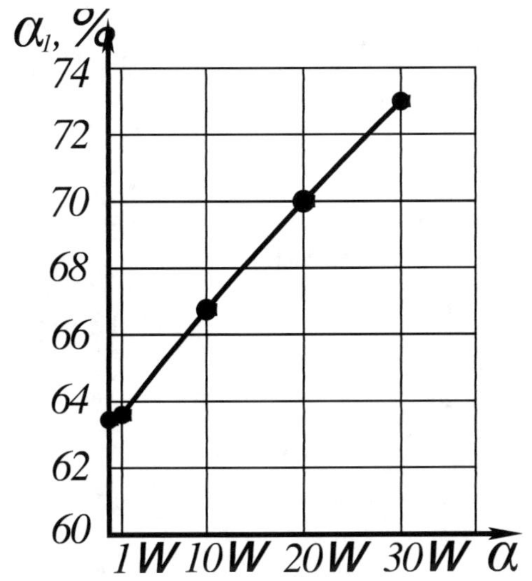

Figure 5. Influence of the full angle α on the measured angle value α_1.

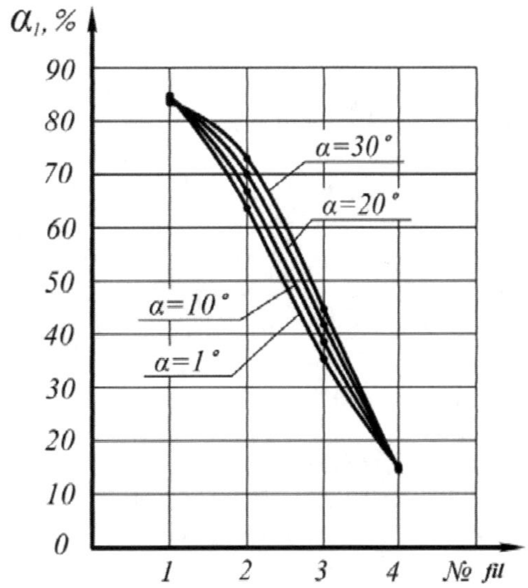

Figure 6. The effect of the distance C_i on the proportion of the angle α_1 in the angle α.

Figure 6 shows the effect of the distance C_i from the left support O_1 to the processed fillet on the fraction of the angle α_1 in the angle α. As can be seen from Figure 6, with an increase in C_i, it decreases sharply. Thus, the proportion of the magnitude of the measured warpage δ_{ms} in the total hogging value δ is not constant and depends on various factors. Therefore, the total hogging value δ by the hogging value δ_{ms} in each specific case should be determined individually. However, according to numerous experiments and practices, it is known that the magnitude of δ_{ms} usually does not exceed several tenths mm (sometimes more than 1 mm), i.e. the value of α_1 does not exceed one degree (table 1). Within these limits, the proportion of α_1 in α ranges from 63.38% to 63.72% (only within 0.34%), i.e. practically it can be considered constant, equal to 63.5% and used for calculations.

The coefficient K in the above stated formula according to the results of real experiments takes different values with a relatively small scattering. However, it remains unclear whether this is natural or a consequence of experimental error. A grapho-analytical study of the "ideal" CS stamping process showed that in this case the coefficient K is a constant value when stamping all four of the fillets.

Thus, the following conclusions were formulated:

1. When hogging (SPD) CS fillets and subsequent measurement of the hogging value of the CS, not the total hogging value δ is traced, but its measured part δ_{ms} (accordingly, not the full angle α, but its part α_1).
2. The share of the measured hogging value δ_{ms} in the total hogging value δ is not constant, but it varies depending on δ, the distance C_i and other factors.
3. The proportion of the measured value of the angle α_1 in the total value of the angle α is also not constant, but it varies depending on the size δ of the distance C_i and other factors.
4. In connection with the foregoing, in each particular case, to obtain accurate hogging values δ and the total hogging angle α corresponding to it, it is necessary to use T-FLEX CAD computer-aided design system.
5. Practically, with small values of δ and α, the fraction of the measured hogging value δ_{ms} in the total hogging value δ varies insignificantly (tenths of a percent) and can be taken as a constant which equals to 63.5% and is used for approximate calculations.
6. The value of the coefficient K in the formula for calculating the hogging value when stamping an ideal shaft is constant.

IOP Conf. Series: Materials Science and Engineering **656** (2019) 012017 doi:10.1088/1757-899X/656/1/012017

3. The results of grapho-analytical research

The experiments were carried out with CS having the following proportions: CP diameter is 47.2 mm, diameters of the main crank necks (CN) are equal to 50.5 mm, the crank radius r = 40.15 mm. CS cheeks have thickness of 20.5 mm on the left and 19 mm. The radius of the CN fillets is 2.5 mm, the radius of the CP fillets is 2 mm. Distance between supports L = 95.5 mm. The distance from the left support to the section A–A b = 100 mm. Material – steel 45.

The experiments were carried out with sequential stamping of fillets. The fillets were processed manually with a spherical head peen (sphere diameter is 4 mm) and a hammer in the sequence 1 – 2 – 3 – 4. The value and direction of the beating vector of the CS were measured in section A–A with the help of an indicator after each fillet was processed (Figure 1). The value and direction of the hogging vector of the CS from the processing of each fillet was calculated as the difference of the beating vectors after and before the processing of this fillet. The results of the experiments are shown in table 2 and in the graph of figure 8.

4. Description of the program

On the basis of theoretical concepts [2] and taking into account the results of grapho-analytical and experimental studies of the process of hogging CS with one CP when stamping fillets, a C# (C Sharp) program was developed using the Microsoft VisualC # 2017 Express programming environment. A block diagram of the algorithm for calculating the hogging value of the CS for stamping of fillets is shown in figure 7.

Block 1 is responsible for entering all the data required for the calculation, such as the mechanical characteristics of the material, the geometric parameters of the CS, the modes of stamping, etc.

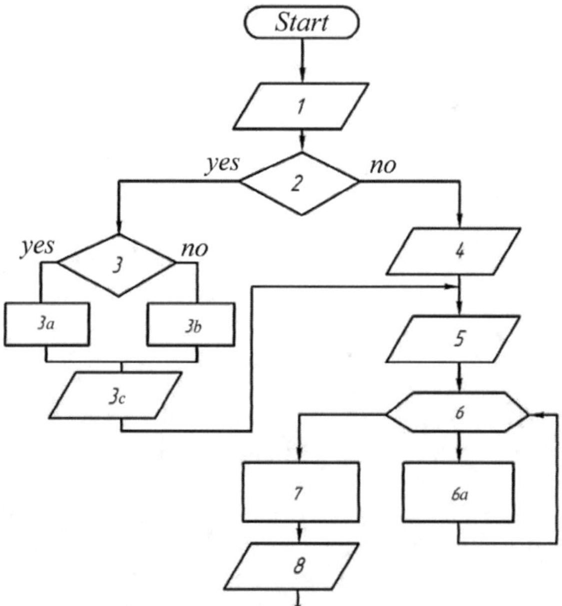

Figure 7. Block diagram of the program.

In Block 2, there is a choice of how to set the values of the coefficient Ψ: automatic calculation or manual input.

In Block 3, the choosing of the formula for calculating the coefficient Ψ from the conditions of the relative position and sizes of CN and CP occurs. In Blocks 3a, 3b the coefficient Ψ is calculated using the selected formula, and then the calculated values are output in Block 3c.

In Block 4 the manual input of the coefficient Ψ is performed if the automatic calculation is not selected.

In Block 5, one must select the numbers of the fillets being hardened.

In Block 6, a cycle is organized to calculate hogging value for the fillets being hardened. In Block 6a, the calculation of the CS hogging value of KV when stamping a single fillet is carried out. The calculation is made according to the above stated formula [2], presented in the working window of the program.

In Block 7, the calculation of the CS total hogging value and the total beating of the CS takes place.

Block 8 is responsible for outputting the results: CS hogging and beating values after processing of fillet by stamping, CS total hogging values, CS total beating values.

In order for the program to calculate the hogging values, beating values from the hardening of single fillets, their total values, as well as to display the obtained values, one must click the “Calculate” button.

As a result, in the “Obtained Results” block, the CS hogging values are displayed after processing each single fillet. In the lower part of this block, the total CS hogging value is displayed. Total hogging value is obtained by vector summing the hogging values from those fillets numbers of which are marked with a check mark.

In the working window of the described program, the characteristics of the CS, as well as the modes of hand-stamping, were entered. The graphs of the experimental and calculated (theoretical) CS hogging values depending on the distance from the fillets to the supports, i.e. from the fillet number, are given in table 2 and figure 8. The discrepancy between experimental and theoretical points averages 10.7%. The result can be considered satisfactory, given that the stamping of the fillets was done manually. In case of manual stamping, it is almost impossible to maintain the mode unchanged (impact energy, step of plastic prints) when processing various fillets. Consequently, the developed program describes the process of CS hogging satisfactorily.

The total amount of CS hogging in section A–A after stamping of all four fillets in the experiment was (–27.5) μm. If this value exceeds the tolerance for the CS beating in section A–A, then it is necessary to adjust the stamping modes. Using the described program, the following combination of stamping modes was selected: fillet 1 – striking energy of the peen E Y = 20 J; fillet 2 – E Y = 22 J; fillet 3 – E Y = 22 J; fillet 4 – E Y = 22 J. With this combination, the CS total hogging value δ_{sum} was (–1) μm, which is acceptable.

5. Comparison of data obtained by the grapho-analytical method as a result of the experiment and calculated by the program

Comparison of the data obtained by the grapho-analytical method, obtained as a result of the experiment and calculated according to the program, is given in table 2. The results of field experiments were used as reference CS hogging values.

Table 2. Comparison of the data acquired.

№ gal.	Experimental hogging values, μm	Grapho-analytical values of hogging, μm	The error of the experimental values,%	Calculated hogging values, μm	The error of the experimental values,%
1	–190	–180.6	5	–171.2	9,9
2	+135	+135	0	+122.7	9.1
3	+72.5	+74.8	3.2	+79.1	9.1
4	–45	–32.3	28.2	–35.7	20.7

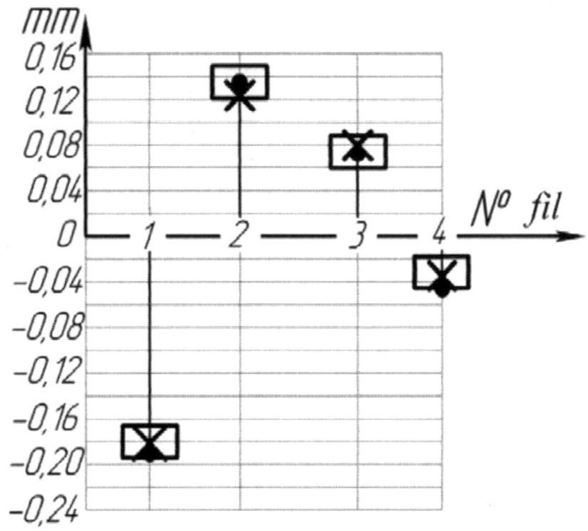

Figure 8. CS hogging graph: "□" – values obtained by the grapho-analytical method; "•" are the values obtained experimentally; "X" are the calculated values.

Thus, with the help of the developed computer program the following is possible to do:

a) at the stage of the design of the strengthening the fillets of CS by stamping operation, to calculate the expected value and direction of hogging, as well as to select such a combination of modes of hardened fillets, at which the total hogging value will be minimal;

b) to make a computer study of the influence of various factors on the hogging value of a CS with one crank neck with any combination of the sizes of their structural elements and from any metal, which as a matter of principle cannot be studied experimentally.

References

[1] Uskov V P 1998 *Spravochnik po Remontu Bazovykh Detalei Dvigatelei* [Basic Engine Repair Parts Handbook] (Briansk: Klincovskaya Gorodskaya tipografiya) p 589

[2] Zajdes S A and Emelyanov V N 2017 *Vliyanie Poverhnostnogo Plasticheskogo Deformirovaniya na Kachestvo Valov* [Influence of Surface Plastic Deformation on Shaft Quality] (Irkutsk: IRNITU) p 380

[3] Emelyanov V 2014 Research on hogging process of crankshafts with five rod journals because of fillets stamping *Journal of Engineering And Technology Research* **2 (2)** 65–69

IOP Conf. Series: Materials Science and Engineering **656** (2019) 012018 doi:10.1088/1757-899X/656/1/012018

Regeneration and utilization of solvents after the process of physical and chemical cleaning of metal optics

S A Filin[1], V E Rogalin[2] and I A Kaplunov[3,4]

[1]Plekhanov National University of Economics, 117997, Stremyannyj per., 36, Moscow, Russia
[2]Institute of Electrophysics and Electric Power, Russian Academy of Sciences, 191186, Dvorzovaya Naberezhnaya 18, St. Petersburg, 191186 Russia
[3]Tver State University, 170100, Zhelyabova Str. 33, Tver, Russia

[4]E-mail: Kaplunov.IA@tversu.rul.ru

Abstract. The possibility of application of a semipermeable polymer membrane in total with the method of rectification purification for the repeated use of freon-114***B***2 and detergent compositions based on it for cleaning metal optics in a closed technological cycle while ensuring the environmental cleanliness of the process and saving solvents has been identified and justified. The investigated adsorption method is effective for cleaning freon-114***B***2 from rosin resins contaminats: among adsorbents (activated carbon, silica gel, anion exchange resin IA-I, silica gel S-80). Silica gel S-80 has the highest adsorption capacity (up to 50 mg / l). It has been shown that regeneration of freon-114***B***2 with the help of a polymer semipermeable membrane after its use for cleaning metal optics allows to reduce the amount of impurities present in it by at least 2 orders of magnitude. Moreover, organic molecules, that have long surfactant chains, oils, fats of organic and mineral origin are almost completely separated from the regenerated solvent.

1. Introduction

For the operation of high-power CO_2 lasers, metallic mirrors are used [1]. They are – a high-precision, expensive products, the working optical surface of which must be operated in the most pure state. The presence of traces of the polishing process and operational contaminants on the surface leads to the degradation of the mirror. To clean the surface of the mirror organic solvents with the brand "especially pure" are used. However, for implementation of the physico-chemical process for cleaning of metal optics in the semi-industrial variant, one, even optimal, choice of washing medium is not enough. It is necessary to organize a closed technological process of purification, including the development of a method for the regeneration of spent solvents (using each time a new portion of expensive solvents for cleaning and disposal, including the environmental factor, involves significant costs). This work is devoted to the study of the process of regeneration of the main component in the compositions used for cleaning metal optics. This is the purification of freon-114***B***2 from technological impurities, which pass into the solvent after the process of cleaning metal optics.

The regeneration of spent solvents is similar to the method of purification of solvents before their use and is usually carried out by the method of rectification. However, in the process of regeneration, it is necessary to remove significant amounts of washed contaminants. When using this method, this can lead to a rapid failure of distillation columns due to the accumulation of significant amounts of contaminants in them and the need for their frequent cleaning, which requires significant costs.

Therefore, in [2, 3], an adsorption purification method is used to regenerate solvents, which complements the distillation purification (at the preliminary stage), and in some cases replaces it. This method is implemented by passing contaminated freons in a condensed form through a layer of highly effective adsorbent: Al_2O_3, SiO_2, coal zeolites with slit pores ~ 5 Å, and zeoline are commonly used. Wherein the water content decreases to 1 • 10-6% by weight, and the decomposition temperature of freons increases accordingly. The disadvantages of the method are: 1) low degree of purification of freon at low productivity (5 kg of pure freon per hour), bulkiness and complexity of equipment during exploitation; 2) when using molecular sieves, activated carbon, aluminum gel, etc. as adsorbents, mainly low molecular weight impurities such as water are removed from solvents (one of the most effective adsorbents for its extraction are synthetic zeolites). In the manufacture of solvents brand "especially pure" the use of this method is possible only in combination with the method of distillation purification, since they may be contaminated with adsorbent material in particular, with anion exchanger AB-17-8 [3]. Therefore, the experiments on the regeneration of freon-114***B***2 were carried out in two stages: 1) removal of rosin and surface-active substances (surfactants) by the adsorption method; 2) distillation purification. Rosin is widely used for the manufacture of polishing pads, and surfactant is a part of the liquid used in the optical processing of metal optics [4]. Using of rosin as model contamination is due to its predominant presence in the impurities that are washed with solvents and the use of surfactants - the difficulty of removal during regeneration compared with other impurities [5, 6].

A characteristic feature of surfactants is the tendency to micelle formation at concentrations exceeding the critical concentration. Most of the studied surfactant adsorption region is the region of the existence of micellar structures, as a result of which the adsorption isotherms of the surfactant are complex [6]. Under dynamic conditions, the capacity of anion exchanger for surfactants depends on the presence of mineral ions (in their absence, the appearance of surfactants (“overshoot”) is observed already in the first portions of the filtrate), the transmission rate of the solution and its concentration. Moreover, at a concentration equal to critical concentration, the adsorption capacity, for example, of cation exchanger KU-2-8, falls. This is probably due to the presence of micelles in the solution, the adsorption of which is difficult.

Techniques for the quantitative determination of various types of surfactants in aqueous solutions are described in [6]. To determine the free surfactant molecules in non-aqueous solutions, the characteristic bands of the spectra related to the stretching vibrations of the bonds of the polar groups of the surfactant molecules are used (the association of surfactant molecules in non-polar liquids, as a rule, is realized through the interaction of these groups).\

2. Experimental technique

As model solutions in the experiments, we used 0.01–0.4 % weight. rosin and 0.01–0.5 % weight. sulfanol surfactant (anionic surfactant, a mixture of isomers of sodium salts) in freon-114***B***2 (due to its maximum content compared with other components in the selected detergent composition). The range of concentrations of contaminants in the model solution is due to the amount of impurities actually present in the solvents before the regeneration process. As adsorbents, activated carbon, silica gel, anion-exchange resin IA-I, ***S***-30 silica gel, a strong base cation exchanger KU-2-8 in the hydrogen form and a strong base anion exchanger ***JRA***-401 in the hydroxyl form were chosen. The use of anion exchangers and cation exchangers is caused by the fact that adsorbents like activated carbon and silica gel absorb mainly low molecular weight impurities and practically do not absorb surfactants [2, 7].

The technique of conducting experiments on adsorption of rosin and surfactant from a solution of freon-114***B***2 under dynamic conditions is as follows. A suspension of the adsorbent in freon-114***B***2 was loaded into a column with a diameter of 11 mm and a height of the absorbing layer of 9 and 15 cm. Since the density of the solvent is higher than that of the solid phase, the adsorbent floated to the surface. The excess solvent was removed from the bottom of the column, and then the adsorbent layer was pressed to the bottom filter with a glass wool swab. Then a model solution was supplied. To determine the content of rosin in freon-114***B***2, the dependence of the optical density of solutions on the concentration of the dissolved substance was used. The determination of abietic acid (which form the main mass of rosin) concentrations, is usually carried out by absorption bands (λ) – 232, 240 and 250 nm [7]. Considering the significant absorption of freon-114B2 in the mid-UV region of the spectrum, λ = 340 nm was used. According to the results of measurements on the SF-4A spectrophotometer, a calibration straight line was built (figure 1), which was later used to determine the concentration of the solution passed through the column (reference solution — 114*B*2 Freon — ("especially pure").

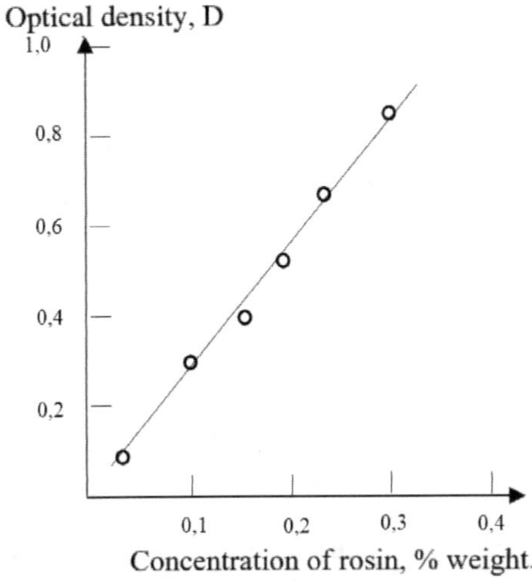

Figure 1. Dependence of the optical density of rosin solution in freon-II4*B*2 on the concentration of rosin.

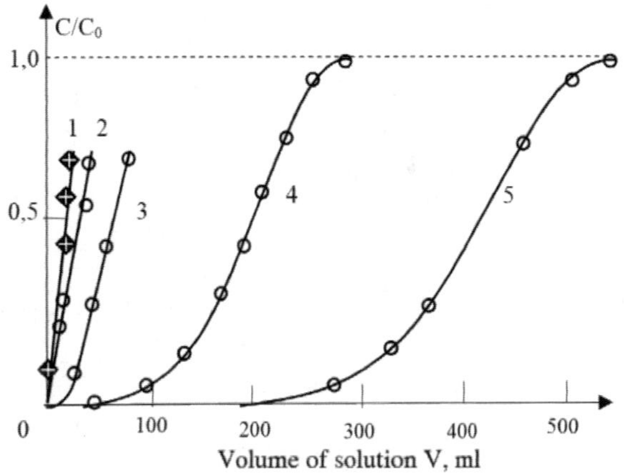

Figure 2. Output curves of rosin adsorption from freon-114*B*2: 1) with activated carbon; 2) with silica gel; 3) with anion exchange IA-1; 4,5) with silica gel *S*-80.

The concentration of surfactant in freon-114*B*2 was determined as follows: an aliquot (15 ml) of the surfactant solution in freon-114*B*2 was placed in a separatory funnel and 25 ml of distilled water was fed into it. The solution was kept for 5–7 minutes. During this time, a portion of the surfactant was extracted into the aqueous phase. After that, the optical density of the obtained aqueous solution was measured on an SF-4A spectrophotometer at λ = 260 nm. This technique has a sensitivity allowing detecting the presence of surfactant in the solution leaving the column at a concentration of up to 10–6 % weight.

3. Results and discussion

The test results on adsorption of rosin from freon-114*B*2 solution in columns are presented in figure 2. It is shown, that when using activated carbon, silica gel and anion-exchange resin IA-1 (curves 1, 2, 3) as adsorbents, the adsorption of rosin from the freon solution is practically absent, the rosin appears in the first portions of the solvent. Silica gel *S*-80 well absorbs rosin from freon-114*B*2, and the absorption value increases with increasing amount of adsorbent in the column (curves 4, 5). The adsorption capacity of silica gel *S*-80, determined from curve 5, is 50 mg/l. When passing surfactant solutions in freon-114*B*2 through columns with adsorbents (silica gel *S*-80, activated carbon, cation exchanger KU-2x2 in hydrogen form and anion-exchanger ***JRA***-401 in hydroxyl form), the presence of surfactant was detected in the first portions leaving the solution columns.

When passing through the column with silica gel *S*-80, which absorbs rosin well, a solution of freon-114*B*2 with a rosin content of 0.1 % wt. and trace amounts (0.005 % wt. surfactant) absorption of rosin was absent. That is, studies have shown the ineffectiveness of the adsorption method of regeneration of freon-114*B*2 from technological contaminations in the presence of surfactants (surfactants are not removed from the solvent, also preventing the removal of rosin).

Therefore, experiments on freon regeneration were carried out using the filtration method through semipermeable polymeric membranes [8]. The advantage of this method is a rather high quality of solvent cleaning and higher productivity than with the adsorption method with minimal energy costs. A significant defect of the method is the lack of universalism: the selectivity of polymer membranes requires the use of a membrane with a specific pore size for a particular type of solvent.

The experimental procedure was as follows. As a material for the manufacture of the membrane, a film of polyethylene terephthalate (PET) with a thickness of 50 mkm (Industry standard 4.023.001-73) was used as the most resistant to freon type solvents. To give this film the maximum defect structure and properties of the polymer membrane against freon-114***B***2, it was pre-soaked for 2 hours in a water-acetone binary solution (43–47% by volume of acetone) with a solubility parameter δ – 33.6 $J^{1/2} cm^{-3/2}$. The choice of this solution for modify the structure of the film is due to the ability of liquids with similar solubility parameters δ to amorphous-crystalline polymers like PET to change the structure of the latter due to crystallization processes. The latter proceed so intensely that the film loses its mechanical strength and is not suitable for further use (significant swelling and partial or complete dissolution of the polymer occur). The intensity of crystallization can be reduced by adding a liquid into the system, which also participates in the formation of the polymer structure, but with a solubility parameter δ, which is significantly different from the solubility parameter δ of the polymer. It is water that is the specific component in the presence of which in the system "PET (δ = 21.9 $J^{1/2} cm^{-3/2}$) – water (δ = 47.8 $J^{1/2} cm^{-3/2}$) – acetone (δ = 19.3 $J^{1/2} cm^{-3/2}$)" a defective porous structure is formed in the polymer.

The exposure time of the PET film in the solution was determined by the time it takes to reach an equilibrium degree of swelling, which is set in a water-acetone solution after 2 h. Similar results on the creation of a defective structure of a PET film gives it processing in a binary solution "water – dioxane" with a solubility parameter (δ = 47.5 $J^{1/2} cm^{-3/2}$) with a dioxane content of 23–27% by volume.

To study the rate of regeneration of freon-114***B***2 we used equipment that includes a mixer for mixing the contaminated solvent; nitrogen trap; a thermostatted separation cell, the case of which consists of two parts, between which a polymer membrane with a working area of 19.2 cm^2 is fixed; backing pump. The principle of its operation is as follows. Before the start of the experiment, the space under the membrane is pumped out with a backing pump to a pressure not exceeding 10 Pa during the experiment. As a model mixture, a solution of freon-114***B***2 with rosin (0.1–0.3 % wt.) and a surfactant of the sulfanol type (0.1 % wt.) were used. This mixture was poured into the upper part of the separation cell. Under the action of pressure drop and its own gravity, it selectively seeped through the pores of the polymer membrane, freeing itself from the bulk of contaminants that are larger than the pores.

The value of the mixture flux density which passed through the polymer membrane over a certain period of time was determined by weighing the amount of the mixture passed. The composition of the mixture was measured on an SF-4A spectrophotometer at λ = 340 nm. The degree of structure defectiveness of the PET film after treatment in binary solutions was determined by IR spectra. The results of the study of the PET membrane modified in binary solutions, degree of defectiveness, the density of the past flow and the concentration of rosin in the model solution before and after passing through the membrane are presented in table 1.

The results of the study of the filtration of a model mixture of rosin in freon-114***B***2 through a semipermeable membrane made of PET showed that modification of the PET film in the "dioxane (25 % vol) - water" solution on both sides allows to increase the density of the passing flow of pure freon-114B2 at 70-430 times, and when modifying the solution "acetone (45 % vol.) – water", – it is 60–130 times compared with the amorphous PET film. Processing the PET film with a binary solution on one side only does not lead to a noticeable increase in the density of the passing stream. The density of the passing mixture when modified in a solution of "dioxane – water" increases 20–45 times and when modified in a solution of "acetone – water" – 25–60 times, compared with the amorphous PET film, depending on the concentration of rosin in the model solution. The concentration of rosin even in the presence of surfactant in an amount of 0.1 % wt in all experiments after filtration through the polymer membrane of PET decreases on average by 2 orders of magnitude, and no surfactant was detected in the output mixture using the methods used in the work. If the concentration of dioxane deviates by 2 % vol from optimal (25 % vol) and acetone to 2 % vol from the optimal (45 % vol), the achieved degree of imperfection of the structure of the PET film decreases and, accordingly, decreases the density of the flow.

Table 1. The dependence of the density of the past flow, the concentration of rosin, the degree of defectiveness.

The method of modification of PET film		Binary solution concentration, % wt	Density of flow, 103 kg/m^2h		Concentration of rosin, % wt.		Degree of defectiveness, %
			Pure freon 114*B*2	Model mixture of rosin in freon-114*B*2	Original model mixture of rosin in freon-114*B*2	Model mixture after filtration	
Amorphous PET		–	14	2	–	–	1–2
Treated with "dioxane -water" solution	On the one hand	25	16	10	0.3	0	6
	From two sides	23	5000	40	0.3	0.030	25
		25	1000	90	0.1	0.025	33
		25	6000	40	0.3	0.010	33
		27	4800	30	0.3	0.030	24
Treated with "acetone -water" solution	On the one hand	45	300	10	0.3	0	7
	From two sides	43	1400	100	0.2	0.030	25
		45	1800	120	0.1	0.030	33
		47	1350	95	0.3	0.030	24

This is probably due to the fact that when the composition of a binary solution deviates from the optimum, its solubility parameter (λ) changes and, thus, an optimal defect structure of the PET film is not achieved. In this case, predominantly pores with a diameter not exceeding 20 Å are formed in the film, through which freon-114B2 (the size of molecules of which is 18 Å) can leak out, being freed from the main mass of contaminants having dimensions of molecules larger than 20 Å. These are mainly molecules of surfactant, fats and oils of organic and animal origin. The contaminants retained by polymer membrane together with the membrane itself are subsequently utilizated.

Recommendation. Regeneration of freon-114B2 and detergent compositions based on it by means of a polymer membrane should be carried out when organizing a closed technological cycle of the physico-chemical process of cleaning metal optics before final purification of solvents by the method of rectification.

4. Conclusion

1. The adsorption method is effective for cleaning freon-114***B***2 from contamination with rosin resins. Among of adsorbents (activated carbon, silica gel, anion exchanger IA-I, silica gel ***S***-80), silica gel ***S***-80 has the highest adsorption capacity (up to 50 mg/l). It has been established that the use of silica gel ***S***-80, activated carbon, cation exchanger KU-2*2 and anion exchanger ***JRA***-401 is not effective when cleaning freon-114***B***2 from surfactants and mixtures of rosin with surfactants. The mechanism of this effect apparently consists in the fact that, at concentrations exceeding critical concentration, large agglomerates are formed in aqueous solutions of surfactants, the adsorption of which is difficult. In non-polar organic liquids, a gradual association of surfactant molecules occurs, and the concentrations at which this process begins may be significantly lower than the critical concentration for aqueous solutions [9].

2. Regeneration of freon-114***B***2 with the help of a polymeric PET membranes after using them to clean metal optics allows reduce the amount of impurities present in it by at least 2 orders of magnitude, and organic molecules with long chains of surfactant type, oils, fats of organic and mineral origin are

separated from the regenerated solvent almost completely. A probable mechanism of this phenomenon is proposed.

3. The use of filtration and distillation purification methods in total allows to reuse the detergent compositions on the basis of freon-114***B***2 for cleaning metal optics in a closed technological cycle, ensuring the environmental friendliness of the process and saving of solvents. The need for pre-treatment is dictated by its advantages: high throughput, low energy intensity and the ability to quickly remove the filtrate and the membrane itself, which is made easily replaceable, and significant losses of solvents when using only the rectification method. Based on the above experimental results and recommendations based on them, the process of physico-chemical cleaning of metal optics was implemented technically.

References

[1] Rogalin V E and Kaplunov I A 2013 Opticheskie svojstva metallicheskih zerkal dlya CO_2 – lazerov [Optical properties of metal mirrors for CO_2 – lasers] *Izvestiya Sochinskogo gosudarstvennogo universiteta* 4–2 (**28**) 120–7

[2] Butenko A I, Birzhivaya V S, Chizhova I I and Kuznecova L P 2005 Regeneraciya othodov rastvoritelej dlya mnogorazovogo ih ispol'zovaniya v proizvodstve [Regeneration of waste of solvents for their reusable use in production] *Tul. ekol. byul* **1–2** 138–40

[3] Filin S A, Rogalin V E and Kaplunov I A 2018 Podgotovka anionita dlya vysokoeffektivnoj ochistki galogenzameschennyh uglevodorodnyh rastvoritelej, ispol'zuemyh dlya ochistki metallooptiki [Preparation of the anionit for highly effective purification of the halogenated hydrocarbon solvents used for cleaning of metal optics] *Zhurnal prikladnoj himii* **91(10)** 1414–19 doi: 10.1134/S0044461818100055

[4] Kaplunov I A, Rogalin V E and Filin S A 2015 Analysis of laser-chemical cleaning of metal surfaces *Non-Ferrous Metals* 1 29–31 doi: 10.17580/nfm.2015.01.07

[5] Kolesnikov V A, Ladygina Yu Sh, Kolesnikov A V, Mec E A and Maslyannikova D V 2019 Osnovnye zakonomernosti elektroflotosorbcionnogo izvlecheniya anionnyh i kationnyh poverhnostno-aktivnyh veschestv iz rastvorov [The main regularities of elektroflotosorbtsionny extraction of anion and cationic surfactants from solutions] *Izvestiya vysshih uchebnyh zavedenij. Seriya: Himiya i himicheskaya tehnologiya* **62(3)** 113–20

[6] Rusanov A I 1992 *Micelloobrazovanie v Rastvorah Poverhnostno-aktivnyh Veschestv* [Formation of Micelles in Solutions of Surfactants] (St. Petersburg: Himiya Publ) 280

[7] *Spravochnik Himika* [Reference Book by the Chemist] 1964 (Leningrad-Moscow: Himiya Publ) Vol **2** 399

[8] Pico J Del 1973 Patent US3762566A Cl. 801d 31/00 *Supported semipermeable membranes and process for preparing same* Oct. 2

[9] Shilova S V, Galiullina G H, Tret'yakova A Ya and Barabanov V P 2015 Micelloobrazovanie anionnogo PAV prirodnogo proishozhdeniya v vodnyh sredah [Formation of micelles anion surfactant of natural origin in water environments] *Vestnik Kazanskogo tehnologicheskogo universitet* **18(16)** 44–46

New way the determination of magnetostrictive parameters composite multiferroics using the magnetoelectric response

D A Filippov[1,6], V M Laletsin[2], Gopalan Srinivasan[3], Jitao Zhang[4], A A Tikhonov[1] and T A Galichyan[5]

[1]Yaroslav-the-Wise Novgorod State University, ul. B. St. Peterburgskaya, 41 173003, Veliky Novgorod, Russia
[2]Institute of Technical Acoustics National Academy of Sciences of Belarus 13, Lyudnikava Praspekt, Vitebsk, BY-210023 Belarus
[3]Physics Department, Oakland University, Rochester, Michigan 48309, USA
[4]Zhengzhou University of Light Industry, Zhengzhou 450002, China
[5]Institute of Mechanics, National Academy of Sciences of Armenia, 24B Baghramyan Ave., Yerevan, 0019 Armenia

[6]E-mail: Dmitry.Filippov@novsu.ru

Abstract. The new way the determination of the thin films magnetostriction characteristics using the magnetoelectric response of composite multiferroic is supposed. The analytical expression for the magnetostriction coefficient using the field dependency of the magnetoelectric voltage coefficient is obtained. The experimental data for the bulk composite lead zirconate titanate – nickel ferrite is presented. It is shown, that the data, obtained by supposed way, are in good agreement with previous data, obtained by another way.

1. Introduction

Composite multiferroics are the perspective materials for the using in the recently developed new direction of electronics – straintronics [1]. The composite multiferroics are of particular interests due to their potential applications for dual H- and E-tunable electronic devices [2]. Dual control in such devices is due to the presence of a magnetoelectric effect (ME) in composite multiferroics, which consists in changing the polarization of the sample in a magnetic field (direct ME effect) or, vice verse, changing the magnetization under the action of an electric field (converse or inverse ME effect). The magnitude of ME effect in composite multiferroics is than several orders higher that its magnitude in monocrystals. The magnetostriction and the piezoelectric module are the more important parameters of these structures. ME effect in composites arises as a result of the elastic interaction between the magnetostriction and piezoelectric subsystems; therefore, the magnitude of the effect directly depends on the magnetostriction and piezoelectric module. These parameters are determined the characteristics of devices based on the ME effect. For example, the change of inductance of the coil by applied electric field is proportional to the product of the piezoelectric coefficient d and the magnetostriction coefficient λ [3–4]. The shift of the magnetic resonance line in an electric field is also proportional to the product $d\lambda$ [5]. The magnitude of the piezoelectric modulus is determined quite accurately, for example, by the resonance-antiresonance method. The determination of magnetostriction has certain difficulties. They arise due to the smallness of the effect in the region of small fields. The small fields region is especially interesting because in this region a nonlinear ME effect appears, on the basis of which it is possible to create a whole range of magnetic field sensors with sensitivity much higher than the sensitivity of Hall sensors [6, 7]. To measure the magnetostriction, as a rule, strain gauges are used, which introduce an additional error in the magnitude of the deformation that occurs under the action of a magnetic field. The difficulties of determining the magnitude of magnetostriction are especially pronounced when measuring the magnetostriction of thin films grown on a piezoelectric substrate, which have proven themselves in devices based on electrical control of magnetic resonance [8, 9]. To estimate the effect in such structures, the value of the magnetostriction coefficient for bulk samples is used. However, as shown by the results of experiments [10] the ME effect in magnetostrictive – piezoelectric structures obtained by electrolysis deposition nickel thin film on piezoelectric substrate has some differences compared to the effect in bulk structures. These features are due to different field dependency of magnetostriction thin films and bulk samples. In this paper, a new method is proposed for determining the magnetostriction of composite multiferroics based on measuring the ME response of the system.

2. Theory: magnetostriction from magnetoelectricity

Magnetoelectric (ME) effect in composite multiferroics is caused by the elastic interaction of magnetostrictive and piezoelectric subsystems. Under the action of a magnetic field, mechanical deformations occur in the magnetostriction phase, which pass through a mechanical coupling into the piezoelectric phase and cause a change in the polarization of the sample, which leads to the appearance of an electrical voltage between the electrodes. As a parameter that numerically determines the ME response of the system, they use the ME voltage coefficient (MEVC) α_E defined as the ratio of the average electric field strength in the sample <E> to the magnitude of the magnetic field H that caused it, i.e. $\alpha_E = \langle E \rangle / H$. The average field value is calculated using the ratio $\langle E \rangle = U / t$, where U is the measured electrical voltage, which occurs on the plates when the sample is placed in an alternating magnetic field H and the constant magnetizing field H_{bias}, and t is the sample thickness. The theory of ME effect in bulk composites at the electromechanical resonance region was first presented in [11, 12]. Using the effective parameters method authors was obtained the expression for MEVC at low frequency region in following form:

$$\alpha_E = \frac{Y d_{33} q_{33}}{\varepsilon_{33} \varepsilon_0}, \tag{1}$$

where Y, d_{33}, q_{33} and ε_{33} are the effective values of the Young's modulus, piezoelectric, piezo-magnetostriction coefficients, and dielectric permeability of the composite, ε_0 is the dielectric constant.

The piezomagnetostriction coefficient d in Exp. (1) is determined by the relation

$$q_{33} = \left.\frac{d\lambda_{\|}}{dH}\right|_{H=H_{bias}}, \tag{2}$$

where $\lambda_{\|}$ is the longitudinal magnetostriction. Since the magnetostriction, and consequently, the piezo-magnetostriction coefficient, depends on the magnetic field, the dependence of the ME response of the system on the bias magnetic field, the so-called field dependence of MEVC, is observed. This dependence allows us to determine the magnetostriction of composite multiferroic. Using equation (1) and equation (2) we get the following expression for magnetostrictive coefficient:

$$\frac{d\lambda_{\|}}{dH} = \frac{\varepsilon_{33}\varepsilon_0 \alpha_E(H)}{Y d_{33}}. \tag{3}$$

Integrating equation (3) with respect to H and using the fact that the parameters Y, d_{33} and ε_{33} are independent of the magnetic field, for the magnetostriction coefficient we obtain the following expression:

$$\lambda(H) = C\int_0^H \alpha_E(H')dH', \tag{4}$$

where $C = \frac{\varepsilon_{33}\varepsilon_0}{Yd_{33}}$ is the constant coefficient for given samples. This coefficient is independent of the magnetic field, but its value depends on the percentage of ferrite and piezoelectric in the composite.

Thus, using relation (3) and the field dependence of MEVC, we can obtain the dependence of the magnetostriction of the composite multiferroic. The value of this coefficient can determine using the measure the value Young's modules, piezoelectric modules and dielectric permeability of composite. Using equation (4), we get for the saturation value of magnetostriction the expression in the form:

$$\lambda_S = C\int_0^\infty \alpha_E(H')dH'. \tag{5}$$

If the value of the saturation magnetostriction is known, then the dependence of the magnetostriction on the magnetic field is easy to determine using the relation:

$$\lambda(H) = \lambda_S A(H) / A_s, \tag{6}$$

where $\lambda = \int_0^H \alpha_E(H')dH'$ and $A_S = \int_0^\infty \alpha_E(H')dH'$ are the integral characteristics of MEVC [13].

Thus, the integral characteristic of MEVC can to obtain the magnetostriction characteristics of composite multiferroics.

3. Details of experiment

Bulk composite ME materials were fabricated by sintering the mixture of nickel ferrite (NFO) and piezoelectric lead zirconate titanate (PZT) powders. With this aim, the samples were prepared in (1 – x) PZT23 – $xNiFe_2O_4$, where parameter x is a weight fraction and equal to 0.1, 0.3, 0.5 and 0.7. The samples were sintered in crucibles with a lead-containing filling for 2 h at a temperature of 1240°C. The cooling rate was not higher than 50°C/h. The electrodes were deposited by means annealing of silver paste at a temperature of 650°C for 20 min. All the samples were shaped as disks with a diameter of 8.8 mm and thickness of 0.8 mm. For polarization of the samples five modes were used: modes 1–4, polarization in the fields with intensity of 1, 2, 3, and 4 kV/mm at room temperature; and mode 5, polarization in the field with intensity of 4 kV/mm at a temperature of 100°C. The polarization of materials was carried out at a temperature of 60-100 ° C for two hours in an electric field of 4 kV / mm, followed by cooling in this field to room temperature for half an hour.

The piezoelectric module $d_{33}$was measured with a YE2730A d_{33} METER instrument. The Young's modulus was defined using the resonance – antiresonance method. The dielectric permeability was determinate from measure of the capacitance, which was measured at a frequency of 1 kHz.

The longitudinal ME effect was studied, when the direction of the electric polarization of the sample and magnetic fields were parallel to each other. The linear ME effect was investigated by measuring the voltage arising on the sample when placed in an alternating magnetic field and a bias magnetic field. The amplitude of the alternating field was 1 Oe at a frequency f = 1 kHz. The magnitude of the bias field varied from 0 to 5000 kOe.

4. Results and discussion

The measurement results of physical parameters of composite for different ferrite content are presented in Table1. As can see from these data, the physical parameters demonstrate nonlinear dependencies on ferrite content. Young's modules increase with increasing ferrite content from

65 GPa for pure PZT to 155 GPa for pure nickel ferrite. The dielectric permeability value for PZT equals 1750; with increasing ferrite content this value decrease from 660 for 10% NFO content to 60 for 70% NFO content. The piezoelectric module value d_{33} for PZT equals 340 pC/N; with increasing ferrite content this value decrease from 200 pC/N for 10% NFO content to 3 for 70% NFO content.

Table 1. The dependence of the parameters of the composite multiferroic on the content of ferrite and piezoelectric.

Contain of ferrite, %	Young's modules Y, GPa	Piezoelectric modules d_{33}, pC/N	Dielectric permeability ε
10	93	198	660
30	115	79	340
50	134	22	140
70	147	3.1	60

In figure 1 the fields dependencies of MEVC and in figure 2 of the curves of magnetostriction, obtained by integration of the MEVC dependencies for different contain of the composite are presented. The integration of MEVC dependencies was performed using the Origin software product.

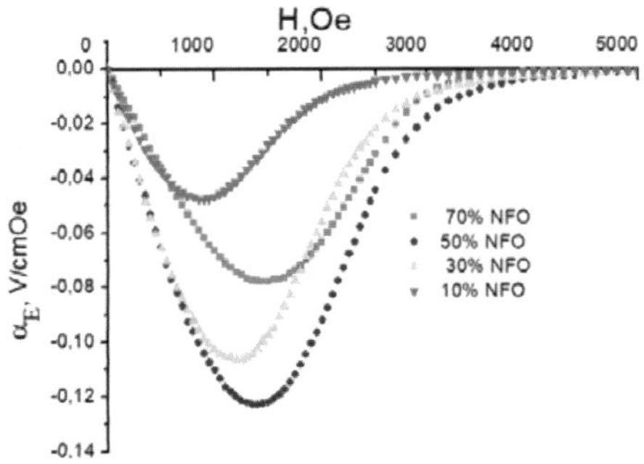

Figure 1. The fields dependencies of MEVC for different ferrite content.

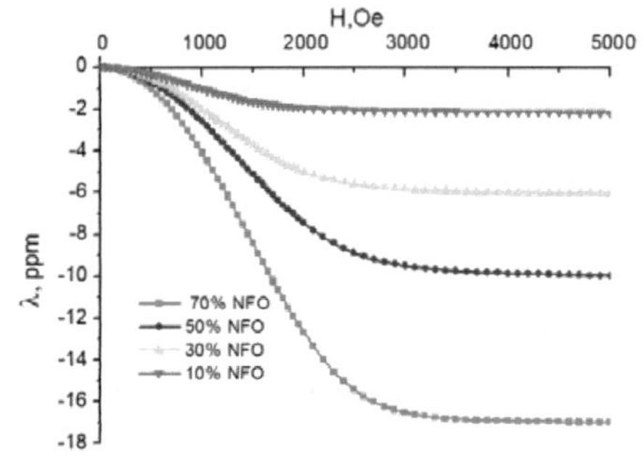

Figure 2. The magnetostriction of NFO-PZT composite multiferroic for different ferrite content.

As can see from figure1, field dependency of MEVC has a maximum. The value of magnetic field H_{max}, corresponding maximum MEVC, depends from the content ferrite in composite and increase with increasing this content. The maximum value of MEVC $\alpha_{E,max}$ depends also from the content ferrite in composite. The maximum of this value $\alpha_{E,max}$=0.125 V/cmOe corresponding to composite with 50% content of nickel ferrite. The curves of magnetostriction, which are shown on figure 2, have typical dependency for nickel contain composite. The magnetostriction has a negative sign and increase with increasing the content ferrite in composite. The values of saturation magnetic field are also increase from 1.8 kOe to 2.8 kOe with increasing the content ferrite in composite from 10% to 70%. The saturation magnetostriction is increase with increasing the ferrite content in composite from –2 ppm at 10% NFO in composite to –17 ppm for 70% NFO in composite. It is good agreement with previously experiments, which shown that for pure polycrystalline nickel ferrite the saturation magnetostriction equals λ_S= –26 ppm.

5. Conclusion

The new way determination of magnetostriction of composite multiferroics is suppoused. It is shown, that the field dependence of MEVC makes it possible to obtain the dependence of the magnetostriction of a composite multiferroic in a wide range of magnetic fields. The experimental results, obtaining with using this way for determination of nickel ferrite-PZT composite magnetostriction are in good agreement with data, obtained previously another way. This way is a perspective for determination magnetostriction thin films on piezoelectric substrate.

Acknowledgment
The work was jointly supported by the Russian Foundation for Basic Research and Belarussian Foundation for Basic Research (Russian project no. 18-52-00021 and Belarussian project no. F18R-300).

References
[1] Bukharaev A A, Zvezdin A K, Pyatakov A P and Fetisov Y K 2018 *Physics Uspekhi* **61** 1175 – 1212
[2] Chung Ming Leung , Jiefang Li, Viehland D and Zhuang X 2018 *J. Phys.* D: *Appl. Phys.* **51** 263002
[3] Lou J, Reed D, Liu M and Sun N X 2009 *Appl. Phys. Let.* **94** 112508
[4] Zhang Jitao *at al* 2019 *J. Phys.* D **52** 165001
[5] Bichurin M I, Kornev I A, Petrov V M, Tatarenko A S, Kiliba Yu V and Srinivasan G 2001 *Phys. Rev.* B **64** 094409
[6] Burdin D, Chashin D, Economov N, Fetisov L, Fetisov Y and Shamonin M 2016 J. *Phys.* D: *Appl. Phys.* **49** 375002
[7] Zhang Juanjuan, Fang Chao and Weng George J 2019 *Proc. R. Soc.* A **475** 2225
[8] Wang Xinjun, Chen Yunpeng, Chen Huaihao, Gao Yuan, He Yifan, Li Menghui, Lin Hwaider, Sun Neville and Sun Nian 2018 *Appl. Phys. Let.* **112** 192903
[9] Zhou Peng *at al* 2019 *Phys. Rev. Appl.* **11** 054045
[10] Laletin V M, Filippov D A, Mozzharov S E and Manicheva I N 2018 *Tech. Phys. Let.* **44** 281
[11] Laletin V M, Stognij A I , Novitskij N N and Poddubnaya N N 2014 *Technical Physics Letters* **40** 969–971
[12] Filippov D A, Laletin V M and Galichyan T A 2013 *Physics of Solid State* **55** 1840–1845

Methods of pulsed local action on magnetic mechanical state of ferromagnetic shape memory alloy

N I Gorbatenko[1], V V Grechikhin[1,2], I S Kraevskiy[1], A V Kudrya[1] and A L Yufanova[1]

[1]Platov South-Russia State Polytechnic University (NPI), 132 Prosvesheniya ul., Novocherkassk 346428, Russia

[2]E-mail: vgrech@mail.ru

Abstract. The task of controlling a positioning system drive with the active element of ferromagnetic shape memory alloy has been considered. The task is to provide the length increment of the active element by a given value from the resultant action – the pulse of magnetic field (or pulse series) with definite amplitude and duration. Methods of solving the task – the methods of tracking and scanning modes of the pulse reversal of the active element – have been suggested. To implement the method of scanning magnetization reversal, it is necessary to introduce a feedback link to reduce the positioning error due to the discrete nature of the effect and have information about the deformation of the element. A specific feature of tracking magnetic reversal is the use of experimental information to exclude overregulation. In testing process, an iterative process has been carried out to correct the mathematical model of deformation, using experimental data. To study the method, a software module has been developed in the LabVIEW graphical programming environment. The data obtained show that application of the method allows to increase the accuracy of controlling positioning systems based on ferromagnetic shape memory alloy.

1. Introduction

A promising direction for the development of positioning systems is the use of ferromagnetic shape memory alloy (FSMA), which has high sensitivity and the ability to change the geometrical dimensions in a wide range under the influence of magnetic, mechanical forces and temperature [1–4]. This allows to increase the accuracy of conversion, expand the range of displacements, and simplify the design of positioning systems. However, there are a number of restrictions on the use of FSMA as active elements. The main ones are the need to make strong magnetic fields, magnetic and mechanical hysteresis, significant effect of temperature change on the FSMA parameters. When making high-speed positioning systems with small weight and size parameters, it is efficient to use local pulsed magnetic fields to control active elements of FSMA. These conditions are met by a positioning system implementing a two-stroke configuration (figure 1). A distributed magnetizing system (*1*) containing *n* pairs of coils without ferromagnetic cores which allows to increase speed, reduce weight and size of the device, as well as two active elements (*2*) made of FSMA, and an actuator (*3*) is used. The mode of pulsed magnetic reversal of active elements of FSMA with the possibility of controlling the magnitude of magnetic field strength to the elements is used [5]. Also, this approach allows to implement the discrete mode of controlling the deformation of FSMA, improve the accuracy of positioning, increase speed, and reduce weight and overall dimensions of the system.

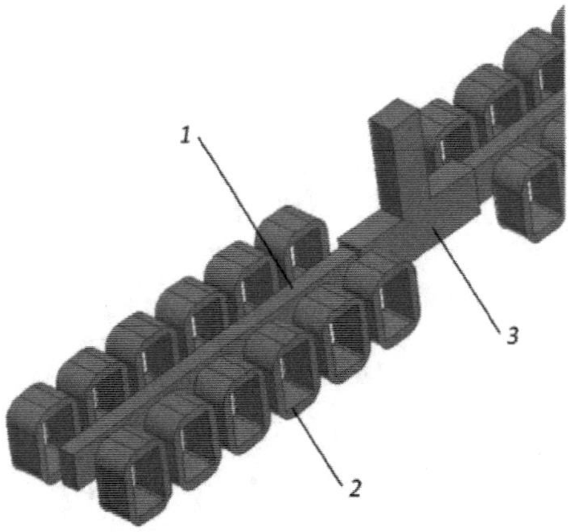

Figure 1. General view of two-stroke positioning system.

There is a close interdependence between the factors affecting FSMA, magnetic intensity H and its deformation ε. Problems of mathematical modeling of deformation of local areas of the active element of FSMA of the positioning system have been discussed in the works [6, 7].

The operation principle of the positioning system drive is based on the variation of linear size (deformation) of the active element of FSMA under the influence of magnetic intensity, the vector of which is perpendicular to the plane of resizing. To ensure large deformations of active elements, a pulsed magnetic field with the intensity up to (350 – 400) kA/m is created.

We formulate the following control problem. It is necessary to deform the active element by a specified value with a permissible control error ε. The task is to provide the length increment of the active element by a specified value from the resultant impact – a magnetic field pulse (or a series of pulses) with definite amplitude and duration. Further we suppose methods of solving the problem.

2. Methodology

It is possible to implement tracking and scanning modes of pulsed magnetization reversal of the active element of FSMA.

The method of scanning conversion supposed by F.E.Temnikov [8] is that the results of conversion of an unknown input signal are compared with those of the known replacing value, i.e., the difference between a scanning function and a defined one is determined with fixing a zero level of the difference. The scanning function changes with a definite cyclical nature. In the positioning system the magnetic field $H_e(t)$ is taken as a scanning function, and the deformation of the active element $\varepsilon(t)$ is taken as a defined function (figure 2).

The duration of conversion cycle T_{tr} is defined by the duration of magnetization reversal t_{rem} and the duration of pause t_p between the control cycles:

$$T_{tr} = t_p\,(1 + Q),$$

where Q is a relative pulse duration.

The magnetization reversal duration t_{rem} in a converter of the magnetic field of scanning balancing with steady stepwise changing of the magnetic field is not constant in the range of displacements and depends on the magnitude of magnetic intensity: $t_{rem}=N_e\,/\,f_0$, where N_e is a numeric code, N_e = ent(H_e / h_k), directly proportional to the magnitude of magnetic intensity H_e, and, accordingly, to ε. A minimal value of relative pulse duration Q is limited by extremely permissible value of heating of magnetizing winding, depending on the magnetization reversal current $i(t)$.

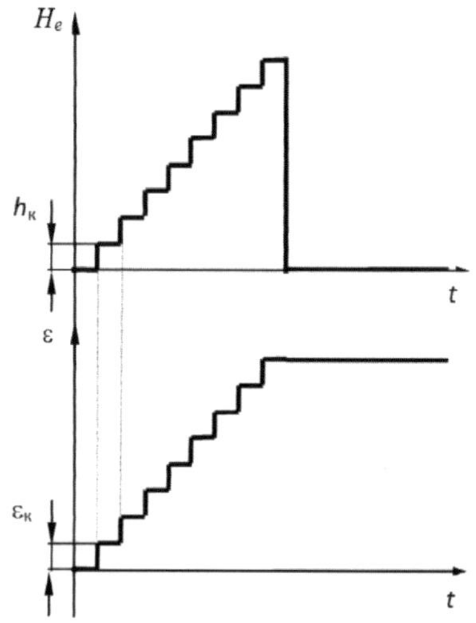

Figure 2. Pulsed magnetization reversal by scanning balancing method.

We determine the conditions, where the aggregate error $\delta_{H_{sum}}$, equal to the sum of quantization error Δh_k and dynamic error $\delta_{H_{dyn}}$ will be minimal. According to [9], the errors Δh_k and $\delta_{H_{dyn}}$ will be equal

$$\Delta hk = \frac{H_e}{N_e}, \qquad \delta_{H_{dyn}} = V_{H_e} t_{rem},$$

where $V_{H_e} = dH_e/dt$, is the speed of changing the field H_e.
In steady stepwise variation law H_e, the time of magnetization reversal is equal to

$$t_{rem} = N_e \Delta t_{st},$$

where Δt_{st} – the time of quantization step working-off.
If $\Delta t_{st}=const$, then the error $\delta_{H_{sum}}$ can be defined by the formula

$$\delta_{H_{sum}} = \Delta h_k + V_{H_e} N_e \Delta t_{st}. \tag{1}$$

Regarding N_e as a continuous variable, we differentiate the expression (1) by N_e and from the extremum condition we define the optimal value N_{eo}, corresponding to the minimum of the aggregate error

$$N_{e0} = \left(\frac{H_e}{V_{H_e} \Delta t_{st}}\right)^{\frac{1}{2}}. \tag{2}$$

With this value N_{eo}, error components $\delta_{H_{sum}}$ are equal, correspondingly

$$\delta_{H_{sum}} = \sqrt{H_e V_{H_e} \Delta t_{st}}. \tag{3}$$

Thus, from (2), (3), knowing maximum values of the field H_e and rates of its variation V_{H_e}, it is possible to determine the values of quantization step h_k and its time of working-off Δt_{st}.

In the pulsed magnetization reversal method discussed above, the selection of parameters determining accuracy and speed of positioning is implemented according to equations (1–3) by maximum values of deformation of the active element and, consequently, by magnetic intensity and the rate of its variation. The deformation, values H_e and V_{H_e} can significantly vary. This means that the use of constant T_{tr} and Δt_{st} will not promote optimal speed and accuracy of positioning.

To implement the method of scanning pulsed magnetization reversal of the active element of FSMA, it is necessary to introduce a feedback link to reduce the positioning error due to the discrete nature of the effect on the active element of FSMA. It is necessary to get information about the deformation of the element that has occurred. A promising direction of measuring the deformation is the use of sensory properties of the active element [10].

Another approach is the method of pulsed magnetization reversal implementing the tracking mode. As a rule, pulses with constant frequency, duration and, consequently, a constant duty cycle Q are used. The duty cycle Q must correspond to the maximum value of displacement. This means that the duty cycle Q has always been overstated, i.e., the converting time is not minimum possible. It is known that the amplitude of current pulse supplied into the magnetizing winding may exceed the value of constant current flowing through the same winding into $\sqrt{Q}$.

In control systems, to obtain, in any sense, the optimal control with unknown characteristics of the controlled process, methods and systems called adaptive are used. If we perform the adaptive regulation of the duty cycle Q in the process of changing the amplitude of compensatory pulses, then the positioning time can be significantly reduced (figure 3).

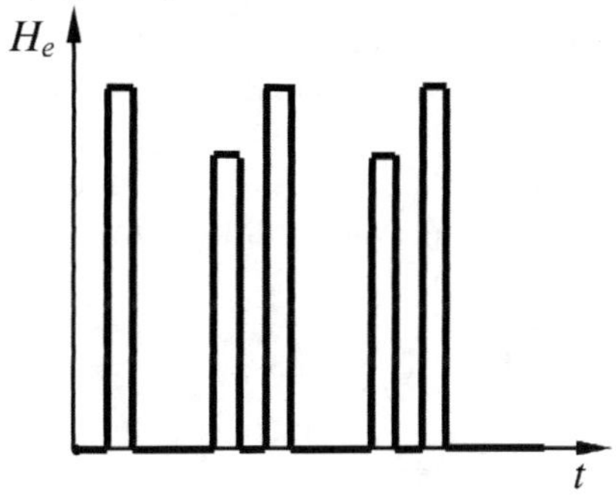

Figure 3. Pulsed magnetization reversal by tracking balancing method.

It should be considered that in a number of cases when implementing the positioning system, an additional control condition may be required, – exclusion of overregulation possibility [11]. We regard this condition in relation to positioning systems with active elements based on FSMA: the actual length L_{act} of the active element after the control action should be strictly less that the specified value of the final length L_{set} by the value of allowable control error ΔL_c (positioning errors) (figure 4).

Displacement control algorithm by the way of deforming the active element based on FSMA includes the following steps: generation of control pulse for magnetization reversal of the active element; measurement of actual length L_{act} of the active element; control error calculation ΔL_c; analysis of the result obtained; determination of parameters of correcting control pulse. Its implementation suggests the preliminary development of control empiric laws. They present experimentally obtained dependences of the active element deformation on various control effects in varied initial conditions determined by the presence of initial deformation of the active element. Various initial deformations of the active element reproduce the result of action of control pulses at previous iterations of the control algorithm.

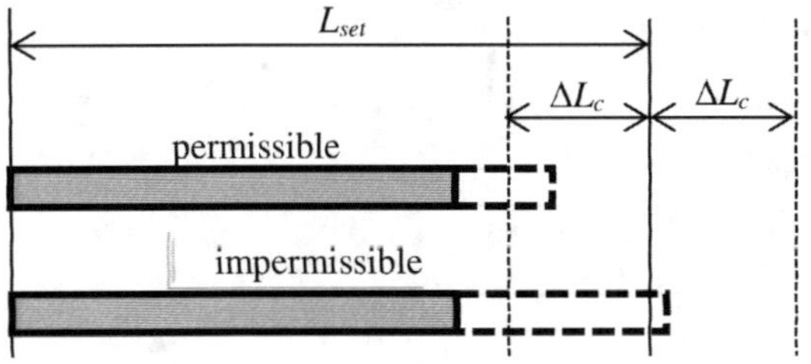

Figure 4. Illustration of additional condition to the active element deformation control.

The specified deformation value of the active element (the specified value of its final length L_{set}) corresponds to the amplitude value of the primary control pulse $U_m^{(1)}$, as well as the table of the amplitude values of the secondary control pulse $U_m^{(2)}$ as a function of the actual length of the active element $L_{act}^{(1)}$ after the primary action:

$$L_{set} = F(U_m^{(1)}, U_m^{(2)} = f\left(L_{act}^{(1)}\right)) \tag{4}$$

IOP Conf. Series: Materials Science and Engineering **656** (2019) 012020 doi:10.1088/1757-899X/656/1/012020

Expression (4) presents the analytical record of empiric law of the active element deformation control with one corrective action.

In general, several corrective pulsed actions are necessary. In this case the analytical record of empiric control with several corrective actions looks like this:

$$L_{set} = F(U_m^{(1)}, U_m^{(2)} = f_1\left(L_{act}^{(1)}\right), U_m^{(3)} = f_2\left(L_{act}^{(2)}\right), \ldots).$$

Implementation of the method of pulsed action on the active element based on FSMA includes the following stages:

– theoretical determination of the primary control pulse amplitude. Herewith, the necessary control pulse amplitude is determined on the base of mathematical model to achieve the specified final length L_{set} of the active element.

– experimental determination of the variation of the results of the primary control action. For this, a series of experiments is carried out, and the variation of values of actual deformation of the active element is determined $\Delta\varepsilon$ (figure 5).

– adjustment of the primary control pulse amplitude in order to exclude situations of overshoot. Herewith, regarding the value $\Delta\varepsilon$, the primary control pulse amplitude is corrected (decreased) so that in a series of experiments the maximum active element deformation would not lead to exceeding the given value of the final length of the active element L_{set}.

– checking the condition of pulse action process completion. It is checked whether the specified control accuracy is ensured as a result of the previous control actions, and, accordingly, whether the required deformation has been performed (figure 6).

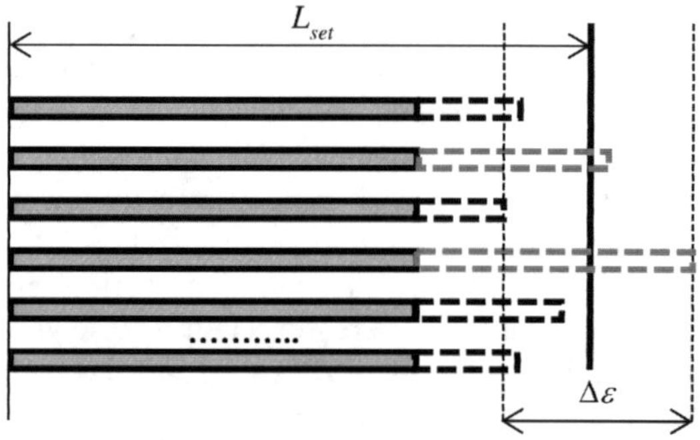

Figure 5. The variation of values of the actual active element deformation.

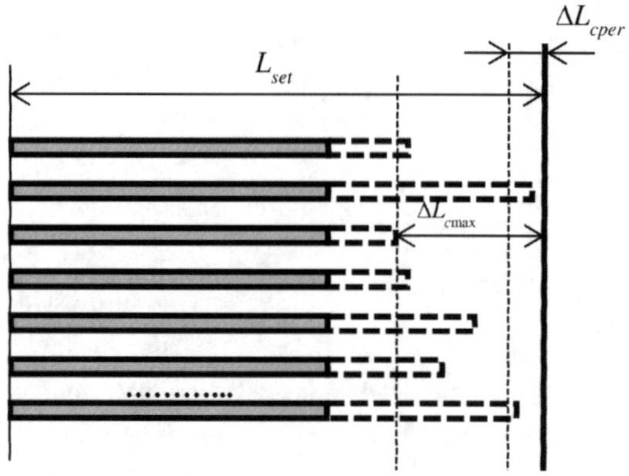

Figure 6. Condition of deformation process completion of the active element.

At subsequent iterations, the actions of the previous stages are repeated, and not one, but several

IOP Conf. Series: Materials Science and Engineering **656** (2019) 012020 doi:10.1088/1757-899X/656/1/012020

series of experiments with the condition of the definite value of the initial deformation of the active element for each series are conducted. These values of the initial deformations are determined by the set of values of the actual final length of the active element detected in an experimental series at the first iteration (after the supply of the primary control pulse). Herewith, the whole set of the mentioned values may be "thinned out" with the accuracy to the permissible control error ΔL_{cper}.

Consequently, with each successive iteration, the parameters of corrective control pulses are selected, due to the actual increment of the actual element at the previous iteration.

3. Results and Discussion

In order to study the supposed method of pulse action on the active element, the computing experiment has been carried out. For its implementation, a software module has been developed in the LabVIEW graphical programming environment.

To enter data into the program, the cluster control element is used. Values of the following parameters are specified: the number of argument values (a control signal amplitude) which will be used to calculate functional dependencies; minimal and maximal amplitudes of the control signal; theoretically minimal and maximal possible increments of the active element length; the exponent parameter in the functional dependence of the active element increment on the control signal amplitude; the specified value of the length increment of the active element; the number of experiments carried out at one definite value of the control signal amplitude determined on the basis of a theoretical model to achieve the given value of the active element length increment; the maximum permissible actual value of the active element length increment.

The program uses the theoretical function of the value of the active element length increment obtained on the basis of mathematical modeling:

$$dL(I) = dL_{min} + dL_{max}\left(1 - e^{-\frac{I-I_{min}}{p}}\right),$$

where dL_{min} – minimally possible increment of the active element length after the control action by the pulse with minimum threshold amplitude I_{min}; dL_{max} – maximally possible increment of the active element length; I – control pulse amplitude; p – exponent parameter.

The graphs of theoretical and actual dependence of the increment size of the active element length on the control signal amplitude have been obtained (figure 7). In Figure: *1* – theoretical dependence of the incremental size of the active element on the control signal amplitude; *2* – actual dependence of the incremental size of the active element length on the control signal amplitude; *3* and *4* – the range of varying the increment of the active element length, depending on the control signal amplitude.

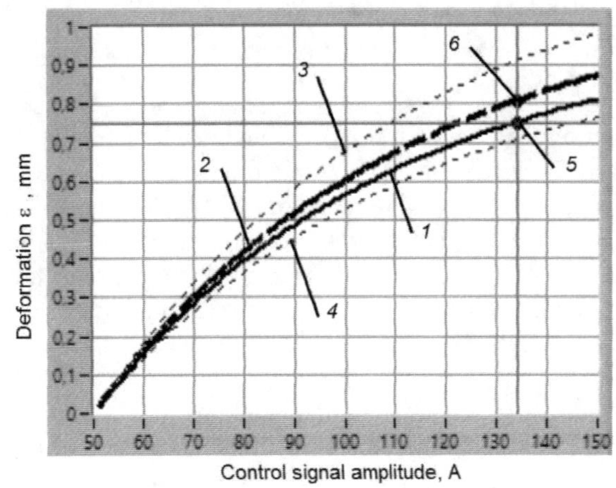

Figure 7. Theoretical and experimental dependences of the active element deformation value on the control signal amplitude.

Point *5* corresponds to the given value of the active element length increment and, accordingly, the necessary control signal amplitude determined based on theoretical concepts. Point *6* corresponds to actual value of the active element length increment after the control action by the pulse with the amplitude determined based on theoretical concepts.

The actual dependence of the increment size of the active element length on the control signal amplitude and the range of its variation (figure 8) which is formed due to processing the data of large complex of experiments have been obtained. The graph contains the following elements: *1* – the set of experimentally obtained values of increments of the active element length after control action; *2* – connecting line for displayed experimental values of increments of the active element length; *3* – the specified value of the active element length increment after the control action; *4* – maximally permissible increment of the active element length after the control action.

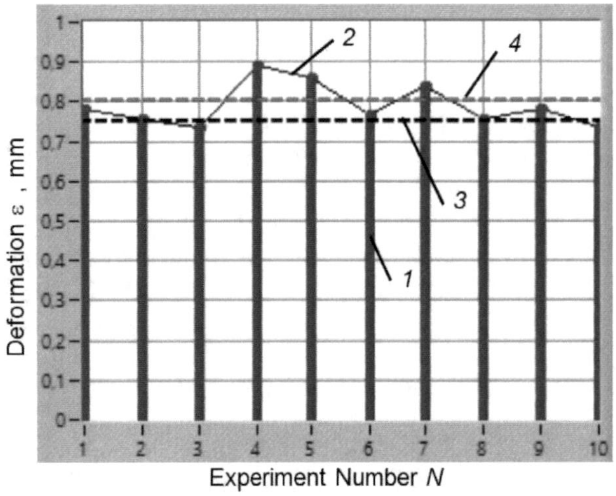

Figure 8. Experimental values of increments of the active element length after the control action with the given amplitude.

The experimental graph (figure 8) corresponds to the theoretical graph (figure 5). Consequently, the developed software module is an instrument which provides the fulfillment of the first, second and third stages of the method of pulse action on the active element based on FSMA described above. Using the graph element (figure 8), according to the third stage, it is necessary to correct the control signal amplitude so that to avoid the situation of exceeding maximally permissible increment of the active element.

The analysis of the results obtained has shown that the use of the pulsed action method allows to deform the active element of positioning system by the given value with the positioning error not more than 0,5μm.

4. Conclusion

The use of FSMA allows to make high-speed positioning systems with small weight and size parameters. To control active elements of FSMA, it is efficient to use local pulsed magnetic fields. It is possible to implement the scanning and tracking modes of pulsed magnetization reversal of FSMA element. If it is necessary to exclude overregulation to obtain optimal speed and accuracy of positioning, it is advisable to use the method of pulsed magnetization reversal of the active element based on experimental information. The goal of the following works is to study the adaptive method of pulsed action on the element of FSMA, using non-searchable self-tuning systems.

Acknowledgements

The study results are obtained with the support of the project №2.7193.2017/8.9 "Development of scientific bases of design, identification and diagnosis systems for highly accurate positioning with application of the methodology of inverse problems of electrical engineering", carried out within the framework of the base part of State job.

IOP Conf. Series: Materials Science and Engineering **656** (2019) 012020 doi:10.1088/1757-899X/656/1/012020

References

[1] Ullakko K at al 1996 *Appl. Phys. Lett.* **69(13)** 1966–1968

[2] Tickle R and James R 1999 *Journal of Magnetism and Magnetic Materials* **195(3)** 627–638

[3] Vasil'ev A N, Buchel'nikov V D, TakagiT, Khovailo V V and Estrin E I 2003 *Physics-Uspekhi* **46(6)** 577–608

[4] Wilson S A at al 2007 *Materials Science & Engineering* R – Reports **56** 1–129

[5] Gorbatenko N I, Grechikhin V V and Shaikhutdinov D V 2015 *Metal Science and Heat Treatment* **11** 609–613

[6] Grechikhin V V, Lozin O I, Kudrya A V, Kudrya N A, Shaikhutdinov D V and Yanvarev S G 2014 *Fundamental research* **11(4)** 744–748

[7] Gorbatenko N, Grechikhin V, Kolomiets A, Kraevskiy I, Kudrya A and Shaykhutdinov D 2017 *International Journal of Applied Engineering Research* **12(23)** 13220–13226

[8] Temnikov F E 1963 *Theory of deployment systems* (Moscow: Gosenergoizdat Publ)

[9] Ornadskiy P P 1980 *Automatic measurements and instruments* (Kiev: High school Publ)

[10] Gorbatenko N I, Grechikhin V V, Kraevskiy I S, KudryaAV and Yufanova J V2018 *IOP Conf. Series: Materials Science and Engineering* **441(1)** 012020

[11] Grechikhin V V, Kraevskiy I S, Lozin O I, Shaikhutdinov D V and Yanvarev S G 2015 *Fundamental research* **12** 671–675

Development of a convolution algorithm for the content of a Web resource for processing by a neural network

S N Gorshkov[1,2], D V Solomonov[1,3], I V Dvoryaninova[1,4] and R M Mihaylichenko[1,5]

[1]North-Caucasus Federal University, Stavropol, Russian Federation

[2]E-mail: gorser96@gmail.com
[3]E-mail: dsolomonov@ncfu.ru
[4]E-mail: Innadv99@mail.ru
[5]E-mail: furiousfaind@gmail.com

Abstract. The article aim to develop a convolution algorithm for the content of Web resources for processing by a neural network. The content of a web resource is represented in a vector form based on a co-occurrence matrix. A convolutional neural network, where the most significant features stand out from them, processes the content vectors of Web resources. The result is a collapsed content of Web resources in the form of a vector of significant features, which is the result.

1. Introduction

Web content is a set of textual and visual information. Such a set cannot be processed in the form in which it is presented to users, so it must first be converted into a form understandable to the computing device, and then be processed using a convolutional algorithm. Neural network processing requires that each word be converted into a form that is understandable to a computing device – a numeric one. For convenience of computations, it is necessary to represent one or several words in a vector form.

When processing content of Web resources, the problem of choosing the correct meaning of a word arises, therefore it is necessary to consider several words at once in order to more accurately determine the context of a fragment of a Web resource. One of the tasks of processing neural networks of texts is the construction of vectors of words related to the field of distributive semantics [4]. This task is based on the assumption that the meaning of a word can be understood by the meaning of its context, by the surrounding words.

2. Development of the algorithm for the vector transformation of the web resource content

Convolution of the content of Web resources is necessary to accelerate the processing of data by a neural network. The most effective means are convolutional neural networks. They convert a sequential set of vectors into a new fixed-length data set, without losing the meaning of the word layout.

When solving the task, first of all it is necessary to determine the method of converting text into a vector. The simplest representation of text in vector form is bag-of-words. In this case, on the basis of the learning corpus, a dictionary is constructed of all the words found in it or n-grams, where n is less than or equal to any predetermined value. The method in a primitive representation has several negative qualities.

When building a dictionary based on words, it is necessary to predict all possible words of Web content, which is impossible, and the volume of the dictionary will be too large. We will calculate the approximate data dictionary, if the dictionary consists of 100,000,000 words, then the vector will consist of 100,000,000 values, 1 vector element in the worst case is an integer value with a size of 4 bytes, which means the size of the vector will be 400,000,000 bytes = 381 MB. A large enough vector and its processing will greatly load the device. In addition to the problem of a potentially large vector size, there is a problem of loss of context and inefficient use of a vector, which will have only one number, and all other zeros.

This problem can be solved using n-grams and a co-occurrence matrix. If you use a 3-character n-gram and form a vector based on the co-occurrence matrix, then the generated dictionary will contain a maximum of 669920 elements, taking into account the special characters $\bar{C}_{158}^{3} = C_{3+158-1}^{3} = \frac{(3+158-1)!}{(158-1)!*3!} = 669920$.

Character n-grams are useful for languages with a rich morphology – the same words can be found in texts, but in different variations (different gender or numeric forms), but the root of the words does not change, and, consequently, the common set of characters.

Each n-gram must be saved to the database, but it is impossible to know in advance which n-grams will be found in the content of the Web resource, since the number and variety of the latter is too large. Therefore, a dictionary in the database will be formed as soon as a new n-gram is encountered in the next content of the Web resource.

Figure 1 shows the algorithm for the formation a database of key phrases.

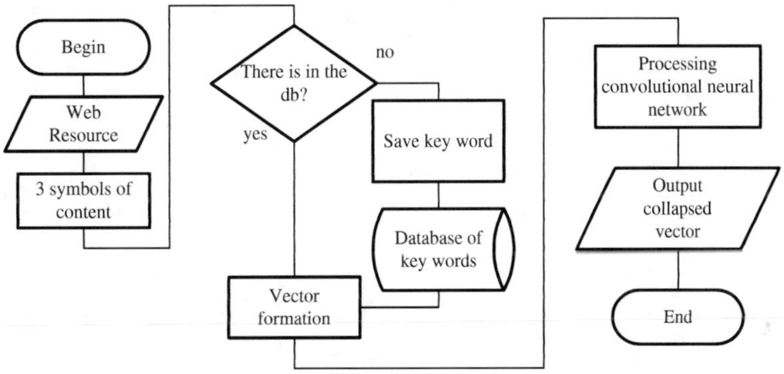

Figure 1. Algorithm for the formation of a database of key phrases.

We describe the structure of the database of key phrases (figure 2). The database will contain 2 tables, the first table stores all known key phrases. The second table contains links between key phrases. The link between key phrases will be considered as a pair of adjacent key phrases in the content of a Web resource, a set of such links will reflect the context.

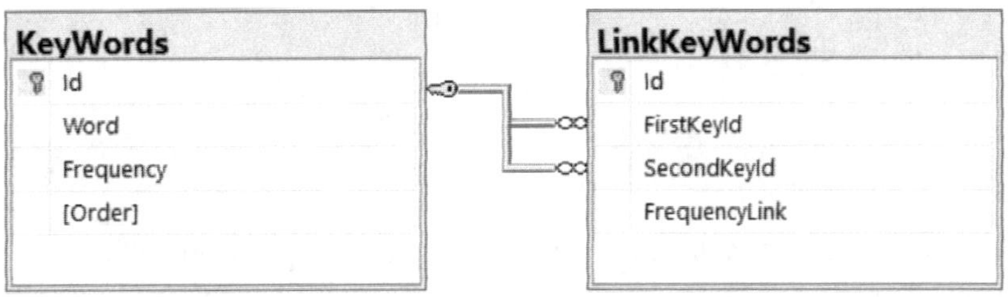

Figure 2. Datalistic database model of key phrases.

IOP Conf. Series: Materials Science and Engineering **656** (2019) 012021 doi:10.1088/1757-899X/656/1/012021

The algorithm of vector formation is shown in figure 3. The algorithm transforms the array of key phrases into an array of corresponding vectors. As an array of key phrases, a sentence or another part of the content of a web resource, broken into symbolic 3 grams.

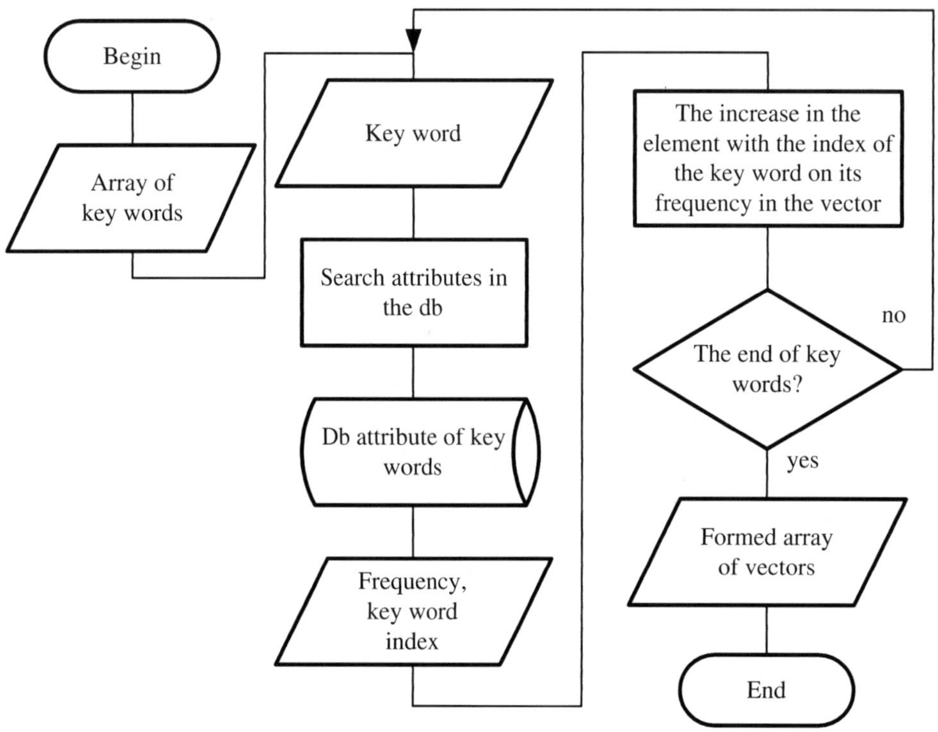

Figure 3. Vector formation algorithm.

As a result of the operation of the algorithm, vectors with a length equal to the size of the ictionary of key phrases are obtained. The vector is filled according to the co-occurrence matrix. For example, for the phrase "neural net consists of neurons" a co-occurrence matrix will be formed.

Table 1. Co-occurrence matrix.

	neu	ral	net	co	nsi	sts	of	ne	uro	ns
neu	0	2	0	0	0	0	0	0	1	0
ral	2	0	2	0	0	0	0	0	0	0
net	0	2	0	1	0	0	0	0	0	0
co	0	0	1	0	1	0	0	0	0	0
nsi	0	0	0	1	0	1	0	0	0	0
sts	0	0	0	0	1	0	1	0	0	0
of	0	0	0	0	0	1	0	1	0	0
ne	0	0	0	0	0	0	1	0	1	0
uro	1	0	0	0	0	0	0	1	0	0
ns	0	0	0	0	0	0	0	0	0	0

Table 1 also presents the vectors for each symbolic 3-gram of the considered example, the vector is a sequence of numbers in the string. 3-gram character "net" will be converted into a vector:
$\overrightarrow{\text{"net"}} = (0\quad 2\quad 0\quad 1\quad 0\quad 0\quad 0\quad 0\quad 0\quad 0).$

IOP Conf. Series: Materials Science and Engineering **656** (2019) 012021 doi:10.1088/1757-899X/656/1/012021

As can be seen from the example, most of the vector is zero, this part of the vector does not carry critical information about the context of the content of the Web resource, but will be involved in further processing, spending more productive resources of the target device. To solve this problem, it is necessary to isolate the main features of the content of the Web resource and form a smaller vector characterizing the content of the Web resource using a convolutional neural network.

3. Development of the algorithm of convolution and selection of significant features of the vector content of Web resources

Features of a convolutional neural network for word processing [9]:

– filter layer has the same width m as the matrix;

– it is allowed to apply simultaneously different filter layers with different heights to highlight additional features.

A convolutional neural network is a very effective method for presenting texts. Due to the complexity of computing and storing data, most n-gram cases, including Google, do not represent n-gram cases with n more than 5. For example, the Google n-gram case from 1,024,908,267,229 words, 95,119,665,584 sentences, 13,588,391 unigrams, 314,843,401 bigrams, 977,069,902 3-grams, 1,313,818,354 4-grams, 1,176,470,663 5-grams after compression in gzip format has a size of 24 GB [1].

A convolutional neural network for representing content of Web resources consists of two layers (figure 4): a convolutional layer and a pooling layer.

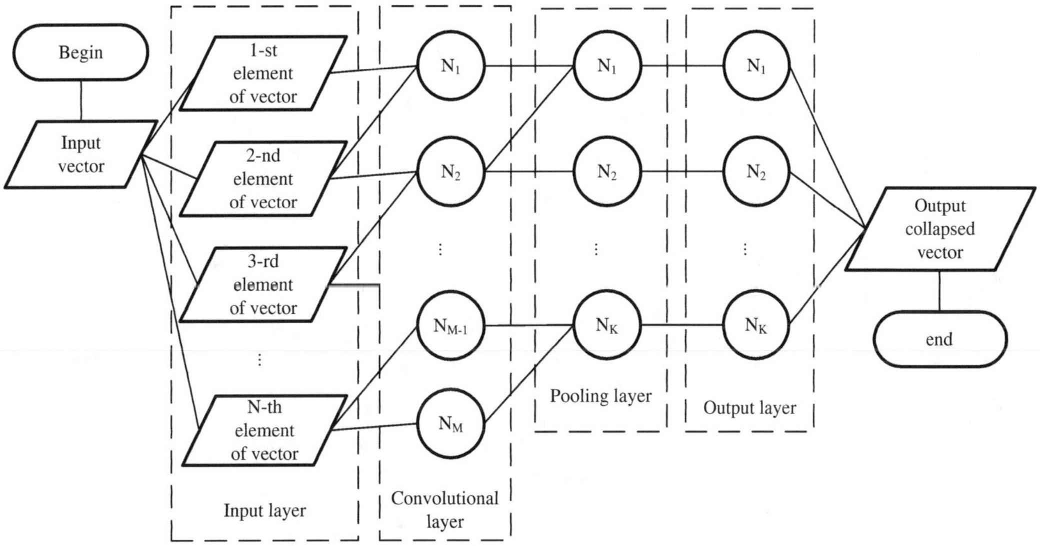

Figure 4. The algorithm of the convolutional neural network for processing the vector content of Web resources.

The convolution layer, in turn, consists of filters. A filter is a neuron, the input of which is formed with the help of windows moving through the text and choosing a certain number of words in succession [2]. At the output of the filter, one vector is formed, which aggregates all the word vectors in it. Then, one vector is formed on the pooling layer, corresponding to the entire sentence, which is calculated as an component-maximum from all the output filter vectors.

Consider the algorithm of the convolutional layer for a simple example (figure 5). In the convolutional layer of the neural network, vectors are processed using predefined filters [7]. The filter size is the number of simultaneously processed word vectors. Figure 5 shows the operation of the size 2 filter. In it, the scalar product with weights of the filter is calculated for each 2 vectors of input data. For example, calculate the first value of the output vector:

$$\begin{pmatrix} 1 & 0 & 0 & 0 \\ 0 & 1 & 1 & 0 \end{pmatrix} * \begin{pmatrix} 0.2 & 0.6 & 0.1 & 0.5 \\ -0.3 & 0.1 & 0.4 & 0.2 \end{pmatrix} =$$
$$1*0.2+0*0.6+0*0.1+0*0.5+0*(-0.3)+1*0.1+1*0.4+0*0.2=0.7.$$

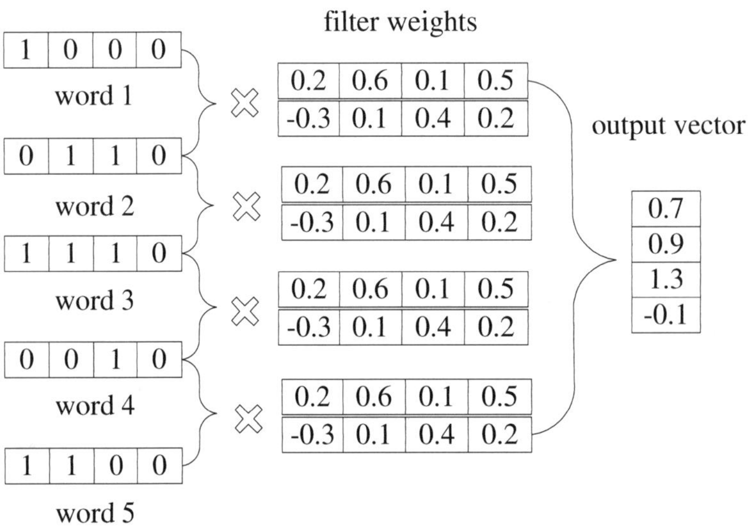

Figure 5. Algorithm convolutional layer convolutional neural network.

Similarly, the remaining values of the new vector are calculated. The resulting vector is transferred to the next layer – the pooling layer. The pooling layer algorithm is depicted in figure 6.

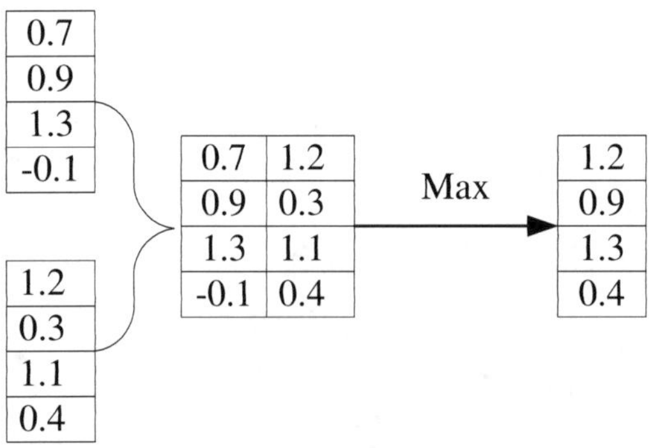

Figure 6. The pooling layer algorithm.

The input pooling layer gets all the vectors of one filter from the previous layer and applies the component-wise maximum value function to them. Figure 6 shows the operation of the pooling layer for vectors $(0.7 \quad 0.9 \quad 1.3 \quad -0.1)$ and $(1.2 \quad 0.3 \quad 1.1 \quad 0.4)$. After processing the layer obtained vector $(1.2 \quad 0.9 \quad 1.3 \quad 0.4)$.

Training a convolutional neural network for the task is based on the method of training without a teacher. Hebb's learning algorithm assumes that the synaptic connection of two neurons is enhanced if both of these neurons are excited.

The algorithm is expressed by equality (1), described in [3]:

$$W_{ij}(t+1) = W_{ij}(t)*(1-\gamma)+\eta*IN(t)*OUT(t), \tag{1}$$

where $W_{ij}(t)$ is the synapse force from neuron i to neuron j at time t, γ is the forgetting factor, is needed to stabilize the learning process, η is the learning factor, the value of which is selected in the interval (0.1), IN(t) – input signal, OUT(t) – output signal. The value of the coefficient γ is a part of the learning coefficient η.

The activation function is a function that calculates the output signal of an artificial neuron. The use of the activation function ReLU significantly increases the rate of convergence of the stochastic gradient descent compared to the sigmoid and hyperbolic tangent. A large gradient passing through the ReLU can lead to such an update of the balance that the neuron is never activated again. One of the modifications ReLU – Leaky ReLU, solves this problem. The activation function of Leaky ReLU is:

$$f(x) = \begin{cases} \alpha * x, & x < 0 \\ x, & x \geq 0 \end{cases}, \quad (2)$$

where α is a small constant.

The principle of choosing the initial values of weights for the layers of the neural network contributes to the speed of learning the network. The Ge initialization method makes it possible to simplify the passage of a signal through a layer, the activation function of which is the ReLU, which compensates for the fact that this function returns zero for half the definition domain. The initialization method of Ge has the form:

$$Var(w) = \frac{2}{n_{in}}, \quad (3)$$

where $\boldsymbol{Var(w)}$ – variance of random weight, $\boldsymbol{n_{in}}$ – number of neurons in the previous layer.

4. Conclusion

As a result, an algorithm for checking the content of Web resources has been developed, which allows you to present any content in the form of vectors that modern computers can easily handle. After the content is verified, the most important features are preserved, initially being the links between the content fragments of the Web resource, which affects the context of the entire content in a minor way. This will allow in the future to effectively clustering the content of Web resources to detect illegal content in Web resources.

References

[1] Thorsten B and Franz A 2006 *Web 1T 5-gram Version 1 LDC2006T13* DVD (Philadelphia: Linguistic Data Consortium)
[2] Yoav Goldberg 2017 *Neural Network Methods for Natural Language Processing* (Morgan & Claypool Publishers)
[3] Osovskii S 2002 *Neural networks for information processing* Translation from polish I D Rudinskii (Moscow: Finance and Statistics Publ)
[4] Information on https://www.osp.ru/os/2017/02/13052221
[5] Uossermen F 1992 *Neurocomputer technology: Theory and practice* (Moscow: Publ “Mir”)
[6] Krug P G 2002 *Neural networks and neurocomputers: A manual for the course "Microprocessors"* (Moscow: MEI Publ)
[7] Information on http://www.wildml.com/2015/12/implementing-a-cnn-for-text-classification-in-tensorflow/
[8] Chipiga A F and Kolkov D A 2006 *Napravlenniy poisk globalnogo ekstremuma mnogomernih functsiy* Basic research
[9] Information on https://neurohive.io/ru/osnovy-data-science/7-arhitektur-nejronnyh-setej-nlp/

Analysis of structures of energy conversion complexes of spacecraft power supply systems in development of their digital control systems

V A Kabirov[1,4], N P Vintonyak[2], S S Tyunin[2], D B Borodin[2], V D Semenov[2] and Yu A Shinyakov[3]

[1]Engineering Center for Electronic Prototyping, Yaroslav-the-Wise Novgorod State University, ul. B. St. Petersburgskaya, 41 173003 Veliky Novgorod, Russia
[2]Department of Industrial Electronics, Tomsk State University of Control Systems and Radioelectronics, Russia, Tomsk Region, Tomsk, Lenina 40 634050
[3]Research Institute of Space Technologies, Tomsk State University of Control Systems and Radioelectronics, Russia, Tomsk Region, Tomsk, Lenina 40 634050

[4] E-mail: kva@vipelec.com

Abstract. The paper discusses the principle of construction of automatic control systems for the energy conversion complex of a spacecraft with a digital control system. It is shown that the proposed digital control system is able to solve all the tasks assigned to it. The results are tested on an experimental model.

1. Introduction

The modern level of energy-converting complexes (ECC) of spacecraft (SC) is characterized by the use of combined control systems (CCS), which include a digital part, providing general control and telemetry for ECC, and an analog part that implements high-speed feedback loops. Modern ECC provide minimum weight and dimensions, high reliability, accuracy and speed. However, the CCS, or rather its analog part, is subject to aging of elements, and it is difficult to implement complex adaptive control algorithms on it. The constant increase in the requirements for the ECC forces us to look for new solutions that will improve the listed quality indicators. One of these solutions, in our opinion, is the implementation of fully digital CS (DCS) of ECC.

Digital control systems (DCS), in which direct digital control of power keys from microcontrollers is implemented, are just beginning to be used in some new products [32]. This is due to the absence of publications and positive examples of implementation proving the necessity and expediency of transferring the ECC to the DCS. In addition, a new digital element base suitable for use in space radioelectronic equipment appears on the market with some lag.

It is known that direct replacement of an analog control system of a pulse converter with a digital one leads to a decrease in its dynamic characteristics and accuracy due to delays and a quantization effect associated with the operation of digital PWM, ADC and digital corrective element [25]. However, the DCS, in comparison with the analog system, has a number of advantages, which are noted in the works [25–28]. Comparison of analog and digital control systems of pulse converters has been given in [28]. Comparing them with combined control systems, we will present the comparison results in table 1.

Table 1. Comparison of analog, digital and combined control systems of pulse converters.

Characteristic	CS Type		
	Analogue CS	DCS	CCS
Overall dimensions	-	+	+/-
Programmability	-	+	+
Ability to change structure	-	+	+/-
Ability to change operation modes	-	+	+
Departure of parameters	-	+	+/-
Possibility of telemetry	-	+	+
Reliability	-	+	+/-
Accuracy	+	-	+
Speed performance	+	-	+
Complexity	+	-	+/-

We briefly comment on some of the characteristics given in table 1.

Programmability and the ability to change is a property of remotely changing the structure of the control system and the algorithm of operation without changing the hardware.

The ability to change modes of operation is a property of a remote change in the course of a laid-down algorithm of operation, for example, in flight in the event of emergency situations.

Departure of parameters – this characteristic is associated with changes in the parameters of elements over time. Analogue CS are strongly affected by aging parameters. Since CCS include the analog part, they are also subject to aging.

The possibility of telemetry is the ability to remotely monitor the operation of the CS.

For ECC of SC, the most important characteristics are: mass and dimensional parameters, the ability to change operating modes, the departure of parameters, the possibility of telemetry, reliability, accuracy and speed of CS. The ability to change the modes of operation and telemetry are fully possessed by the ECC implemented using the CCS. Improvement of the remaining characteristics can be achieved only with the use of the DCS.

Improving the weight and dimensions of the DCS is achieved due to a higher degree of integration of the digital element base with respect to analog one. The use of digital microcircuitries of a high degree of integration allows creating systems on a chip [29].

Modern spacecrafts have an active lifetime (ALT) up to 15 years. The departure of parameters during the aging of elements leads to a significant deterioration in the accuracy of measured parameters. For example, super-precision in terms of long-term stability, thin-film resistors C2-29C have a resistance stability ± 0.25% for 25,000 hours of operating time. Such resistors are used in measuring circuits and corrective links. In digital control systems, corrective links of feedback loops are implemented using digital signal processing methods and, as a result, have higher temporal stability indicators.

The accuracy of the digital control system is determined by the level of discretization of the analog signal. Modern high-speed ADCs, for example, ADC128S102QML-SP with a sequential approxima-

IOP Conf. Series: Materials Science and Engineering **656** (2019) 012022 doi:10.1088/1757-899X/656/1/012022

tion, have a digit capacity of up to 12 bits, a signal conversion rate of 1 MHz and a maximum conversion error of ± 2 LSB. The accuracy of such ADC is ± 0.05% of the full scale of measurement, which is five times as accurate as the stability of a single resistor. Therefore, it can be argued that the accuracy of the DCS for unattended systems with a large ALT is significantly higher than the accuracy of analogue and combined CS.

The reliability of the control system is determined by the failure rate of its constituent elements and the amount of failures. For example, the failure rate of resistor C2-29B is 20 × 10-9 1 / h, and the failure rate of programmable logic integrated circuits (PLIC) is in the range from 10.3 × 10-91 / h to 74.6 × 10-9 1 / h, depending on design standards CMOS technology [30]. Comparisons of these values show that the failure rate of a single resistor and an PLIC containing thousands of logic elements are comparable. Therefore, the implementation of CS on digital element base allows to reduce the number of discrete elements and, accordingly, to increase the reliability.

The implementation of control systems on a digital element base requires an analysis of the tasks of control systems in a selected ECC structure, such as uniform current division in energy-converting ECC modules, ECC reliability, ensuring the output voltage stabilization performance when developing disturbing influences, and ensuring the energy balance of primary sources and load with a minimum number of charge cycles / discharge of energy storage devices (batteries).

2. Structure of the DCS and the uniformity of the division of currents in the energy-converting modules of ECC

The increase in power of the power supply system (PSS) of spacecraft (SC) is conducted primarily due to the parallel connection of energy conversion modules (ECM) of the energy-converting complex (ECC), which is confirmed by block diagrams of ECC, leading manufacturers of platforms for spacecraft [1, 2, 3, 4,5]. The parallel inclusion of the ECM provides the possibility of technical implementation of the modular structure of the ECC, which allows to increase the efficiency and durability of the ECC. However, the implementation of, for example, a discharge device (DD) in the form of parallel-connected switching power supplies (modules), poses the task of uniformly distributing the current between the modules to ensure uniform thermal conditions, which ensures an increase in the lifetime of the switchgear.

The task of uniform current distribution between parallel-connected pulse converters can be solved in various ways, and is considered in the works [6–11, 24]. In [6], the most complete classification, review and comparison of methods for uniform current distribution between parallel-connected pulse converters is given. In [7], the issues of uniform current distribution are interpreted in terms of the foundations of the theory of electrical circuits and the theory of automatic control, which makes them physically more understandable and makes it possible to outline additional ways to solve them. Obviously, not all solutions considered in [6, 7] satisfy the requirements imposed on a spacecraft power supply system (SC PSS).

The analysis showed that, at present, for the SC ECC, two variants of control organization of parallel operating converters are most widely used. The first version of the organization of control in accordance with the work [6], corresponds to the class of combined systems: “Systems with a falling output characteristic, united by regulation on the external current loop” (Outer Loop Regulation & Voltage Droop Via Output Current) [8, 9, 10, 11, 24]. Let's name it for brevity ACS (automatic control system) with the falling characteristic presented in figure 1. These systems are also characterized by the adjustment of the reference voltage by signals from the internal and external current loops.

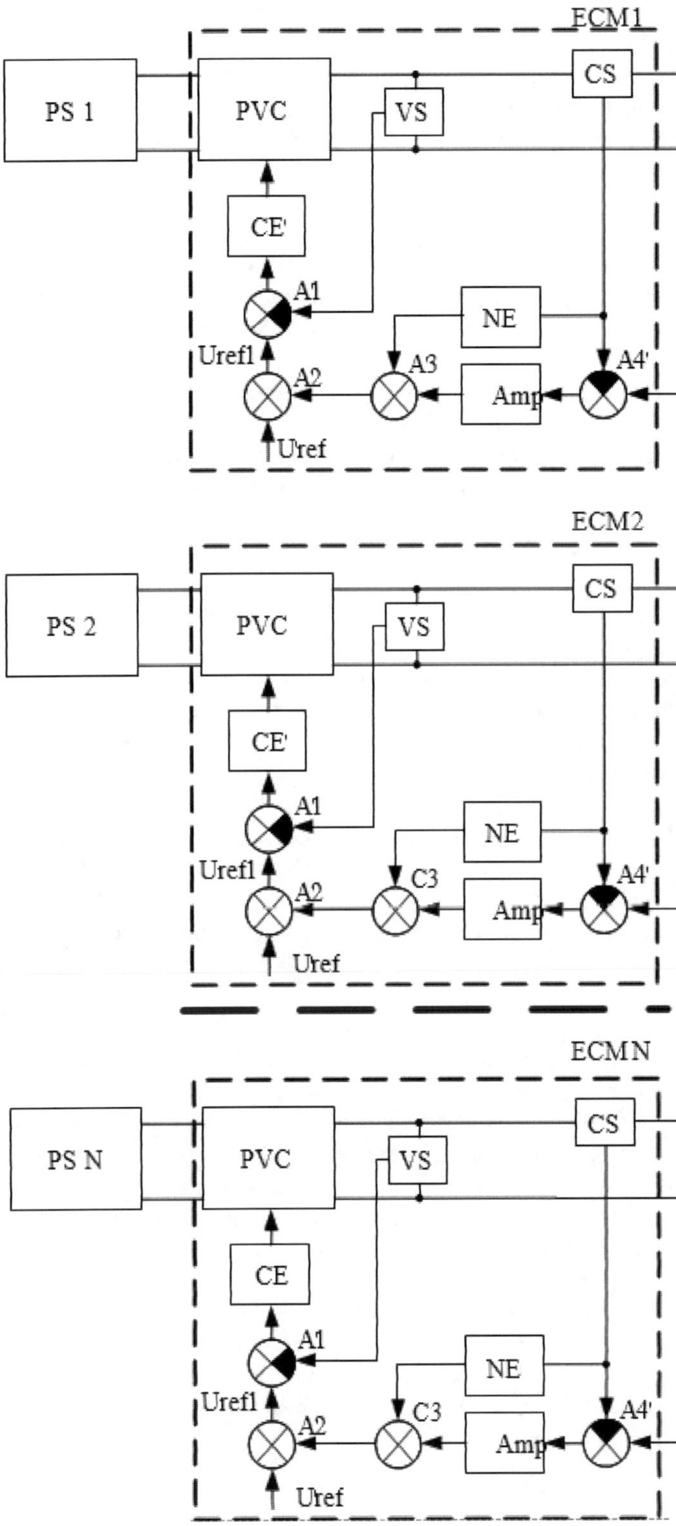

Figure 1. Structural scheme of the ACS with a falling characteristic.

IOP Conf. Series: Materials Science and Engineering **656** (2019) 012022 doi:10.1088/1757-899X/656/1/012022

ACS with the falling characteristic consists of N of energy-converting modules (ECM1-ECMN) connected by output in parallel to the load L via the CSS current sensor, the signal of which is reduced to the average current of one ECM. At the input, all the ECM are connected to their own power sources PS1 - PSN. In general, all the ECM at the input can be connected to one or several PSs, depending on their power. All energy-converting modules ECM1..ECM N, through the CSS current sensor are covered by common feedback on average current, which, together with all ECM, forms an external circuit for correcting the reference voltage. All energy-converting modules ECM1..ECM N, through the CSS current sensor are covered by common feedback on average current, which, together with all ECM, forms an external circuit for correcting the reference voltage. Each ECM has its own voltage feedback loop, an independent reference voltage Uref and an internal circuit for adjusting the reference voltage. The feedback loop for the voltage of each of the ECM includes a voltage sensor VS, an adder A1, a corrective short circuit element CE and a pulse voltage converter PVC. The internal circuit of the correction of the reference voltage is formed by the current sensor CS, the adder A4, the nonlinear element NE, the amplifier Amp, the adders A3, A2 and includes a voltage feedback loop subordinate to the internal circuit of the current reference voltage correction. The external circuit for the correction of the reference voltage includes the current sensor CSS and the internal circuits for the correction of the reference voltage for the current of all the ECMs, which are also subordinates to this circuit.

The internal circuit of the correction of the reference voltage allows to provide a falling output characteristic of an ECM with a short-circuit current limitation, and an external one allows the currents to be equalized between the energy-converting modules of the ECM1 .. ECM N.

The alignment of the output currents ECM1..ECMN by external circuit correction of the reference voltage is as follows. If the output current of a single ECM exceeds the average value of the load current, the reference voltage of the $Uref_1$ voltage stabilization circuit decreases, leading to a decrease in the output voltage and a decrease in its current.

Power sources with a falling output characteristic can work in parallel, the equality of the currents flowing through the sources will be determined by the slope of their output characteristics. The signal from the external current sensor CSS additionally ensures the equalization of the currents of each pulse converter.

The second version of the organization of control by parallel-working converters, according to [6], corresponds to the ACS class with a subordinate current loop (Inner Loop Regulation) [12], shown in figure 2.

The structure of the ACS with a subordinate current loop also consists of N of energy converting modules (ECM1..ECMN), connected in parallel to the load L. Each ECM is also powered by its own power source PS1 .. PSN. The parallel operation of pulsed power sources is provided by two feedback loops: internal current subordinate circuits for each ECM and external voltage loop. The internal sub-ordinate feedback loop for current is implemented by a current sensor CS, an adder A1, a correcting link for the current CLC, and a pulsed voltage converter PVC. The external feedback loop for voltage is implemented by the voltage sensor VS, the main error amplifier MEA, and a subordinate loop cur-rent of each of the ECM. The main error amplifier MEA consists of the adder A2, the correction link for the voltage CLV and the reference voltage source Uref.

Internal subordinates of current circuit transform each of the ECM into a controlled current source. The driving signal for the output voltage at the load L is the common reference voltage source U_{ref}, and the driving signal for the current I_3 of each PVC is the output signal of the correction link in the voltage CLV. The equality of the currents of all the ECM is ensured by the accuracy of stabilization of their internal current feedback loops.

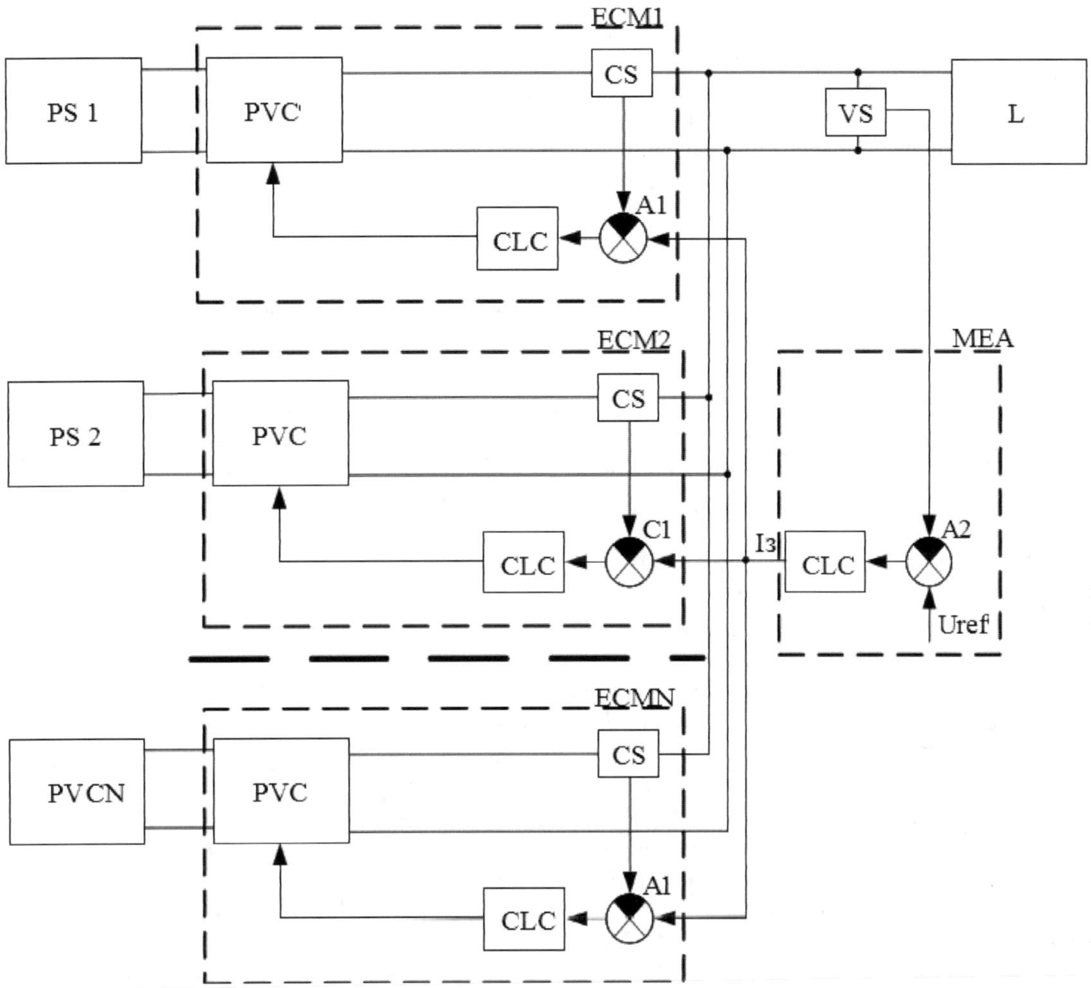

Figure 2. Structural scheme of the ACS with a subordinate current loop.

In [6], it is noted that the lack of a ACS with a falling characteristic is a possible instability in transient conditions and, because of this, the magnitude of the gain of the internal open-loop feedback circuit due to the voltage of the ECM is limited. This has a negative effect on the accuracy of stabilization of the output voltage of an ECM when exposed to any of the disturbing factors. In these structures, the accuracy of the division of currents between the modules is in conflict with the accuracy of the stabilization of the output voltage.

With high demands on the accuracy of the output voltage, a definite preference is given to ACS with a subordinate current loop. The issues of ensuring the accuracy and dynamics of the ACS with a subordinate current loop are the subject of numerous studies, e.g. [13,14].

3. Structure of DCS and ensuring the reliability of ECC

The disadvantages of ACS with a subordinate current loop are low structural reliability, which is explained by the fact that in this structure the voltage feedback channel, which includes the voltage sensor VS and the main error amplifier MEA, is common to the entire system. The failure of any of the elements of this channel leads to the failure of the entire system.

The main disadvantage of ACS with a subordinate current loop is eliminated by the implementation of multi-channel negative voltage feedback. To do this, all the channels, each of which includes its own voltage sensor and the main error amplifier MEA are connected to the input in parallel. Their outputs are connected to the "majority" element (ME) [15] with the required number of inputs, which selects one signal from three (all) signals of each of the negative voltage feedback channels and transmits this signal to all the ECM. In this case, the common element of the entire system becomes the ME, which should have a high degree of reliability.

In foreign literature, the unit combining all channels of voltage feedback, called the Main Error Amplifier (MEA) while the ME is not included in this unit. The algorithm for selecting the ME signal (one of three or more) and its scheme in the ECC review are not discovered.

In work [16], four algorithms for the selection of the ME signal are distinguished: the majority choice, the choice by the median, the choice by the mean value, and the choice by the principle of pluralism.

Majority choice of a signal corresponds to the algorithm of choosing a signal by majority vote.

The choice of the signal by the median corresponds to the algorithm for sorting the values of the input signals in ascending order and choosing such an input signal whose value will be in the middle.

The choice of the signal by the mean value corresponds to the algorithm for determining the average value from the values of all input signals and the choice of such an input signal, the value of which has the minimum deviation from this average value.

The choice based on the principle of pluralism corresponds to the algorithm for choosing such an input signal, the value of which was most often encountered during the previous operating time of the ME. For a single sample of input signals, the algorithm of choice according to the principle of pluralism will give the same result with the majority choice, for many samples the result will be different. In [16], it is argued that the algorithm of choice on the basis of pluralism has the highest probability of choosing the right value from a set of samples. In this case, the sample is understood as the values of analog signals at a particular point in time. Its disadvantage is the need to memorize the prehistory of the samples and complex technical implementation.

Of the four algorithms considered for selecting one signal from among a variety of input analog signals, in our opinion, only two of them are suitable for the structure of spacecraft power supply system - this is the choice by the median and the choice by the mean value.

The implementation of the selection algorithm by the median at the level of principal schemes is presented in the works [17, 18, 20, 21, 22, 23].

Thus, in the ECC, which is presented in [15], the only non-reserved functional unit is the ME, which reduces the reliability of the ECC control system. At the same time, in the work [15] the possible technical solutions for the reservation of this ME are not indicated.

One of the options for improving the reliability of the ECC control system, due to the degree of reservation of the ME, is to increase the number of MEs to the number of modules connected in parallel, and to construct the MEs in each of the modules. In this case, the reliability requirements for the ME will be significantly reduced. The proposed version of the control system with the reservation of the ME is presented in figure 3.

The ACS scheme with the reserved ME repeats the ACS structural scheme with a subordinate current loop. The difference is that in each of the energy-converting modules of ECM1..ECMN, the majority element ME is included, and the total voltage feedback is implemented by the three main error amplifiers MEA1..MEA3. It should be noted that in each MEA there is its own reference voltage source U_{ref}, and the variation in the magnitude of this voltage will affect the divergence of the MEA output signals during their operation.

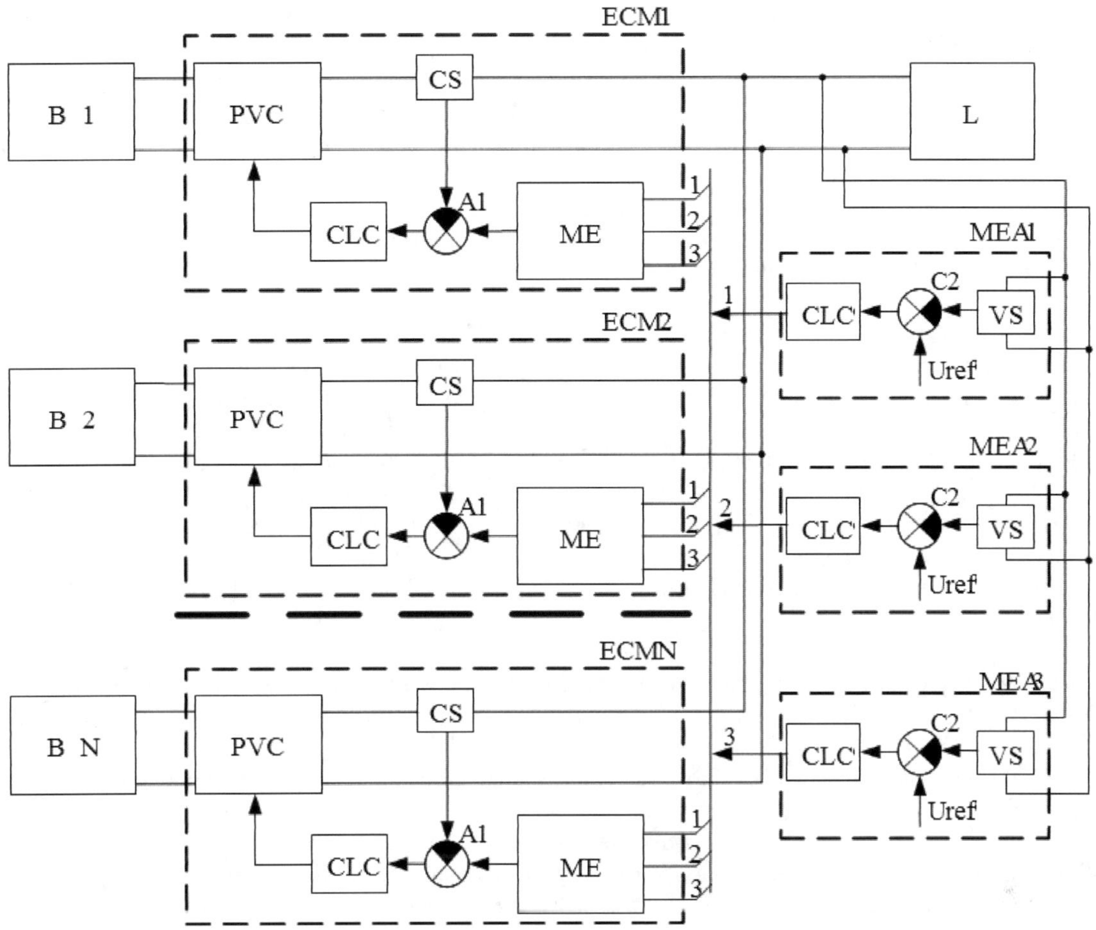

Figure 3. Structure of scheme ACS with reversed ME.

4. DCS structure and speed

The decrease in the dynamic characteristics of the stabilization circuit of the ACS, due to the transition to digital control systems, can be compensated for by the simpler implementation of adaptive control algorithms in the DCS [33].

In paper [31], one adaptation algorithm is proposed, which assumes stabilization of the loop gain coefficient of the open-loop circuit for the voltage of the ECC which varies, depending on the number of concurrently working modules of the discharge devices of the ECC. The positive influence of the chosen adaptation algorithm on the dynamic characteristics of the converter and the output impedance is noted. It is shown that the output impedance of the system does not change when the number of working modules of the discharge devices changes.

However, the loop gain coefficient depends not only on the number of working modules of the discharge device in the ECC, but also on the parameters of the primary power sources. For example, the coefficient of loop amplification of an open circuit by voltage when operating on the current branch of a solar battery (SB) is proportional to the SB current. And the coefficient of loop amplification of the subordinate current circuit is proportional to the voltage of the battery (B). These parameters of power sources and energy storage devices in the course of work vary over a wide range from 0 to I_{max} for SB, and from $U_{\text{minАБ}}$ to $U_{maxАБ}$ for B. Accordingly, a change in these parameters also leads to a significant change in the loop gain factors, and hence to a change in the dynamic properties of the system. At the same time, the corrective links of the analog and combined control systems ensuring the stability of the automatic control system are adjusted to the mode of the SB maximum current and the maximum voltage of the B and have the maximum speed and the specified stability margin. By reducing these characteristics, the performance of the ACS is reduced and a decrease in the stability margin is possible.

The implementation of the stabilization algorithms for loop gains in the DCS will allow us to obtain the dynamic characteristics of the ECC, which in some modes are no worse than those of analogue CSs, while in others they are significantly superior.

5. ECC structure and energy balance

One of the main tasks of the control system of ECC is to maintain the energy balance of primary sources and loads with a minimum number of charge / discharge cycles of the battery (B). At the same time, the ECC should maximize the energy generated by the solar battery (SB), transferring it to the load and to the B, in order to maintain the B in the maximum charged state. And only if the power of the load exceeds the power of the SB, the battery B can be discharged. This task is performed by the zone control method based on the error signal (the foreign term "Three-Domain Control"), which is implemented by a buffer amplifier [19].

Figure 4 shows the structural scheme of the ACS of ECC with DCS, which provides a solution to all the problems considered.

The proposed ACS of ECC with DCS is composed of N of modules of voltage stabilization MVS1..MVSN, five-channel digital controle module DCM, power sources SB1..SBN, energy storages B1..BN, five-channel digital communication interface CI which connects all control modules of ACS.

Each MVS includes: bit channel (BC), charger channel (CC) and energy conversion channel of SB (CSB), majority element ME with five inputs, and five-channel digital communication interface receiver (DCI).

All energy conversion channels consist of a pulse voltage converter (PVC) and buffer amplifier (BA). Energy conversion channels CC and CB, besides a buffer amplifier and PVC, also have their current sensor (CS), adder A and correcting link for the current CLC, implementing the subordinate charge / discharge current control loop of the B.

The function of reserved MEA is performed by five channels of digital control module (CDCM1..CDCM5). Each channel of DCM (CDCM1.. CDCM5) consists of a voltage sensor VS, analog-to-digital converter ADC, adder A, voltage reference source Uref, correction link for the voltage CLV and a single channel digital communication interface transmitter (DCI).

The operation principle of ACS of ECC with DCS is as follows. Output voltage of ECC is transmitted to the voltage sensor into all five channels of DCM (CDCM1...CDCM5). Further, this voltage is transformed by the ADC into a numeric form. Each channel of DCM independently calculates the required signal of control action (SCA1..SCA5) in the numeric form and transmits it through the transmitter DCI via five-channel communication interface CI into all receivers of DCI of all MVS. The DCI transmitters add a checksum to the packets of transmitting SCA to eliminate errors during data transmission. Signals of control action (SCA1..SCA5) are received by the DCI receivers of each MVS and transmitted to the inputs of the majority element ME, which selects one of the incoming SCAs and transmits it to the buffer amplifiers of all the MVS energy conversion channels. Buffer amplifiers of energy conversion channels determine the sequence of operation of these channels, realizing the zone principle of control by the error signal.

The control algorithms incorporated in the buffer amplifiers guarantee the absence of the mode of simultaneous operation of the charger channel (CC) and the bit channel (BC), by the fact that these channels operate alternately. This ensures the minimum number of charge / discharge cycles of the battery (B) and the maximum efficiency of using the energy generated by the SB.

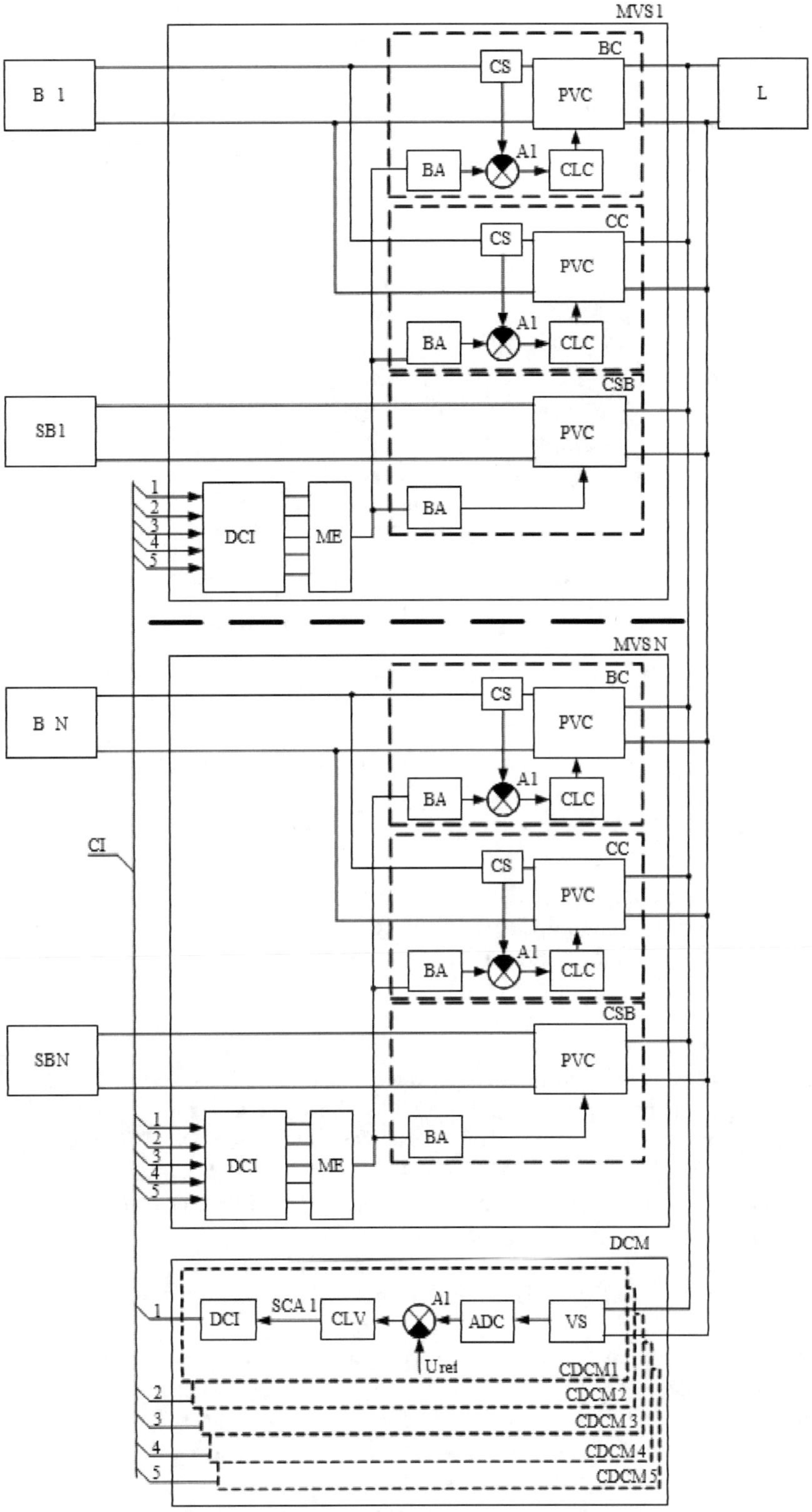

Figure 4. Structural scheme of ACS of ECC with DCS.

In the particular case, all ECC modules can be implemented in separate structural units, which makes it easy to increase the output power and increase the degree of reservation of the ECC. In this case, all the blocks designed to control all the pulse voltage converters of all the MVS channels can be implemented in a single digital integrated circuit of the large integrated circuit (LIC), programmable logic integrated circuits (PLIC) or on a specialized microcontroller. Such a microcircuit will play the role of the MVS control system.

Figure 5 shows the model of the ECC with two energy-converting modules MVS and a five-channel digital control module DCM.

Figure 5. Model of ECC with DCS.

6. Conclusions

– The proposed structural scheme of ACS of ECC with DCS and its implementation in the model showed the possibility of using digital control systems in the ECC.

– The use of digital control systems fundamentally allows to solve all the tasks imposed on the SC ECC.

– The undoubted advantage of digital control systems is the possibility of implementing complex corrective links that adapt to external disturbing influences due to changes in illumination and SB degradation, voltage changes on the battery (B), as well as a wide range of load changes and failure of both ECC modules and other nodes of power supply system of SC.

References

[1] Chris Greenwood, Steve Lenhart, Bruce Inenaga, Cacey Jennings, Aaron Mendelsohn, Michael Staley and Robert Vaughan 2002 SS/L'S Superpower subsystem development and application on the 1300 family of spacecraft *20th AIAA International Communication Satellite Systems Conference and Exhibit*

[2] Abbas A Salim In-Orbit Performance of Lockheed Martin's Electrical Power Subsystem for A2100 Communication Satellite" *AIAA-2000-2809*

[3] Garrigos A, Carrasco J A, Blanes J M and Sanchis-Kilders E 2006 A power conditioning unit for high power geo satellites based on the sequential switching shunt series regulator *IEEE MELECON Electrotechnical Conference* pp 1186–1189

[4] Ich D Ngo 2009 Evolution of Solar Array Shunt Regulators for Boeing Satellites *7th International Energy Conversion Engineering Conference*

[5] Soubrier L 2011 Trehet High Power PCU for Alphabus: PSR100V *Proceedings of the 9th European Space Power Conference* (France, Saint Raphael) ESA SP-690

[6] Shiguo Luo, Zhihong Ye, Ray-Lee Lin and Fred C and Lee A 1999Classification and Evaluation of Paralleling Methods for Power Supply Modules *IEEE Power Electron. Specialists Conf.* pp 901–908

[7] Yuehui Huang and Chi K Tse 2007 Circuit Theoretic Classification of Parallel Connected DC–

DC Converters *IEEE Transactions on Circuits and Systems I*: Reg. Papers Vol 54 **5** pp 1099–1108
[8] Chernyshev A I, Kazantsev Yu M, Patlakhov E N 1982 *Multichannel power supply system with uniform current distribution* (Certificate of authorship 985773 of the USSR) Bul **48**
[9] Gordeev K G, Shinyakov Yu A, Cherdantsev S P and Elman V O 1986 *Multichannel power supply system with uniform current distribution (*Certificate of authorship 1206763 of the USSR) Bul **3**
[10] Emelyanov P, Merunko A V, Manernov B A, Elman V O and Lenshin A V 1990 *Multichannel power supply system with uniform current distribution (*Certificate of authorship 1534450 of the USSR) Bul **1**
[11] Emelyanov P, Merunko A V, Manernov B A, Chernyshev A I and Elman V O 1991 *Multichannel power supply system with uniform current distribution* (Certificate of authorship 1658136 of the USSR) Bul **23**
[12] Soubrier L, Besdel P, Daubresse T and Trehet E 2008 High performance BDR for the PCU of Alphabus *Proc. of the '8th European Space Power Conference'* (Germany, Constance) ESA **SP-661**
[13] Ming Fu and Donglai Zhang 2017 Design and Analysis of a Conductance Compensator for Keeping Constant Bandwidth and Output Impedance in Average Current Mode Control *IEEE Transactions on Power Electronics* Vol 32 **1**
[14] Choi B 1998 Comparative Study on Paralleling Schemes of Converter Modules for Distributed Power Applications *IEEE Transactions on Industrial Electronics* Vol 45 **2**
[15] Otero J, Lapena E and Simon P 2011 Modular PCDU product line for LEO applications *Proc. '9th European Space Power Conference'* (France, Saint Raphaël) ESA **SP-690**
[16] Blough D M and Sullivan G F 1990 A Comparison of Voting Strategies for Fault-Tolerant Distributed Systems *IEEE symposium on Reliable Distributed Systems* pp 136–145
[17] Wright W H, Verdes P, Biess J J, Park C 1971 *Shunt regulation electric power system* (Patent 3600599 USA)
[18] Namazi A, Askari S and Nourani M 2008 Highly reliable A/D converter using analog voting *Proc. of Intern. Conf. on Comp. Design (ICCD)* (Lake Tahoe, A) pp 334–339
[19] Capel A, O'Sullivan D and Marpinard J.-C. 1988 High-power conditioning for space applications *Proc. of the IEEE* Vol 76 **4** pp 391
[20] Kozlov R A, Kucherov V K and Tatarinov K A 1982 *Three-channel analog majority element* (Certificate of authorship 3243815 of the USSR) Bul **35**
[21] Andreev D V 2003 The median identifier (Patent 2204164 of the RF) Bul **13**
[22] Terekhin B Yu, Vihareva T I and Koldin V N 1985 *Three-channel majority amplifier* (Certificate of authorship 3691585 of the USSR) Bul **30**
[23] Dorofeev V B 1999 *Majority device* (Patent 2141130 of the RF)
[24] Sustin B P, Ivanchura V I, Chernyshev A I and Islyaev Sh N 1994 *Spacecraft power supply systems* (Novosibirsk: VO "Nauka", Siberian Publishing Company) p 318
[25] Radu Etz, Stefan R Daraban, Dorin M Petreus and Adina R Rusu 2010 A Comparison between Digital and Analog Control for a Buck Converter *IEEE Electronics Technology International Spring Seminar* (Warsaw, May 2010)
[26] Maksimovic D, Zane R and Erickson R 2004 Impact of Digital Control in Power Electronics *Proc. of ISPSD* pp 13–22
[27] Yan-Fei Liu and P C 2005 Sen Digital Control of Switching Power Converters *IEEE Conference on Control Applications Toronto*
[28] Ahmad H and Bakkaloglu B 2008 A DC-DC digitally controlled buck regulator utilizing first-order Σ-Δ frequency discriminators *Conf. Proc. IEEE Applied Power Electronics Conference and Exposition APEC*, art. **4522745** pp 346–352
[29] Shagurin I 2009 Systems on a chip. Features of implementation and development prospects *Electronic components* (Electronica Publishing House) **1**

[30] Stroganov A 2007 Assessment of the durability of LSI based on the results of accelerated testing *Technologies in the electronics industry* **3**
[31] Ming Fu, Donglai Zhang and Tiecai Li 2017 Design and Analysis of a Conductance Compensator for Keeping Constant Bandwidth and Output Impedance in Average Current Mode Control *IEEE Transactions on Power Electronics* Vol 32 **1**
[32] Romanenko A S 2013 *Redundant equipment for regulation and control of the power supply system of a small spacecraft* Dis. Cand. Techic. Sciences (Tomsk: TUSUR Publ) p 197
[33] Lopatin A A 2006 *Indirect adaptive control of a DC drive:* Dis. Cand. Techic. Sciences (Krasnoyarsk: KGTU Publ)

Morphology of temperature field of heat pipe in conditions of photon

V Karachinov[1,2], D Petrov[1] and D Evstigneev[1]

[1]Yaroslav-the-Wise Novgorod State University, ul. B. St. Petersburgskaya, 41 173003 Veliky Novgorod, Russia

[2]E-mail: Polnovo@yandex.ru

Abstract. Within the framework of computer experiment with the use of finite element method the technique of visualization of nonstationary temperature field morphology in the design of heat pipe having a symmetrical structure in the presence of localized heat flow source of a certain shape is developed. The features of the structure of isothermal surfaces and lines are visualized and investigated for the uniform starting mode model in the range of power values of the heat flux source induced by photon heating P = 0 ÷ 5W on both sides of the center of the pipe.

1. Introduction

Photon heating, as a physical phenomenon, represent certain scientific and practical interest among researchers engaged in the development of new diagnostic methods in various industries and medicine, as well as technologies that protect objects, for example, from exposure to laser radiation [1–4]. The known technological advantages of photon heating, including varying the size and shape of the heating zone, obtaining high energy density (power), the demanded monochromatic radiation and the simple implementation of the “flying spot” principle have provided it with wide application in the processes of modifying materials and non-destructive thermal control of layered, composite, cellular and other products [1, 5–9]. Highest rates of informativeness and reliability of the new method of quality control of containment with localized phase transitions, including heat pipes (HP), through the use of non-contact optical methods of heat supply and temperature measurement [10, 11].

Currently, HPs are products of mass production and they are widely used in special cooling systems for various electronic means, including computer technology [12]. In this regard, high requirements are imposed on their reliability, which stimulate the development of methods for studying thermal modes of HP. However, in the literature devoted to research in the field of design, testing and application of HPs, there is no information on the morphology of temperature fields during photon heating of HPs [13].

Therefore, this paper presents the results of a study aimed at creating a method of computerized visualization of the morphology of a nonstationary thermal field of TTs under the conditions of using photonic heating of HPs with a symmetric structure.

2. Research methods

When developing the technique, was chosen the flat aluminum HP with capillary-porous structure of which was implemented as a net fuze of aluminum wire as the basic design. Ammonia served as a coolant. The dimensions of the geometric model of HP were, respectively (figure 1): length L = 200mm; width B = 20 mm; thickness H = 4 mm, and the shell wall thickness was assumed to be 0.7 mm [12].

Figure 1b shows the structure of the thermal model of the HP. The central location of a flat circular heat source (HS) (2), induced by external luminous flux, allows using known heat transfer mechanisms to create reversible heat fluxes Q, which move in the material of the body, fuze, steam channel and cause a change in time of the corresponding temperatures (T_k, T_f, T_p). Thus, a diagnostic signal is formed within the surface of the HP casing in the form of thermal contrasts, carrying information about morphological changes in temperature fields.

When stating the problem of analyzing unsteady heat transfer in the axial section of the "surface heat dissipation – heat pipe" system, the following basic assumptions were made:

1) The thermophysical characteristics of the medium in the area of the fuze HP are calculated as effective taking into account the volume fractions of each component. Such approach is widely used in solving problems in porous media and has performed well in calculating the thermal conductivity of fairly complex systems. At the same time, the effective thermal conductivity coefficient (λE) remains constant both for the condensation zone and for the TT evaporation zone [12, 14]: A subsection
Some text.

$$\lambda_E = \lambda_L \frac{\lambda_L+\lambda_S-(1-\varepsilon)(\lambda_L-\lambda_S)}{\lambda_L+\lambda_S+(1-\varepsilon)(\lambda_L-\lambda_S)} \quad (1)$$

where $\lambda_L$and λ_S - coefficient of thermal conductivity of the liquid and the fuze skeleton; ε - fuze porosity (for aluminum mesh is 0.6 [14]).

The thermophysical characteristics of the materials and elements of the heat pipe are given in table 1.

Table 1. Property of material.

System element		Material or substance	Density, ρ, kg/m^3	Specific heat, C, J/(kg·K)	Thermal conductivity, λ, W/(m·K)
Case		Aluminum	2689	951	237,5
Fuze	Capillary structure	Aluminum	1691,5	2484	60,25
	Refrigerant	Ammonia			
Refrigerant		Ammonia	0,8	2160	0,022

2) We consider that the processes of heat release in the condensation zone do not directly affect the formation of the temperature field in the evaporation zone. In this case, the dynamics of heat transfer in the HP is determined by the process of thermal conductivity in the wall of the body and the fuze, as well as their heat capacity. Moreover, the coolant vapor (ammonia) is an ideal thermal bond, the heat capacity of which is close to zero.

3) HP is considered as a system of bodies interconnected by an ideal thermal bond, which is in heat exchange with the environment and instantaneous regularization of the thermal regime. Then, to calculate the non-stationary thermal mode of the HP, the known dependence of the rate of temperature change for the regular mode of the system of bodies can be used [15]:

$$m = \psi_C \frac{S_C\alpha_C}{C_{HP}} \quad (2)$$

where ψ_C - coefficient of uneven temperature of the heat removal zone; C_{HP} - heat capacity of the heat pipe, j / c; S_C - area of the condensation zone m^2; α_C - heat transfer coefficient of the condensation zone W / (m · K). The accepted assumptions do not impose fundamental restrictions on the generality of the problem statement and reflect a rather real operating mode of the heat pipe.

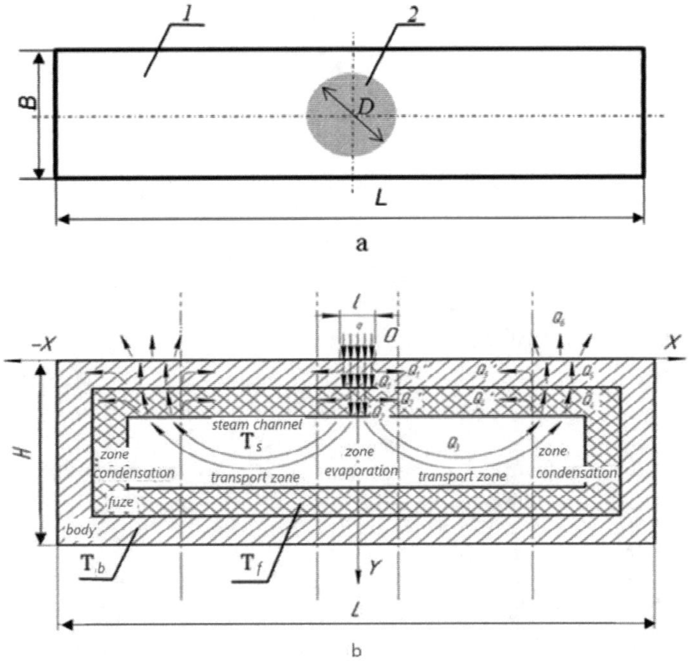

Figure 1. Geometric (a) and thermal (section along the axis) (b) HP type models with a round HS. 1 – Case; 2 – HS.

It should be noted that free (natural) convection at normal pressure and thermal radiation [14, 15] were chosen as the main mechanisms for dissipating thermal energy from the surface of a horizontally oriented HP body to an air (gas) medium with temperature Tc in the thermal model. When constructing a mathematical model, the non-stationary heat conduction problem in a plane-parallel formulation was considered. For this case, using the original heat conduction equation (3) [15]:

$$C\rho\frac{\partial T}{\partial t}=\lambda_x\frac{\partial^2 T}{\partial x^2}+\lambda_y\frac{\partial^2 T}{\partial y^2}+\lambda_z\frac{\partial^2 T}{\partial z^2}+q_0\cdot \tag{3}$$

The system of equations (4) was compiled describing the heat transfer in the fuze, the steam channel and the body of the HP [13,15]:

$$\left.\begin{aligned}
&C_1\rho_1\cdot\frac{\partial T_1}{\partial t}=\lambda_1\left(\frac{\partial^2 T_1}{\partial x^2}+\frac{\partial^2 T_1}{\partial y^2}\right);\quad 0<x<(L-\delta_{ct}-\delta_f);\ (\delta_{ct}+\delta_f)<y<(H-\delta_{ct}-\delta_f)\\
&C_2\rho_2\cdot\frac{\partial T_2}{\partial t}=\lambda_2\left(\frac{\partial^2 T_2}{\partial x^2}+\frac{\partial^2 T_2}{\partial y^2}\right);\ 0<<(L-\delta_{ct});\ \ (\delta_{ct})<y<(H-\delta_{ct})\\
&C_3\rho_3\cdot\frac{\partial T_3}{\partial t}=\lambda_3\left(\frac{\partial^2 T_3}{\partial x^2}+\frac{\partial^2 T_3}{\partial y^2}\right);\quad 0<<L\ ;\ 0<y<\delta_{ct}
\end{aligned}\right\} \tag{4}$$

where C – isobaric heat capacity, J/(kg · K); T – temperature, K; x, y – coordinates, m; λ – coefficient of thermal conductivity W/(m · K); ρ – density, kg/m^3; t – time, s; L – heat pipe length, m; H – height of the heat pipe, m; δ_{wp} – wall thickness of the heat pipe, m; δ_f – fuze thickness, m. The indices in (4) correspond to the following bodies: 1 – steam channel; 2 – fuze; 3 – heat pipe body.

For the developed thermal model under conditions of uniform start-up of HP's, the following initial and boundary conditions were used [14, 15]:

1) at the initial moment of time for the edges of all bodies included in the model, the temperature was fixed:

$$T_{i,t=0} = T_C = \text{const}; \quad (5)$$

2) for all HS edges, taking into account the surface isothermality, a condition of the first kind was set:

$$T = T_s; \quad (6)$$

3) for the inner ribs forming the vapor channel, the fuze and the body, the boundary conditions of the second and fourth kind were set:

$$\frac{\partial T_i}{\partial n} = \psi_i \quad (7)$$

$$-\lambda j \cdot \frac{\partial Ti}{\partial y} = -\lambda_2 \cdot \frac{\partial T_3}{\partial y}; \quad (8)$$

4) on the external fins (casing) of the HP model, a condition of the third kind was defined, which describes both convective and radiant heat exchange with the environment:

$$\lambda_p \frac{dT}{dn} = -\alpha_\kappa(T - T_c) - \beta(T^4 - T_c^4) \quad (9)$$

where β – a value equal to the product of the Stefan-Boltzmann constant ($\sigma_0 = 5.7 \cdot 10–8\ W / m^2K^4$) and the coefficient of radiation of the surface of the probe material; α_κ – convection heat transfer coefficient. The numerical values of the coefficient α_κ were obtained from preliminary calculations using the well-known similarity method [15].

The removal of heat flux through conductive connections of HP fasteners was not taken into account. The system of equations (4–9) was solved by a numerical method (finite element method) on a PC. Using the software "ANSYS 13.0" [15], a series of calculations was performed, which made it possible to estimate the degree of influence of the power of the induced HS of round shape on the field characteristics of the thermal model of the HP. The grid step was chosen optimal, based on the available resources of the PC, and was 1.4 mm for the whole model and 0.5 mm in the field of HS. Studies were performed at an ambient temperature of TC = 300K, the duration of the HS was 120 s, and its power changed in the range of values of P = 0 ÷ 5 W.

3. Research results and discussion

It is known that most HPs are designed for isothermal operation at a certain temperature and transmitted power [12–14]. At the same time, in practical use, virtually every HP goes through a starting mode and its duration, as well as the nature of the temperature field change depends on the design features of the pipe, on the methods of heat supply and cooling, and on the time variation of the input power.

The research results showed that when using the model of uniform starting mode HP, the nature of the change in the maximum surface temperature in the interval of time values 0 ÷ 120s in the area of localization of HS is close to linear (figure 2). This character of dependences is preserved with an increase in the values of the power induced by radiation on the surface of the HS HP. This, in turn, made it possible, during a computer experiment, to study the conditions for removing the HP to the operating mode, the nature of the temperature distribution along the pipe, the factors that impede the launch, and others. Thus, using the method of field characteristics [16], the morphology features of the nonstationary temperature field of HPs with an HS of circular shape were visualized. Typical examples of simulation results are shown in figure 3.

Their analysis showed that in the thermogram mode (figure 3a) for a fixed point in time, the morphology of the HP temperature field on either side of the HS is represented by a system of symmetric isothermal surfaces (zones). At the same time, it should be noted that only near the HS, the shape of the zone quite well repeats the shape of the round HS.

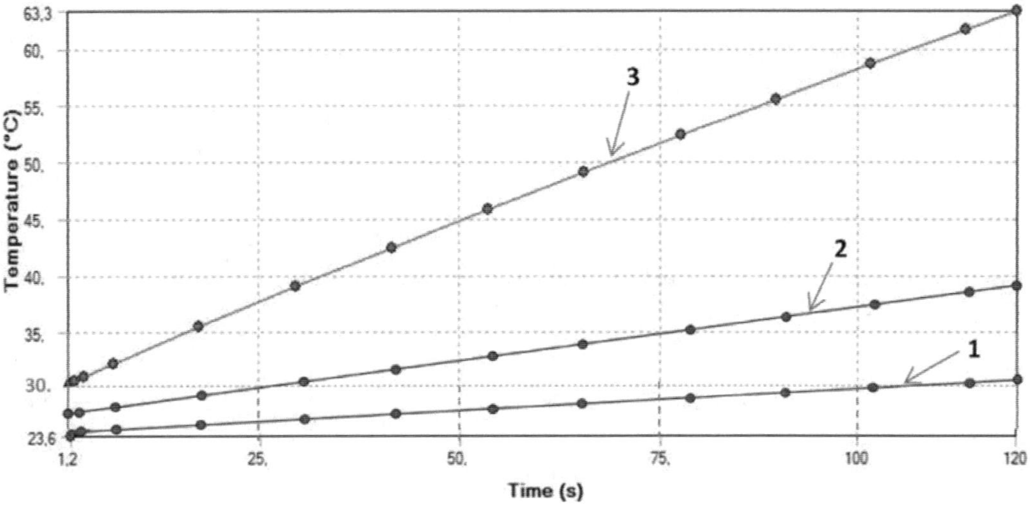

Figure 2. The change in the maximum surface temperature of ammonia HP in time. Surface HS power: 1 – 1W; 2 –2W; 3 – 5W.

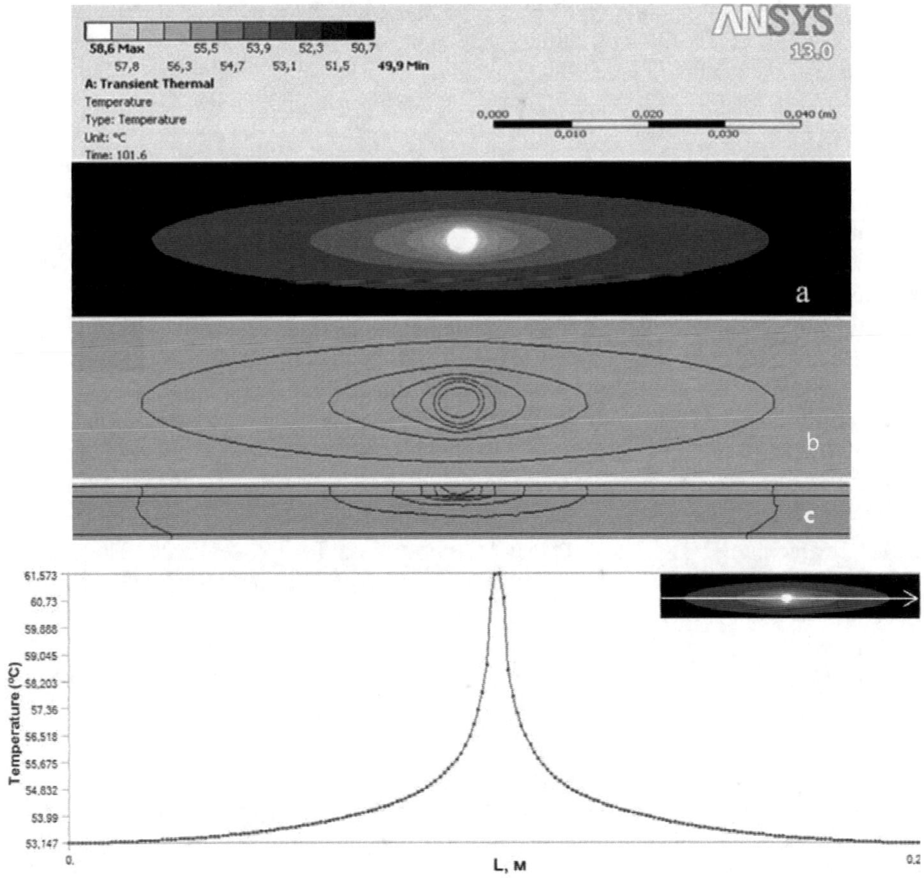

Figure 3. Morphology of the temperature field in ammonia HP. Input data: HS power P = 5 W; a – thermogram mode; b, c – contour mode; g – mode selected circuit (along the axis of the HP).

As they move away from HS, the shapes of the zones acquire an almost equally curved convex profile, which is formed due to the external heat exchange of the HP surface [13, 14]. Additional calculations showed that with increasing HS power, the length of the zones only increased, and their morphological features did not undergo significant changes. It is quite clearly seen that in the isoline mode (figure 3b), the isotherms continuously repeat the boundaries of the corresponding zones. In the course of the research, the effect of the front lag of the corresponding isotherm was found on the opposite to the HS surface of the HP (figure 3c). This effect, in our opinion, is due to the peculiarities of the transport of heat flow from the HS in the transverse direction of the HP [12, 14].

The nature of the temperature change along the length of the HP (along the selected contour) reflects quite well the graph in figure 3d. In the absence of defects specially introduced into the HP model, it has strict symmetry about the center of the pipe.

4. Conclusion

According to the computer experiment, in the model of a uniform start-up of ammonia HP, the shape of isothermal lines near the induced HS radiation reproduces the shape of this source quite well, and the effect of the front of the corresponding isotherm on the opposite to HS surface of the HP occurs.

Reference

[1] Rykalin N, Angles A, Zuev I and Kokora A 1985 *Laser and electron beam processing of metals* (Moscow: Mechanical Engineering Publ) p 496
[2] Napartovich A 1991 *Handbook of laser technology* (Moscow: Energoatomizdat Publ) p 544
[3] Libenson M 2007 *Laser-induced optical and thermal processes in condensed media and their mutual influence* (SPb: Science Publ) p 423
[4] Eichler U and Eichler G 2008 *Lasers. Execution, management, application* (Moscow: Technosphere Publ) p 440
[5] Karachinov V, Votinov M, Emelyanova G and Seleznev B 1982 *The Effect of Laser Radiation on the Characteristics of the Si-SiO2 System* (*III-All-Union Conference "The Effects of Ionizing Radiation and Light on Heterogeneous Systems" Proc.* (Kemerovo) p 183
[6] Dvurechensky A, Kachurin G, Nidaev E and Smirnov L *1982 Impulse annealing of semiconductor materials* (Moscow: Science Publ) p 208
[7] Djibouti Z 2006 *The mechanisms of pulsed photon annealing in semiconductors with covalent and mixed bonds: abstract of thesis* (Moscow) p 74
[8] Nosov P, Shirankov A, Tretyakov R and Grigoriants A 2016 Heating of optical elements from high-purity quartz glasses by radiation from high-power fiber lasers *Instrument making* Vol 59 **12** 1028-1033
[9] Vavilov V 2012 *Infrared thermography and thermal control* (Moscow: Spectrum Publ) p 570
[10] Karachinov V *Method of heat pipe quality control RF* Patent No. 2456524 of the IPC F28D 15/02
[11] Petrov A, Karachinov V, Kiliba Y, Petrov D, Ionov A and Evstigneev D 2018 *TV quality control method heat pipes from the side of the "shadow" surface* (Minsk, Belarus) p 198–201
[12] Karachinov V 2007 *Protection of RES in extreme conditions. Special cooling devices* (Veliky Novgorod: NovSU) p 176
[13] Lux A and Matveev A 2007 Investigation of high-performance ammonia heat pipes of energy-saving thermal control systems for large-sized spacecraft structures *Samara State University Bulletin. Natural science series* Vol 6 401–418
[14] Dan P and Ray D 1979 *Heat pipes* (Moscow: Energy Publ) p 272
[15] Lykov A 1978 *Heat and mass transfer* (Moscow: Energy Publ) p 480
[16] Chigarev A, Kravchuk A and Smalluk A 2004 *ANSYS for engineers: Ref. allowance* (Moscow: Mashinostroenie-1 Publ) p 512

The influence of the temperature conversion rate on the deformation processes of the shape memory effect and transformation plasticity

E A Khlopkov[1,2,7], P N Zhurbenko[3], A A Tikhomirov[4], G A Volkov[5], G Z Zatulsky[6] and Yu N Vyunenko[4]

[1]Peter-the-Great Saint Petersburg Polytechnical University, Saint Petersburg, 195251, Russia
[2] Yaroslav-the-Wise Novgorod State University, Veliky Novgorod, 173003, Russia
[3] Joint-Stock Company «Proletariy», Surazh, 243500, Russia
[4] OOO «OPTIMIKST LTD», Saint Petersburg, 195426, Russia
[5] Saint Petersburg State University, Saint Petersburg, 198504, Russia
[6] Physics-technological Institute of Metals and Alloys of the National Academy of Science of Ukraine, Kiev, 03142, Ukraine

[7]E-mail: hlopkovelisey@mail.ru

Abstract. The influence of the temperature conversion speed on the mechanical characteristics of metals with shape memory is mentioned in connection with the dissipative properties of this materials class. This paper presents the different cooling and heating regimes research results of ring-shaped force elements and spiral drives demonstrating the possibility of this parameter influence on their deformation characteristics in the temperature range of martensitic transformation.

1. Introduction

In the study of the shape memory effect (SME) and plasticity of transformation (PT), the development of which is due to thermoelastic martensitic transformations, until recently did not notice a significant effect on the deformation properties of the rate of temperature change [1]. However, in the process of developing drives for the multiple operation of technological devices operating in the mode of the SME and reversible shape memory (RSM), noticeable changes in the characteristics of the shape change due to the heating and cooling modes were found.

Investigation of the dependence of the deformation phenomena of the SME on the heating rate was carried out on spiral drives made of CuZn18Al7 wt.% (figure 1, a) [2] and ring-shaped force beam elements (RFBE) from an alloy of TiNi55.4 wt.% (figure 1, b) [3]. The diameter of the copper-zinc-aluminum alloy wire was 4 mm. RFBE was made of a two-millimeter wire. The height of the spiral of twelve turns was ~ 50 mm under the condition that the turns completely fit. The ring of the power element of titanium nickelide consisted of five turns and its outer diameter was ~ 63 mm. Samples of CuZnAl drives were annealed at 1073 K for 30 min and then quenched in water. The outer diameter of the helix was ~ 20 mm. The drive from titanium nickelide was annealed for 20 min at ~ 800 K and quenched in water.

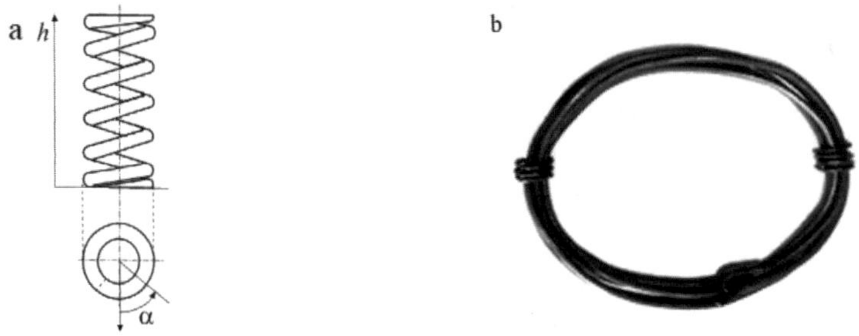

Figure 1. The appearance of the spiral (a) and RFBE (b).

The dependence of the deformation effects of the plasticity of the transformation on the change in cooling modes was studied on an ring-shaped force element made of a TiNi55.4 wt.% alloy from a two-millimeter wire. The ring consisted of one coil and had the shape of an elongated oval. The sample was annealed at 773 K for 20 min. Then, it was subjected to a three-fold thermal cycling with the transfer of material from the fully martensitic state to the austenitic state and back.

2. Results and discussion

When the helix was heated, a change in its height and rotation of the upper turn relative to the base was observed. Figure 2, a shows the temperature dependence of the height of the spiral. When heated from room temperature to 315 K, the geometrical characteristics of the CuZnAl actuator did not change. In the range of temperatures from 315 K to 335 K, the height of the helix increased by ~ 130%, reaching 112 mm. Further, up to 400 K, the shape of the helix was stable. The distance between the coils reached ~ 6 mm. The cooling of the spiral led to a significant restoration of the original geometric dimensions. In the temperature range from 313 K to 298 K, the drive height decreased to ~ 63 mm. Such an evolution of height did not change when the heating regimes changed. A different situation with the spiral torsion when heated. The rotation of the upper turn relative to the base (figure 2, b) is reflected in two graphs. Curve 1 shows how the angle of torsion α changes when the helix was placed from room conditions (temperature ~ 296 K) in a thermostat heated to 400 K. Curve 2 reflects the rotational process at a heating rate of the helix ~ 1 K / min. If in the first case, all deformation processes take place in 6 minutes. At the same time, the swing of the turn was only 3°. In the second case, the development time of the rotational motion lasts ~ 25 minutes. The angle of rotation at the same time reached 7°.

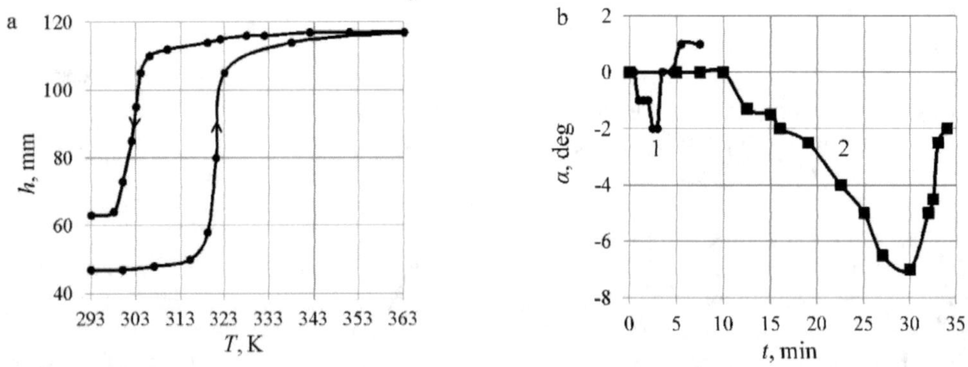

Figure 2. The change in the height of the spiral as a result of the SME (a) and the angle of rotation α for different heating modes (b): 1 – at T = 400 K, 2 – at a rate of change T ~ 1 K / min.

The study of the dependence of RFBE on changing the heating rate was carried out on a special LIND dynamometer, the circuit of which is shown in figure 3. At room temperature, a pair of "metallic muscles" was brought into force interaction with the counterbody. A steel spiral spring was chosen as the counterbody. The dynamometer was placed in a thermostat chamber. With an increase in temperature, the control point displacement $\Delta = d(0) - d(t)$ ($d(0)$ is the initial characteristic size of the RFBE and d (t) is the characteristic size at the current time t) was monitored. Knowing the value of Δ, one can determine the level of the generated force: $F(t) = F0 + K1 \cdot \Delta$ ($F0$ is the strength of the initial interaction of RFBE with the counterbody, K1 = 100 N / mm is the rigidity coefficient of the counterbody).

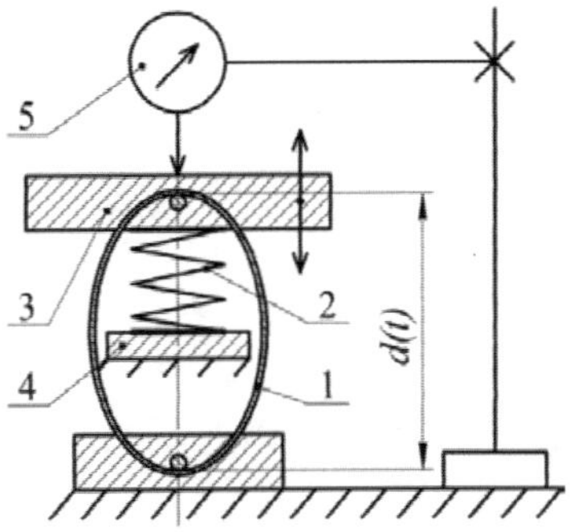

Figure 3. Loading RFBE in a LIND dynamometer: 1 – RFBE, 2 – counterbody, 3 – movable plate, 4 – regulating fixed plate, 5 - measuring system, $d(t)$ – characteristic size of RFBE.

The RFBE heating modes are represented by temperature change diagrams in Figure 4. The initial force interaction of the "metallic muscles" with the counterbody corresponded to F0 = 706 N. As a result of LIND's placement in the heated thermostat (curve 1, figure 4), the development of the shape memory effect occurs monotonously. Forming begins after 2 minutes from the moment of placement in the thermostat, and ends at 12 minutes.

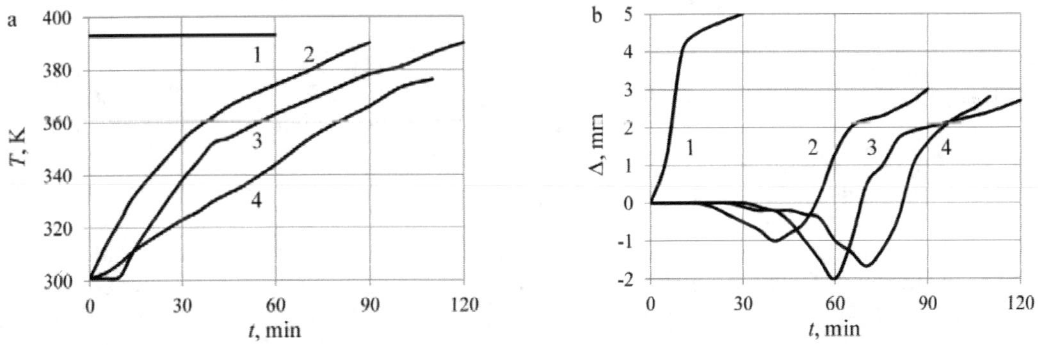

Figure 4. The dependence of the thermostat temperature change (a) and the distance between moving LIND plates (b) on the time in the heated state at 1 – T = 393 K and at heating rates: 2 – 1 K / min, 3 – 0.83 K / min, 4 – 0.68 K / min.

If the dynamometer is heated with a thermostat (curves 2–4, figure 4, a), then we observe a deformation counter-flow, that is, the initial stage of shaping the RFBE is opposite to the direction of the SME. With an average rate of temperature change T in the thermostat ~ 1 K / min (curve 2, figure 4, a) the maximum offset of the opposite flow of 1 mm was noted at 42 minutes (curve 2, figure 4, b). At the same time, the interaction force of RFBE with the counterbody decreased to 606 N. A decrease in heating rates to 0.83 K / min (curve 3, figure 4, a) increases the displacement to 2 mm (curve 3, figure 4, b). The maximum is marked at the sixtieth minute. The minimum value of the force acting on the RFBE was found to be 506 N. At a heating rate of 0.68 K / min (curve 4, figure 4, a), the maximum countercurrent was observed at 70 minutes (curve 4, figure 4, b). However, its value decreased to ~ 1.67 mm. And the smallest force of interaction of RFBE with a counterbody increased relative to the same parameter noted in the third experiment, and was ~ 539 N.

Deformation processes depending on the cooling rate were studied using a spring dynamometer (figure 5). The oval of the force element was heated in the unloaded position to 400 K. This temperature is sufficient for transferring the ring material to the austenitic state. Under these conditions, a sample with d1 = 73 mm was brought into contact with a stretched spring with a force of F0 = 6.4 N. Then, during the cooling process, the ring was changed to form (figure 6). The oval ring was pulled along the line of action of the load. The control point offset is represented for four cooling modes. When the sample cooled with a thermostat (curve 1, figure 6), the form change began at ~ 100 min at 320 K. This process lasted about 200 minutes to a temperature of 300 K. The ring cooling rate at this stage was ~ 0.1 K / min. The value of d1 increased by Δ = 7.3 mm, where $\Delta = d(t) - d1$. Unlike heating, when d1 decreases as a result of the SME, in this situation the magnitude of this geometric parameter increases. The force effect $F(t) = F0 - K2 \cdot \Delta$ (K2 = 0.6 N / mm) decreased to 2.2 N.

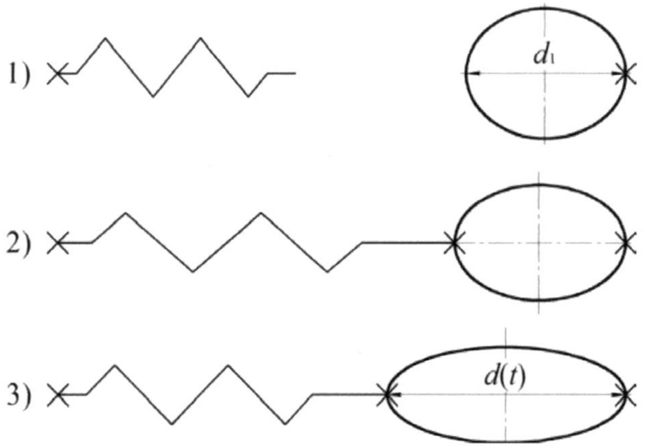

Figure 5. Load diagram of the power elements in the dynamometer: 1 – the initial state of the RFE with a size of d_1, 2 – after loading in the austenitic state, 3 – the sample after the deformation of the plasticity of the transformation with the dimension *d (t)*.

When the dynamometer is removed from the thermostat and cooled at room temperature of 299 K (curve 1, figure 6), the form change is observed from 8 to 60 minutes. The value of d_1 increased by 6.5 mm, and the force effect on the oval decreased to 2.75 N. At the same time, the average cooling rate in the range of direct martensitic transformation is 0.4 K / min.

When cooled in cold rooms with temperatures of 280 K (curve 3, figure 6) and 270 K (curve 4, figure 6), the change in the geometric parameters of the power element began at the 4th and 3rd minutes, respectively. In the third cooling mode, d1 increased by 5 mm, and the force interaction with the spiral spring decreased to 3.4 N. During the fourth mode of lowering the temperature, the change in d_1 was only 3.5 mm, and *F (t)* = 4.4 N. The rates of temperature change were respectively 1.7 K / min and 2.9 K / min. In this case, *t* was equal to 10 minutes. Change of the form occurred in the minimum time interval. In all four experiments, the heating of the annular force elements led to a complete restoration of the original shape.

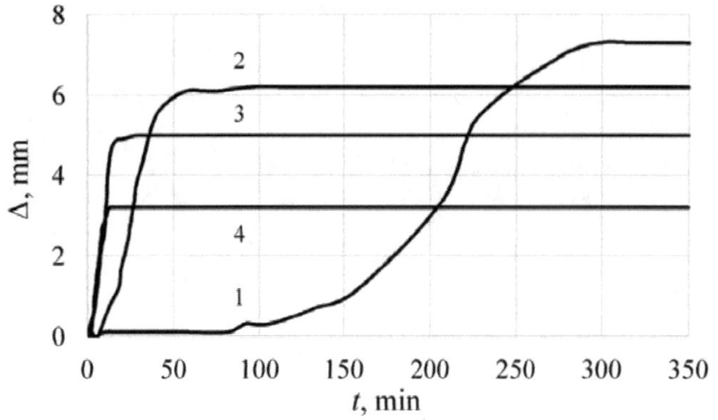

Figure 6. The effect of cooling rate on oval shaping: 1 – cooling with a thermostat; isothermal cooling: 2 – at 300 K, 3 – at 280 K, 4 – at 270 K.

To analyze the causes of the results obtained, a numerical experiment was performed. A calculation is made of the evolution of the temperature field in a cylinder when its surface is cooled from 400 K to M_f = 313 K (the temperature of the end of the direct martensitic phase transition) and to a temperature 40 K below M_f. The calculations were performed using the heat equation [4] in cylindrical coordinates (1):

$$\rho \cdot c(U) \cdot \frac{\partial U}{\partial t} = k \cdot \left(\frac{\partial^2 U}{\partial r^2} + \frac{1}{r} \cdot \frac{\partial U}{\partial r}\right) \tag{1}$$

where ρ is the density of the material, k is the coefficient of thermal conductivity, U is the temperature, r is the radial coordinate, t is the time. The heat capacity c (U) outside the transformation temperature range is a constant c0, and in the temperature range the lattice rearrangement is approximated by a quadratic function (2):

$$(U) = c_1 \cdot \frac{(U - M_s)\cdot(M_f - U)}{(M_f - M_s)^2} + c_0 \tag{2}$$

where M_s is the onset temperature of the direct transformation (in the numerical experiment 323 K). The value of the constant c_1 is found from equation (3):

$$Q_{tr} = \int_{M_s}^{M_f} c_1 \cdot \frac{(U - M_s)\cdot(M_f - U)}{(M_f - M_s)^2} dU \tag{3}$$

where Q_{tr} is the latent heat of transformation.

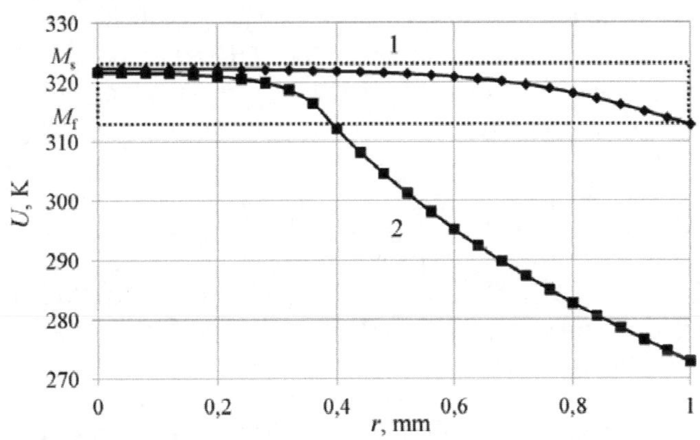

Figure 7. The evolution of the temperature field with a direct martensitic transformation in a cylinder r = 1 mm with a sharp change in surface temperature from 400 K to M_f (1) to M_f – 40 K (2).

In figure 7, it can be seen that the supercooling of the cylinder causes the heterophase zone to move from the sample surface to the center by a narrow band. At the same time, at the M_f temperature on the cylinder surface, the heterophase state V_{get} covers almost the entire volume of cylinder V. At the same time, the lifetime of this state is much longer (figure 8). Similar results were obtained for heated cylinders.

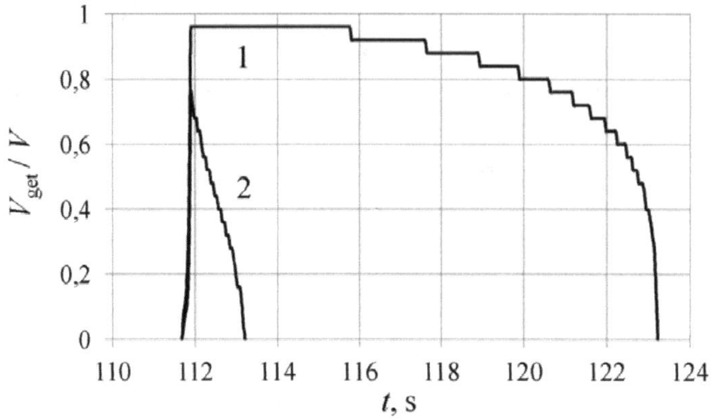

Figure 8. Evolution of the direct martensitic transformation zone V_{get} / V in a cylinder r = 1 mm over time with a sharp change in surface temperature from 400 K to M_f (1) to $M_f - 40$ K (2).

3. Conclusion

Thus, an extensive heterophase state and its duration provide an improvement in the deformability of the force element during the cooling process. When heated, this is done only partially. The phenomena of countercurrent compete with the change in shape due to the effect of shape memory. The advantage in the development of deformation processes is on the side of the SME.

When creating structures made of titanium nickelide, it is necessary to take into account the technological conditions of their operation, due to the temperature conditions of equipment operation.

References

[1] Belyaev S P, Volkov A E, Ermolaev V A *et al* 1998 *Shape Memory Materials: Reference Book* ed VA Likhachev (St. Petersburg: NIIKh SPbGU)

[2] Vyunenko Yu N, Zatulsky G Z, Komissarov V N, Kulikov A A and Soldatova J V 2000 Shape memory effect in CuZnAl samples of spiral form *Tambov University Reports. Series: Natural and Technical Sciences* **5** 281–282

[3] Vyunenko Yu N 2010 Mathematical modeling of deformation processes and experience of technological application of the SME *Basic Problems of Material Science* **7** 28–31

[4] Vyunenko Yu N 2003 The mechanism of shape memory effect caused by the residual stress field evolution *Materialovedenie* **12** 2–6

The influence of the thermomechanical conditions and the geometric parameters of TiNi alloy force elements on the deformation processes of two-way memory effects

E A Khlopkov[1,2,5], N N Belousov[3], T A Makarova[1], Ya Yu Brovkina[1], A O Mikhailov[4] and Yu N Vyunenko[4]

[1] Peter the Great Saint Petersburg Polytechnical University, Saint Petersburg, 195251, Russia
[2]Yaroslav-the-Wise Novgorod State University, ul. B. St. Petersburgskaya, 41 173003 Veliky Novgorod, Russia
[3]Donetsk Institute for Physics and Engineering named after A.A. Galkin, Donetsk, 83114
[4]OOO «OPTIMIKST LTD», Saint Petersburg, 195426, Russia

[5]E-mail: hlopkovelisey@mail.ru

Abstract. The paper presents data of the two-way memory effect deformation processes dependencies of the various TiNi force elements from different temperature and force modes.

1. Introduction

Studies of metallic materials with shape memory effect (SME) have shown the possibility of controlling their deformation properties due to thermal cycling with partial transfer of material from the austenitic to martensitic state and vice versa [1]. This was used to improve the operation of ring force beam elements (RFBE) in small-sized SheR presses [2]. Obtaining the first ring welded structures allowed the use of ovals as power elements not only in the mode of the SME, but also in the mode of reversible shape memory (RSM) [3]. In this connection, there is interest in regulating the deformation-strength properties of welded rings operating in the (RSM) mode, due to incomplete martensitic transformation. For wave-shaped power elements, the form-changing of reversible shape memory with complete phase transitions was investigated.

2. Results and discussion

First of all, we studied the thermal cycling of ring elements from temperature M_m to A_{sup} and back, where $M_s \le M_m \le M_f$, $A_{sup} > A_f$ (M_s and M_f are the temperatures of the beginning and end of the direct martensitic transformation, respectively, A_f is the temperature of the end of the reverse phase transition).

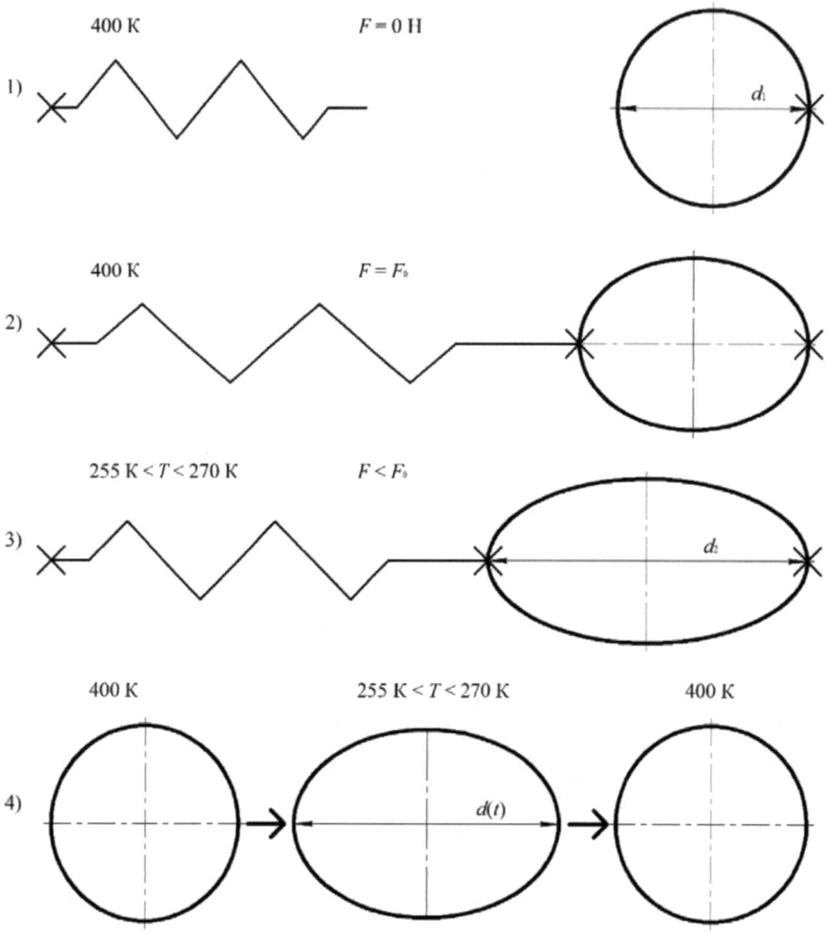

Figure 1. Scheme of increasing the reversible memory of the RFE form: 1 RFE – before loading, 2 – RFE at the time of loading in the austenitic state, 3 – RFE after deformation of the direct transformation plasticity, 4 – implementation of the reversible memory of the RFE form, d_1 – the initial size of the RFE, d_2 – the size of RFE during deformation of plasticity of transformation, $d(t)$ – RFE size during deformation of RSM.

The deformation of the sample under the conditions of plasticity of the transformation was performed using a coil spring (figure 1). Welded rings with a diameter of ~ 62 mm were made of TiNi alloy. Cooling occurred under isothermal conditions at temperatures of 270 K refrigerating chambers – within the interval [M_s; M_f], 255 K – below M_f (Figure 1, position 3). Heating RFE was carried out without load in the thermostat. The temperature in the chamber was 430 K (above A_f) (figure 1, position 4). When heated, the original form was restored completely. At the next stage of heating, the cooling of the RSM was $\Delta_{TWME} = d(t) - d_1$ (figure 1, position 4). After the first thermal cycle described, several preparatory process heat cycles were made. At the stage of lowering the temperature, the sample was under load, and during heating the SME developed in the free ring. Then, observations of the deformation process of the reversible shape memory were repeated. Three rings with different levels of the initial value of the training effort in each cycle: 8.2 H, 18.8 H, 32.7 H were tested.

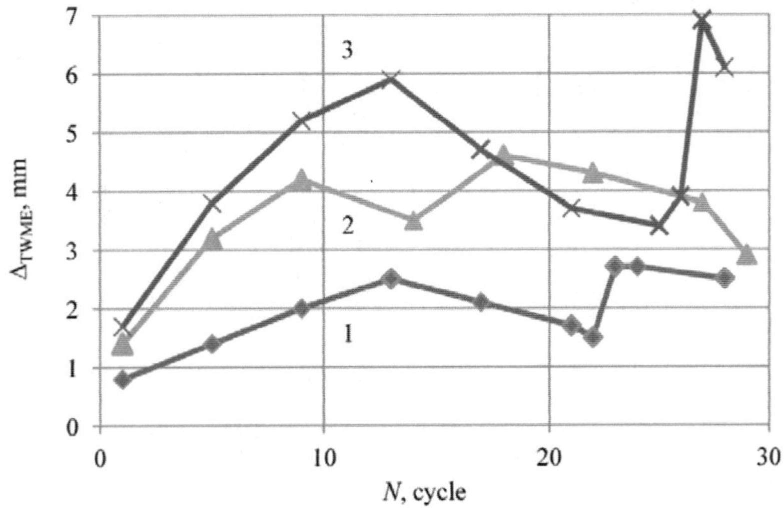

Figure 2. Cyclic change in the value of the reversible shape memory during preparation with a force F: 1 – 8.2 H, 2 – 18.8 H, 3 – 32.7 H and a minimum temperature of the thermal cycle 270 K.

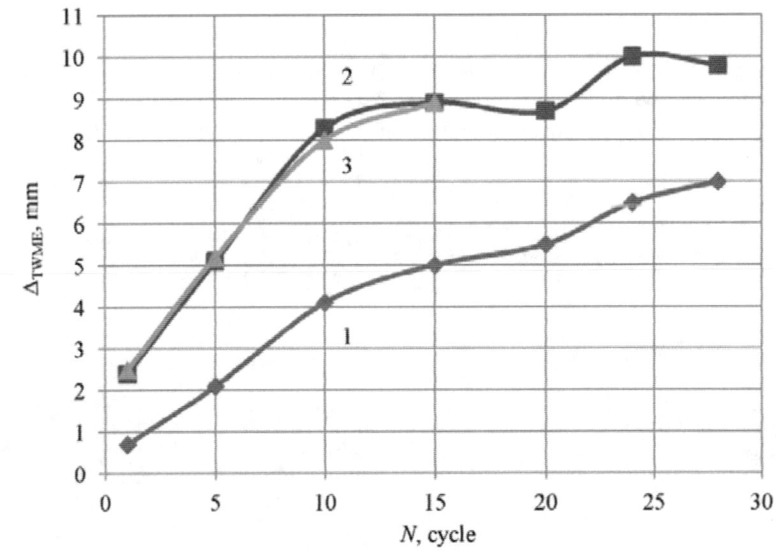

Figure 3. Cyclic change in the value of the reversible shape memory when preparing with a force F: 1 – 8.2 H, 2 – 18.8 H, 3 – 32.7 H and the minimum temperature of the thermal cycle is 255 K.

Figures 2 and 3 show how during the RSM under the specified temperature variation regimes the size of the rings along the loading line changed during thermal cycling. In the works [3, 4], the authors noted an increase in the reversible shape memory in nickel-titanium products during thermal cycling in the full transformation mode, it is also confirmed by the curves in figure 3. At an effort of $F = 8.2$ H, a nearly linear increase in the RSM values was observed for 28 cycles. An increase in the force effect to 18.8 H led to the Δ_{TWME} saturation, starting from the 10th cycle. And with a load of 32.7 H, the characteristics of the shape change practically coincided with the corresponding process values at

F = 18.8 H. In the case shown in figure 2, when the plasticity deformation of the transformation during thermal cycling developed in the range with a lower temperature limit within the interval of the direct phase transition, the growth of the reversible shape memory in all cases, stopped after reaching a certain value. Then there was a decline in the values of the deformation characteristics of the RSM. Further thermal cycling of the ring-shaped power elements resulted in a non-monotonic change in Δ_{TWME} at the cooling stage.

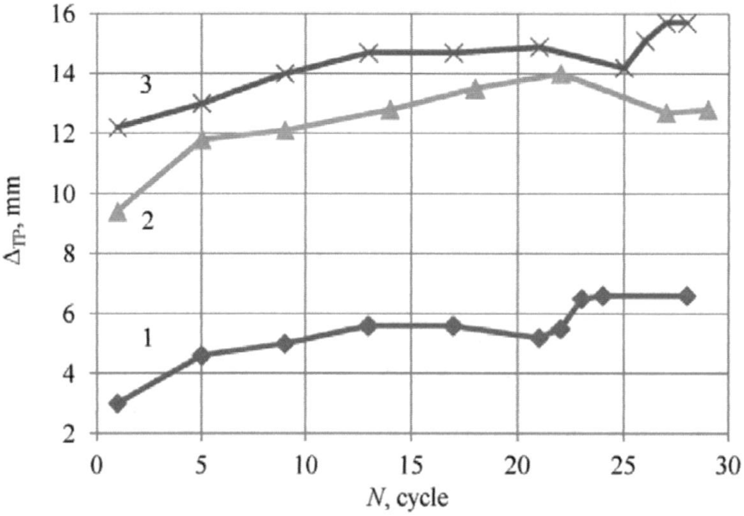

Figure 4. Cyclic change in the strain of plasticity of the direct transformation during preparation with a force F: 1 – 8.2 H, 2 –18.8 H, 3 –32.7 H; and minimum temperature of the thermal cycle 270 K.

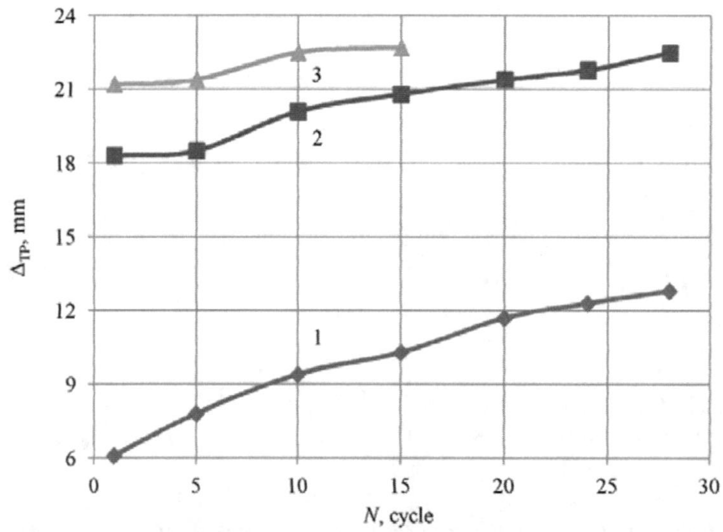

Figure 5. Cyclic change in the strain of plasticity of the direct transformation during preparation with a force F: 1 – 8.2 H, 2 – 18.8 H, 3 – 32.7 H and the minimum temperature of the thermal cycle is 255 K.

The pattern of shape changes in annular force elements caused by deformations of the plasticity of the transformation during thermal cycling $\Delta_{TP} = d_2 - d_1$ is shown with the curves in figures 4 and 5. It should be noted that the reversals of the reversible shape memory (figure 4) occurred with the growth of the shape changes of ovals due to the plasticity of the direct transformation.

After 16 thermal cycle the sample which was under load of 32.7 H collapsed in the welding zone, being in the austenitic state. Since the second research program was carried out on this power element, the total operating time of the ring before failure was 73 triggers. This result requires an improvement in the technology of welding nickel-titanium rings.

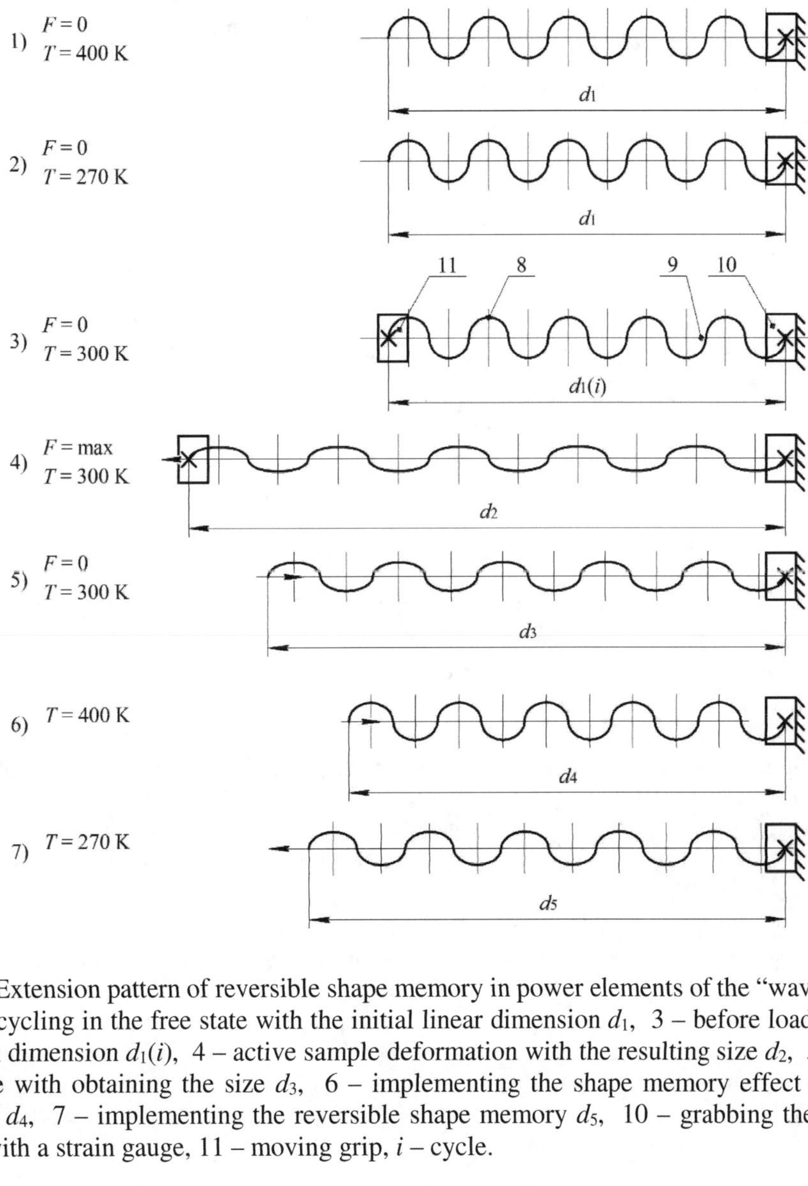

Figure 6. Extension pattern of reversible shape memory in power elements of the "wave" type 8: 1,2 – thermal cycling in the free state with the initial linear dimension d_1, 3 – before loading along the line 9 with dimension $d_1(i)$, 4 – active sample deformation with the resulting size d_2, 5 – unloading the sample with obtaining the size d_3, 6 – implementing the shape memory effect with forming dimension d_4, 7 – implementing the reversible shape memory d_5, 10 – grabbing the deformation machine with a strain gauge, 11 – moving grip, i – cycle.

As an alternative to spiral drives [5], capable of changing the characteristic size by 130% (helix height), a wave-like structure was investigated (figure 6). To study the characteristics of the reversible shape memory, the shape change scheme was chosen, shown in figure 6. The power elements with five repetitive S-shaped links with a total length of 94 mm were made of a 2.5 mm diameter TiNi alloy wire. For the selected alloy, the characteristic temperature is $A_s = 350$ K, $A_f = 370$ K, $M_s = 310$ K and $M_f = 280$ K, where A_s is the onset temperature of the reverse martensitic transformation. The active deformation of the sample along the loading line 9 was carried out using the test setup described in [6] (the process is shown in figure 6). After the shape changing the sample was unloaded. As a result, a new form was acquired, due to plastic deformation of the material of the structure (5, figure 6). Then it was heated in a thermostat (6, figure 6). A temperature of ~ 430 K (above A_f) was maintained in his chamber. The reverse transfer of the actuator material was carried out in a refrigerating chamber at a temperature of 270 K (7, figure 6). At this stage of the experiment, the value of the RSM $\Delta_{TWME} = d_5 - d_4$ was monitored. After the completion of the specified cycle of thermomechanical processing, the sample with the characteristic size d_5 was subjected to the next loading. Then technological operations of unloading, heating and cooling were repeated.

Figure 7 shows the interdependence of the force effect on the wave-like structure and its deformation under loading up to 70 H.

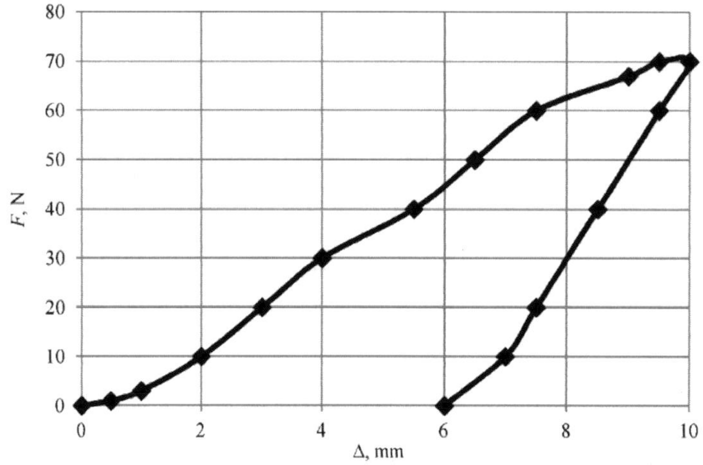

Figure 7. Dependence of the shape changes Δ on the applied load F and the subsequent unloading of the power element of the "wave" type, obtained on the deformation machine DTV.

Curve 1 in figure 8 shows the increase in the characteristic of the shape changing of the power element $\Delta = d_2 - d_1(i)$ from cycle to cycle. On the x-axis it is indicated the value of the maximum force F of the corresponding stage of loading. The first loading of a wavy structure with a force of 70 H caused an increase in Δ to 10 mm. Deformations of reversible shape memory under these conditions were not found. An increase in the force effect to 80 H led to an increase in the characteristic size of the "wave" to 20 mm. The magnitude of the displacements due to the reversible shape memory was 2 mm. A further increase in the deforming force led to an almost linear increase in the characteristic parameter d_2 and Δ_{TWME}. Thermomechanical processing of the sample according to the scheme shown in figure 6 allowed to get the movement of the wave-shaped power element in the RSM mode 19 mm by the 7th cycle. After obtaining this level of deformation parameters, the sample was transferred from the martensitic state to the austenitic state and back 10 times without load. At the same time, no degradation of the characteristics of the reversible shape memory was found. The temperature dependence of shape changing RSM Δ_{TWME} is shown in figure 9.

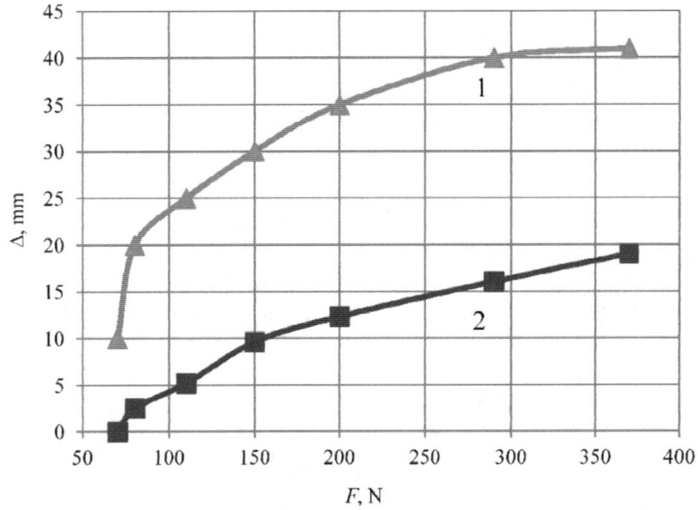

Figure 8. The dependence of the shape changing of the active deformation (1) and the deformation processes of the RSM (2) from the maximum applied load *F*.

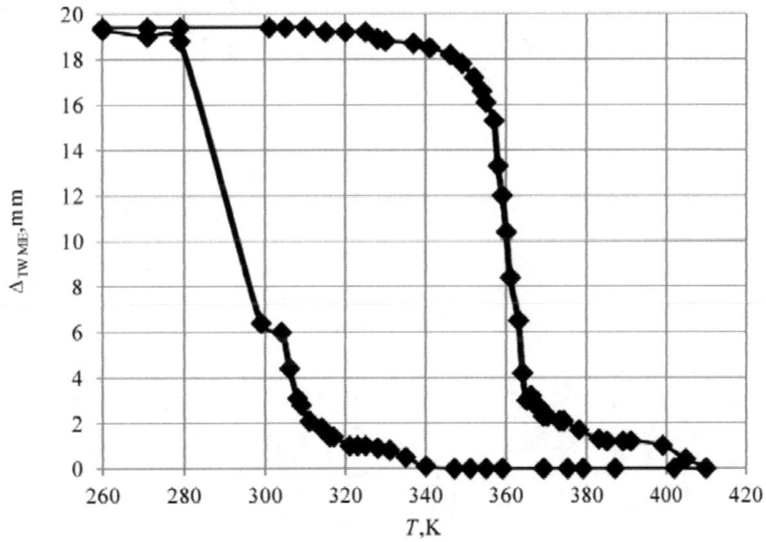

Figure 9. Temperature dependence of the shape changing of RSM Δ_{TWME} during heating and cooling.

3. Conclusion

Thus, the obtained results and the data [3] indicate the need to ensure the completeness of the direct martensitic transformation to achieve controlled shape changing of the power elements at the expense of the RSM in process plants. The required level of movement of the reversible shape memory can be obtained by choosing the level of power effects. The experiments have also shown the need to improve the quality of welded joints.

To ensure a high level of reversible shape memory changes in technological processes, it is advisable to use wave power elements.

IOP Conf. Series: Materials Science and Engineering **656** (2019) 012025 doi:10.1088/1757-899X/656/1/012025

References

[1] Belyaev S P, Resnina N N and Anshukova K O 2013 Work performance in TiNi alloy during heat cycling in the temperature range of B2 ↔ R martensitic transformation *Technical Physics Letters* **39** 284–286

[2] V`yunenko Yu N, Volkov G A and Khlopkov E A 2018 Temperature factor to control deformation–power behavior of ring-shaped bundle force TiNi elements *Technical Physics* **63** 1167–1170

[3] V`yunenko Yu N and Belousov N N 2018Two-way shape memory effect in ring-shaped designs *Actual Problems of Strength* 482–484

[4] Andronov I N, Verbakhovskaya R A, Danilov A N and Korepanova V S 2010 Martensitic type of two-way shape memory in the device for restoring the systolic size of the mitral valve ring, *Zavodskaya Laboratoriya* **76** 38–42

[5] V`yunenko Yu N, Zatulsky G Z, Komissarov V N, Kulikov A A and Soldatova J V 2000 Shape memory effect in CuZnAl samples of spiral form *Tambov University Reports Series: Natural and Technical Sciences* **5** 281–282

[6] Khlopkov E A, Tikhomirov A A, Latyshev K A, Brovkina Ya Yu, Volkov G A, Smirnov I V, Buldakov P Yu and Vyunenko Yu N 2018 Two-stage character of ring-shaped bundle force elements from TiNi alloys deformation *Tambov University Reports Series: Natural and Technical Sciences* 303–306

IOP Conf. Series: Materials Science and Engineering **656** (2019) 012026 doi:10.1088/1757-899X/656/1/012026

Mechanical properties of weld joints of TiNi alloy wire used in force elements

E A Khlopkov[1,2,5], D V Kurushkin[1], E S Ostropiko[2,3], V M Khanaev[4], S B Sapozhkov[2] и Yu N Vyunenko[4]

[1] Peter the Great Saint Petersburg Polytechnical University, Saint Petersburg, 195251, Russia
[2] Yaroslav-the-Wise Novgorod State University, ul. B. St. Petersburgskaya, 41 173003 Veliky Novgorod, Russia
[3] Saint Petersburg State University, Saint Petersburg, 198504, Russia
[4] OOO «OPTIMIKST LTD», Saint Petersburg-Novosibirsk, 195426, Russia

[5]E-mail: hlopkovelisey@mail.ru

Abstract. The paper presents data on the mechanical testing of argon-arc weld joints of the TiNi 55.42 % wt. wires diameter of 2 mm. The mathematical modeling results of the temperature fields evolution of the weld zone during cooling are given.

1. Introduction

Currently, such functional materials as alloys with shape memory effect (SME) are widely known [1]. These metals begin to find their application in the space industry and medicine [2–4]. However, in modern engineering technology [5–10] there are practically no technical solutions implemented with the help of unique properties of materials with SME. The exception is ring force beam elements (RFBE, "metal muscles") of nickelide titanium [11]. These compact drives are actuators of the press used in the manufacture of anti-vibration layered compositions. For ten years, "metal muscles" demonstrate trouble-free operation in the process. In this case, their mechanical properties were investigated depending on various technological conditions [12–15]. However, the use of RFBE in work in the reversible shape memory (RSM) mode was undertaken only with the advent of hard TiNi wire joints [16]. The authors [16] managed to obtain the strength of the weld of a wire with a diameter $d = 2$ mm equal to 66% of the starting material. In the works [17], the authors were able to make welded joints capable of withstanding a force of up to 80% of the applied to the starting material before breaking. However, this level of strength was obtained when welding wire was with d less than 1 mm.

The aim of this work was to search for ways to improve the quality of welded joints of wire from TiNi alloy to create RFBE, capable of operating in the RSM mode. For experimental research, a wire of $d = 2$ mm from TiNi55, 42 wt. percentage. Welding was carried out in two modes by argon-arc welding.

2. Results and discussion

In the first case, a welded joint was obtained using the Svarog TIG P AC / DC source in an *Ar* atmosphere. A voltage of 12 V and a current of 10 A were set at the source. Figures 1 and 2 show the plots of the interdependence of deformations and applied stresses, which make it possible to estimate the strength characteristics of a wire sample from the source material and welded samples. Mechanical tests were performed on an Instron universal machine, under conditions when the material was in the martensitic state. The traverse speed was 1 mm / min.

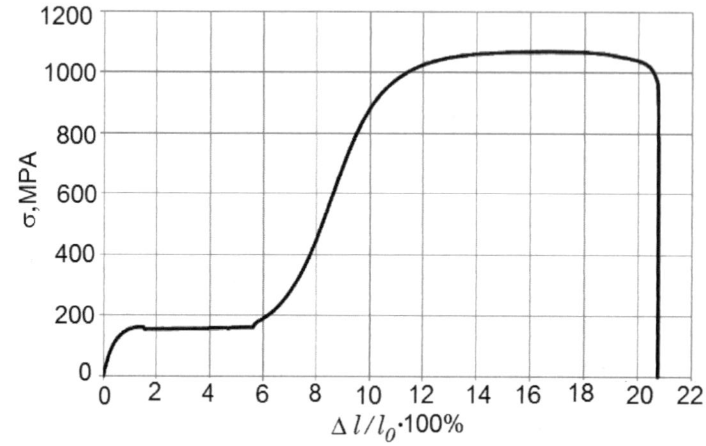

Figure 1. Strength properties of the original TiNi sample.

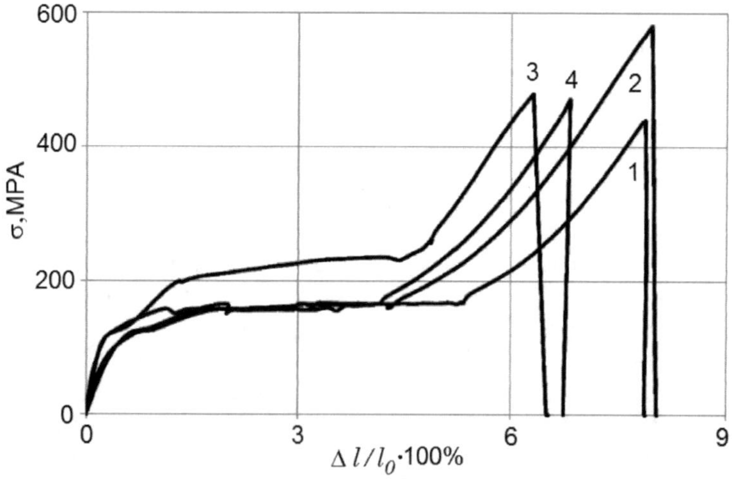

Figure 2. Strength properties of welded TiNi samples: 1 – mode 1 with heat treatment, 2 – mode 2 with heat treatment, 3 – mode 2 without heat treatment, 4 – mode 2 with heat treatment and thermal cycling.

Up to a stress of 120 MPa, the deformation $\varepsilon = \Delta l/l_0$ accumulated according to Hooke's law up to 0.4% (figure 1). A further increase in stresses up to 160 MPa translates the process of deformation into the plastic flow regime. This process is maintained up to $\varepsilon = 6\%$. Further, the magnitude of the deformation increases with σ according to a law close to linear. Exceeding the level of 900 MPa transfers the deformation process into submission to the law of plastic flow. Achieving a voltage level of 1060 MPa and exceeding ε of 20% leads to the destruction of the sample.

Curve 1 in figure 2 shows that in the first technological mode, it was possible to obtain a welded specimen with a breaking load of 40% of the corresponding value of the source material (figure 1).

Before mechanical testing, annealing was performed at T = 773 K for 30 min, then cooled with a furnace. This heat treatment (HT) was performed to eliminate residual stresses in the material of the weld zone.

The second welding mode was performed on the EWM Force TIG 552 apparatus. The voltage was set at 18.5 V, the current strength being 45 A. Ne was used as a protective atmosphere in the technological operation. Curve 2 in figure 2 shows that in this mode it was possible to make a welded joint with a strength of up to 55% of the original sample. Before the breaking tests, the welded core was heat-treated the same as the starting material. Without annealing, the welded specimen collapsed at a load two times smaller than the limiting force effect on the solid piece of wire (curve 3, figure 2).

The fourth sample, the compound of which was made in the second mode, besides annealing, was subjected to thermal cycling through the interval of direct and reverse transformation. For this, the sample was alternately placed in a thermostat chamber with a temperature of 100 °C and a refrigeration chamber with a temperature of (–2) °C. The exposure time in isothermal conditions was equal to 15 minutes This made it possible to transfer the material completely from the austenitic state to the martensitic state and back. The performed procedure did not increase the strength of the sample (curve 4, figure 2). The destruction took place practically under the same load as the third one. However, the deformability was higher. At lower stresses, the sample showed a greater elongation than a similar one without heat treatment (curve 3, figure 2).

To analyze the characteristics of the state of the wire material in the welding zone, a numerical experiment was performed in which the evolution of the temperature field was studied. It was assumed to simplify the calculations that heat exchange with the environment does not occur. As a result of the immediate impact of electric current, the material in the welding zone is heated to 1000 °C. The temperature distribution along the wire corresponds to curve 1 in figure 3. On a wire segment from 0 to 0.5 mm, the temperature is 1000 °C, the rest of the wire material is at room temperature 25 °C. Further, the evolution of the temperature field was calculated using the one-dimensional heat conduction equation (1):

$$\rho \cdot c(U) \cdot \frac{\partial U}{\partial t} = k \cdot \frac{\partial^2 U}{\partial x^2} \tag{1}$$

where k is the coefficient of thermal conductivity, U is the temperature, t is the time, x is the linear coordinate, ρ is the density of the material. The heat capacity of the material $c\ (U)$ depended on temperature. This made it possible to take into account the latent heat of martensitic transformation. Distribution of the latent heat of the phase rearrangement of the crystal lattice quadratic function was approximated (2):

$$c(U) = c_0 + c_1 \cdot \frac{(U - A_s)\cdot(A_f - U)}{(A_f - A_s)^2} \tag{2}$$

where c_0 is the heat capacity of the single-phase state. This value was considered constant and the same for low-temperature and high-temperature states. Respectively, A_s and A_f are the temperature of the beginning and end of the reverse transformation. For the model experiment, these values were assumed to be 60 °C and 70 °C. The second constant c_1 was determined from the relation (3):

$$\int_{A_s}^{A_f} c_1 \cdot \frac{(U - A_s)\cdot(A_f - U)}{(A_f - A_s)^2} \mathrm{d}U = Q_{\mathrm{tr}} \tag{3}$$

where Q_{tr} is the latent heat of the martensitic transformation.

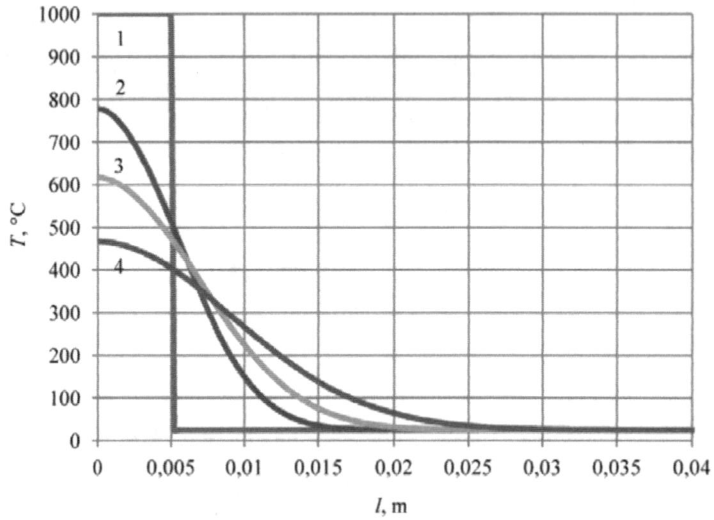

Figure 3. The evolution of the temperature field in the welding zone in the absence of phase transformation under cooling conditions at the moment of time: 1 – 0 s, 2 – 5 s, 3 – 10 s, 4 –20 s.

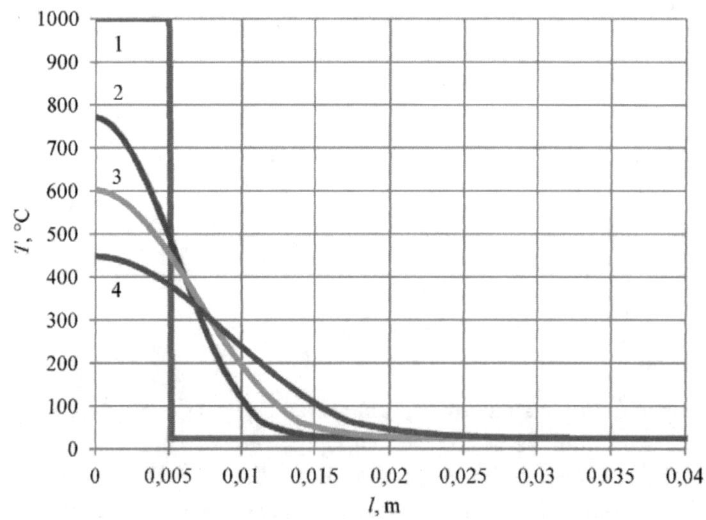

Figure 4. The evolution of the temperature field in the welding zone with the presence of phase transformation during cooling at the time point: 1 – 0 s, 2 – 5 s, 3 – 10 s, 4 – 20 s.

To simplify the problem, we assume that the direct transformation proceeds in the material below room temperature. For comparison with materials that do not undergo transformations, the problem of the evolution of the temperature field was solved when, in equation (1), $c(U) = c_0$ over the entire temperature range from 25 °C to 1000 °C. Figure 3 shows the distribution of heat at different times in the absence of phase transformations. Comparing with the curves similar in time in figure 4, we see that the phase transition with the latent heat of transformation significantly speeds up the process of cooling the welding zone. Figure 5 shows that with increasing Q_{tr}, the rate of temperature decrease in the interval from 0 to 0.5 mm increases.

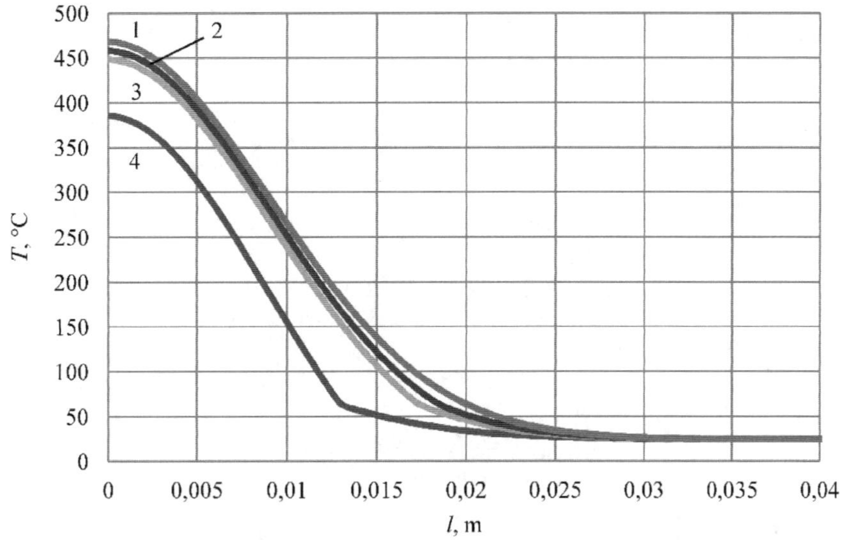

Figure 5. Temperature field distribution in the welding zone at the moment of time 20 s with the latent heat of transformation Q_{tr}: 1 – 0 kJ / kg, 2 – 20 kJ / kg, 3 – 40 kJ / kg, 4 – 200 kJ / kg.

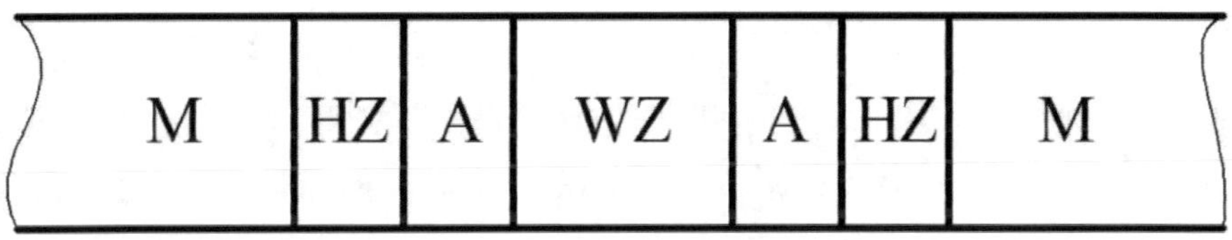

Figure 6. Welded joint areas in the TiNi alloy: WZ – joint area, A – austenitic section, HZ – heterophase state of the material, M – martensitic region.

The analysis of the obtained curves shows that in the wire segment, in the final analysis, there are three sections with a different structure (figure 6). The austenitic state of the material is formed in the adjacent welding zone. With sufficient length, the austenitic state will be followed with a heterophase zone, in which the ratio of austenite and martensite will change as the distance from the zero point changes. Further, the alloy will remain in the original martensitic state. In this way, a clearly heterogeneous material of the simplest welded structure was obtained. It can be assumed that the rate of cooling of the weld zone may introduce additional features in the structure of the material of the construction. Such a gradient system will require not only additional heat treatment in the form of annealing, but also for structures of multiple action, as M.A. Khusainov repeatedly noted [18]. It will be necessary at least fifteen times thermocycling of the annealed part with the transfer of material from martensite to austenite and back to stabilize the temperature ranges of phase transitions.

3. Conclusion
Thus, the welding technology will require the development of special heat treatment regimes, focused on the upcoming technological conditions of operation of the fabricated structure.

Acknowledgments
The reported study was funded by RFBR, project number 19-38-90285.

References
[1] Likhachev V A, Kuzmin S L and Kamentseva Z P 1987 *The shape memory effect* (Leningrad: Leningrad State University) p 216
[2] Belyaev S P, Volkov A E, Ermolaev V A at al 1998 *Shape Memory Materials: Reference Book* (St. Petersburg: NIIKh SPbGU)
[3] Kravchenko Y D, Likhachev V A, Razov A I, Trusov S N and Cherniavsky A G 1966 Testing of shape memory alloys in the erection of large-scale structures in open space *Technical Physics* **41** 1167–1171
[4] Gyunter V E, Kotenko V V, Polenichkin V K and Itin V I 1985 Use of alloys with shape memory in medicine *Soviet Physics Journal* **28** 433–437
[5] Anukhin I V, Anukhin V I, Lyubomudrov S A and Murashkin S L 2015 Thermal imaging in selecting the cutting conditions for high-temperature intermetallic alloys *Russian Engineering Research* **35** 544–548
[6] Kolodyazhniy Y Y, Lyubomudrov S A and Makarova T A 2016 Quality assurance issues of hard-processing aluminum alloy parts fabrication for aircraft construction and engine-building *Journal of Engineering and Applied Sciences* **11** 3019–3023
[7] Kozar I I , Kolodyazhniy D Yu, Radkevich M M Tsimko TA 2014 A mathematical model of error for intractable alloy turning *Nauchno-tekhnicheskie vedomosti SPbGPU* **2** 194–201
[8] Tarasov S B Stepanov SN 2007 Development of the reference and high precision instruments and devices for replacing foreign-made equipment *Nauchno-tekhnicheskie vedomosti SPbGPU* **4** 29–33
[9] Nikitkov N V, Kovelenov N Y and Shabalin D N 2012 Calculation mode cuttings under diamond cutting stocking up from hard frail material Nauchno-*tekhnicheskie vedomosti SPbGPU* **3** 155–159
[10] Zhukov E L, Kozar I I and Olodyazhniy D Y 2016 Problems of ensuring quality of a surface layer when producing components from hard-to-process heat resistant alloys *Acta Metallurgica Slovaca* **22** 128–132
[11] V'yunenko Yu N, Volkov G A and Khlopkov E A 2018 Temperature factor to control deformation–power behavior of ring-shaped bundle force TiNi elements *Technical Physics* **63** 1167–1170
[12] Khlopkov E, Volkov G and V'yunenko Yu N 2017 Specific features of the behavior of TiNi force elements in thermocycling *Materials Today: Proceedings* **4** 4879–4883
[13] Kiselev A Yu, Belousov N N, Khlopkov E A and V'yunenko Yu N 2018 The effect of the size factor on the functional properties of shape memory alloy ring-shaped force elements Materials *Research Proceedings* **9** 24–27
[14] V'yunenko Yu N 2013 Mechanical characterictics of the power element of materials with SME *Tambov University Reports. Series: Natural and Technical Sciences* **18** 2023–2024
[15] Khlopkov E A and Vyunenko Yu N 2018 Influence of cooling rate on the deformation processes associated with direct martensitic transformation in TiNi alloy *Materials Research Proceedings* **9** 28–31
[16] Vyunenko Yu N and Belousov N N 2018 Two-way shape memory effect in ring-shaped designs *Actual Problems of Strength* 482–484
[17] Nishikawa M, Tanaka H, Kohda M, Nagaura T and Watanabe K 1982 Behaviour of welded part of Ti-Ni shape memory alloy Journal *de Physique Colloques* **43** 839–844

[18] Khusainov M and Volnyanskaya O Y 2001 Influence of the preliminary thermocycling mode on the deformation-force parameters of the shape memory alloy *Modern Problems of Strength* P **2** 123–127

Justification of rejection values of defect characteristics in ultrasonic testing of welded joints

N N Konovalov[1,2] and A A Petrov[2,4]

[1]Yaroslav-the-Wise Novgorod State University, ul. B. St. Petersburgskaya, 41 173003 Veliky Novgorod, Russia
[2]LLC "RASEc plus", Ryazansky pr., 32 109377, Moscow, Russia

[3]E-mail: knn5@mail.ru
[4]E-mail: PetrovAlerxA@yandex.ru

Abstract. Probabilistic models of substantiation of rejection values of the measured characteristics of defects during ultrasonic testing of welded joints using phased antenna arrays that reduce possible test errors are considered. It is shown that the most important for usage are the reliability indicators of non-destructive testing, taking into account the sub-rejection. The recommended application areas for evaluating the reliability D* criteria of ultrasonic testing using alternative and quantitative indicators for selecting rejection values of measured defect characteristics have been determined.

1. Introduction

Most often, non-destructive testing of welded joints is subject to hazardous production facilities [1]. To carry out their control, methodological documentation is needed, including, in particular, rejection indicators that meet certain standards for the admissibility of welding defects. Moreover, the choice of rejection indicators has a significant impact on the reliability of non-destructive testing.

In accordance with GOST (Russian National Standard) 20911-89 "Technical Diagnostics. Terms and definitions" "reliability of technical diagnostics (control of the technical condition) is the degree of objective correspondence of the results of diagnosis (control) to the actual technical condition of the object. With regard to non-destructive testing quantitatively the reliability of the control is calculated as the probability of making error-free decisions when assessing the quality (condition) of controlled object [2–4].

2. Analysis of reliability indicators of non-destructive testing

To determine the reliability indicators of non-destructive testing, the values of possible control errors are estimated: errors of the 1st kind (supplier risk – α, over rejection) and errors of the 2nd type (consumer risk – β, under-rejection). The reliability of the control can be calculated by the formulas [4, 5]:

$$D_\alpha = 1 - \alpha,$$
$$D_\beta = 1 - \beta, \qquad (1)$$
$$D_\Sigma = 1 - (\alpha + \beta).$$

Federal rules and regulations “Basic requirements for non-destructive testing of technical devices, buildings and structures at hazardous production facilities”, approved by order of the Federal Service for Environmental, Technological and Nuclear Supervision No. 490 of 21.11.2016, have established that the choice of non-destructive testing technologies and means should be carried out on the basis of the condition for ensuring the detection of unacceptable defects. Thus, the applied indicators of the reliability of control should, first of all, be focused on reducing the values of possible control errors of the 2nd kind. Taking this into account, the indicator D_Σ is of particular interest, which, since under-rejection is more dangerous than over rejection, protects the interests of the supplier well and not fully the consumer, and D_β, which protects only the interests of the consumer [4, 5]. Two types of determination of indicators of reliability are possible: by alternative and quantitative characteristics.

2.1. The choice of rejection values of the characteristics of defects in assessing the reliability of control indicators for an alternative criterion

In assessing the reliability of an alternative criterion, a matrix is used (table 1), comparing the test method with the reference one, which is recommended to open the controlled welds.

Table 1. Matrix for assessing the reliability of control by the number of checked items (on an alternative criterion).

The number of controlled items		Reference method (opening)	
$n_\Sigma = n_g + n_n + n_\alpha + n_\beta$		Fit	Unfit
Trial method	Fit	Twice fit n_g	Under rejection n_β
	Unfit	Over rejection n_α	Twice unfit n_n

Reliability can be calculated by the formulas:

$$D_\alpha = (n_\Sigma - n_\alpha)/n_\Sigma = 1 - n_\alpha/n_\Sigma$$

$$D_\beta = (n_\Sigma - n_\beta)/n_\Sigma = 1 - n_\beta/n_\Sigma \quad (2)$$

$$D_\Sigma = (n_\Sigma - n_\alpha - n_\beta)/n_\Sigma = 1 - (n_\alpha + n_\beta)/n_\Sigma$$

With the most common of the physical types of non-destructive testing, ultrasonic testing of the boundary corresponding to the standard-acceptable defect size h_n and the adopted rejection level of the ultrasonic signal amplitude A_{br}, divide the correlation field “defect size – ultrasonic signal amplitude” into four areas corresponding implicitly to four cells of the validity matrix: n_g, n_n, n_α, n_β (figure 1).

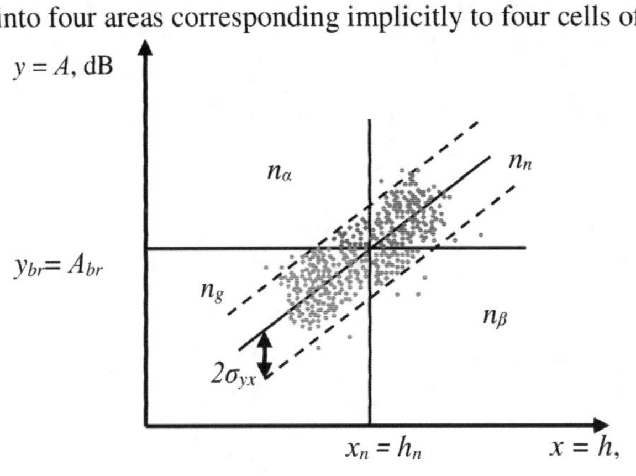

Figure 1. Graphic model of point estimate of the reliability of ultrasonic testing on an alternative basis (distribution of experimental points), n_g, n_n are twice fit and twice unfit elements of welded joints, respectively; n_α, n_β are over rejection and under rejection respectively.

The most significant internal defects in butt welded joints are lack of penetration. They are often found, have a significant impact on the carrying capacity of welded joints and their occurring is rather difficult to prevent [4].

With traditional ultrasound testing, basic information about the detected defect is usually obtained on the basis of an analysis of the corresponding amplitude of the echo signal. At the same time, there is a good correlation between the amplitude of the echo signal and the penetration depth located in the lower part of the butt joints [4]. Considering that the use of phased array antenna technology also makes it possible to estimate the real sizes of defects [6], it is advisable to investigate not only the correlation fields "the amplitude of the echo signal A is the penetration depth h", but also the correlation fields "measured penetration depth h_i" and "real non penetration depth h_r".

Schemes for determining the rejection values of the amplitude of the echo signal and the size of the defect (depth of non penetration) based on the assessment of the reliability of control on an alternative criterion are proposed. The rejection values determine how the ordinates of the points of intersection of the values of the normative – acceptable size of the defect with the lower limit of the scatter of the correlation dependence "the measured characteristic of the defect – size of the defect".

Thus obtained rejection values of the amplitude of the echo signal A_{br} (figure 2a) and the measured depth of non penetration h_{br} (figure 2b) practically ensure the absence of under rejection. Analysis showed that to determine the rejection values of the measured characteristics of defects, it is advisable to use the ranges of variation ± 2σ. The use of ranges of ± 3σ realizes the rejection values of the measured characteristics of defects, which without thoroughly increase the errors of the 1st kind (over rejection).

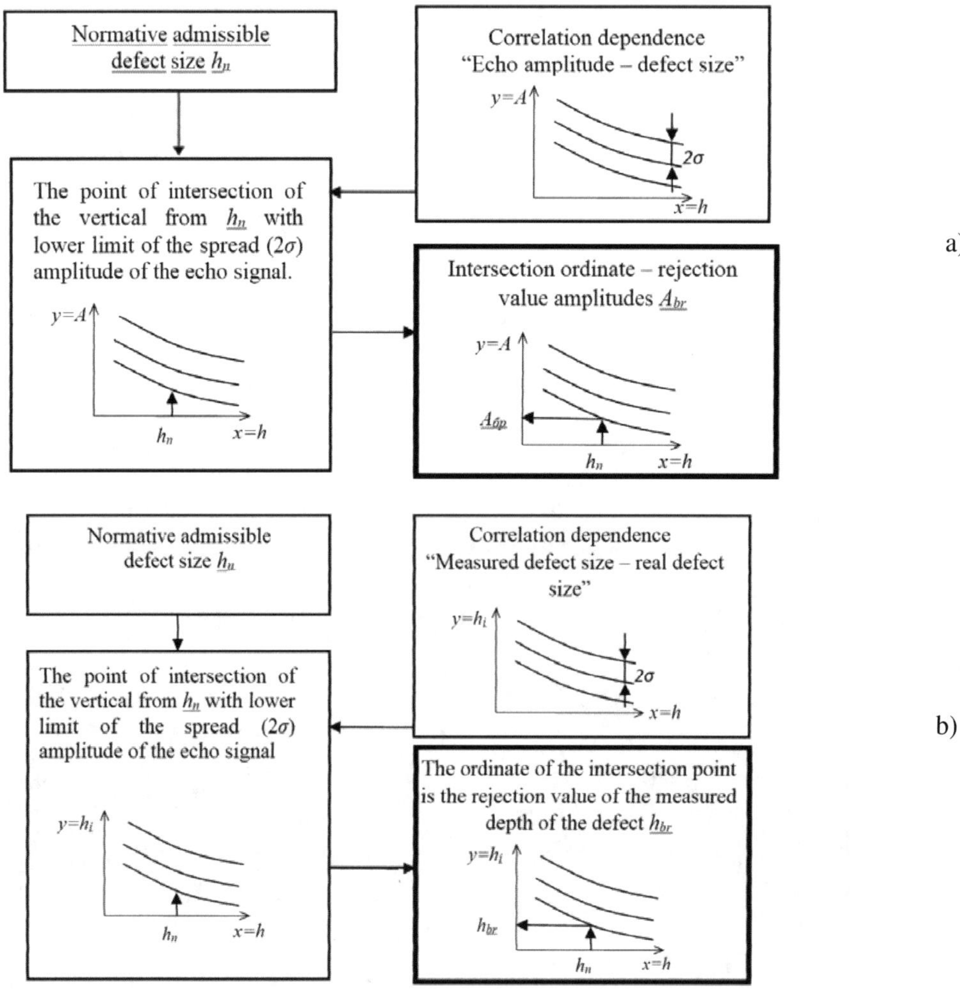

Figure 2. Schemes for determining the rejection value of the amplitude of the echo signal A_{br} (a) and

the rejection value of the depth of non penetration h_{br} (b) based on the use of the assessment of the reliability of control on an alternative criterion.

As an example, an assessment of the reliability of ultrasonic testing on an alternative criterion of a one-sided butt-welded joint made of low-carbon steel with a thickness of 10 mm was made (rejection sensitivity S_{eqv} = 2 mm^2, the allowable depth of penetration h_n = 2 mm) (figure 3). The control was performed by Harfang flaw detector with a phased antenna array of the PE-5.0M3208P type with a frequency of 5 MHz, consisting of 32 elements with a pitch of 0.8 mm. The antenna array was located on a standard prism of rexolite T1-WOD with a width of 28 mm and a tilt angle of 35°.

The rejection value of the amplitude of the echo signal by the correlation dependence "the amplitude of the echo signal A – the depth of the non penetration h" must be reduced by 11 dB, which practically ensures that there is no under rejection on signal amplitude (figure 3a). Similarly, on the correlation dependence "measured depth of non penetration h_i – real depth of non penetration h_p", with a rejection value of 1.4 mm depth, there will be practically no non penetration on depth of non penetration (figure 3b).

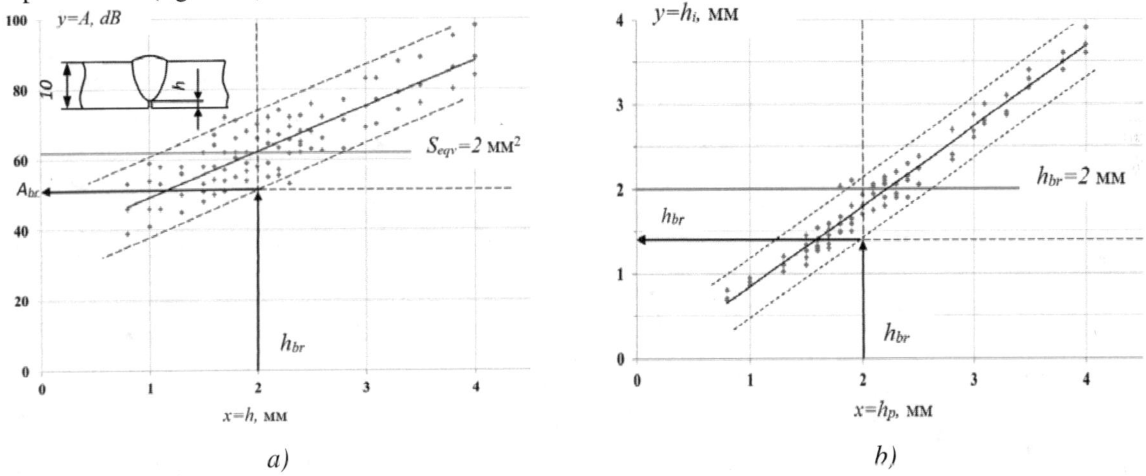

Figure3. Determination of the rejection value of the amplitude of the echo signal by the correlation dependence "the amplitude of the echo signal A – depth of non penetration h" (a) and the rejection value of the depth of the non penetration according to the correlation dependence of "measured depth of non penetration h_i and is the real depth of the penetration h_p (b).

2.2 The choice of rejection values of the characteristics of defects in assessing the reliability of control indicators on a quantitative criterion

A scheme for determining the rejection value of the size of the defect based on the assessment of the reliability of control on quantitative basis (figure 4) is proposed. The assessment of the reliability of quantitative basis is associated with the modeling of the conditional boundary of the division of production into fit and unfit and finding errors of the 1st and 2nd kinds. To obtain graphic image of the complete model for assessing the reliability indicators based on quantitative basis, it is necessary to build three functions: the operational characteristic of the control G (h) = 1-W_Σ (h) or the total detection function of defects W_Σ (x) = 1-G (x), the distribution defects function on depth f_p (h), accuracy function of the evaluation of depth of defects f_b (h). The usage of these functions makes it possible to determine the errors of rejecting: the probability of not detecting unacceptable $\Delta\chi_\beta$ defects, the probability of under rejecting $\chi_{\beta B}$, the probability of over rejecting χ_α [7, 8].

The composition of functions f_p (h) W_Σ (h) = f_{on} (h) is the probability distribution of defects that are detected during ultrasound testing. The steeper the function W_Σ (h) is and located to the left, the more the functions f_{on} (h) and f_p (h) will coincide. During development of technology and the choice of means of ultrasonic testing, they try to ensure maximum detection of defects, and the function $f_{on}(h)$

shows the possibilities of control. The composition of the functions f_p (h) $(1\text{-}W$ (h) $= f_{ob}$ (h) allows you to determine the errors in estimating the size of defects.

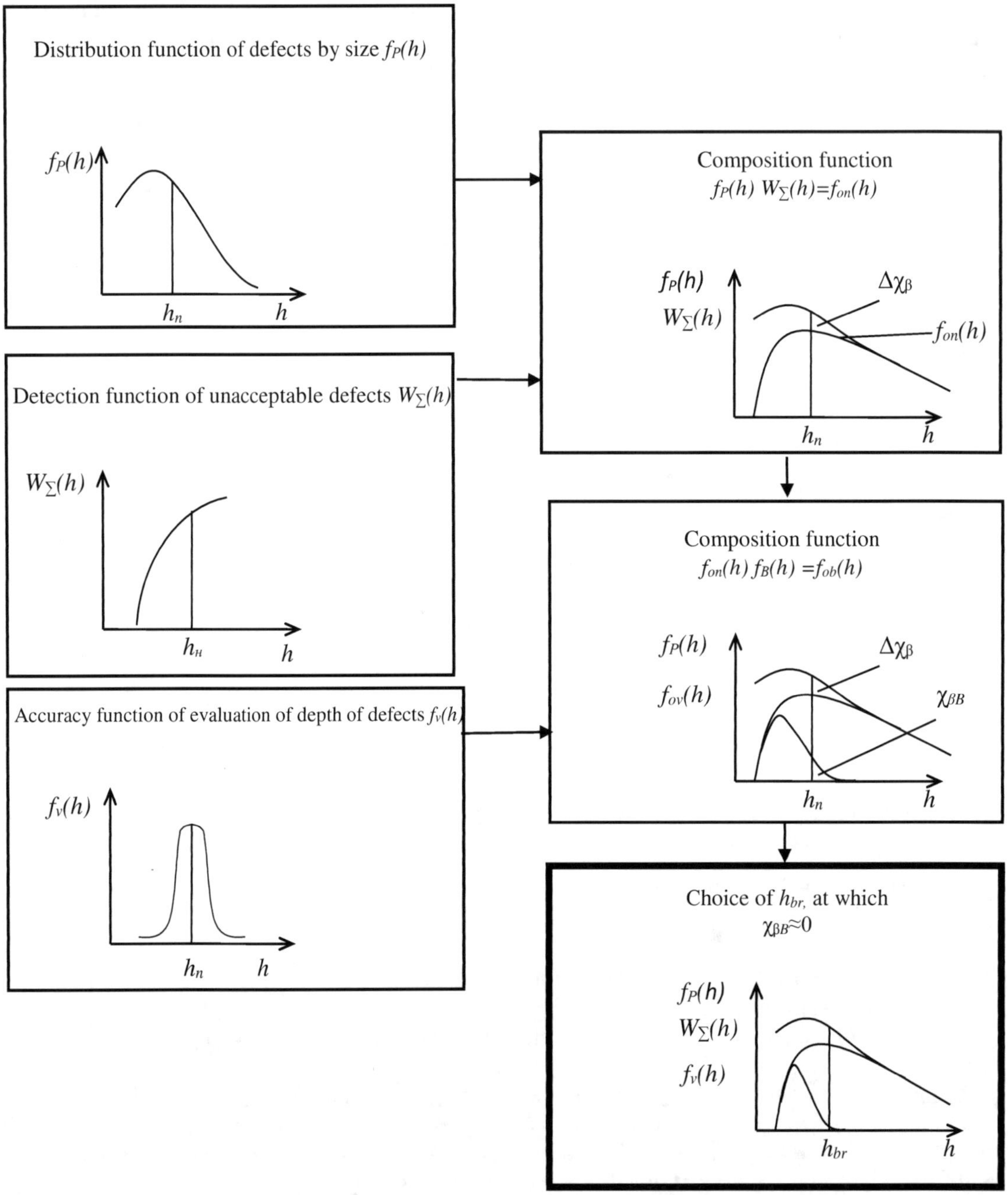

Figure 4. The scheme for determining the rejection value of the size of the defect based on the use of a graphic image of the complete model for assessing the reliability of a quantitative criterion.

IOP Conf. Series: Materials Science and Engineering **656** (2019) 012027 doi:10.1088/1757-899X/656/1/012027

The rejection value of the size of the defect is determined by the location of the accuracy function of estimating the size of defects f_v (h). The function f_v (h) should be located in such a way as to minimize the value of $\chi_{\beta B}$.

Graphic image of the complete reliability assessment model for example for which the reliability of ultrasonic testing on alternative criterion was estimated above in this article (see figure 3) is shown in figure 5. The numerical values of screening errors, that is, the corresponding areas on the graphic image of the full assessment model reliability, calculated by calculating the areas on the graphic image using scale paper.

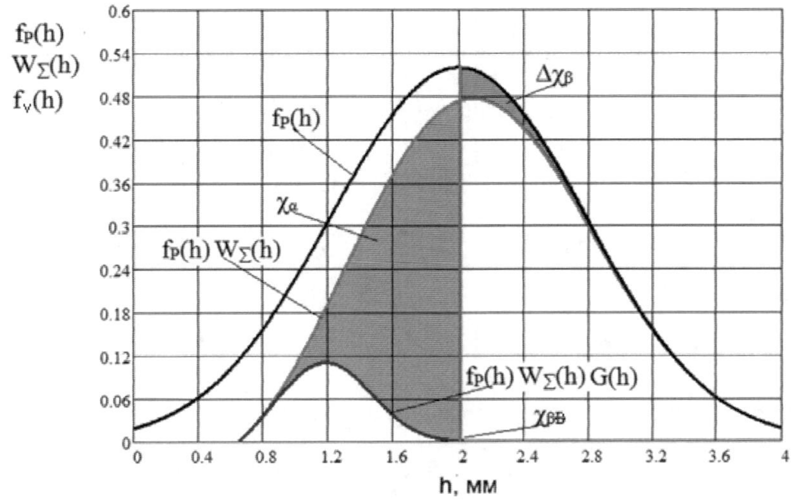

Figure 5. Graphic images of the complete model of ultrasonic testing reliability assessment for a regulatory-admissible defect h_n = 2 mm.

2.3. Analysis of rejection values of characteristics of defects obtained during assessing the reliability of control indicators for the alternative and quantitative criteria

Reliability indicators obtained during evaluating by alternative criterion and during evaluating by quantitative criterion have sufficiently close numerical values (table 2).

Table 2. Comparison of indicators of reliability at assessment on alternative criterion and quantitative criterion.

	Criterion of evaluation	Confidence factor	
		D_Σ	D_β
Ultrasonic test using phased antenna array	Alternative	0.863	0.91
	Quantitative	0.82	0.952

With a standard-permissible defect depth h_n = 2 mm to ensure under rejection $\chi_{\beta B}\approx 0$, the rejection value of depth of defect, determined from graphic image of complete model of assessment of reliability, is h_{br} = 1.8 mm. The rejection value of the depth of the defect, obtained with an alternative approach, is numerically smaller and amounts to h_{br} = 1.4 mm. Due to the scatter of the obtained rejection values of the depth of the defect with different approaches in its determination, it is advisable when assigning the rejection value to be guided not only by increasing the requirements for the quality of the welded joint, but also take into account the material costs of the defect correction.

Bear in mind that the use of different methods for assessing the reliability of control requires the use of different sets of necessary data and has differences in what rejectionable measured characteristics of defects can be obtained (figure 6).

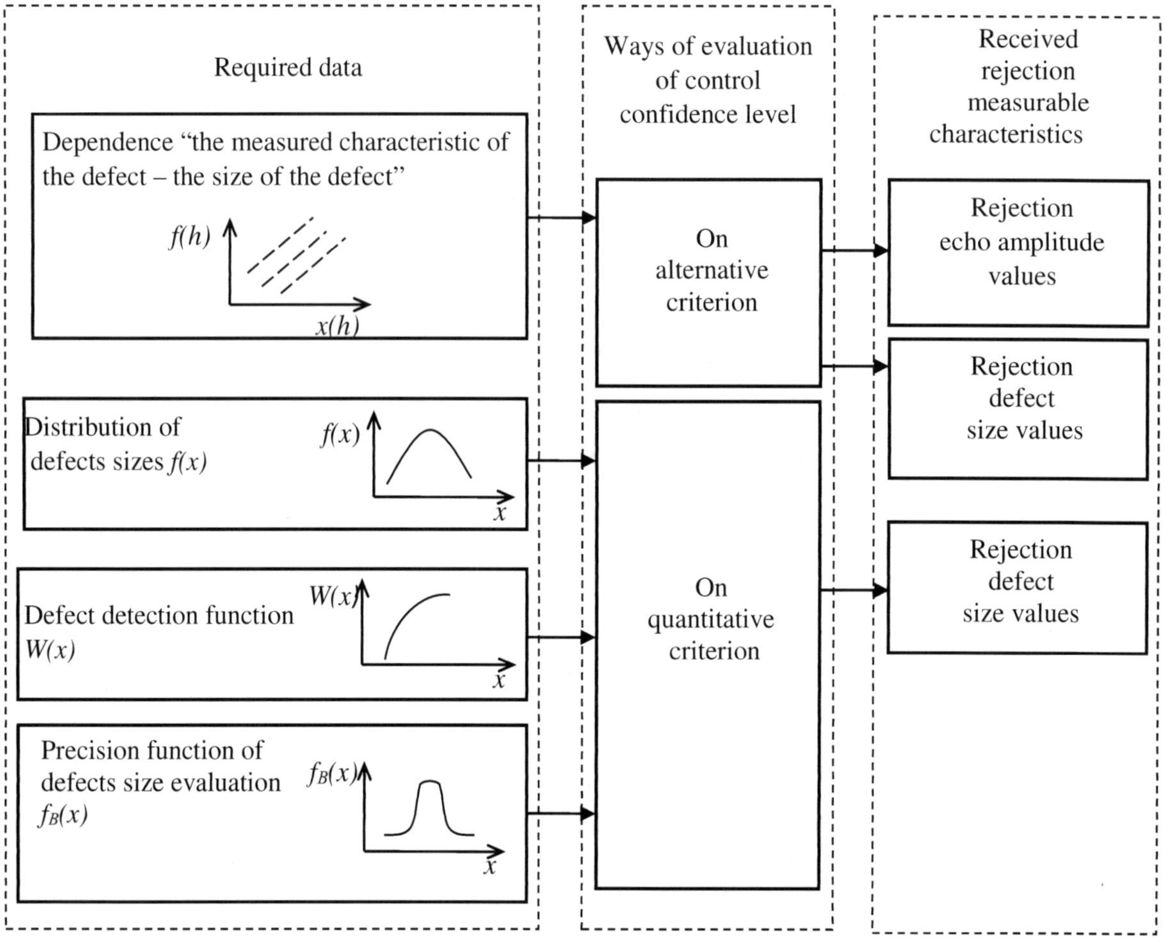

Figure 6. Comparison of methods for assessing the reliability of control for determining the rejection values of the measured characteristics of defects according to the necessary data and obtained rejection measured characteristics of defects.

Evaluation of reliability of ultrasonic testing on alternative criterion allows selection of the rejection values of the amplitude of echo signal and size of defect, ensuring reduction of under rejection. This requires only knowledge of correlation dependencies "amplitude of echo signal – size of defect" and "measured size of defect – real size of defect." Assessment reliability on the basis of quantitative criterion allows to select only rejection value for the size of defect, which ensures reduction of under rejection. This requires knowledge of distribution of defects sizes, their detectability and accuracy of estimating sizes of defects.

3. Conclusion

Advantage of evaluating reliability of ultrasonic testing on alternative criterion to determine the rejection values of measured characteristics of defects is the ability to assess the rejection values of amplitude of echo signal and size of defect, which provide reduction of under rejection. Usage of assessment of indicators of reliability of control on alternative criterion is advisable for operational control of product, for example, during operation. Assessment of reliability by quantitative criterion makes it possible to select only rejection value of the size of defect, which ensures the reduction of under rejection. Assessment of reliability of control on quantitative criterion provides additional important information on the distribution of defects in size, the detectability of defects and accuracy of estimation of the size of defects, which is appropriate for regulating the technology of manufacturing (repairing) products and the technology of their ultrasonic testing.

References

[1] Konovalov N N 2004 Requirements for the quality of welded joints *Welding production* **11** p 45–46

[2] Ivanov V I, Konovalov N N, Kotelnikov V S, Kotelnikov V V and Musatov V V 2005 Issues of accident risk assessment using technical diagnostics *Contr. Diagnostics* **3** p 12–20

[3] Konovalov N N, Dergachov A N, Usachov I I and Arefiev A A 2016 Two-parameter estimation of reliability of ultrasonic testing *19th World Conference for Non-Destructive Testing Book of abstracts* June 13–17 (Germany, Munich) p 23–24

[4] Konovalov N N 2004 Rationing of defects and reliability of non-destructive testing of welded joints *M FGUP NTC "Industrial Safety"* p 128

[5] Konovalov N N 2004 Reliability of ultrasonic testing of welded joints with constructive incomplete welds *Non-destructive testing* **9** p 38–41

[6] Meleshko H V, Petrov A A, Shitikov B S, Mezintsev E D 2012 Detection and estimation of the sizes of reflectors of various types of flaw detectors with phased antenna arrays *Vestnik MEI* **2** p 91–94

[7] Scherbinsky B G 2014 *Technology of ultrasonic testing of welded joints* 3-rd izd., rev. and add (Saint-Peterburg: CVEN) p 495

[8] Scherbinsky B G 2012 On the possibility of patternless prior validation of ultrasonic testing techniques *Non-destructive testing* **11** p 14–31

IOP Conf. Series: Materials Science and Engineering **656** (2019) 012028 doi:10.1088/1757-899X/656/1/012028

Creating an algorithm for improving the working surfaces of tools, using computer modeling

A I Kruglov[1], Y M Zubarev[1], D V Zhelnov[2] and I V Shvetsov[3,4],

[1]Saint-Petersburg State Polytechnic University Peter The Great, Saint Petersburg, Russia
[2] Department of Expertise and Certification of the Union "St. Petersburg Chamber of Commerce and Industry", RSaint Petersburg, Russia
[3]Yaroslav-the-Wise Novgorod State University, ul. B. St. Peterburgskaya, 41 173003, Veliky Novgorod, Russia

[4]E-mail: Igor.Shvetsov@novsu.ru

Abstract. The issue in question relates to machine building, particularly to machining of metals. We used the approach and software environment allow comparison with other aforementioned approaches and software environments to more accurately simulate the stress state of the material of the tool and, thereby, to increase accuracy of the design parameters of the tool for specific operating conditions. Introduction of automated systems of metals is directly related to the device status monitoring process of metal cutting, work process and increase its efficiency in the conditions of limited use of staff. Tasks of further improvement of machine-building production are inextricably linked with the effective use of technological equipment. Improving the efficiency of CNC machine tools is limited "hard" daemon setting the specified modes of the metals, when not taken into account peculiarities of processing of a concrete detail and lays the modes for the most demanding conditions.

1. Introduction

Reliability, efficiency, and environmental safety of hydraulic turbines, pumps and other energy-converting machinery depend to a large extent on the quality of materials of bearing assemblies and end seals. Metallic antifriction materials used now cannot work without lubricating with the mineral oils, this complicates the construction of bearing assemblies and does not exclude the oil getting into the environment. Domestic power engineering still lags behind foreign companies in terms of the use of composite antifriction materials. Plastic deformation, shear and destruction of the crystal lattice during the machining of materials is accompanied by the formation of a substructure with a high concentration of defects, the problem of thermal and temporary stability of defective structures arising. This structural stability must be the guarantee of preserving the properties of products, especially it concerns their work during the whole period of their operation in real conditions. Despite the difference in technological solutions, all methods of machining materials have common features distinguishing them from static or low-speed deformation and relief cracking. The most important of them are the following: high stress gradient between deformable and non-deformable materials in the small volume of the transition zone, short duration of impact of external forces, the formation of a substructure with high density of defects of a crystal lattice, and, in some cases, the absence of significant macroscopic deformation [1]. Under the conditions of high-speed deformation of the material, an adiabatic shear occurs – the formation of local zones of increased plastic deformation, the energy of which is converted into heat, this decreasing the yield limit with increasing temperature. When heat is emitted under the conditions of chip formation at a higher rate than its removal into cooling medium due to heat conductivity, heat is concentrated near strips or shear planes, this leading to the occurrence of the so-called adiabatic shear bands. During shear deformation, the transition from uniform deforming to adiabatic shear is determined by the magnitude and speed of deformation. Temperature in the region of the adiabatic shear depends on the magnitude and speed of the deformation and thermal physical characteristics of the material. Under the conditions of high – energy action, one of possible explanations of behavior of materials has been given on the basis of appearance of "highly excited states" in crystals. Under these conditions, the behavior of materials becomes non-linear, hydrodynamic nature of plastic flow is possible, a metastable structure and phase appear. In accordance with this theory, highly excited crystal becomes essentially a superposition of several structures, and the number of destroyed structural states far exceeds the number of atoms, this state of a substance being called atom-vacancy. In extreme conditions, when the process of chip formation occurs, the dynamic balance between crystalline and amorphous phases is even more probable. Abnormally intensive flows of defects occurring in these conditions cause much higher rates of transferring the substance. In general, the concept of structural transformations in the conditions of high-energy impact is formulated like this: into the crystal "energy is introduced, accumulative in the form of excited atom-vacancy conditions, which is then released as defect flows of standard or pulsed nature [1].

2. Results and Discussion

Techniques of examining high-speed processes which allow fixing structural changes in the process of chip formation are not quite perfect yet, and from the point of view of flow of various processes in material processing they are missing. Therefore, various aspects of phenomena of the interaction of a substance with thermal fields, plastic deforming and destruction of materials have not been fully understood due to the lack of complete theoretical and experimental data on the diagnostics of changes of the substance condition at atomic levels. Decomposition processes of oversaturated solid solutions, ordering and disordering of alloys, diffusion, self-diffusion, and mass transfer of substances, as well as the choice and recrystallization of deformed macro- and microobjects of metals have not been sufficiently studied. The study of these processes will allow more efficient use of mechanical processing, process control, and integrated methods of processing, as well as of the knowledge we have about physical phenomena [1].

The normal load on a cutting tool is a function of component cutting forces, and physical mechanical characteristics of the materials of the tool and the blank depend on the temperature in the cutting zone. But the actual contact area of the tool with the blank can be defined as the product of the number of contact points by their average size. The increase in contact area with the increase in wear leads to the growth of the number of contact points resulting in the increased heat flow. At the same time, the temperature in the contact zone affects the amount of heat flow and the nature of diffusion processes [1].

In the study of thermal processes, the dependences obtained by schematization and simplification of actual processes of thermal distribution are used. In general, these simplifications are as follows:

– heat sources are considered to be either concentrated or distributed according to the corresponding law which allows to describe quite easily the process of heat distribution;

– body shapes or components of the system are simplified;

– thermal physical quantities (heat conductivity coefficient, thermal diffusivity, specific heat capacity, and linear expansion coefficient) are taken in some cases as independent on temperature, what somewhat distorts the actual process, but significantly simplifies mathematical expressions.

Study and description of main patterns of flowing of various processes during machining allow one to approach the explanation of mass transfer in the conditions of chip formation, formulate the main principles of intensification of different processes during metal processing. The study of the processes of high-speed deforming and destruction of materials allows us to judge, on the one hand, the dynamics of the substance condition change in substructural and atomic levels by the nature of the processes

IOP Conf. Series: Materials Science and Engineering **656** (2019) 012028 doi:10.1088/1757-899X/656/1/012028

course, and, on the other hand, to develop qualitatively new physical principles for intensifying the methods of material processing, the control of processing, and the diagnostics of the state of a cutting tool. Based on a comprehensive study of the relationships of deformation, post-deformation and material destruction processes, structural changes in the conditions of high speed processing, a scientifically determined approach to the use of the results obtained in making new control methods is possible.

One of the problems is the selection of friction pair – a metal alloy and a method of its surface treatment. The sufficiently high surface hardness of carbon plastics (compared to, e.g., to fluoroplastic-4), special physical and chemical properties require high mechanical characteristics of the response pair, which leads to increased cost of the product [2, 3, 4].

In mechanical engineering, technologies of changing physical and mechanical state of the part surface are widely used. Figure 1 presents a diagram illustrating various technological measures aimed at changing the physical mechanical (physical chemical) condition of both the whole material from which the friction pair element is made and of a thin surface layer. Experiments on metal samples are still crucial for calculating the properties of a material, checking the model, adjusting it to solve the problems of increasing wear resistance. Due to the use of computer simulating, they decrease dramatically in volume, complexity and cost [2, 3, 4].

Making block-diagrams of programs for calculating thermal fields by the finite element method greatly simplifies and reduces the cost of making the improved surface layer of the tool and opens up possibilities for computer modeling of the quantities of these surfaces without significant expenses for laboratory physical experiments.

Figure 1 presents a diagram illustrating the methods that allow changing physical and mechanical state of the surface layer of friction pairs.

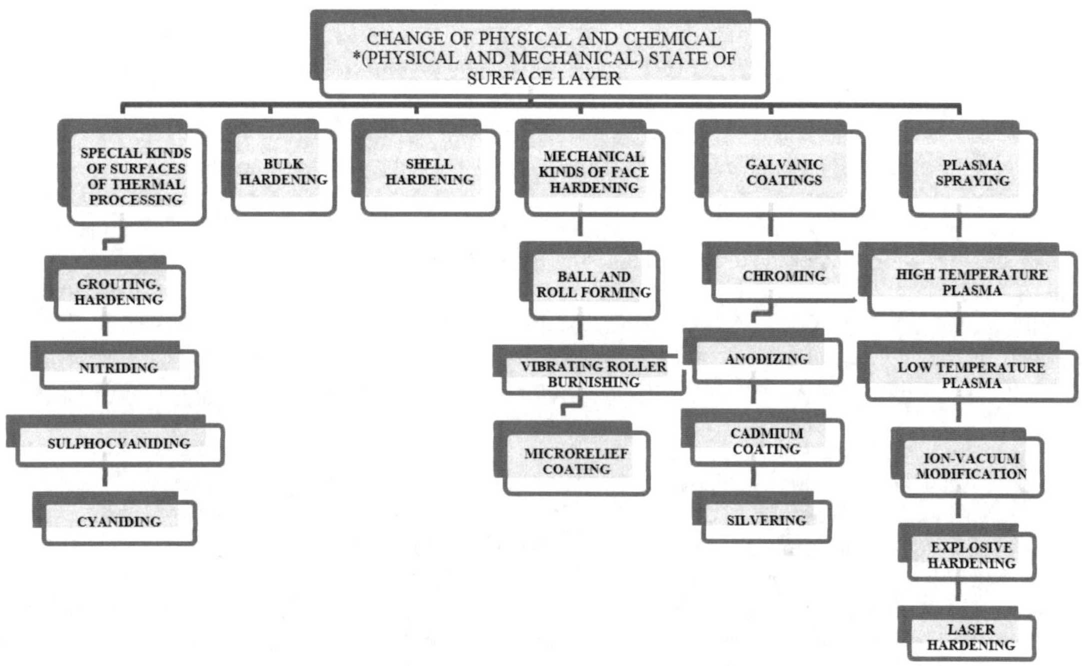

Figure 1. Diagram illustrating methods that allow changing physical mechanical state of the surface layer of friction pairs.

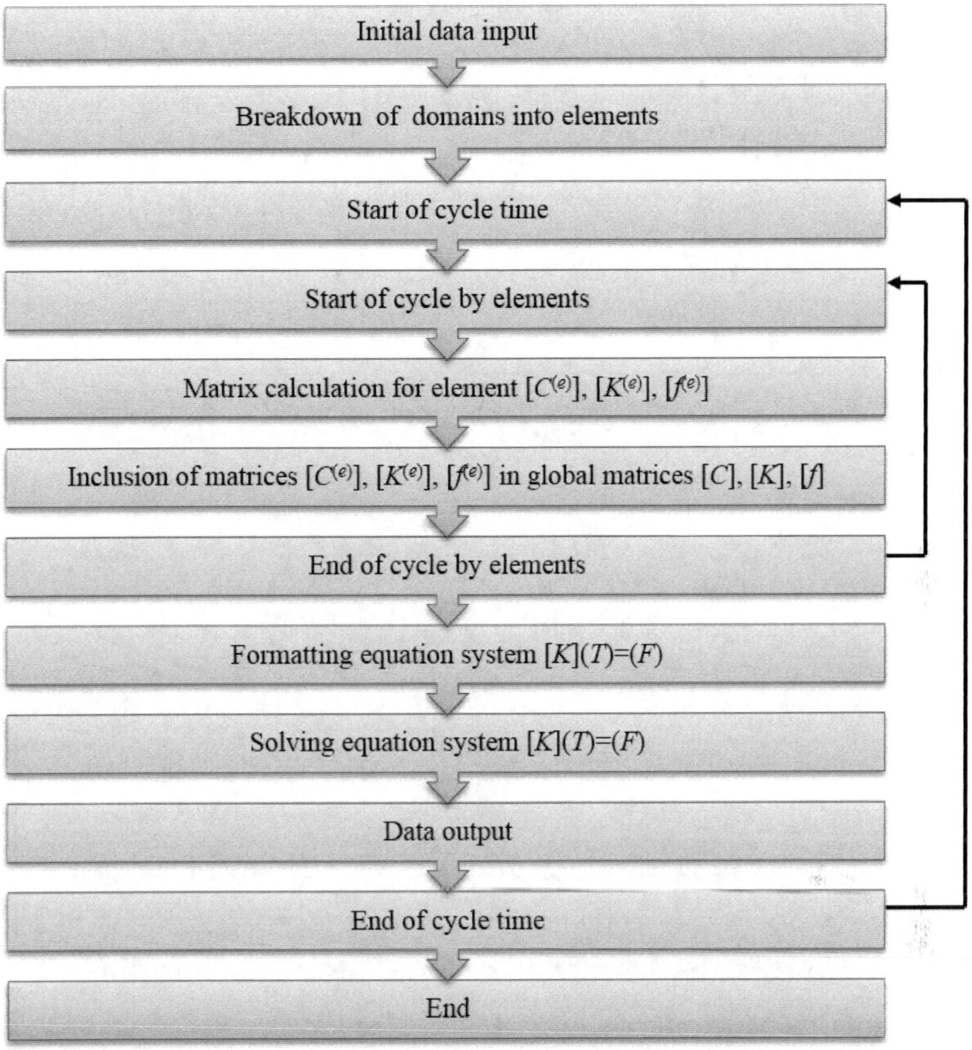

Figure 2. Block-diagram of the program implementing the calculation of thermal fields by the finite element method.

The given methods can be divided into the following groups:

– a group of technologies associated with the bulk quenching – or thermal treatment aimed at increasing the material hardness;

– technological operations aimed at the surface layer hardening, e.g., by high-frequency currents;

– technological operations associated with special kinds of surface thermal processing: cementing followed by hardening and tempering, or methods of nitriding, cyaniding, sulphocyaniding, etc.;

– technological operations associated with plating with galvanic coatings: chroming, anodizing, cadmium coating, coppering, silvering, etc.;

– mechanical technological operations of hardening kinds of surface processing: regular microrelief coating; vibrating roller burnishing; roll and ball forming, etc.;

– methods of plasma spraying which may be divided into two groups: high temperature plasma; low temperature plasma;

- methods of ion-vacuum surface modification, or they are also called ion implantation;
- methods of explosive surface hardening;
- methods of laser surface hardening.

One of the most promising methods is the method of plasma spraying which allows modifying the material surface to required technical parameters, such as hardness, roughness, residual surface stresses.

However, application of plasma spraying is accompanied by the number of difficulties, such as the occurrence of a defective layer of extending residual stresses, and microfissures.

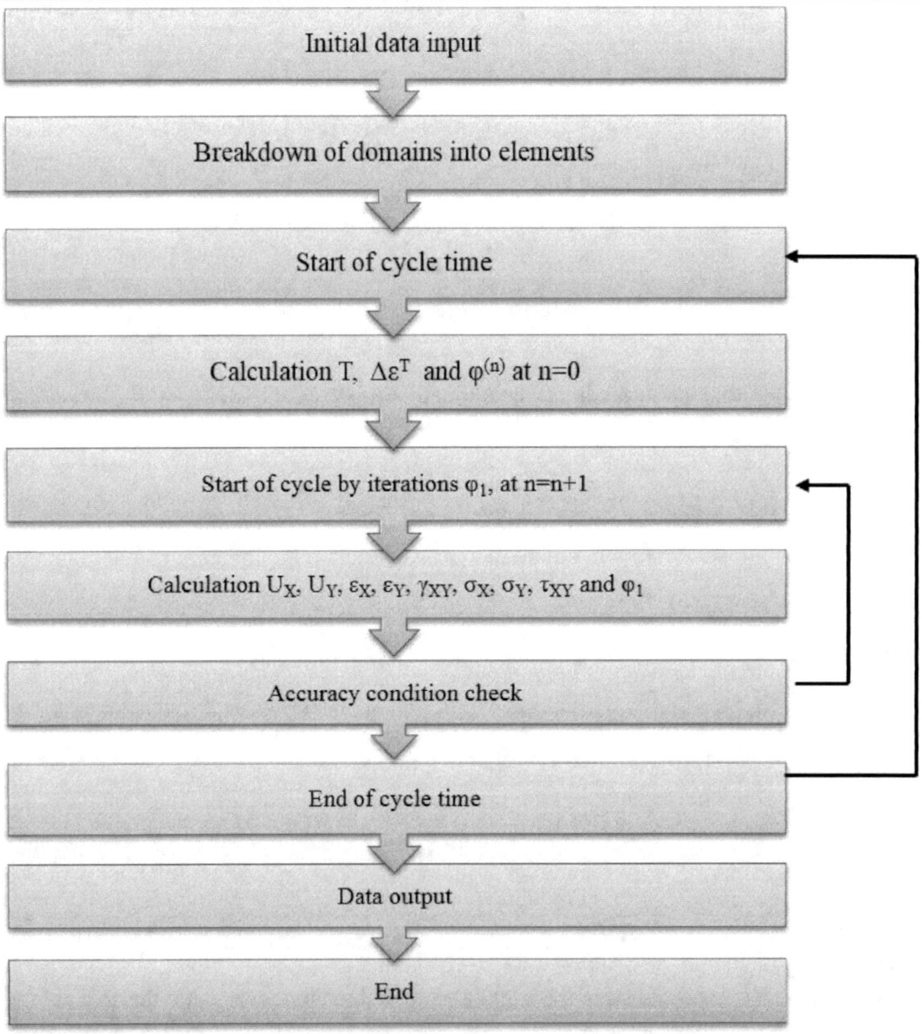

Figure 3. Block-diagram of the program which implements the implementing calculation of flat stress state.

The given defects occur if the spraying technology is wrong or it is not observed. Development of hard facing technology is accompanied by examining on natural samples. To reduce the cost of examinations, it is possible to use computer simulating for hard facing processes. Compared to routine "natural" experiments, computer simulating requires preliminary efforts to create models in the form of software supply. However, the following experiments on the model prove to be more operative, cheap and effective.

For the purpose of rational spraying parameters, it is possible to use modern computers and application package implementing calculations by the finite element method. The main idea of the finite element method is that any analog quantity such as temperature, pressure or displacement can be approximated by a discrete model that takes place on a set of piecewise continuous functions defined on a finite number of subdomains.

Piecewise continuous functions are defined by the way of values of analog quantity in the finite number of points in the considered domain. To define the thermal deformation fields, the shape of an object for which calculations will be made is first set. Then the shape is broken down into elements. It is thought that on each element all characteristics of the material such as temperature, stress, deformation, displacements, as well as physical chemical properties of the material do not change, or change linearly. The calculation of thermal fields is made according to the algorithm (figure 2), and the calculation of stresses is made according to the algorithm (figure 3). Particularly, the description of physical, chemical and mechanical processes has been presented in the literature [1, 5, 6].

3. Conclusion

In the presented work the materials were compiled on the basis of the calculations made and computer simulating of the surface hardening process at the “Automata” department in the SPbPU, which allow us to recommend the given programs for application of wear resistance coatings in mechanical engineering.

References

[1] Shvetsov I V 2017 *Fiziko-himiko-mekhanicheskieprocessy v proizvodstve: monografiya* [Physico-chemical and mechanical manufacturing processes: monograph] (Veliky Novgorod: NovSU Publ) p 100

[2] Volkov V Y 2014 *Teoriya I opisanie programmy modelirovaniya: uchebno-metodicheskoe posobie* [Theory and description of the simulation program: textbook] (Moscow; Berlin: Direkt-Media) p 85

[3] Kurkin S A, Khovov V M, Aksenov Yu N *at al* 2002 *Komp'yuternoe proektirovanie I podgotovka proizvodstva svarnyh konstrukcij:* ucheb. posobiedlyavuzov [Computer aided design and preparation of production of welded structures: Proc. manual for schools] (Moscow: MGTU im. N. Uh. Bauman Publ) p 464

[4] Kukudzhanov V N 2008 *Komp'yuternoe modelirovanie deformirovaniya, povrezhdaemosti I razrusheniya neuprugih materialov I konstrukcij*. Uchebnoe posobie [Computer simulation of deformation, damage and fracture of inelastic materials and structures. Textbook] (Moscow: MIPT Publ) p 215

[5] Shvetsov I V, Aleinikova M A, Zubarev S M and Nikulenkov O V 2016 Problems of dynamics and control during fine machining of body parts *Research Journal of International Studies* **12** 215–217

[6] Shvetsov I V, Belyakov V N, Zhelnov D V, Letenkov O V and Nikulenkov O V 2018 Diagnostirovanie rabotosposobnosti rezhushchego instrumenta [Diagnostics of Wheel Life in Grinding] *Russian Engineering Research* Vol 38 **8** 610–613

IOP Conf. Series: Materials Science and Engineering **656** (2019) 012029 doi:10.1088/1757-899X/656/1/012029

Impact of thermocycling type on nature of changing reversible deformations in titanium nickelide

S V Kryuchkov[1,3], N P Bogdanov[1], V G Malinin[2], V L Savich[1] and M N Konovalov[1]

[1] Ukhta State Technical University, Ukhta city, Komi Republic, Russia, Pervomaiskaya street, 169000
[2] Orel State Agrarian University, Orel city, Russia, Generala Rodina street, 69, 302019

[3]E-mail: kruk06@yandex.ru

Abstract. For cyclic functioning of mechanisms for various functional purposes based on materials with a shape memory effect (SME), it is necessary to possess skills of forming the characteristics of a cyclic shape memory, i.e. the shape memory effect implemented upon heating and plasticity of direct transformation observed upon cooling of the material under load. For this purpose, a systematic experimental research of the impact of a type (mode) of thermocycling of samples of titanium nickelide under torsion on the nature of changing reversible deformations in the process of thermocycling was performed. To assess quantitatively the effectiveness of a thermocycle, the value of a thermocycle completeness coefficient was introduced which is determined as the ratio of the value of the deformation restored for a cycle to the value of the deformation accumulated for a thermocycle. Three modes of loading were studied.

1. The material was loaded in a martensitic state to a certain level, then it was heated through the interval of a reverse martensitic transformation, in the austenitic state it was unloaded and cooled to its original state.
2. The material was loaded in a martensitic state, then heated through the interval of a reverse martensitic transformation to an austenitic state and cooled to its original state.
3. The material loaded in a martensitic state was unloaded. Then it was heated through the interval of a reverse martensitic deformation to an austenitic state in a free state, where it was loaded and the loaded material was cooled to a martensitic state. For all thermomechanical modes of loading upon heating, the value of the deformation restored and accumulated upon cooling was defined experimentally.

It has been experimentally found that the highest value of the restored deformation is implemented in the third loading mode, and the lowest one – in the first one. The intermediate deformation is implemented in the second mode.

The highest value of the deformation determined by the effect of the direct transformation plasticity takes place in the third mode.

1. Introduction

Alloys with a shape memory effect (SME) are capable to demonstrate the reversible shape changing during thermocycling through the intervals of martensitic transformations under a mechanical stress [1]. Herewith, at the stage of heating the SME itself occurs as a rule, and it is implemented in the direction opposite to the acting loading, and at the cooling stage the accumulation of the deformation in the direction of the acting force, or a direct transformation plasticity (DTP), is observed. In the work, the systematic study of the impact of loading mode type, the values of acting stresses – τ and the number of thermocycles – N on the values of deformations restored when heated, through the interval of reverse martensitic transformation, and accumulated in the process of the direct transformation plasticity has been made.

In the work [2], the example of the TN-1 alloy has shown that, if the thermocycling is made according to the second mode by the way of thermocycling under loading through the intervals of martensitic transitions, then the value of the SME deformation implemented upon subsequent heating will be essentially higher that that after an isothermic loading and subsequent unloading at temperature $- T_D$. Herewith, the value of the ratio $\frac{\gamma_p^s}{\gamma_p^i}$ depending on the value $-T_D$ and the level of acting loadings varies from 1.4 to 27.0. All above stated testify that the most effective way of forming the reverse deformation, and, particularly, of the shape memory effect includes thermocyclic variants of impacting on the material under loading.

The purpose of the present work is to perform a systematic experimental study of the impact of a certain type of thermocycling titanium nickelide samples on the nature of changes in reversible deformations, as well as to make an analytical model to quantify this impact.

2. Materials and methods of the experiment

The TN-1 (53,5-56,5 Ni, 43,2-46,5 Ti% by weight) alloy with the characteristic temperatures of martensitic transitions $M_S = 326$ К, $M_F = 298$ К, $A_S = 365$ К, $A_F = 395$ К was studied. Smooth cylindrical samples, with the length and the diameter of the working part 33 and 4 mm correspondingly, were exposed to low temperature annealing in 823 K for 1 hour in the state before the study. Next, the experiments in three modes detailed in [3, 4] were performed. In all modes, the thermocycling was performed within the interval of temperatures $T_{min} - T_{max}$. Minimal temperature matched the martensitic state at $T_{min} = 291$ K, and maximal temperature matched the austenitic state at $T_{max} = 520$ K. In all experiments two complete thermocycles within the interval of temperatures $T_{min} \leq T \leq T_{max}$ were performed according to the following three modes of thermocyclic impact:

The first mode: heating from the martensitic state to the austenitic one under the constant stress $\tau_S = \tau$; unloading in the austenitic state to the stress $\tau_0 = 0$; cooling from the austenitic state to the martensitic one in the unloaded state; loading in the martensitic state to the stress $- \tau_S$, where γ_p – the deformation of the shape memory, figure 1.

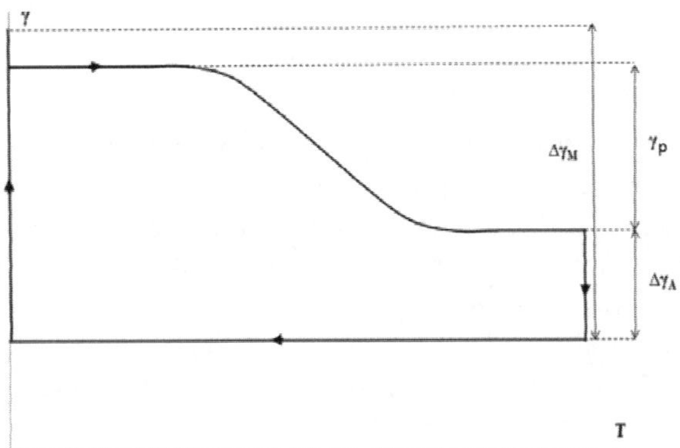

Figure 1. First mode circuit.

The second mode: Heating from the martensitic state to the austenitic one and cooling from the austenitic state to the martensitic state under constant stresses $\tau_S = \tau_0 = \tau$, where τ_S and τ_0 –stresses at the stages of cooling and heating correspondingly, where γ_{tp} – the deformation of transformation plasticity, figure 2.

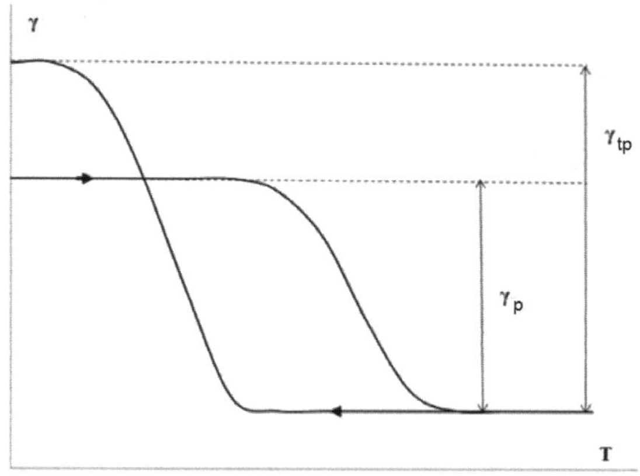

Figure 2. Second mode circuit.

The third mode: Unloading in the martensitic state from stress $\tau = \tau_0$ to $\tau = 0$; heating without loading to the austenitic state; loading in the austenitic state to the value of stresses $\tau_0 = \tau$; the next cooling to the martensitic state under stress, figure3.

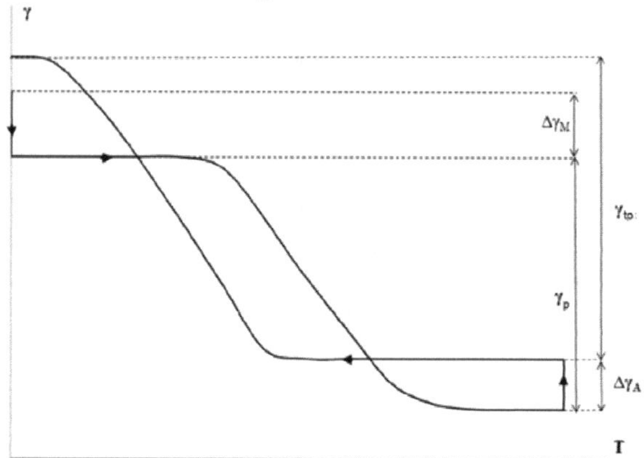

Figure 3. Third mode circuit.

Experiments were made according to the techniques detailed in [5]. Thermocycling was made with the average speed of changing temperature $|\dot{T}| \cong 0.08$ degree/sec. Herewith, the measuring of the value of the temperature T was made with the help of torque chromel-copel thermocouple with accuracy to 2°C. At the heating and cooling stages, the steering angle of cross section of the sample φ was measured. Herewith, the accuracy of measuring the angle φ was 0.009 rad, and that of the torque M– 0,01H· m. Values of the integral characteristics M and φ allowed calculating the estimated values of shear stresses and deformations referred to the outer fiber of the working part of the sample to the approximation of an ideally-plastic hinge. Herewith, the following working hypotheses were used in the basis of estimating:

1. Hypothesis of flat sections is performed: the transverse section of a beam, flat and perpendicular to the longitudinal axis, remains flat and perpendicular to the longitudinal axis after being deformed.
2. Radius drawn from the center of the beam's cross section remains a straight line (does not bend) after the deformation.
3. The distance between the cross sections does not change after the deformation. The beam's axis does not bend, the diameters of the cross sections do not change.

These suggestions allow finding estimated values of shear stresses and deformations:

IOP Conf. Series: Materials Science and Engineering **656** (2019) 012029 doi:10.1088/1757-899X/656/1/012029

Estimated value of the shear stresses, as we have already noted, was performed to the approximation of ideally plastic body and referred to the outer fiber of the sample according to the formula $\tau=12\ M/\pi d^3$, where M– a torque, d– the diameter of the sample's working part.

Estimated shear deformation in the outer fiber was found by the formula: $\gamma=(d\cdot\theta/2)\cdot 100\%$, where d – the diameter of the sample's working part. Herewith, the error of measuring the shear deformation did not exceed 0.05%. Herewith, the thermocycling was performed for the first and third mode of thermocycling under two stresses $\tau = 300$ and 400 MPa. Let's note that the less stress was approximately equal to $\sigma_{0.2}/\sqrt{3}$, and the higher one – to $\sigma_{top}/\sqrt{3}$ for the alloy TN – 1. Besides, in the second mode, the thermocycling was performed at $\tau = 100;\ 200;\ 300$ and 400 MPa.

3. Results of the experiments

Figures 1–3 demonstrates the functional modes of thermomechanical impact on the samples.

In figure 1, the deformation γ_p is restored when heated, the deformation $\Delta\gamma_A$ is back additionally in the austenitic state under the next isothermal unloading, the deformation does not change with the following cooling from the austenitic state to the martensitic one, the isothermal additional loading of the deformation in the martensitic state gives accumulation of the deformation $-\Delta\gamma_M$.

In figure 2, the heating stage corresponds the deformation restoration by the value γ_p, and the cooling stage – the accumulation by the value γ_{tp}.

In figure 3, the deformation $\Delta\gamma_M$ is restored in the martensitic state under isothermal unloading, during the following heating in the free state the deformation is restored by the value γ_p, the following isothermal additional loading in the austenitic state gives accumulation of the deformation by the value $\Delta\gamma_A$, and during the cooling in the unloaded state the deformation γ_{tp} is accumulated. To quantify the experimental results, a coefficient of thermocycle **completeness** was introduced – as the ratio:

$$k_c = \frac{\gamma_h}{\gamma_{sf}};\ (1)$$

where γ_{sf} and γ_h – the accumulated deformation and the deformation restored for a thermocycle.

In the first mode: at $\tau_s = \tau; \tau_0 = 0; \gamma_h = \gamma_p + \Delta\gamma_A; \gamma_{sf} = \gamma_0 + \Delta\gamma_M = \Delta\gamma_M; \gamma_0$ – the deformation implemented during the cooling at $\tau_0 = 0$, which is equal to zero in the described experiments; $\Delta\gamma_A$ – the deformation removed under the isothermal unloading in the austenitic state; $\Delta\gamma_M$ – the deformation accumulated under the isothermal loading in the martensitic state. Then:

$$k_c = \frac{\gamma_h}{\gamma_{sf}} = \frac{\gamma_p+\Delta\gamma_A}{\gamma_0+\Delta\gamma_M} = \frac{\gamma_p+\Delta\gamma_A}{\Delta\gamma_M};(2)$$

In the second mode: at $\tau_s = \tau_0 = \tau;\ \gamma_h = \gamma_p; \gamma_{sf} = \gamma_{td};$ then

$$k_c = \frac{\gamma_h}{\gamma_{sf}} = \frac{\gamma_p}{\gamma_{tp}};\ (3)$$

In the third mode: at $\tau_s = 0; \tau_0 = \tau; \gamma_h = \Delta\gamma_M + \gamma_p; \gamma_{sf} = \Delta\gamma_A + \gamma_{tp};\ \Delta\gamma_M$ – the deformation returned into the martensitic state under the isothermal unloading; $\Delta\gamma_A$ – the deformation accumulated in the austenitic state under the isothermal additional loading.

$$k_c = \frac{\gamma_h}{\gamma_{sf}} = \frac{\Delta\gamma_M+\gamma_p}{\Delta\gamma_A+\gamma_{tp}};\ (4)$$

Experimental dependences of the deformations conditioned by the SME, DTP, as well as of the values k_c of the number of thermocycles for the second loading mode are presented in figure 4–6, which show that γ_p increases monotonously with the number of thermocycles, reaching its saturation at N = 4, figure 4.

The deformation conditioned by the DTP decreases monotonously during the thermocycling process, coming to saturation at N = 4, figure 5. In figure 6, there is a dependence of completeness coefficient on the number of thermocycles for the second mode, which shows that k_c increases monotonously with the number of cycles. Herewith, we can observe that the higher is the level of acting stresses, the lower is the completeness coefficient.

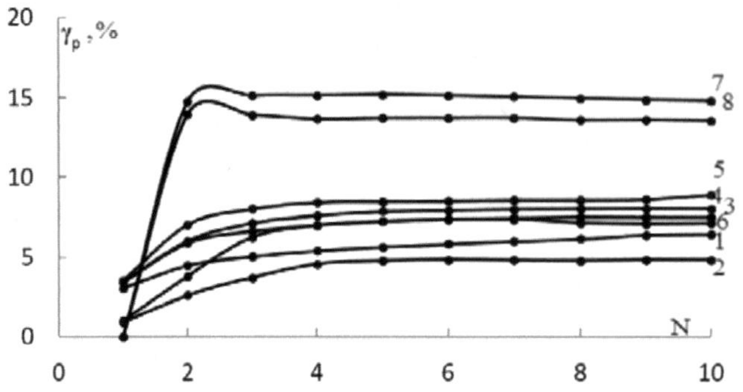

Figure 4. Dependences of memory deformation on the number of thermocycles in the first – (1, 2), curve 1 atτ = 300 MPa, curve 2 at τ = 400 MPa; in the second (3, 4, 5, 6) (curves 3at τ = 100 MPa; curve 4 at τ = 200 MPa; curve 5 at τ = 300 MPa; curve 6 at τ = 400 MPa) and in the third mode (7, 8) (curve 7 at τ = 300 MPa; curve 8 at τ = 400 MPa).

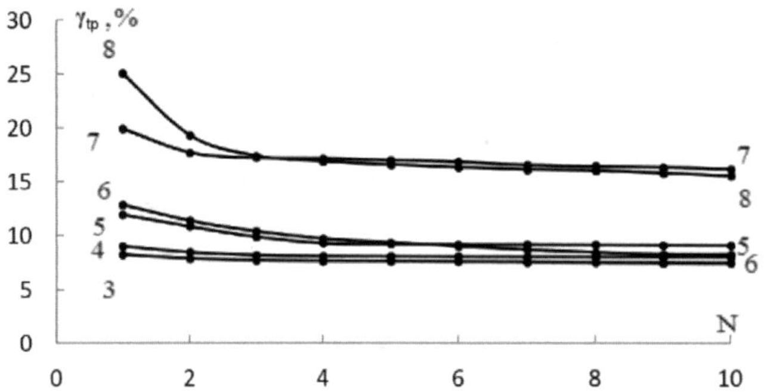

Figure 5. Dependences of direct transformation plasticity on the number of thermocycles in the second (3, 4, 5, 6) (curves 3 at τ = 100 MPa; curve 4 at τ = 200 MPa; curve 5 at τ = 300 MPa; curve 6 at τ = 400 MPa) and the third thermocycle (7, 8) (curves 7 at τ = 300 MPa; curve 8 at τ = 400 MPa).

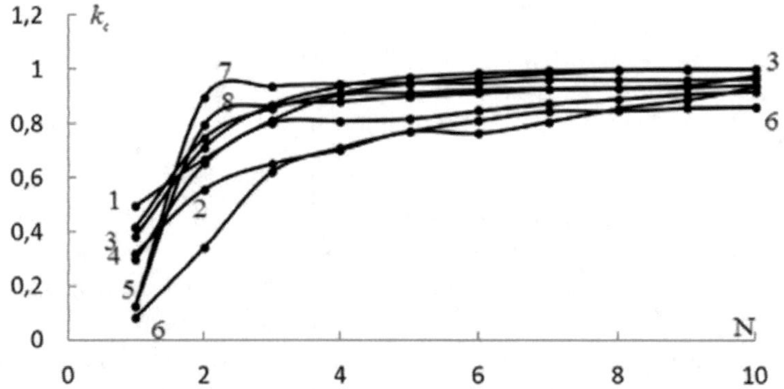

Figure 6. Dependences of completeness coefficient on the number of thermocycles in the first – (1, 2) in the second (3, 4, 5, 6) (curves 3 at τ = 100 MPa; curve 4 at τ = 200 MPa; curve 5 at τ = 300 MPa; curve 6 at τ = 400 MPa) and in the third mode (7, 8) (curves 7 at τ = 300 MPa; curve 8 at τ = 400 MPa).

Figures 4–6 presents the impact of the number of thermocycles on the values γ_p, γ_{tp} and k_c for the levels of acting stresses 300 and 400 MPa for the first and third mode, as well as for the levels of acting stresses 100, 200, 300 and 400 MPa for the second mode.

Herewith, the following is justified: the highest values $-\gamma_p$ are implemented in the third mode – heating in the free state, cooling under stress – and reach 15–16%, the intermediate value is implemented in the second mode – thermocycling under stress – and reach 8–9%, and the lowest value in the first mode – heating under stress, cooling in the free state – reach 6%. Herewith, it is important to note that in all cases the value γ_p for 300 MPa was higher than at 400 MPa, which is connected with the fact that, under stresses 400 MPa, reversible components of the deformation are suppressed due to irreversible components of deformations because of the effect of thermocyclic creep implemented in the materials [6, 7], so suppression of the SME occurs in the transition from stresses 300 MPa to stresses 400 MPa. Herewith, the following regularities are present: $\frac{d\gamma_p}{dN} \geq 0$; $\frac{d^2\gamma_p}{dN^2} \leq 0$, figure 4.

Figure 5 presents the dependences of deformations conditioned by the effect of a direct transformation plasticity (DTP) on the number of thermocycles for the first and third thermocycling mode. Here, it is important to note that in the second thermocycle the DTP effect itself is absent, as the circuit in figure 2 shows. Figure 5 shows that the DTP deformation is higher under stress 300 MPa than 400 MPa. It is characteristic that the DTP deformation for the third mode is about two times higher than for the second mode. The value of the DTP deformation decreases monotonously with the number of cycles.

Figure 6 presents the dependences of the completeness coefficient on the number of thermocycles. For all modes the monotonous growth of the deformation with the number of cycles is characteristic. Herewith, $\frac{d\gamma_p}{dN} > 0$; $\frac{d^2\gamma_p}{dN^2} < 0$ was characteristic. Herewith, the following was justified: the most complete cycle was observed in the third mode, curves 5–6, and the least complete cycle – in the first mode, curves 1–2. Intermediate values of the completeness coefficient occur in the second mode, curves 3–4.

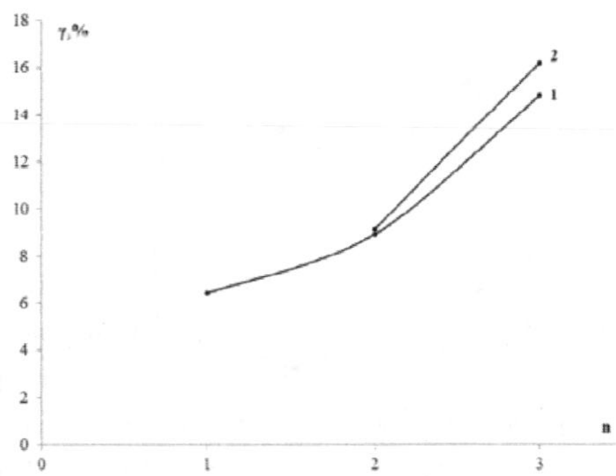

Figure 7. Dependences γ_p curve (1) and γ_{tp} curve (2) on the number of a testing mode. τ = 300 MPa, at N=2.

Figure 7 presents the dependences of the deformations conditioned by the SME and DTP on the number of a loading mode which show that the values of the SME deformation are minimal in the first mode, intermediate in the second mode, and maximal in the third loading mode. Deformation conditioned by the DTP is absent in the first mode, and in the third mode it is two times higher than in the second one.

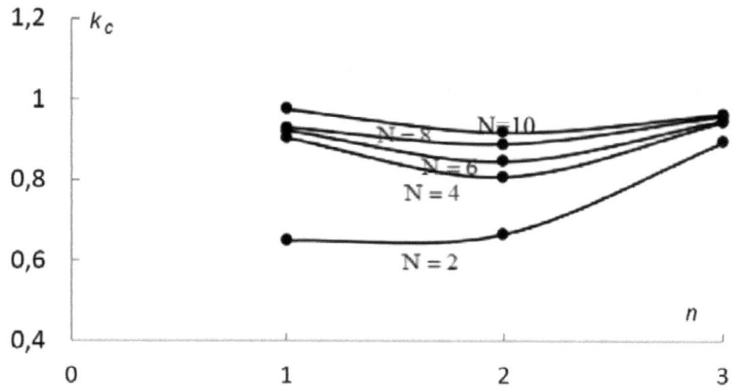

Figure 8. Dependences of completeness coefficient on the mode number, for N=2, 4, 6, 8 and 10.

Figure 8 presents the dependences of the completeness coefficient on the mode number for different thermocycles. The course of curves show that at the initial thermocycling stage at N=2 the completeness coefficient is minimal in the first, intermediate in the second, and maximal in the third thermocycling mode. It is associated mostly with the fact that in the first thermocycles the SME has not been yet formed, and this is especially noticeable in the first mode where the SME is being formed mainly due to the isothermal additional loading in the martensitic state. As thermocycles repeat, the SME continues to form and k_c increases with the number of cycles, this matching the ascending family of dependences of thermocycle completeness on the mode number, figure 8. Failure of the value k_c for the second mode is conditioned by the fact that the thermocyclic creep [6] implemented in the process of thermal changes more suppresses the SME in those thermocycles, where there is a stress both at the heating stage and at the cooling stage. This determines the course of extreme curves in figure 8, at N=2, 6, 8, 10.

4. Conclusions

Summarizing the above stated, we may conclude the following: the impact of thermocycling type on the nature of changes of reversible deformations in titanium nickelide conditioned by the SME and DTP, as well as on the completeness coefficient of the thermocycle – k_c was studied experimentally in the thermocycling process.

It has been found that, regardless of the mode of deformation thermocyclic impact, the conditioned SME as well as the thermocycle completeness coefficient increase monotonously during the thermocycling, the deformations conditioned by the DTP decreasing monotonously.

It has been shown that the deforming effects of the SME and DTP are maximal in the third thermocycling mode, while the minimal deformation associated with the SME is implemented in the first thermocycle.

From the above stated it follows that, for functional elements of constructions working under minor, close to zero, stresses the use of the third thermocycling mode is the most effective.

For forceful functional elements under major acting stresses, the first thermocycling mode is the most favorable, as in this case the high completeness of the cycle will be observed with the sufficiently high values of SME deformations.

References

[1] Likhachev V A, Kuzmin S L and Kamentseva Z P 1987 *Shape memory effect* (Leningrad: LSU Publ) p 216

[2] Andronov I N and Verbakhovskaya R A 2009 Shape memory effect of TN-1 alloy during the isothermal and non-isothermal deforming *Plant laboratory. Methods of mechanical testing*

Vol 75 **8** 52–54

[3] Andronov I N, Verbakhovskaya R A, Ovchinnikov S K and Severova N A 2007 Effects of reversible shape memory and thermocyclic return of deformation in TN-1 alloy *Plant lab.* Vol 73 **2** 64–67

[4] Andronov I N and Ovchinnikov S K 2005 Effects of reversible change of titanium nickelide during thermocycling under stress *Deformation and destruction of materials* **5** 28–30

[5] Andronov I N, Bogdanov N P, Churilina I V and Bogdashova N S 2016 Impact of loading modes on reversible and irreversible deformations in TN-1 alloy during thermal changes through intervals of martensitic transitions under torsion conditions *Deformation and destruction of materials* **1** 19 – 23

[6] Andronov I N, Kuzmin S L and Likhachev V A 1983 Thermocyclic creep of copper-manganous alloys associated with FCC ↔ FCT transformations *News of Higher Educational Institutions. Non-ferrous Metallurgy* **3** 84 – 88

[7] Andronov I N and Kryuchkov S V 2005 Computer modeling of deformational properties of martensitic inelasticity during thermocycling *Mater. of the 2nd Intern. Seminar "Physical and Mathematical Modeling of Systems": Modeling of physical processes in condensed media in systems of numerous* particles (Voronezh: Voronezh. State Techn. Univ. Publ) pp 3–12

[8] Belyaev S P, Resnina N N and Volkov A E 2006 Influence of irreversible plastic deformation on the martensitic transformation and shape memory effect in Ti-Ni alloy *Materials Science and Engineering* A Vol 438–440 627–629

[9] Andronov I N, Ryabkov Yu I, Bogdanov N P, Severova N A, Danilov A N and Churilina I V 2014 Irrevesible deformations and effect of super-plasticity of TN-1 alloy during thermocycling through intervals of martensitic transitions under loading *Deformation and destruction of materials* **9** 19 – 26

IOP Conf. Series: Materials Science and Engineering **656** (2019) 012030 doi:10.1088/1757-899X/656/1/012030

The dynamic of cooling of cast iron in the cristllizer in condition of the semi-continuous production process bimetallic band "steel-cast iron"

V P Lihoshva[1,3], A V Shmatko[1,4], V V Savin[2,5], L A Savina[2,6] and A N Tymoshenko[1,7]

[1]Physics-technological institute of metals and alloys of the National academy of science of Ukraine, Vernadsky Avenue, 34/1 03680, Kiev, Ukraine
[2]Immanuel Kant Baltic Federal University, A Nevsky st, 14 236016, Kaliningrad, Russia

[3]E-mail: plazer_v@mail.ru
[4]E-mail: blacknorfolk@gmail.com
[5]E-mail: VVSavin@kantiana.ru
[6]E-mail: LSavina@kantiana.ru
[7]E-mail: Marschal@i.ua

Abstract. The dynamics of hard crust formation is investigated in the liquid melt by contact with the working walls of the copper crystallizer under conditions of continuous production of bimetallic products such as "band" at different speeds of the process. The optimal speed is identifying to achieve maximum performance with the satisfactory quality of the finished products. Recommendations for increase speed of carrying out process are offered.

1. Introduction

The competitiveness of most industrial enterprises is largely determined by the performance and reliability of the equipment used and depends on the number of technological breaks and repairs (scheduled and emergency). Parts, operating in conditions of intense wear, determine the share whose replacement leads to downtime of the production equipment and technology. In particular, an urgent problem is to increase the wear resistance of the working bodies of machines and mechanisms of the mining industry operating in conditions of abrasive and impact abrasive wear. A promising direction for solving the problem of reducing the intensity of wear is the use bimetallic products [1, 2]. The main advantage of double-layer and multi-layer parts is the achievement of a set of properties of the material basis and the working layer, which can't be achieved by using monolayer products. Thus, the use of working bodies on the basis of bimetallic compounds with the necessary indicators of wear resistance of the deposited metal in conjunction with the specified structural properties of the base metal increases the total resource.

At the moment, there are many ways to obtain multilayer structures. The most widespread were processed of surfacing (plasma, laser, gas-powder, etc.) [3–7]. Despite its advantages, these methods have one common drawback-low it is process productivity. Modern technology requirements this is a high performance production, these include the technology for continuously in-situ blanks. Taking into account the disadvantages and advantages of modern processes of production of bimetallic and multilayer products, a hybrid technology of semi-continuous production of bimetallic construction of the "band" type was proposed [8, 9]. Its main point is the connection of the liquid of the melt preheated to the temperatures required for the diffusion of their connection with the substrate (workpiece) of a different material in the mold [10]. One of the key points of obtaining products using the above method is the crystallization of the melt.

In continuous casting machines, much attention is paid to optimizing the cooling rate of the melt, since it is an important factor in predicting the quality of products [11]. Copper crystallizers of both vertical and horizontal types are used for the production of continuous cast billets at modern metallurgical plants [12, 13]. With the help of the crystallizer, the finished product is formed by increasing the crust of the solidifying metal on the border of the "wall of the crystallizer – melt". According to [13], the value of the hard crust at the exit of the crystallizer is normalized (depending on the cross-sectional configuration) to avoid breaking the molten metal through the hardened wall under the action of the liquid metal head. The main task of the crystallizer is to provide intensive heat removal from the melt. During operation, the crystallizer removes 15–30% of accumulated heat from the liquid metal [13].

An important influence on the crystallization process has a gas gap, which is formed by metal shrinkage during phase transitions and further cooling. The thickness of the gas gap is unstable and can change along the perimeter and height of the crystallizer.

2. Results and Discussion

As part of the development of the foundry-plasma technology for bimetallic products, a numerical simulation of the melt cooling process was carried out to predict the crystallization front, taking into account the influence of a constant gas gap, in order to formulate recommendations on the choice of speed for the practical conduct of the process.

In order to optimize the calculation time, the processes of heat and mass transfer in the filling device were not taken into account in the model. As the material of the workpiece used steel DIN 1.0402 the overall dimensions of which are 70x10 (WxH, mm). As a fill layer took grey cast iron DIN GG35 (WxH=50x20, mm). Since the technology of obtaining a bimetal strip is a semi-continuous process, the length of the steel substrate and the filled layer was taken conditionally infinite.

The design of the crystallizer was used for the calculation without taking into account the thermal processes associated with the heat transfer of the cooling liquid.

According to the scheme (figure 1) cooling of molten cast iron was calculated as it progresses in the U-shaped crystallizer taking into account the contact with the steel billet. It was assumed that the pre-heated workpiece to the minimum temperature necessary for the diffusion compound (800 °C). Calculations were carried out for steel strip feed rates in the range of 2–20 mm/s. The temperature of the walls of the crystallizer, not in contact with the workpiece material and the melt, was set equal to the ambient temperature.

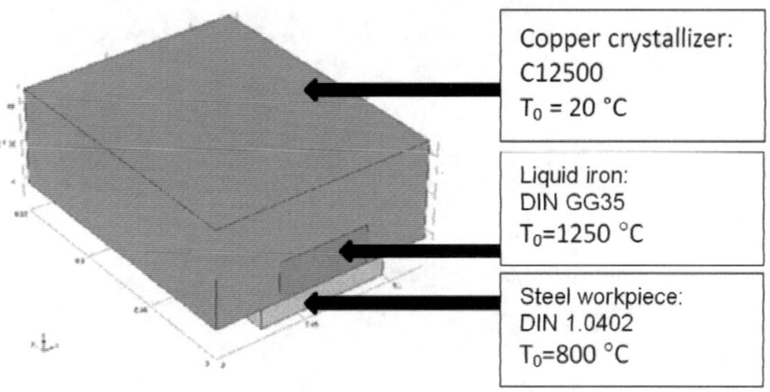

Figure 1. Schematic formulation of the problem of liquid cast iron cooling in the crystallizer.

The model of calculation of cooling of a bimetallic strip at contact with walls of the crystallizer taking into account processes of heat transfer includes the solution of the following equations

Continuity equations:

$$\frac{\partial \rho}{\partial t} + \nabla \cdot (\rho u) = 0 \quad (1)$$

The Navier-Stokes equation:

$$\rho \frac{\partial u}{\partial t} + \rho u \cdot \nabla u = -\nabla p + \nabla \cdot \left(\mu\left(\nabla u + (\nabla u^T)\right) - \frac{2}{3}\mu(\nabla \cdot u)I\right) + F, \quad (2)$$

where ρ – density (kg/m^3);
u – velocity vector (m/s);
p – pressure (Pa);
μ – dynamic viscosity (Pa·s);
F – vector of body force (N/m^3).

Heat transfer equation for liquid phase:

$$\rho C_p + \rho C_p u \cdot \nabla \mathrm{T} = \alpha_p T\left(\frac{dp_a}{dt} + u \cdot \nabla p_a\right) + \frac{\tau}{S} + \nabla \cdot (k \nabla T) + Q, \quad (3)$$

where C_p – specific heat at constant pressure (J/kg·K);
T – absolute temperature (K);
τ – tensor of viscous stresses (Pa);
S – tensor of rate of deformation (1/s);
Q – heat source (W/m^3).
q – heat flow by conduction (W/m^2);
$q = -h(T_{ext} - T)$
T_{ext} - ambient temperature (K);
k – thermal conductivity (W/m·K);
p_a – absolute pressure (Pa);
α_p – coefficient of thermal expansion (1/K).

Heat transfer equation for the solid phase:

$$\rho C_p \frac{\partial T}{\partial t} = -(\nabla \cdot q) - T\frac{\partial E}{\partial t} + Q, \quad (4)$$

where E – elastic component of the entropy (J/(m^3·K)).

The following dependence of the liquid phase fraction on the temperature was used to determine the crystallization front:

$$B = \frac{(T - T_{m.av.} + \partial T')\cdot a \cdot b}{\partial T'} + c, \quad (5)$$

where B – proportion of the liquid phase;
$T_{m.av.}$ – average temperature interval of crystallization: $T_{m.av.} = \frac{T_l + T_s}{2}$;
T_l – liquidus temperature, °C;
T_s – solidus temperature, °C;
$\partial T'$ – crystallization half-interval, °C: $\partial T' = \frac{T_l - T_s}{2}$;
a, b, c – logical variable:
if $(T \leq T_{m.av.} + \partial T')$ then $a = 1$;
if $(T > T_{m.av.} + \partial T')$ then $a = 0$;

if $(T \geq T_{m.av.} - \partial T')$ then $b = 1$;
if $(T < T_{m.av.} - \partial T')$ then $b = 0$;
if $(T > T_{m.av.} + \partial T')$ then $c = 1$;
if $(T \leq T_{m.av.} + \partial T')$ then $c = 0$.

The calculations took into account the maximum gas gap, which was determined from the equation:

$$\delta = \gamma(1 - B)l\varepsilon, \quad (6)$$

where l – value of the linear size, mm;
ε – linear shrinkage of cast iron, %.

Since we assume that there is no gas gap at the interface between the steel substrate and the cast iron layer, given the presence of a diffusion connection between the steel workpiece and the cast iron layer, the γ coefficient is introduced into the equation, which determines the proportion of linear shrinkage at the boundaries of the cast-iron layer. Thus, the gas gap on the surfaces contacting with the crystallizer was taken 0.25 mm and 0.4 mm on the vertical and horizontal borders, respectively.

The heat flux through the air gap can be determined from the equation:

$$q_{air} = (\lambda_{air}\delta)\left(T_{surf} - T_{cr}\right) + \varepsilon\sigma_0(T_{surf} - T_{cr}), \quad (7)$$

where λ_{air} – thermal conductivity of air in the gap;
δ – value of the gas gap;
T_{surf} – temperature of the contact surface of the melt;
T_{cr} – surface temperature of crystallizer;
ε – reduced emissivity factor of contact surfaces;
σ_0 – Stefan-Boltzmannconstant.

The choice of the optimal speed was carried out for the following reasons:
- formation of a hard crust of the required value to exclude the possibility of breakthrough and spilling liquid cast iron at the exit of the crystallizer;
- ensure maximum productivity process;
- ensuring satisfactory quality of finished products.

Based on the analysis of the results obtained, the speed interval was divided into three groups, taking into account the recommendations of applicability in the practical conduct of the process (table 1).

The feed rates of the workpiece 2–8 mm/s are characterized by complete cooling of the melt supplied to solidus temperatures at the time of exit from the crystallizer. Since there is no liquid phase at the crystallizer outlet inside the cast iron layer volume, the risk of shrinkage defects is reduced due to liquid metal feeding. It should be noted that the smaller the length of the liquid hole, the less likely the product shrinkage porosity and shrink holes. For figure 2 the values of the liquid phase in the longitudinal section at the center of the bimetal structure are presented, which give an idea of the crystallization front for velocities up to 8 mm/s. The necessary heat sink, which provides the values of the hardened crust (figure 2), provided by increasing the contact time of the supplied liquid metal with the working walls of the crystallizer. Since productivity is an important factor in industrial applications, results are preferred when choosing the optimal speed is preferable $V_{cast} \approx 8$ mm/s (table 1, gr. II).

With a further increase in the feed rate of the strip, and as a consequence, and the melt flow rate in the crystallizer, incomplete crystallization of the cast iron layer is observed at the time of exit from crystallization (figure 3). Moreover, the higher the speed, the greater the proportion of liquid and liquid-solid phase in the volume, which in turn can lead to a breakthrough of the metal and the appearance of shrinkage defects due to the inability to compensate for shrinkage of the liquid metal. Thus, speeds V_{cast} > 10 mm/s are unsatisfactory when the length of the mold is 150 mm.

Table 1. The fraction of the liquid phase in the cross section at the exit of the crystallizer for the velocity range.

№ / V, mm/s	I	II	III	*liquid phase drier, B*
2				
4				
6				
8				
10				
12				
14				
16				
18				
20				

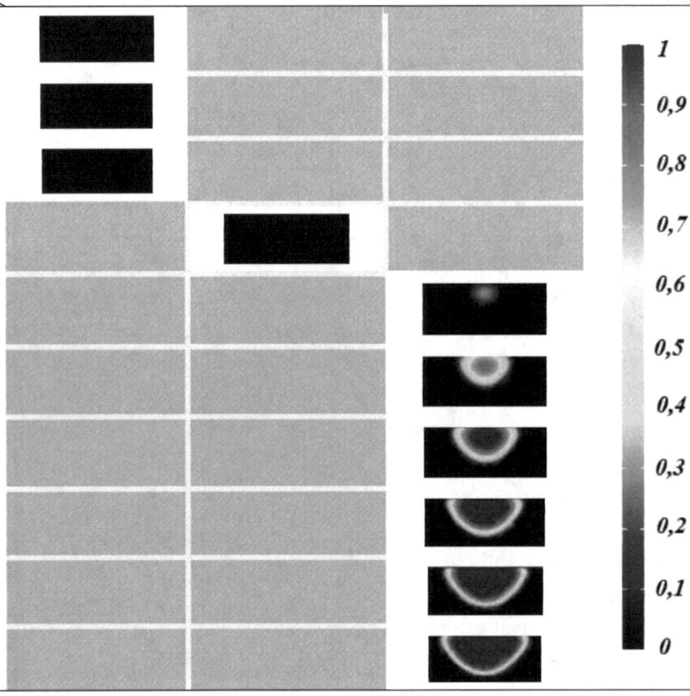

where I – satisfactory result; II – optimal result; III – unsatisfactory result.

If it is necessary to increase the productivity, it is necessary to increase the length of the crystallizer or take measures to reduce the effect of the gas gap on the heat transfer processes from the melt to the walls of the crystallizer. To do this, in the crystallizers of modern CCM, a change in the flow section in the direction of reduction is applied to neutralize the effect of shrinkage processes as cooling and advancement in the crystallizer [14]. In the case of a horizontal semi-continuous casting machine, the situation is complicated by the difference in cooling conditions at the boundaries of the heat exchange "cast iron-wall crystallizer" and "cast iron – steel substrate."

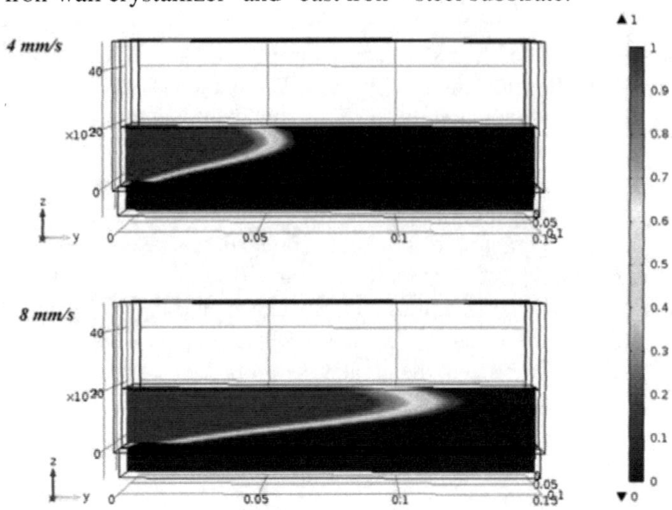

Figure 2. Volume fraction of the liquid phase in the longitudinal section for feed rates of the workpiece 4 mm/s and 8 mm/s, respectively.

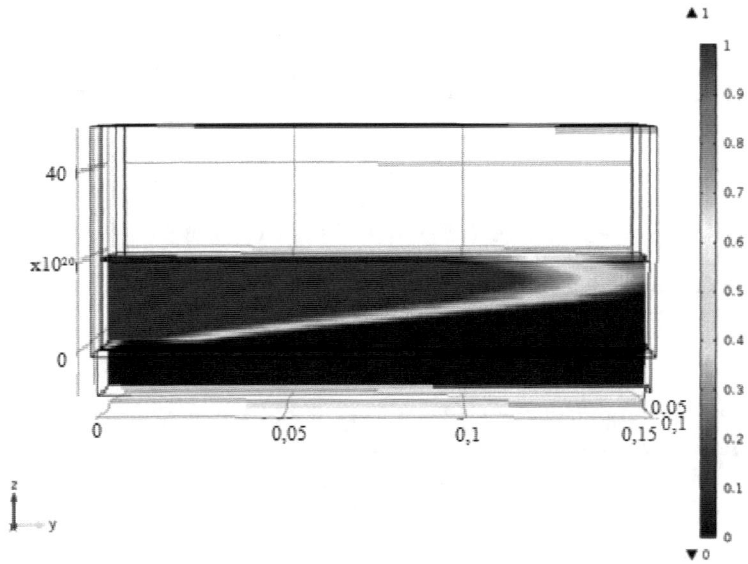

Figure 3. Volume fraction of the liquid phase in the longitudinal section for the feed rate of the workpiece 10 mm/s.

3. Conclusions

Taking into account the results of numerical simulation of the cooling process of cast iron in a semi-continuous process of obtaining bimetallic structure "band" formulated recommendations for the choice of speed.

Mass-flow rate V_{cast} = 2…6 mm/s is undesirable because of the poor performance of the process. The optimal value is $V_{cast} \approx$ 8 mm/s. If it is necessary to increase the speed, it is necessary to increase the contact time of the cast iron layer with the wall of the crystallizer by increasing its length, or take measures to reduce the influence of the air gap on the heat transfer processes from the melt by changing the flow section to compensate for the shrinkage processes.

Acknowledgments
Research was supported by the Ministry of science and higher education of the Russian Federation, unique project identification number RFMEFI57817X0252.

Reference

[1] Wrobel T 2011 Bimetallic layered castings alloy steel – gray cast iron *Science and engineering* **2** 118–125

[2] Posypajko I Ju and Socenko O V 2011 Povyshenie iznosostojkosti smennyh detalej promyshlennyh smesitelej [Increasing the wear resistance of replaceable parts of industrial mixers] *Metall i lit'e Ukrainy – Metal and casting in Ukraine* **1** 32–35

[3] Hasui A and Morigaki O 1985 *Naplavka i napylenie* [Surfacing and spraying] ed V N Popova (Moscow: Mashinostroenie) p 240

[4] Gologoskij E G 1998 *Mehanizirovannye sposoby naplavki i napylenija detalej stroitel'nyh, dorozhnyh i kommunal'nyh mashin [Mechanized methods of surfacing and spraying parts of construction, road and municipal machines]* (Moskow: MIKHiS)

[5] Tkachev V N, Fishtejn B M, Kazincev N V and Aldyrev V A 1970 *Indukcionnaja naplavka tverdyh splavov [Induction hardfacing of hard alloys]* (Moscow: Mashinostroenie)

[6] Steen W M and Watkins K G 1993 Coating by laser surface treatment *Journal de Physique* **3** 581–590

[7] Pauleau Y 2006 *Material Surface Processing by Directed Energy Techniques* (London: Elsevier) Retrieved from https://www.elsevier.com/books/materials-surface-processing-by-directed-

energy-techniques/pauleau/978-0-08-044496-3

[8] Lihoshva V P, Najdek V L, Karichkovs'kij P M, Pelikan O A, Glushkov D A and Nadashkevich R S 2010 *Ukrainian Patent UA №54486* (Kiev: Derdavne pidpriemstvo "Ukrains'kij institut intelektual'noї vlasnosti")

[9] Lihoshva V P, Aftandiljanc E G, Pelikan O A, Timoshenko A M, Nadashkevich R S and Rejntal' O O 2012 *Ukrainian Patent UA №74270* (Kiev: Derdavne pidpriemstvo "Ukrains'kij institut intelektual'noї vlasnosti")

[10] Lihoshva V P and Shmatko O V 2017 Bezperervnij livarno-plazmovij metod otrimannja bimetalevih ta bagatosharovih konstrukcij [Continuous casting-plasma method of obtaining bimetallic and multilayer structures] *XIII Mezhdunarodnaja nauchno-prakticheskaja konferencija «LIT''E. Metallurgija. 2017»* [XIII Internat. Scien.-pract. conf. "Casting. Metallurgy. 2017»] (Zaporozh'e: Nacional'naja metallurgicheskaja akademija Ukrainy) pp 166–167

[11] Allazadeh M R 2012 Cooling rate optimization of as-cast consciously cast steel *Iranian J. of Material Science and Engineering* Vol 9 **(3)** p 16

[12] Brovman M O 2004 Perspektivah razvitija nepreryvnogo lit'ja metallov [About the perspectives for the development of continuous casting of metals] *Nacional'naja metallurgija - National metallurgy* **5** 66-70

[13] Bulanov L V, Korzunin L G and Parfenov E P 2003 *Mashiny nepreryvnogo lit'ja zagotovok Teorija i raschet* [Machines for continuous casting of blanks Theory and calculation] (Ekaterinburg: Ural'skij centr PR i reklamy)

[14] Kozyrev N A, Gizatulin R A and Valuev D V 2014 *Nepreryvnaja razlivka stali i splavov: uchebnoe posobie* [Continuous casting of steel and alloys: a tutorial] (Tomsk: Izdatel'stvo Tomskogo politehnicheskogo universiteta)

IOP Conf. Series: Materials Science and Engineering **656** (2019) 012031 doi:10.1088/1757-899X/656/1/012031

Superhydrophobic coating based on silicone resin SILRES® MSE 100

V I Loganina[1]

[1]Penza State University of Architecture and Construction, Street Titov,28, 440028, Penza, Russia

E-mail: loganin@mail.ru

Abstract. The anti-icing properties of coatings based on the developed composition are considered. It was shown, that after wetting during 72 hours, the coatings based on the proposed composition retained a superhydrophobic effect. The wetting angle on the mortar substrate was more than 150 degrees. Adhesion of coatings to the substrate after wetting amounted to 1 point. Information is given about the results of evaluation of the adhesion of ice to a superhydrophobic metal surface. It is shown, that the force of detachment of a drop of water on a superhydrophobic metal surface is 3 times less. It ensures easier rolling of a drop of water from the surface and its anti-icing properties. It is established, that the adhesion of ice to the superhydrophobic surface is less compared to the hydrophilic one. It is shown, that there is no difference in the rate of ice evaporation on a hydrophilic and superhydrophobic surface.

1. Introduction

In order to protect the roofs of buildings from icing, various anti-icing systems are used. Super-hydrophobic coatings are promising as an anti-icing protection due to their unique properties:

– a significant (more than 3 times) slowing down of the process of droplet freezing on a super-hydrophobic surface [1–3];

– short duration of the contact of the superhydrophobic surface with a falling drop [4];

– small hysteresis of wetting angle - no more than 10–15 degrees [5, 6];

– low adhesion of ice to the surface [7].

Analysis of scientific and technical literature suggests, that the manufacture of superhydrophobic materials is possible with a combination of hydrophobic material and surface topography [8]. Despite the large number of proposed compositions, the problem of anti-icing remains relevant. In some cases coatings after a certain period of operation lose the superhydrophobic effect [9]. Thus, the development of anti-icing coatings for roofs of buildings is an extremely important task.

Based on the foregoing, a formulation for anti-icing coating was developed [10, 11]. When developing the composition, silicone resin SILRES® MSE 100 was used as a binder, and aerosil R 972 was used as a filler. The results of the conducted research showed, that after wetting for 72 hours a coating based on silicone resin SILRES® MSE100, 5% and 10% Noah concentration retained superhydrophobic effect – the wetting angle on the mortar substrate was more than 150 degrees. The adhesion of coatings to the substrate (estimated by the method of lattice cut in accordance with GOST 15140-78 “Paint materials. Methods for determination of adhesion ") after wetting amounted to 1 point.

2. Results and Discussion

For an experimental study of the anti-icing properties of coatings, metal profiled roofing sheets "MR-20" with polymer coating were used as substrates.

Under the action of a negative temperature (–18 °C) when freezing, a drop of water takes on a different shape – a pointed tip appears (figure 1). Obviously, this is due to the vertical expansion of ice in combination with the surface tension on the remaining liquid [15–17].

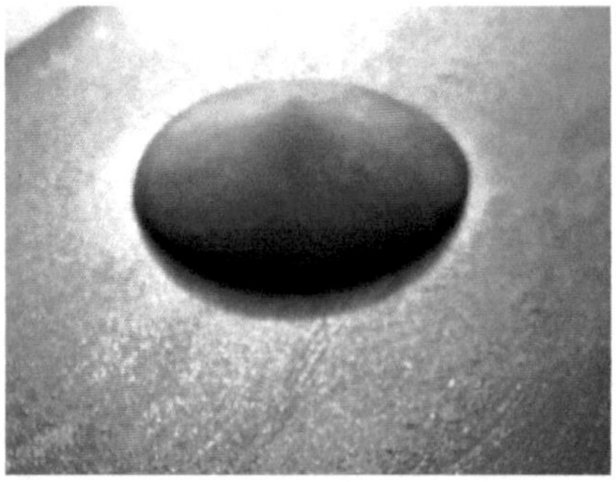

Figure 1. Form a drop of water when frozen on a superhydrophobic surface.

It was established, that the amount of ice on the untreated surface, (the plate is located at an angle of 30 degrees when moistened) is 0.59 kg/m^2, and on a superhydrophobic surface – 0.15 kg/m^2. During the subsequent moistening of the plates with the same amount of water and its freezing, no difference in the increase in the amount of ice was revealed. The increase in the amount of ice during subsequent wetting and freezing under the given experimental conditions averages 0.109 kg per 1 m^2.

For assess the kinetics of freezing a drop of water on a superhydrophobic surface, the following experiment was conducted. A drop of water was placed on the superhydrophobic surface of the mortar substrate, which was placed in a freezer at a temperature of –18 °C. Studies of the dynamics of freezing drops on the surface were performed using a thermal imager TESTO 875-1. The research results are given in table 1.

Table 1. The temperature at the surface of the water droplet and coating.

Time, minute	Temperature, ° C	
	water drop tops	on the surface of the coating
1	20	23.5
3	18.4	22
6	16.7	19.8
9	12.7	15.0
12	10.0	11.9
15	7.3	8.1
18	6.0	5.0
21	4.7	3.0
24	1.7	0.5
27	–0.3	–1.8
30	–1.0	–2.5
33	–1.7	–3.0
36	–1.8	–3.5

An analysis of the experimental data (table 1) shows, that in the first stage of freezing (up to 15 minutes) the temperature of the surface of the coating is higher, than the temperature of the water drop, i.e. heat is transferred from the surface of the coating to a drop of water. On the 18th minute of freezing, the temperature of a water drop becomes 1°C lower, than the surface temperature of the coating. In the future, the temperature difference increases.

The results of studies of the temperature distribution on the surface of a water droplet indicate that its distribution is uneven. So, after 27 minutes of cooling, the temperature at the top of the droplet surface is + 0.3 °C, below –0.6 °C, after 30 minutes cooling –1.0 °C and –1.6 °C, respectively, after 33 minutes –1.4° C and –2.0 ° C, respectively.

To compare the adhesion of ice to a hydrophilic and superhydrophobic surface, the following experiment was conducted. Metal profiled roofing sheets "MR-20" were placed in a freezer with a temperature of -10 ° C and a relative humidity of 90%. Part of the metal plates were previously treated with anti-icing composition. After curing the coating, they were moistened and placed in a freezer. It was established, that after freezing at –10 °C, the removal of ice from a superhydrophobic surface at room temperature at a plate tilt angle of 90 degrees begins after 1 minute 55 seconds, and after 3 minutes ice fell from the plate (figure 2).

The adhesion of ice to the metal hydrophilic surface is higher compared to the super-hydrophobic surface: at a plate tilt angle of 90 degrees, ice does not slip, after 6 minutes of being at room temperature, the ice on the metal plate melted (figure 3).

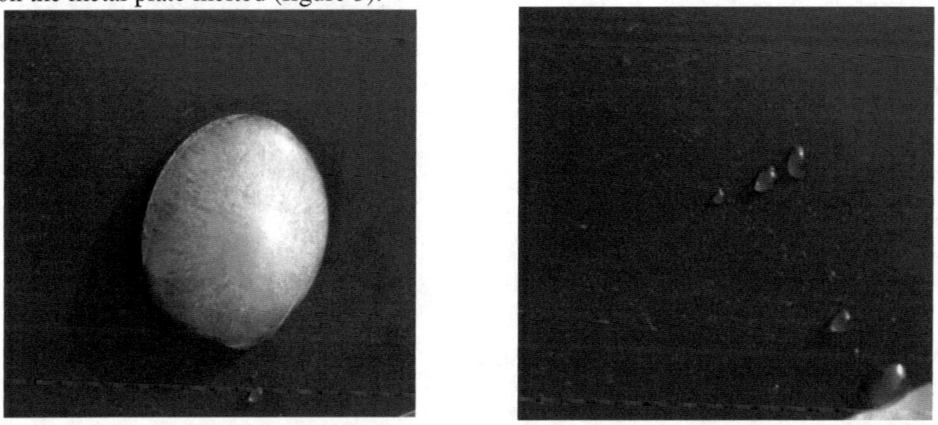

a) b)

Figure 2. Photo image of ice rolling down from superhydrophobic surface. a – the beginning of ice rolling; b – after rolling.

At a freezing temperature of –5 °C, ice on a super-hydrophobic surface began to roll at room temperature after 52 seconds, after 1 minute 35 seconds he himself fell.

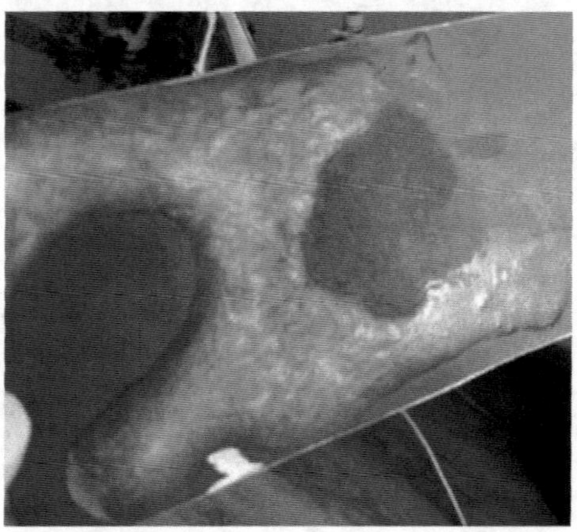

Figure 3. Image of rolling ice from a hydrophilic surface.

When the angle of inclination of the metal plate was 30 degrees in the process of moistening and subsequent freezing, ice rolling from the superhydrophobic and hydrophilic surface was not observed. However, in the process of thawing, the untreated plate had a continuous film of water on the surface, and on super-hydrophobic – the formation of individual droplets, i.e. manifestation of the hydrophobic effect.

Ice evaporation from a hydrophilic and superhydrophobic surface in a freezer was calculated. According to the experimental conditions, the relative humidity of the air in the freezer was 90%. It has been established that during evaporation there is practically no difference in the evaporation rate from the hydrophilic and superhydrophobic surface. So, after 1 hour, the evaporation rate from 1 m^2 of a super-hydrophobic surface is 15.12 g/hour, and from a hydrophilic surface – 14.99 g/hour. After 2 hours, the rate of evaporation from the surface decreases and amounts to 14.88–14.89 g/h with 1 m^2. After 6 hours in the closed space of the freezer (volume 196 **l**) a dynamic equilibrium is established between the processes of evaporation and condensation. The rate of condensation becomes equal to the rate of evaporation; from this point on, the amount of ice ceases to decrease. Differences in the values of the evaporation rate of ice from the hydrophilic and superhydrophobic surfaces were not recorded.

3. Conclusion

The results of the research indicate that the proposed composition forms a coating characterized by anti-icing properties that remain in service. The adhesion of ice to the metal hydrophilic surface is higher compared to the superhydrophobic surface, which makes it easier to remove from the superhydrophobic surface.

References

[1] Shilova O A, Proskurina O I, Antipov V N, Khamova T V, Esipova N E, Pugachev K E, Ladilina E Yu and Kruchinina I Yu 2014 Sol-gel synthesis and hydrophobic properties of anti-friction coatings for use in high-speed mini-turbine generators *Physics and Chemistry of Glass* **40(3)** 419–425

[2] Boynovich L B and Emelyanenko A M 2008 Hydrophobic Materials and Coatings: Principles of Creation, Properties, and Application *Successes of Chemistry* **77(7)** 619–638

[3] Lakshmi R V and Bharathidasan T 2011 Fabrication of superhydrophobic and oleophobic sol–gel nanocomposite coating *Surface & Coatings Technology* **257(24)** 7

[4] Nosonovsky M and Bhushan B 2009 Superhydrophobic Surfaces and Emerging Applications: Nonadhesion, Energy, Green Engineering *Current Opinions Coll. Interface Sci.* **14** 270–280

[5] Shirtcliffe N J, McHale G, Newton M I and Perry C C 2003 Intrinsically Superhydrophobic Organosilica Sol-Gel Foams *Langmuir* **19(14)** 5626–5631

[6] Venkateswara R A, Latthe S S, Nadargi D Y, Hirashima H and Ganesan V 2000 Preparation of MTMS based transparent superhydrophobic silica films by sol-gel method *J. Colloid Interf. Sci.* **332(2)** 484–490

[7] Thorpe A A, Smith J R and Peters V 2000 Poly (methylpropenoxyfuoroalkylsiloxane) s: a class of fuoropolymers capable of inhibiting bacterial adhesion onto surfaces *J. Fluor. Chem.* **104** 37–45

[8] Rajiv S, Kumaran S and Sathish M 2019 Long-term-durable anti-icing superhydrophobic composite coatings *J. of applied polymer science* **136(7)** 47059

[9] Sinebryuhov S L, Sidorova M V, Egorkin V S, Nedozorov P M, Ustinov A Yu and Gnedenkov S V 2011 Anticorrosive, anti-friction coatings on magnesium alloys for aviation *Bulletin of the Far East Branch of the Russian Academy of Sciences* **5** 95–105

[10] Loganina V I, Kislicyna S N and Sergeeva K A 2019 The choice of polymer concentration in the development of a formulation for anti-icing coating *Bulletin PGUAS: construction, science and education* **1** 19–22

[11] Loganina V I 2019 Study of the kinetics of freezing of a drop of water on a superhydrophobic surface of coatings *Vestnik MGSU* **14(1)** 311–317

Structural and analytical mesomechanics of fracture

V G Malinin[1], N A Malinina[1,3], V V Malinin[1] and O A Malukhina[2]

[1]Orel State Agrarian University, General Rodin ul. 69, 302019, Orel, Russia
[2]Yaroslav-the-Wise Novgorod State University, ul. B. St. Petersburgskaya, 41, 173003, Veliky Novgorod, Russia

[3]E-mail: malinina2006@yandex.ru

Abstract. Developing methodology of structural-analytical theory of strength, this article outlines methods for constructing a variant of defining equations of structural-analytical mesomechanics of failure, based on a rational synthesis of the main achievements of continuum mechanics, physics of plasticity and strength of solids, and other related disciplines. Loaded material is considered as a multi-level system in which micro, meso and macro levels are organically interconnected. Each scale level is characterized by its mechanisms and laws of deformation, the evolution of structure and damaging.

A method for constructing a two-level fracture criterion is being developed, which takes into account the hierarchical nature of the development of damaging and mutual influence of structural concentrators zones (SCZ) and macro stress concentrators. In order to take into account the influence of SCZ at the mesoscale level, experimentally justified tensor parameters of the magnetomechanical effect are introduced: magnetic distortion tensor and tensor intensity of intrinsic magnetic field of dispersion that occurs in the SCZ of the loaded body. Possible directions for the development of fracture mesomechanics are discussed.

1. Introduction

It is known that the ability of material to resist nucleation and development of cracks strongly depends on the organization of defective material structure, the presence of SCZ, which necessitates the development of fracture mechanics in order to introduce parameters characterizing the structural state of the material based on technical diagnostics methods, which led to the emergence of a new scientific direction – mesomechanics of destruction. For the named direction, attempts to link the criterion of fracture mechanics with the parameters of meso and microstructure of the material are characteristic. In this case, there is an urgent task of identifying effective parameters that allow controlling the real state of structural concentrators, and interaction with macro concentrators in order to formulate the criterion of destruction reflecting the influence of the heterogeneity of the stressed state.

To solve the problem of establishing relationship between microstructure and macro characteristics with the goal of formulating the criteria for fracture, a very promising approach is that combining the method of orientation averaging and the method of continuous approximation [1–3]. This approach allows us to formulate the criteria for the destruction of representative volume characterizing the mechanical properties of material point at macroscale level, based on information on the micro and mesostructure of local region of the material in the process of loading.

It is fundamentally important that the method of orientational averaging allows one to take into account different spatial orientation of structural elements at micro and mesoscale levels [2], and to naturally model their evolution in the process of deformation up to destruction [1, 2]. The formal algorithm of the orientation averaging method involves the following procedure. Let the structural-mechanical properties of representatively macro-volume of solid body be characterized by response $\{\Pi\}$ when certain physical field $\{\Sigma\}$ is applied to it. It is assumed that this field acts uniformly on each structural element at micro and mesoscale level within the representative macro volume. The response $\pi(\Omega)$ of each structural element is a function of the field, the structural and mechanical properties of micro and meso volumes, and such is determined by the orientation of the corresponding structural elements relative to active field $\{\Sigma\}$:

$$\pi(\Omega)=\pi(\{\Sigma\};\{a_i\};\{\Omega\}). \tag{1}$$

Performing averaging the parameters $\pi(\Omega)$ over all orientations (superposition is assumed), we obtain the desired functional dependence:

$$\{\Pi\}=\int_{\{\Omega\}} f(\Omega)\pi(\Omega)(\{a_i\};\{\Sigma\};\{\Omega\})d^3\Omega. \tag{2}$$

where $f(\Omega)$ is density distribution function according to the orientations of the representative volumes of structural elements. Function $f(\Omega)$ must satisfy the natural normalization conditions. With this approach, it is possible to obtain analytical relationships that reflect adequate relationships between different types of deformation, damage to the structure, on the one hand, and the stresses, temperatures, radiation, electrical, magnetic fields, etc. that generate them, – with another.

Mathematical side of the problem is connected with the development of averaging methods and continuous approximation methods for hierarchically structured environments of various scale levels, formally represented by formulas (1) and (2). The physical aspect of the problem is related to the formulation of the relations of type (1) for the corresponding physical fields. The problem of specifying dependencies (1) is fundamental in this approach and should be solved taking into account the physical content of the problem, as well as reflecting the main achievements obtained in the mechanics of deformable solid. This aspect is considered in detail in [1, 3].

2. Method for constructing criterion for the destruction of material in high gradient stress fields

When constructing fracture criterion, we will use the methods of structural-analytical mesomechanics [2] supplementing them with hypotheses reflecting the factor of substantial non-locality of structural-mechanical properties associated with the influence of inhomogeneity of macrostresses in the vicinity of stress concentrators [4]. Let us consider the method for constructing two-level criterion for destruction of material under a macrohomogeneous stress state.

The representative volume of macroscale level will be given by laboratory basis *xyz*. In this basis, stress and strain tensors σ_{ik}, deformation ε_{ik} and temperature *T* are identified [1–3]. The representative volume of the structural element will be specified by the local basis *lmn* using the geodetic coordinate system $\{\Omega\}$ [1]. In the local bases of cut and separation, dyad $n_i l_k$ on the plane of separation with the normal *n* is specified, and the direction of the formation of slit cracks is given by the unit ort l, Figure 1. Diagram presented in figure 1 at $r=r_m$ depicts macro point M, which is located at a distance r_m from the tip of macro concentrator, and its surroundings are characterized by structural elements in the form of local points of a sphere with a radius r_m which is considered to be structural characteristic of material (analogous to parameter ρ_0 in the M.Ya. Leonov – K.N. Rusinko model).

IOP Conf. Series: Materials Science and Engineering **656** (2019) 012032 doi:10.1088/1757-899X/656/1/012032

In theory of strength within the framework of the Neuber-Novozhilov-type integral criterion, it is advisable to use orientation space shown in figure 1 at $r = r_m$. For the case of using the gradient approach, it is sufficient to represent the orientation space $\{\Omega\}$ as figure 1 at $r_m = 1$. As noted in [1-3], in the case of using the Cauchy stress tensor, because of its symmetry in the analytical representation of structural elements, instead of sphere, it is enough to consider hemisphere with radius $r = r_m$ or $r_m = 1$, It significantly, reduces the counting time in numerical simulation.

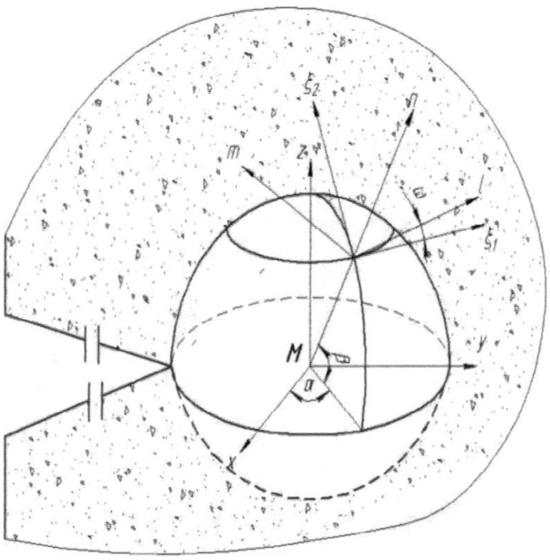

Figure 1. Diagram of orientation coordinate system $\{\Omega\}$.

Note that for the case of macro homogeneous net stretching, you can specify structural elements on one-fourth of the hemisphere. The presented variants of the analytical display of structural elements relate to tactical techniques.

Figure 2 presents the scheme of the orientational coordinate system $\{\Omega\}$, and also shows the matrix of direction cosines α_{ik}, which translates laboratory basis *xyz* into local coordinate system *lmn*.

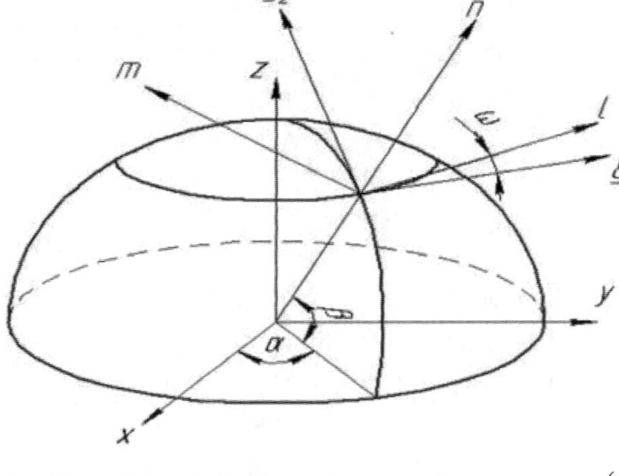

Figure 2. The scheme of orientation coordinate system $\{\Omega\}$: *xyz* is laboratory basis; *lmn* is local basis: $0 \le \alpha \le 2\pi$; $0 \le \beta \le \pi/2$; $0 \le \omega \le 2\pi$.

$$\alpha_{ik} = \begin{pmatrix} \alpha_{11} & \alpha_{12} & \alpha_{13} \\ \alpha_{21} & \alpha_{22} & \alpha_{23} \\ \alpha_{31} & \alpha_{32} & \alpha_{33} \end{pmatrix}; \; \alpha_{ik} = \begin{pmatrix} -\sin\alpha\cos\omega - & \sin\alpha\sin\omega - & \cos\alpha\cos\beta \\ -\cos\alpha\sin\omega\sin\beta & -\cos\alpha\sin\beta\cos\omega & \\ \cos\alpha\cos\omega - & -\cos\alpha\sin\omega - & \sin\alpha\cos\beta \\ -\sin\alpha\sin\beta\sin\omega & -\sin\alpha\sin\beta\cos\omega & \\ \cos\beta\sin\omega & \cos\beta\cos\omega & \sin\beta \end{pmatrix}. \quad (3)$$

Small solid angle $d^3\Omega$ for considered orientational space α, β, ω is determined by the formula

$$d^3\Omega = \cos\beta \, d\alpha \; d\beta \; d\omega. \quad (4)$$

Note that in the laboratory basis *xyz*, the parameters of the macroscale level are set, including the tensor of idealized σ_{ik} and effective stresses $\overset{*}{\sigma}_{ik}$, and in the local basis *lmn*, the corresponding parameters of the structural elements, including the effective stress tensor τ^*_{ik}.

The density of structural damages $\pi(\Omega)$, oriented inside the solid angle $d^3\Omega$ in the direction given for normal tearing cracks $n_i = n_i(\alpha, \beta)$, and for shear cracks by dyad $n_i l_k = f(\alpha, \beta, \omega)$ will be characterized by parameter $\pi(\Omega) \cdot d^3\Omega$.

Fracture criterion in the zone of stress macroconcentrator, as in the case of a macrohomogeneous stress state [3], must take into account the necessary criterion conditions of kinetic and force character at the structural and macroscale levels. However, the specificity of the structural-mechanical state of material associated with sharp localization of macrostresses in the vicinity of a macro concentrator requires when building a fracture model to take into account the influence of the type of stress state, the degree and nature of localization of macrostresses in the stress concentrator zone [4].

Let us formulate the criterion of destruction, considering that destruction will occur when the maximum damage in the most loaded structural element reaches a critical value, i.e.

$$\{\Pi\} = \int_{\{\Omega\}} f(\Omega)\delta(\Omega - \Omega_0)\pi(\Omega) d^3\Omega = const; \max \pi(\Omega_0) = 1. \quad (5)$$

Here, $\delta(\Omega - \Omega_0)$ is Dirac delta function; Ω_0 is coordinates of structural element in which damaging $\pi(\Omega)$ reaches maximum. As argument of functional $\pi(\Omega)$, you can use effective normal stress $\tau^*_{nn}(\Omega)$, i.e.

$$\pi(\Omega_0) = \tau^*_{nn}(\Omega_0) / \tau_0 = \tau^*_{33}(\Omega) / \tau_0, \quad (6)$$

where τ_0 is critical stress by separation equals to ultimate strength of material under the net tension $(\sigma_в)$ of standard sample. In this case, criterion (5), (6) for a complex stress state coincides in form with the criterion of maximum tensile stresses, formulated in terms of the effective stresses of the laboratory basis *xyz*, i.e.

$$\max \tau^*_{nn}(\Omega) = \sigma^*_1. \quad (7)$$

IOP Conf. Series: Materials Science and Engineering **656** (2019) 012032 doi:10.1088/1757-899X/656/1/012032

Note that, in (6) and (7), the kinetic aspect of preparing material structure for macrofracture is not taken into account.

3. Tensor characteristics of intrinsic dispersion of magnetic field

Important task is the development of experimental-theoretical models reflecting the basic laws of the evolution of structural stresses based on the registration of parameters of intrinsic magnetic field of scattering (SIMF). It is necessary to introduce appropriate conceptual representations and mathematical tools. In this case, by measuring magnetic field strength H*i*, one can determine the distribution of H*i* in terms of coordinates of investigated article.

To test the existence of SIMF in the sample as vector object after the material was deformed, experiments were performed on thin steel plates. Plates were used without concentrators and with various macro concentrators in the form of central round and elliptical holes, side cuts of various configurations from steel 3, steel 45, steel 09G2S.

Serially manufactured specialized control device SIMF and a three-component fluxgate sensor developed by the company "Energodiagnostika" were used. In particular, a stress concentration meter magnetometric IKN-1M-4 was used, with a microprocessor-based recording device. H*i* measurements were performed at various points of the sample, and at each point the components of the vector were scanned, H*i* rotating the sensor at angles of 30°, 45°, 60° and 90°. Each experiment was repeated three times. The measurement results confirmed the H*i* invariance, i.e. the law of conservation of the first rank tensor was fulfilled when the coordinate system was rotated

$$H'_i = \alpha_{ik} H_k , \tag{8}$$

where α_{ik} are guiding cosines.

According to estimate of $(Hi\,Hi)^{1/2}$ invariant, the measurement error did not exceed (1-3)%, and for individual components, spread did not exceed 5%. The obtained results on the distribution of vector field H*i* allowed to introduce into consideration the model of the SIMF distortion tensor as a gradient of intensity vector H*i* according to the formula

$$H_{ik} = \nabla_i H_k , \tag{9}$$

where ∇_i is the Nabla operator [1].

Processing the experimental data using formula (2) allowed us to obtain experimentally grounded mathematical object that characterizes the heterogeneity of the SIMF. The next step was to experimentally check the parameters of H_{ik} (9) in order to preserve H_{ik} as a second-rank tensor when coordinate system is rotated by the angles of: 30°, 45°, 60° and 90°. The results of processing the data obtained showed satisfactory compliance with the law of conservation:

$$H'_{ik} = \alpha_{ip} \alpha_{kq} H_{pq} . \tag{10}$$

Scatter in the values of the first, second, and third H_{ik} invariants was within 15%, which confirmed the tensor nature of the parameters H_{ik}. It should be noted that these experimental studies required the creation of original equipment, each experiment lasted for 4-5 hours. Within three years, a large number of experiments were carried out, which confirmed the tensor character of the introduced parameter H_{ik}, called the distortion of the intrinsic magnetic field of scatter.

The next step was the introduction of symmetric and antisymmetric tensors on the basis of the H_{ik} distortion tensor:

$$\mathrm{H}_{ik}^{s}=\frac{1}{2}\left(\nabla_i \mathrm{H}_k+\nabla_k \mathrm{H}_i\right),\ \mathrm{H}_{ik}^{a}=\frac{1}{2}\left(\nabla_i H_k-\nabla_k H_i\right). \quad (11)$$

Introduction of H_{ik}^{s} allows the development of symmetric structural mechanics in accordance with the traditional symmetric mechanics of deformable solid. Introduction of parameter H_{ik}^{a} opens up the possibility of developing structural models that take into account moment structural stresses and corresponding twist-bending.

4. Effective stresses in high gradient fields of macrostresses

When analyzing the strength of material in high gradient force fields, the question of choosing an equation for effective stress is of particular importance. The effective stress must take into account the influence of structural concentrators and their evolution under load. This circumstance can be taken into account using vector-invariant intensity of intrinsic magnetic field of scattering S_i [3], measured directly in the vicinity of the macroconcentrator.

Another important aspect is the need to take into account the influence of type of stress state, the nature and degree of heterogeneity of the stress state in the zone of stress concentrator on the structural and mechanical properties of material. These circumstances can be reflected in the fracture criteria using the achievements obtained in the development of gradient methods [4].

Critical analysis of the main results obtained in gradient approach, taking into account the structural and mechanical parameters of metal magnetic memory method, made it possible, based on the methods of structural and analytical mesomechanics [1–3], to formulate the following equation for effective stresses τ_{ik}^{*} acting in the structural elements experiencing the opening of cracks by separation and cut:

$$\tau_{ik}^{*}(\Omega)=\delta_{i3}\delta_{k3}\tau_{33}^{*}(\Omega)+\left(\delta_{i1}\delta_{k3}+\delta_{i3}\delta_{k1}\right)\tau_{31}^{*}(\Omega). \quad (12)$$

taking into account criteria (5)

$$\max \tau_{33}^{*}(\Omega_0)=\sigma_1^{*};\ \max \tau_{31}^{*}(\Omega_0)=\tau_{\max}^{*}. \quad (13)$$

For effective stresses $\tau_{ik}^{*}(\Omega)$ in the framework of criterion (5), we formulate the equations for σ_1^{*} and $\tau_{\max}^{*}$. Under non-uniform stress state in the SCZ, to formulate the criterion of brittle fracture by separation, taking into account the equivalence principle [3], we will compare with the ultimate strength of the material $\sigma_в$ the maximum effective stress σ_1^{*}. Taking into account the achievements obtained in the gradient approach [4], the following equation is formulated for the calculation σ_1^{*}:

$$\sigma_1^{*}=\sigma_1/\left(1-\beta+\sqrt{\beta^2+Lq_1q_2}\right);\ \beta={}^{L}\!/_{d}, \quad (14)$$

$$q_1=S_i^{\max}/S_i^{cp}, \quad (15)$$

$$q_2=\left(grad\,\sigma_1\cdot Sign\ grad\,\sigma_1\right)/\sigma_1, \quad (16)$$

where q_1 is relative vector intensity invariant S_i [3] of the intrinsic magnetic field of scattering, characterizes at the structural level the damageability of metal at hardening stage; q_2 is relative gradient of the first principal stress, is found from the elastic solution of corresponding boundary value problem [4]; *L* is parameter, has the dimension of length and characterizes the effective size of structural stress concentrators The designations *max* and *cp* in (15) indicate the maximum and average values S_i in the vicinity of macro concentrator; *d* is the characteristic size of macro stress concentrator.

In the case of development of transverse or longitudinal slides in the stress concentration zone, to formulate the criterion of shear failure, based on the equivalence principle, we will compare with the shear strength limit τ_S maximum effective stress value τ^*_{max} that is less than maximum shear stress τ_{max} (taken as equivalent to shear failure criterion):

$$\tau^*_{max} = \tau_{max} / \left[1 - \beta + \sqrt{\beta^2 + Lq_1q_3}\right] \tag{17}$$

$$q_3 = (grad\ \tau_{max} \cdot sign\ grad\ \tau_{max}) / (\tau_{max} \cdot sign\ \tau_{max}), \tag{18}$$

where q_3 is the relative gradient of the maximum shear stress τ_{max}.

It should be noted that introduction of product of parameters $L \cdot q_1 \cdot q_2$ and $L \cdot q_1 q_3$ into relations (14) – (18) reflects the influence of interaction of structural stresses arising in SCZ and gradient of macroscopic stresses in the vicinity of macro concentrator on the structural and mechanical properties of the material in high gradient stress fields.

5. Two-level structural and analytical failure criterion

We formulate a variant of two-level fracture criterion, suggesting that the formation and development of crack in the vicinity of a macroconcentrator requires the fulfillment of fracture criterion at both the macroscale and structural (micro and mesoscale) levels [1–3].

Let us first consider the case of destruction of article with a macroconcentrator by mechanism of separation. We assume that separation by breaking in the neighborhood of the local macro-point will occur when two conditions are fulfilled: the achievement of the ultimate strength $\sigma_в$ by effective stresses σ_1^* and the mandatory fulfillment of the fracture criterion at the structural level, i.e. when the parameter q_1 reaches its value q_{np}.

The latter means, according to [1–3], the emergence in this structural element of critical defect structure characterized by the limiting state of the zones of structural concentrators. Then, taking into account the condition that the initial propagation of macrocrack will occur at the site of the maximum normal stress $\max \tau_{nn}$, i.e. of the first main stress and will occur only when the two conditions mentioned are fulfilled, we formulate the final macroscopic fracture criterion by breaking Π_0^M in the form:

$$\Pi_0^M = X(\sigma_1^* - \sigma_B) X(q_1 - q_{np}) = 1, \tag{19}$$

где where X(x) is the Heviside function.

Note that in general case, the maximum values of effective stress σ_1^* and the structural parameter q_1 can be reached at the point that does not coincide with the point of the maximum value of the first main stress. Therefore, when calculating the strength using criterion (14) - (19), it is necessary to check criterion (19) not only at the apex of concentrator, but also in other points of the body.

In accordance with equation (19), if $\Pi_0^M = 1$, then the body will collapse, and it will not be destroyed if $\Pi_0^M = 0$. Thus, macroscopic destruction of body with macro stress concentrators will occur when the critical level is reached and effective stresses σ_1^* (in this case, the force nature of the fracture is reflected) and the parameter that characterizes the kinetic nature of the fracture. Note that parameter q_1 is a constant of material and characterizes the degree of structural heterogeneity and the maximum deformation capacity of metal in the zones of structural stress concentrators.

Now suppose that in vicinity of macroconcentrator the stress-strain and structural-kinetic state correspond to the fracture variant by the development of transverse or longitudinal shear cracks. For the formulation of the fracture criterion, as in formulation of criterion in the event of breakdown by separation, we apply the hypothesis that macroscopic destruction in the stress concentration zone will occur if fracture criteria are met at both structural $(q_1 = q_{np})$ and macroscale levels, i.e. when the effective shear stress $\tau_{\max}^*$ reaches value equal to shear strength τ_S. Then, the final criterion of destruction by cut Π_c^M at macroscale level will look like:

$$\Pi_C^M = X\left(\tau_{\max}^* \, sign\, \tau_{\max}^* - \tau_S\right) X\left(q_1 - q_{np}\right) = 1 \tag{20}$$

Note that, as in case of fracture criterion with normal break (19), the maximum values of effective stress $\tau_{\max}^*$ and structural parameter q_1 in the fracture criterion by slice (20) can be reached at the point not coinciding with the point of the maximum value of idealized tangential stress; therefore, it is necessary to check criterion (20), both at the top of the concentrator and at other points in the body. Criteria of destruction (19) and (20), reflecting various cases of limiting state, can be combined and write generalized criterion of destruction in the form:

$$\Pi_\Sigma^M = \mathrm{X}\left(q_1 - q_{np}\right)\left[\mathrm{X}\left(\sigma_1^* - \sigma_\mathrm{B}\right) + \mathrm{X}\left(\tau_{\max}^* \, sign \tau_{\max}^* - \tau_S\right)\right] \geq 1. \tag{21}$$

The formulated version of structural-analytical criterion of destruction is fundamentally different from existing criteria of destruction. The main difference is that structural-analytical criterion contains two necessary conditions for final destruction, namely, the kinetic criterion of destruction at the structural level, i.e. requirement for the parameter q_1 to reach the limiting value q_{np}, which reflects the creation of critical defect structure, and the force criterion at macroscale level by tear-off or shear mechanism.

An important point in this approach is the natural combination of methods of fracture mechanics and methods of technical diagnostics of the structural-mechanical state of the material, which allows you to create models that take into account the mutual influence of internal structural stresses in the SCZ and macrostresses in the vicinity of the macroconcentrator on the strength properties of materials. It is characteristic that to describe the strength properties of a material around macroconcentrators, it is necessary to introduce two complex parameters: structural-mechanical (Lq_1q_2) and structural-geometric ($\beta = L/d$) content. The introduced complexes reflect the mutual influence of structural concentrators and stress macroconcentrators on the strength properties of material.

In the structural-mechanical complex Lq_1q_2, one of parameters is a characteristic of the representative sizes of the structural heterogeneity of materials (L), the second reflects the influence of structural stress gradients (q_1), and the third parameter characterizes the influence of macrostress gradients (q_2). The structural-geometric complex β is the ratio of the characteristic size of the structural heterogeneity L to the representative size of the macro concentrator d. An important element of the formulated criterion is also the fact that strength characteristics, namely, tensile strength under uniaxial tension $\sigma_в$, shear strength τ_c and yield strength σ_T, as well as the invariant structural parameter q_{np}

6. Comparison of experimental studies with theoretical calculations

This section provides analysis of the adequacy of fracture criteria for calculation of limit state of thin steel plates with stress concentrators. The advantage of structural-analytical gradient fracture criterion is substantiated. For convenience of analysis, the results of calculating limiting state of plates with macroconcentrators according to gradient fracture criterion and experimental data are presented in Table 1. Note that the experimental data were obtained from the results of testing at least 3 samples for each type of concentrator and Table 1 shows the average values.

Table 1. Comparison of calculation results for structural-analytical gradient fracture criterion with experimental data.

Concentrator type	Calculation method	Theoretical concentration coefficient α_T	Average value of stress σ, MPa		Relative difference by σ
			Calculation	Experiment	
Circle *D*=12 mm	Theory	2.4	460.2		4.8%
	ANSYS	2.508	482.5	483.3	0.16%
Circle *D*=18 mm	Theory	2.1	368		8%
	ANSYS	2.31	437.4	400	8.5%
Ellipsis *l*=12 mm	Theory	3.8	480		1%
	ANSYS	3.16	434.8	475	8.4%
Ellipsis *l*=18 mm	Theory	4.4	368.2		3.9%
	ANSYS	3.37	388.5	383.3	1.3%
U-shaped l=12 mm	Theory	5	284.6		24.1%
	ANSYS	3.16	336.7	375	10.2%
U-shaped *l*=18 mm	Theory	5.9	162.3		32.8%
	ANSYS	2.76	252.5	241.7	4.2%
V-shaped *l*=12 mm	ANSYS	3.79	301	341.7	11.9%
V-shaped *l*=18 mm	ANSYS	3.53	234.6	233.3	0.05%

Prediction made by the structural-analytical gradient fracture criterion is in good consistent with experimental data. The deviation of the calculation results from the data obtained during tests is in the range from 0.5% to 10%, which practically coincides with variation in experimental studies [6].

7. Conclusion

It should be noted that the pronounced specificity of structural concentrators along grain boundaries takes place in materials with submicro and nanostructure. The introduced tensor characteristics of SIMF make it possible to monitor the change in the grain size during plastic deformation up to submicro and nanostructured states [6].

The developed approach takes into account multi-level and multi-stage nature of destruction. The emergence of microcracks is necessarily preceded by the formation of zones of structural concentrators with critical defective structure, which is characterized by diagnostic parameters of the magnetic memory method in the form of invariant (12) and integral criterion, which together with the field of effective stresses τ_{ik}^{*} determine the formation of structural (macro and meso) cracks. Cracks create structural damage that can grow under the right conditions. When such property is acquired by large number of structural cracks, prerequisites appear for macroscopic destruction, which will occur only if the body is sufficiently damaged.

References

[1] Likhachev V A and Malinin V G 1993 *Structural analytic theory of strength* (Saint-Petersburg: "Science" Publ.) p 471

[2] Malinin V G and Malinina N A 2005 Structural analytic mezomekhanics deformed solid body *Phisics mezomekhanics* **23** 31–45

[3] Malinin V V 2011 Structural-mechanical approach in fracture mesomechanics: Appendix to *Engineering Journal Reference Publ. in Engineering journal: Spravochnik* **6** 52–56

[4] Legan M A 1994 Determination breaking load places and directions of development by means of gradient approach *PMTF* **5/35** 117–124

[5] Dubov A A and Dubova A A 2006 *Method of magnetic memory and control devices* (Moscow: "Tisso" Publ) p 332

[6] Golenkov V A, Malinin V G and Malinina N A 2009 *Structural and analytic mezomekhanics and its applications* (Moscow: "Mechanical engineering" Publ.) p 635

IOP Conf. Series: Materials Science and Engineering **656** (2019) 012033 doi:10.1088/1757-899X/656/1/012033

Structural-analytical mesomechanics of a solid body with reversible martensitic transformations

V G Malinin[1], G V Malinin[2], N A Malinina[1,4] and O A Malukhina[3]

[1] Orel State Agrarian University, ul. Generala Rodina, 69 302019, Orel, Russia
[2] PJSC «Tupolev», nab. Akademika Tupoleva, 17 105005, Moscow, Russia
[3]Yaroslav-the-Wise Novgorod State University, ul. B. St. Peterburgskaya, 41 173004, Veliky Novgorod, Russia

[4]E-mail: malinina2006@yandex.ru

Abstract. In the given article the structurally-analytical mesomechanics for environments with martensitic mechanism of mass transfer is developed. Wave equations are derived to describe the deformation process initiated by martensitic transformations at meso and macro-scale levels.

1. Introduction

This article presents the analysis of structural and mechanical aspects of the problem of structural and analytical mesomechanics of multilevel media with reversible phase transformations of martensitic type and physical justification of the need to use the effective field method as a tool to take into account the interaction of deformation processes, the evolution of damage and reversible martensitic transformations at various scale and structural levels. Particular attention is paid to the factors of inheritance of defective structures in reversible transformations of austenite to martensite, as well as the development of a hierarchy scenario of martensitic transformations and plastic deformation. This scenario allows us to provide a holistic perception of the methodology of structural and analytical mesomechanics.

2. Analysis of structural and mechanical aspects of the problem

The task of creating the theory of deformation of materials with shape memory effect (SME) is to build a model of a deformable solid body, which should take into account its complex and multi-level structural organization, various deformation mechanisms, including dislocation slip, diffusion, martensitic transformations and evolution of the material damage on the one hand and generating their thermal effects and mechanical stresses on the other.

The task of solid state physics in this regard is to analyze the micro-scale level and build models of mechanical behavior of crystals, which take into account specific physical mechanisms and their influence on the corresponding parameters of the governing equations, taking into account the structural organization of crystals. The physics of plasticity and fracture allows to analyze crystallography and thermodynamics of phase transformations, to describe the laws of motion of structural imperfections in a loaded solid body, using the methodology of the theory of defects, in particular, the apparatus of the theory of dislocations. Significant progress has been made in the study of dislocation plasticity of crystals. A large array of experimental data obtained using a well-developed experimental technique gives a broad idea of the mechanism of formation of elementary acts and laws of plasticity. Here not only the structural and physical mechanisms of inelastic deformation processes are understood but also effective methods of calculation are created. At the same time, the physical theory of plasticity of crystals did not reach engineering level, having kept the value only for an explanation and the description of elementary acts of deformation and destruction or close to them. In the framework of this approach the basic mechanisms of the motion of defects on the microscale level and the quality of the interpretation of many laws of macrodeformation and macroresults are studied. Nevertheless, the analysis of the stress-strain state of the macroscopic system as a whole is beyond the capabilities of the microscopic approach of the defect theory [1–3].

The mechanics of the deformed solid provides for the creation of analytical models that allow to predict the mechanical behavior of real macroscopic products. In this case, the problem is solved by means of phenomenological hypotheses formulated on the basis of experimental data on the mechanical behavior of macroscopic samples under given loading programs and involving the basic laws of mechanics: dynamic equilibrium equations, continuity equations for deformations, balance conditions for temperature and defining equations of state of phenomenological nature. The calculation of the stress-strain state of a solid body performed within the framework of the considered approach doesn’t take into account the architecture of the internal structure of the material, the processes of martensitic transformations of crystal lattices and real deformation mechanisms. As a result, the theory of defects and phase transformations usually is not used in the models of solid mechanics.

It should be noted that the mechanics of plasticity of materials (experiencing dislocation not elasticity) is widespread in engineering practice and has a good analytical base. However, its substantial progress is very modest. Being openly phenomenological, it describes only the regularities, on which the analytical relation is calibrated. The predictive power of the equations of plasticity mechanics in relation to complex methods of mechanical, thermal and other methods of influence on the material is unsatisfactory. It should be noted that the fundamental principles of plasticity mechanics used in the derivation of defining relations, such as the Drukker and Odquist postulates, the hypothesis of the existence of yield surfaces, a single deformation curve, established at the time on the basis of the analysis of experimental study of the behavior of objects such as iron or copper, do not stand up to criticism with respect to a number of new materials, in particular to materials with shape memory effect or to complex (non-trivial) modes of deformation. For example, in titanium nickelide (the most well-known representative of the material having EPF), deformation hardening can be determined not by the length of the deformation path, as in steel, but depends on the final value of the deformation. In the same object the deformation can initiate the release, not the absorption of energy, etc. [1, 3].

The above-mentioned examples, the number of which can be increased, rightly raise the question of the causes of the situation and possible solutions to the problem. We will analyze the denoted issues.

The reasons for the absence of the physical theory of plasticity on the engineering level are quite obvious. In addition to microstructural acts of plasticity, the laws of which are well studied by analyzing the behavior of single dislocations or their simplest formations, large-scale processes play an important role. In complex ensembles of defective formations, powerful collective effects come into power. This leads to the fact that the properties of the ensemble are not identical to the properties of single dislocations that make up the ensemble. Strong inside ensemble interactions create a complex collective properties. In large-scale ensembles, the properties of self-organization of the structure can come to the fore, which in terms of synergetics should be considered as dissipative [2, 3]. The material undergoes a variety of kinetic phase transitions, the control parameter of which is not only temperature, but also other variables, for example, the total density of dislocations. Moreover, in complex structures, in addition to translational plasticity, rotational plasticity is inevitably excited and characteristic turbulence occurs. One more large-scale level is involved in the process. As the analysis of experimental data shows, in real highly plastic objects of loading is accompanied by mass transfer of matter at several structural and large-scale interacting levels [2].

This analysis allows us to draw the most important conclusion that the consistent physical consideration of the problem of plasticity requires the correct consideration of numerous ways to implement the elementary act not only on the lower deformation floor but also on each of the subsequent scales of structures, their properties and laws of evolution as well as the nature of the inter-level mutual influence and interaction between structures of one type. It is obvious that macroscopic properties are formed at all stages of the mass transfer process and cannot be reduced to only one of them [1 – 4].

It is necessary to formulate such a theory of deformation that would be based on strict physical postulates, it means taking into account real physical processes and at the same time would allow to solve practical problems of engineering nature. Numerous attempts to build such a theory have been made long time ago but the hope to move directly from the micro-scale to the macro-scale area with the help of various methods of orientation and statistical averaging have not been successful [2].

As it is noted in [2, 4], the current situation is determined by two fundamental circumstances: first, a consistent and correct description of the evolution of the complex stochastic distribution of martensitic lamellae and their ensembles faces with insurmountable mathematical difficulties and secondly, self-organization of dislocation ensembles and martensitic lamellae leads to a new quality – in a continuous environment there is a move of larger than the dislocation of meso defects [2]. For the transition from the microstructural level to the macroscopic level, it is necessary to take into account the contribution of the evolution of the intermediate meso structural level where the move of the corresponding meso defects dominates, providing the formation of translational-rotational deformation modes.

The formulation of a general algorithm for constructing a model of physical mesomechanics to describe the deformation of a solid body with any internal structure, taking into account the possibility of transformation of the structure during the phase transformations and for arbitrary modes of its loading is an important problem. This article is devoted to one of the possible solutions to this problem based on the development of methods of structural and analytical mesomechanics [4] for materials with shape memory effect.

3. The script hierarchy of scale and structural levels

Deformation of a loaded object deformed by initiation of reversible martensitic reactions can be represented as a multilevel relaxation process. We will briefly consider the main stages of the origin and development of mass transfer processes in reversible phase transformations of martensitic nature. For convenience, figure 1 shows the scheme of scale and structural levels, adopted during the construction of the model.

According to the data of physical studies, the first loss of shear stability occurs at the micro level, in the local areas of the crystal, where the nuclei of the martensitic phase occur. Microconcentrators of stresses arising at the structural inhomogeneities cause martensitic transformation of the crystal lattice in certain crystallographic planes and directions. Such local structural transformation appears as the emergence and development of martensitic nuclei in the austenitic matrix. The indicated stage is shown in figure 1A, where the shaded areas are schematically represented by martensite nuclei. Since the influence of microconcentrators is short-range, the resulting distortion transformations with a good approximation can be characterized by translational modes of deformation.

In the process of deformation the density of the forming martensitic crystals increases, the variants of martensitic transformations that do not coincide with the favorably oriented directions in the field of external stresses due to the loading trajectories in the stress space are involved.

The mentioned circumstances cause the need for self-organization of martensitic domains in accordance with the given boundary conditions, that means with the directions of the principal axes of the strain tensor at each moment of the deformation process. As a result, there are self-encoded groups of martensitic crystals forming dissipative structures within the initial structure of the material, in accordance with figure 1b. Such dissipative structures contribute to the appearance of translational-rotational deformation modes according to the shift+rotation scheme and are classified as mesoscale level 1.

At the considered scale and structural level, the hydrodynamic characteristics of the substance, depending on the history of thermomechanical action and the type of stress-strain state, are clearly manifested [3, 4].

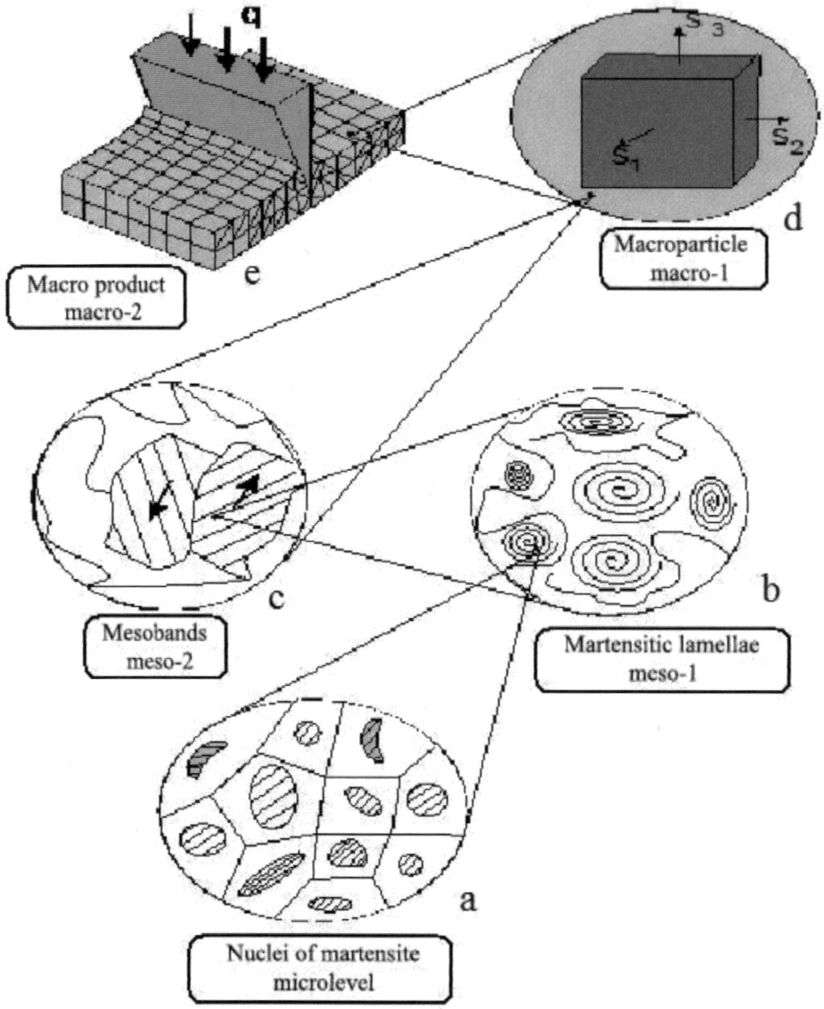

Figure 1. Scheme of scale and structural levels of a deformable object made of materials with a martensitic mechanism of mass transfer, the shaded areas represent the nuclei of the martensitic crystals: a – micro, b – meso-1, c – meso-2, d – macro-1.

With the development of martensitic reactions, the density of structural imperfections of a larger scale level increases and, at a certain critical value of meso concentrators, when the distortion of the transformation at the meso-1 level reaches the limit value, the shear stability is lost in the local zones of the next mesostructural level, meso-2, figure 1c. Thus, structural changes in areas of considerable length and in arbitrary directions, not necessarily coinciding with crystallographic variants of shifts at martensitic transformations, become possible. Extended band-pass deformation structures appear, they originate on meso stress concentrators and spread over long distances through many structural elements regardless of the crystallographic orientation. At this scale level, the stochastic features of the formation of the translational-rotational field of the deformable material are of particular importance [3, 4].

Self-organization of complex stochastic processes in various volumes of the large-scale level meso-2 leads to the formation of a non-trivial tensor properties at the scale level of particulate substances, i.e., at the level of macro-1, figure 1d. Inhomogeneous distribution of macrodeformation in the loaded actual product requires analysis of the stress-strain state in the framework of solution of the boundary task of mechanics of deformable solids, that means the consideration of the large-scale macro-2, figure 1e.

It should be noted that the scheme of structural and scale levels of the deformation process given in figure 1 under the conditions of phase transformations of martensitic type, is very conditional, however, it allows to properly represent the sequence and the main stages of modeling, which in the following sections are formulated in the form of the corresponding connected defining relations. A brief overview of mathematical models of martensitic transformations initiated by phase deformations and structural stresses is given further.

4. Model of martensitic transformations process at the micro-scale level

Characteristic averaging volume $V_0 >> a^3$), where a - is the lattice parameter [2].

Micro-level deformation β_{ik}^{Φ} in martensitic reactions is identified with distortion *of* transformation D_{ik}. The kinetics of the accumulation of the martensitic phase (Φ) is shown in the figure 2 [2, 3]:

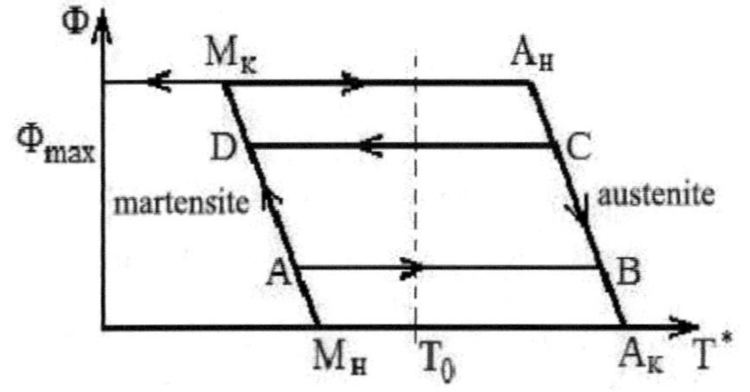

Figure 2. The scheme of kinetics of the martensite change ($\Phi = V_{0\Phi}/V_0$) in the local field of the orientational space, depending on the effective temperature T* when martensitic reactions are direct (A→M) and reverse (A←M)

$$\dot{\Phi} = -\dot{T}^* \begin{Bmatrix} H(1-\Phi_\Sigma)H(-\dot{T}^*)H[M_H - \Phi(M_H - M_K) - T^*](M_H - M_K)^{-1} + \\ + H(\Phi)H(\dot{T}^*)H[T^* + \Phi(A_K - A_H) - A_K](A_K - A_H)^{-1} \end{Bmatrix};$$

$$\dot{\beta}_{ik}^{\Phi} = \dot{\beta}_{31}^{\Phi}(\delta_{i3}\delta_{k1} + \delta_{i1}\delta_{k3}); \dot{\beta}_{31}^{\Phi} = D_{31}\dot{\Phi};$$

$$D_{ik} = D_{31}(\delta_{i3}\delta_{k1} + \delta_{i1}\delta_{k3}); \dot{T}^* = \dot{T} - \frac{T_0}{q_0} D_{ik}\tau'_{ik};$$

$$\tau'_{ik} = \tau'_{31}(\delta_{i1}\delta_{k3} + \delta_{k1}\delta_{i3}); \tau'_{ik} = \tau_{ik} - \psi_{ik}; \tau_{ik} = \alpha_{pi}\alpha_{qk}\sigma_{pq}, \psi_{ik} = \alpha_{pi}\alpha_{qk}\rho_{pq} \quad (1)$$

In the formulae (1) T^* – effective temperature, D_{ik} – tensor distortion of the phase origin, T_0 – the temperature of thermodynamic balance; q_0 – thermal effect of reaction; τ_{ik} – voltage induced by external loads; ψ_{ik} – structural strains; δ_{ik} – Kronecker delta, H – Heaviside function, Φ_Σ – the total phase at the macro level; M_H, M_K, A_H, A_K – characteristic temperatures; α_{ik} – derectional cosines that transform local basis in the global

5. Model of the process of martensitic transformations at the mesoscale level V_{m1} MESO-1

($V_0 << V_{m1} << V_{m2}$), where V_{m2} – characteristic volume meso-2 [2-5].

Deformation at the meso-level -1 L_{ik}^{Φ} in martensitic reactions it is identified with meso-1 distortion B_{ik}^{Φ}. The governing equations in the considered scale and structural levels obey the conservation equations of hydrodynamic type. The modification of these equations is represented by the proportion (2).

$$B_{ik}^{\Phi} = B_{31}^{\Phi}\delta_{i3}\delta_{k1}, \dot{B}_{31}^{\Phi} = -\overline{\nabla}_\Omega \cdot \overline{I}_\beta + \sigma_\beta, L_{ik}^{\Phi} = L_{31}^{\Phi}\delta_{i3}\delta_{k1}, L_{31}^{\Phi} = B_{31}^{\Phi}\Phi_{m1}^{0};$$

$$\Phi_{m1}^{0} = \int_{0}^{t} \int_{-\Gamma}^{\Gamma} \int_{-Д}^{Д} \psi(S_{\Gamma})\varphi(S_{Д})\dot{\Phi} dS_{\Gamma} dS_{Д} dS;$$

$$\left\langle \dot{\beta}_{31}^{\Phi} \right\rangle = \int_{-\Gamma}^{\Gamma} \int_{-Д}^{Д} \psi(S_{\Gamma})\varphi(S_{Д})\dot{\beta}_{31}^{\Phi} dS_{\Gamma} dS_{Д};$$

$$\vec{I}_{\beta} = A_{1\Phi}\overline{\nabla}_{\Omega}\left\langle \beta_{31}^{\Phi}\right\rangle; \left\langle \beta_{31}^{\Phi}\right\rangle = \int_{0}^{t}\left\langle \dot{\beta}_{31}^{\Phi}\right\rangle dS;$$

$$\sigma_{\beta} = K_{1\Phi}\left\langle \dot{\beta}_{31}^{\Phi}\right\rangle + l_{1\Phi}(\overline{\nabla}_{\Omega} \times \vec{a}_{\Phi}) \cdot \vec{e}_{n}$$

$$B\vec{a}_{\Phi} = \left\rangle \dot{\beta}_{31}^{\Phi} \right\langle \vec{e}_{l}; l_{1\Phi} = \eta_{1\Phi} \cdot \left(\frac{mod\, \nabla_{\Omega}\left\langle \dot{\beta}_{31}^{\Phi}\right\rangle}{\nabla_{\Omega}^{2}\left\langle \dot{\beta}_{31}^{\Phi}\right\rangle} \right) \tag{2}$$

In the formulae (2) $B_{ik}^{Ч}$ – transformation distortion on meso-1; vector $\vec{I}_{\beta}$ characterizes the density of the distortion flow on meso-1; parameter σ_{β} describes the production of an internal source of distortion $B_{31}^{Ч}$ due to the initiation in meso amount V_{m1} statistical ensemble of velocity distortions of the martensitic transformation ($\left\langle \dot{\beta}_{31}^{\Phi}\right\rangle$) and the circulation flow of the deformation field (component $l_1(\nabla_{\Omega} \times a_{\Phi}) \cdot \bar{e}_n$). Parameter l_l characterizes the size of the emerging vortex; $K_{1\Phi}$, $A_{1\Phi}$, $\eta_{1\Phi}$ – the constants of a substance; ∇_{Ω} – the operator "nabla" for the orientation of the coordinate system.

6. Model of the process of martensitic transformations at the mesoscale level V_{m2} MESO-2

($V_{m1} \ll V_{m2}$), where V_{m2} – characteristic volume meso -2 [3–7].

Distortion rate $\dot{\varphi}_{ik}^{Ч}$ and the intensity of structural strains $q_{ik}^{Ч}$ at the considered scale level we will describe a system of integro-differential equations that takes into account the condition of preserving the continuity of the environment. Within the development of the model the continuity condition of the environment means the need to harmonize the sliding velocity of the considered bands and shifts in the random meso bands of deformation. Using a technique developed in statistical mechanics of continuum, the following analytical relations can be formulated for meso-2:

$$\dot{\varphi}_{ik}^{\Phi} = \dot{\varphi}_{31}^{\Phi}\delta_{i3}\delta_{k1};$$

$$\dot{q}_{31}^{Ч}(\Omega_i) = \int_{\{\Omega_i'\}} f(\Omega_i')R(\Omega_i, \Omega_i')\tfrac{1}{2}[\alpha_{m3}(\Omega_i')\alpha_{n1}(\Omega_i') + \alpha_{m1}(\Omega_i')\alpha_{n3}(\Omega_i')]\overline{D}\left(\dot{\varepsilon}_{mn}^{Ч}\right)d\Omega_i';$$

$$\dot{\varphi}_{31}^{\Phi} = \int_{\{\Omega_{am}'\}} f(\Omega_{am}')A_m(\Omega_{am}, \Omega_{am}')\tfrac{1}{2}[\alpha_{m3}(\Omega_{am}')\alpha_{n1}(\Omega_{am}') + \alpha_{m1}(\Omega_{am}')\alpha_{n3}(\Omega_{am}')] \times$$

$$\times \overline{D}(\overset{*}{\sigma}_{mn})d\Omega'_{am} + \int_{\{\Omega_{ma}'\}} f(\Omega_{ma}')M_A(\Omega_{ma}, \Omega_{ma}')\frac{1}{2}[\alpha_{m3}(\Omega_{ma}')\alpha_{n1}(\Omega_{ma}') +$$

$$+ \alpha_{m1}(\Omega_{ma}')\alpha_{n3}(\Omega_{ma}')] \times \overline{D}(\varepsilon_{mn}^{\Phi})d\Omega'_{ma};$$

$$\overline{D}\left(\overset{*}{\sigma}_{ik}\right) = \overset{*}{\sigma}_{ik} / \overset{*}{\sigma}_{i}; \overset{*}{\sigma}_{ik} = \sigma_{ik} - \rho_{ik}; \overset{*}{\sigma}_{i} = \frac{1}{\sqrt{2}}\left(\overset{*}{\sigma}_{ik}\overset{*}{\sigma}_{ik}\right)^{1/2};$$

$$\overline{D}\left(\dot{\varepsilon}_{ik}^{\Phi}\right)=\dot{\varepsilon}_{ik}^{\Phi}/\dot{\varepsilon}_{i}^{\Phi};\dot{\varepsilon}_{i}^{\Phi}=\frac{1}{\sqrt{2}}\left(\dot{\varepsilon}_{ik}^{\Phi}\dot{\varepsilon}_{ik}^{\Phi}\right)^{1/2} \quad (3)$$

In the formulae (3) $\overline{D}\left(\overset{*}{\sigma}_{ik}\right)$, $\overline{D}\left(\dot{\varepsilon}_{ik}^{\Phi}\right)$ – guiding tensors of effective stresses $\left(\overset{*}{\sigma}_{i}\right)$ and phase deformation rates $\dot{\varepsilon}_{ik}^{\Phi}$; $f(\Omega_i)$ and $f(\Omega_i')$– density distribution by orientation $\{\Omega_i\}$and $\{\Omega_i'\}$; $A_m(\Omega_{am},\Omega_{am}')$, $M_a(\Omega_{ma},\Omega_{ma}')$ and $R(\Omega_i,\Omega_i')$–impact functions that make sense of structural pliability, structural shape memory and structural relaxation, respectively; $q_{31}^{\Phi}(\Omega_i)$– structural stress intensities ρ_{ik}; $\{\Omega_i\}$– areas of orientation space where shifts occur; $\{\Omega_i'\}$– spheres of impact [2].

7. Model of the process of martensitic transformations at the macro-scale level V Macro level-1

(V_{m2}<<V<<R^3), where R – the characteristic size of macro-production.

At the macro level, in the results of orientation averaging, the defining relations are obtained in the traditional form adopted in the mechanics of a deformable solid [2–7].

$$\Phi_{\Sigma}=\int_{\{\Omega_0\}} f(\Omega_0)\Phi_{ml}^{0}(\Omega_0)d\Omega_0;$$

$$\dot{\varepsilon}_{ik}^{\Phi}\{V_0\}=\frac{1}{2}\int_{\{\Omega_0\}} f(\Omega_0)(\alpha_{i3}\alpha_{k1}+\alpha_{i1}\alpha_{k3})\left\langle\dot{\beta}_{31}^{\Phi}(\Omega_0)\right\rangle d\Omega_0;$$

$$\dot{\varepsilon}_{ik}^{\Phi}\{V_{ml}\}=\frac{1}{2}\int_{\{V_{ml}\}} f(\Omega_{ml})(\alpha_{i3}\alpha_{k1}+\alpha_{i1}\alpha_{k3})\dot{L}_{31}^{\Phi}(\Omega_{ml})d\Omega_{ml};$$

$$\dot{\varepsilon}_{ik}^{\Phi}\{V_{m2}\}=\frac{1}{2}\int_{\{\Omega_i\}} f(\Omega_i)[\alpha_{i3}(\Omega_i)\alpha_{k1}(\Omega_i)+\alpha_{i1}(\Omega_i)\alpha_{k3}(\Omega_i)]\dot{\varphi}_{31}^{\Phi}d\Omega i.$$

$$\dot{\rho}_{ik}^{(i)}=\frac{1}{2}\int_{\{\Omega_i\}} f(\Omega_i)[\alpha_{i3}(\Omega_i)\alpha_{k1}(\Omega_i)+\alpha_{i1}(\Omega_i)\alpha_{k3}(\Omega_i)]\dot{q}_{31}^{\Phi}d\Omega i; \quad (4)$$

$$\dot{\varepsilon}_{ik}^{\Phi}\{V_{m2}\}=A_{ikpq}^{\Phi}\overset{*}{\sigma}_{pq}+M_{ikpq}^{\Phi}\varepsilon_{pq}^{\Phi};$$

$$\overset{*}{\sigma}_{ik}=\sigma_{ik}-\rho_{ik};\ \dot{\rho}_{ik}^{(i)}=R_{ikpq}^{(i)}\dot{\varepsilon}_{pq}^{\Phi}. \quad (5)$$

Here, $\varepsilon_{ik}^{\Phi}\{V_0\}$, $\varepsilon_{ik}^{\Phi}\{V_{ml}\}$, $\varepsilon_{ik}^{\Phi}\{V_{m2}\}$ – macroscopic strain tensors characterizing the contribution of mass transfer processes at micro, meso-1 and meso-2 structural and scale levels, respectively. The correlations (1)-(5) are discussed in more detail in [4].

8. Mesomechanics model taking into account the wave character of deformation

As it is known [4], the peculiarity of the meso bands motion in the deformable solid is the complex nature of the interaction of effective shear stresses with the subsystem of structural imperfections, which in the given boundary conditions (determined by the direction of the main axes of the stress state in the stress space at "soft" loading trajectories or corresponding directions in the deformation space under rigid loading conditions) leads to the wave nature of the plastic flow spread.

We take as a generalized coordinate the shear component of the distortion tensor φ_{31}^{Φ} at the meso-2 scale level [4]. It is necessary to enter the tensor parameter to construct the wave equation $\Delta_{\Omega}\varphi_{31}^{\Phi}\delta_{i3}\delta_{k1}$ (Δ_{Ω}– Laplace operator). However, it is more appropriate instead $\Delta_{\Omega}\varphi_{31}^{\Phi}\delta_{i3}\delta_{k1}$ to consider the

IOP Conf. Series: Materials Science and Engineering **656** (2019) 012033 doi:10.1088/1757-899X/656/1/012033

appropriate analogue, by component $\left\langle B_{31}^{\Phi}\right\rangle$ [4], as a parameter $\Delta_{\Omega}\left\langle B_{31}^{\Phi}\right\rangle\delta_{i3}\delta_{k1}$. This technique is inconsistent for the transfer processes described at the same scale level. In our case of multilevel analysis, this approach is natural and methodically justified, since it allows us to organize additional coherence of the processes of evolution of deformation structures at the micro -, meso - and macro-scale levels. Taking into account the abovementioned, we introduce a tensor parameter d_{ik} in the form:

$$d_{ik} = \Delta_{\Omega}\left\langle B_{31}^{\Phi}\right\rangle\delta_{i3}\delta_{k1} = d_{31}\delta_{i3}\delta_{k1} \tag{6}$$

The wave equation for the shear strain component at the meso structural level of meso-2 can be represented by the following equation:

$$\begin{aligned} m_2\ddot{\varphi}_{31}^{\Phi}(\Omega)+m_1\dot{\varphi}_{31}^{\Phi}(\Omega)+m_0\varphi_{31}^{\Phi}(\Omega)= A_B d_{31}(\Omega)+ \\ +\int f(\Omega')A(\Omega,\Omega')\frac{1}{2}[\alpha_{m3}(\Omega')\alpha_{n1}(\Omega')+\alpha_{m1}(\Omega')\alpha_{n3}(\Omega')]\overline{D}\left(\overset{*}{\dot{\sigma}}_{mn}\right)d\Omega'_{am} + \\ + \int_{\{\Omega_2''\}} f(\Omega'')M_A(\Omega,\Omega'')\frac{1}{2}[\alpha_{m3}(\Omega'')\alpha_{n1}(\Omega')+\alpha_{m1}(\Omega')\alpha_{n3}(\Omega')]\overline{D}\left(\varepsilon_{mn}^{\Phi}\right)d\Omega'_{ma} \end{aligned} \tag{7}$$

Here Δ_{Ω} – Laplace operator for hydrodynamic orientation space $\{\Omega_2\}$, m_0, m_1, m_2, A_B – the material constants. Availability is dependent on the orientation of coordinates and time, the components of the right side of the equation (7) indicates a strong attenuation of the waves of plastic flow in a structurally heterogeneous environment. The specificity of the initiation of meso structure vibrations in the deformation process allows us to characterize it as an active excitable environment.

The available experimental data [4] convincingly testify about the auto wave nature of shifts of meso bands of deformation at the meso structural level. Equation (7) is a natural generalization of equation (3) and takes into account the wave nature of the deformation process at the meso structural level.

Taking into account the description of auto wave processes at the scale level of macro-1, leads to the appearance in the equation for the calculation of the components of the tensor of inelastic deformation $\varepsilon_{ik}^{H}\{V_{m2}\}$ of derivatives, both of the first and second order, which naturally describes the oscillating nature of the spread of the traveling pulse in the active elastoplastic environment.

Having performed the orientation averaging of the relations (7) of the mesoscale level, we will obtain the desired wave equations at the macroscale level in the form:

$$m_2\ddot{\varepsilon}_{ik}^{\Phi}\{V_{m2}\}+m_1\dot{\varepsilon}_{ik}^{\Phi}\{V_{m2}\}+m_0\varepsilon_{ik}^{\Phi}\{V_{m2}\}=A_B D_{ik}\{V_{m2}\}+A_{ikmn}^{\Phi}\overset{*}{\sigma}_{mn}+M_{ikmn}^{\Phi}\varepsilon_{mn}^{\Phi} \tag{8}$$

Here parameter $D_{ik}\{V_{m2}\}$ is calculated with a formula:

$$D_{ik}\{V_{m2}\}=\frac{1}{2}\int_{\{\Omega_2\}} f\{\Omega_2\}[\alpha_{ik}(\Omega)\alpha_{kl}(\Omega)+\alpha_{i1}(\Omega)\alpha_{k3}(\Omega)]d_{31}(\Omega)d\Omega_2 \tag{9}$$

Equations (8) include tensor parameters characterizing the evolution of the structure at the macro-scale level: A_{ikmn}^{Φ}, M_{ikmn}^{Φ} and ε_{mn}^{Φ}, $\overset{*}{\sigma}_{mn}$. Here A_{ikmn}^{Φ} the kinetic coefficients of the structural compliance, the kinetic coefficients of the structural shape memory, $\overset{*}{\sigma}_{mn}$ the effective stress. The indicated parameters are functionals depending on the history of loading in time and in the space of stresses (deformations), and are calculated by the method described in [4] in the derivation of the defining relations for the macro-scale level.

In the framework of this model, the basic functional and mechanical properties of materials with shape memory effect are predicted taking into account the wave character of deformations initiated by martensitic transformations.

To solve the practical problems of engineering mechanics [4], additional equations are required tot take into account the spatial arrangement of volumes V (consideration of the scale level of macro-2), i.e. the traditional static and geometric equations of the boundary value problem of mechanics as well as the initial and boundary conditions for the corresponding variables

9. Formulation of the boundary task at the macro-scale level of macro-2

Analytical relations (1) – (9) should be considered as defining equations for structurally inhomogeneous environment in macropoint, see figure 1d. To solve the practical problems of engineering mechanics, it is necessary, of course, to formulate additional equations that take into account the spatial arrangement of the volumes V1, that is the consideration of the scale level of macro-2, see figure 1e, dynamic and static properties of the environment, etc., as well as the initial and boundary conditions for the corresponding variables. The above mentioned is reduced to the formulation of the following equations of the boundary value problem of mechanics:

9.1. The dynamic equation of balance for voltage:

$$\Delta_i \sigma_{ik} = \rho \ddot{u}_k \tag{10}$$

Where ρ – density of environment; u_k – displacement vector;

9.2. The condition of continuity for the deformation:

$$e_{ksr} e_{qmt} \nabla_s \nabla_t \varepsilon^e_{rm} = -e_{ksr} e_{qmt} \nabla_s \nabla_t \varepsilon^H_{rm} \tag{11}$$

where e_{ksr} – Levi-Civita symbol; $\varepsilon^H_{ik} = \varepsilon^T_{ik} + \varepsilon^\Phi_{ik}$;

93. Balance condition for temperature:

$$\dot{T} = \frac{K_{ik}}{\rho \cdot c} \nabla_i \nabla_k T - \frac{q_0}{\rho \cdot c} \dot{\Phi}_\Sigma \tag{12}$$

where K_{ik} – tensor of thermal conductivity coefficients; c – specific heat. The second term in the right part (12), containing $\dot{\Phi}_\Sigma$, takes into account the thermal effect of the martensitic transformation reaction. In the case of isotropic heat transfer at the macro level the thermal conductivity tensor is equal to:

$K_{ik} = k_0 \delta_{ik}$

where k_0 – scalar coefficient of thermal conductivity.

For the final formulation of a specific boundary task, equations for voltage (10), deformations (11) and temperatures (12) must be supplemented with the corresponding boundary and initial conditions.

10. Conclusion

It is necessary to point out the following important aspects. In contrast to the traditional methodology of continuum mechanics, when the components of strain tensor are introduced as derivatives of the displacement field of points of the continuum, in this approach, they are specified through appropriate micro - and meso - characteristics, taking into account physical regularities in the formation of martensite crystals at the micro-level and processes of self-organization of the martensitic structures at the mesoscale level. In the physical sense, the formulated analytical relations take into account not only the translational and rotational nature of mass transfer at the mesoscopic level, but also the cross-effects of the interaction of elements of the medium of different scale levels.

References

[1] Likhachev V A, Kuz'min S A and Kamentseva Z P 1987 *Shape memory effect* (Leningrad: LGU Publ) p 216
[2] Panin V E 1998 Basics of physical mesomechanics *Physical mesomechanics* **1** 5–22
[3] Likhachev V A and Malinin V G 1993 *Structural analytic theory of strength* (Saint-Petersburg: Nauka Publ) p 471
[4] Golenkov V A, Malinin V G and Malinina N A 2009 *Structural-analytical mesomechanics and its applications* (Moscow: Maschinostroenie Publ) p 635
[5] Malinin V G and Malinina N A 2005 Structural-analytical mesomechanics of a deformable solid body *Physical mesomechanics* **23** 31–45
[6] Malinin V G, Malinina N A and Malinin G V 2002 Mechanical characteristics of materials with the effect of the memories of the form under complex mode isothermal loading *Appendix to Engineering journal: Spravochnik* **10/1** 27
[7] Malinin V G, Malinina N A ans Malinin G V 2002 Mechanical characteristics of materials with the effect of the memories of the form under complex mode isothermal loading *Appendix to Engineering journal: Spravochnik* **10/2** 25

IOP Conf. Series: Materials Science and Engineering **656** (2019) 012034 doi:10.1088/1757-899X/656/1/012034

Structural and analytical mesomechanics of alloys with shape memory effect in the concept of interacting levels of deformation and destruction

GV Malinin[1,4], V G Malinin[2], N A Malinin [2] and O A Malukhina[3]

[1]PJSC "Tupolev", Academician Tupolev Embankment, 17, 105005, Moscow, Russia
[2]Oryol State Agrarian University, ul. General Rodina, 69, 302019, Orel, Russia
[3]Yaroslav-the-Wise Novgorod State University, ul. B. St. Peterburgskaya, 41 173003, Veliky Novgorod, Russia

[4]E-mail: malinin2002@yandex.ru

Abstract. This article presents the results of the development of structural-analytical mesomechanics based on the concept of interacting multi-level (micro, meso and macro) plastic flow, the evolution of structural damage and martensitic phase transformations. It presents a system of constitutive relations, which take into account the mutual influence of plastic deformations, processes of destruction and structural phase transformations of martensitic nature for alloys with shape memory effect (SME).

1. Introduction

In this article, on the basis of [1 6], the results on the development of methods of structural-analytical mesomechanics of materials with the SME are presented. Unlike previous works on the development of structural-analytical mesomechanics, in this article, for the first time in a related and multi-level formulation, the defining relations for the macroscale level are derived that take into account the possibility of simultaneous multilevel plastic flow, reversible martensitic transformations, micro-destruction and their mutual influence.

Let us consider the problem of calculating the functional and mechanical properties of materials with an SME undergoing shear inelastic deformations, micro destruction processes, and martensitic-type structural and phase transformations. In accordance with the methodology of [1–6], the behavior of the material will be considered at three scale levels (micro-meso-and macro). At the micro level, deformations and stresses are averaged over small volumes of a crystal and are the characteristics of the movements and interactions of atoms. Here, the main acts of mass transfer are realized in the concepts of solid-state physics. At the meso level, deformations and stresses are characterized by significant spatial gradients and they develop under essentially nonequilibrium conditions, which determines the dominant role of translational-rotational deformation models, the description of which requires the use of laws of nonequilibrium thermodynamics. And, finally, at the macro level, deformations and stresses are obtained by averaging over a large number of interacting micro and mesoregions and are interpreted in the ideas of engineering mechanics of materials. Let us consider the method of constructing the defining relations for the calculation of micro-meso-and macro-scale properties during plastic deformation of materials with an SME.

IOP Conf. Series: Materials Science and Engineering **656** (2019) 012034 doi:10.1088/1757-899X/656/1/012034

2. The effective field method

In the process of elastoplastic deformation, as a rule, the nucleation and motion of dislocations is inhomogeneous, differently in different grains, inhibition occurs at grain boundaries and other obstacles. As a result, microcracks can be opened in the most loaded structural concentrators. The deformation created in such conditions is incompatible and leads to the formation of internal or intrinsic stress fields, which in turn influence the further deformation process. The nature of the internal stress fields is such that they weaken the stresses from external forces σ_{ik} in the plastic shear zone ($a\rho_{ik}$) and strengthen them where the shift is inhibited. A similar situation occurs in the crystal zone, where the process of destruction is formed, microcracks arise and appear (br_{ik}). If phase transformations are initiated, that are strongly localized under the conditions considered, then there arises an intrinsic stress field (ρ_{ik}^{Φ}), due to the heterogeneity of deformations of martensitic nature. This is equivalent to the statement that the deformation forms three regions in crystals in which the processes of plastic shears, the formation of microcracks and the formation of martensitic transformations are interconnected. To describe the processes under consideration, following the methods of structural-analytical mesomechanics [1, 6], we write out formulas for the effective stresses that act in each of the crystal regions.

In the region of the crystal, where plastic shifts occur, it is reasonable to present the effective stresses $\overset{*}{\sigma}_{ik}$ [6] in the form of:

$$\overset{*}{\sigma}_{ik} = \sigma_{ik} - a\rho_{ik} + br_{ik} + \rho_{ik}^{\Phi}. \tag{1}$$

Effective stresses $\overset{**}{\sigma}_{ik}$ [6], acting in the zone of initiation of cracks, will look like:

$$\overset{**}{\sigma}_{ik} = \sigma_{ik} + a\rho_{ik} - br_{ik} + \rho_{ik}^{\Phi}. \tag{2}$$

In the zone of initiation of phase transformations of the martensitic type, two subregions are formed: one in which the effective stresses ${}^{-}\sigma_{ik}^{\Phi}$ decrease due to plastic shears, and the other where the effective stresses ${}^{+}\sigma_{ik}^{\Phi}$ increase in comparison with σ_{ik}. Following [5, 6] you can write:

$$\begin{aligned} {}^{-}\sigma_{ik}^{\Phi} &= \sigma_{ik} - a\rho_{ik} + br_{ik} - \rho_{ik}^{\Phi}, \\ {}^{+}\sigma_{ik}^{\Phi} &= \sigma_{ik} + a\rho_{ik} - br_{ik} - \rho_{ik}^{\Phi}; \end{aligned} \tag{3}$$

where a and b are permanent; ρ_{ik}, r_{ik} and ρ_{ik}^{Φ} are oriented stress tensors determined at the macroscale level.

3. Hereditary phenomena in reversible martensitic transformations

Plastic deformation of austenite and martensite leads to hardening associated with an increase in yield strengths in austenite τ^{SA}, martensite τ^{SM}, generation of fields of own stresses ρ_{ik}, r_{ik} and ρ_{ik}^{Φ}. Since austenite can turn into martensite (and vice versa), the natural question arises about the inheritance of strengthening factors. Within the framework of this approach, it is not possible to discuss the inheritance process in detail, therefore, we assume that the yield strengths of both phases exist and develop independently of each other. Own stress fields ρ_{ik}, r_{ik} and ρ_{ik}^{Φ} appear during plastic deformation of any of the phases and are common to them, i.e. act in representative volumes regardless of the phase composition. Thus, inheritance is assumed without changing these fields during martensitic transformations.

4. Microscale level

Plastic deformation of austenite and martensite leads to hardening related to the micro level analysis, we introduce the volume of averaging $V_0>>V_a$, where V_a is the atom volume. Following the methodology of structural-analytical mesomechanics and using the same notation as in [6], we formulate the defining relations for the description of the corresponding deformation fluxes and parameters characterizing the evolution of structure and damage at the micro level:

$$ {}^{(m)}\dot{\beta}^{y}_{ik} = {}^{(m)}C_{ikpq}\dot{\tau}_{pq},\ {}^{(m)}\dot{\beta}^{T}_{ik} = {}^{(m)}\gamma_{ik}\dot{T}; \tag{4} $$

$$ {}^{(m)}\dot{\beta}^{H}_{ik} = {}^{(m)}\dot{\beta}^{H}_{31}(\delta_{i3}\delta_{k1} + \delta_{k3}\delta_{i1}),\ {}^{(m)}\dot{\beta}^{H}_{31} = {}^{(m)}\dot{\beta}^{t}_{31} + {}^{(m)}\dot{\beta}^{a}_{31}; \tag{5} $$

$$ {}^{(m)}\dot{\beta}^{t}_{31} = A_t \exp\left[-\frac{u_t - \gamma_t \tau'_{31}\operatorname{sign}\tau'_{31}}{kT}\right](\tau'_{31}\operatorname{sign}\tau'_{31})^{n}\operatorname{sign}\tau'_{31}; \tag{6} $$

$$ {}^{(m)}\dot{\beta}^{a}_{31} = {}^{(m)}A_a(\dot{\tau}'_{31} - \dot{\tau}^{s}_{0}\operatorname{sign}\tau'_{31})H(\tau'_{31}\operatorname{sign}\tau'_{31} - {}^{(m)}\tau^{s})H(\dot{\tau}'_{31}\operatorname{sign}\tau'_{31} - {}^{(m)}\dot{\tau}^{s}_{0}); \tag{7} $$

$$ \dot{\tau}^{s} = {}^{(m)}A^{-1}_{a}\dot{\beta}^{a}_{31}\operatorname{sign}\dot{\beta}^{a}_{31} - {}^{(m)}\chi\dot{T} - {}^{(m)}A_s \exp\left[-\frac{{}^{(m)}u_s - {}^{(m)}\gamma_s\tau'_{31}\operatorname{sign}\tau'_{31}}{kT}\right]\times $$
$$ \times({}^{(m)}\tau^{s} - {}^{(m)}\tau_0 + {}^{(m)}\chi T)^{m} H({}^{(m)}\tau^{s} - {}^{(m)}\tau_0 + {}^{(m)}\chi T); \tag{8} $$

$$ {}^{(m)}\dot{\tau}^{s}_{0} = {}^{(m)}\dot{\tau}^{s} - {}^{(m)}A^{-1}_{a}\dot{\beta}^{a}_{31}\operatorname{sign}\dot{\beta}^{a}_{31};\ {}^{(m)}\tau^{s}_{0}(s^{\tau}) = {}^{(m)}\tau_0 + s^{\tau};\ \tilde{A}^{\tau}_{min} \le s^{\tau} \le \tilde{A}^{\tau}_{max}; \tag{9} $$

$$ \tau'_{ik} = \tau'_{31}(\delta_{i3}\delta_{k1} + \delta_{k3}\delta_{i1}),\ \tau'_{ik} = \tau_{31} - \alpha_{\rho}\psi^{(\rho)}_{31} + \alpha_{r}\psi^{(r)}_{31} + \alpha_{\Phi}\psi^{(\Phi)}_{31} + \nu^{\Sigma}_{31}, $$
$$ \tau_{31} = \alpha_{p3}\alpha_{q1}\sigma_{pq}\left[\left(1 - \frac{\Pi_{\Sigma}}{\Pi_0}\right)^{\varsigma}\right]^{-1}; \tag{10} $$

$$ \psi^{(p)}_{ik} = \alpha_{pi}\alpha_{qk}\rho_{pq},\ \psi^{(r)}_{ik} = \alpha_{pi}\alpha_{qk}r_{pq},\ \psi^{(\Phi)}_{ik} = \alpha_{pi}\alpha_{qk}\rho^{\Phi}_{pq}, \tag{11} $$

$$ \dot{\pi}^{0} = \left[H(\hat{\tau}_{33} - \tau^{0})\delta\left(\frac{\beta}{\beta^{0}} - 1\right)\frac{\dot{\beta}}{\beta^{0}} + H(\beta - \beta^{0})\delta\left(\frac{\hat{\tau}_{33}}{\tau^{0}} - 1\right)\frac{\dot{\hat{\tau}}_{33}}{\tau^{0}}\right]\times $$
$$ \times\left(1 + \alpha^{0}\frac{\beta}{\beta^{0}}\right)H(1 - \pi^{0}) + \alpha^{0}H(\beta - \beta^{0})H(\hat{\tau}_{33} - \tau^{0})\frac{\dot{\beta}}{\beta^{0}}; \tag{12} $$

$$\dot{\pi}^{0} = \left[H(\bar{\tau}_{31}\,\mathrm{sign}\,\bar{\tau}_{31} - \tau^{c})\delta\left(\frac{\beta}{\beta^{c}} - 1\right)\frac{\dot{\beta}}{\beta^{c}} + H(\beta - \beta^{c})\delta\left(\frac{\bar{\tau}_{31}\,\mathrm{sign}\,\bar{\tau}_{31}}{\tau^{c}} - 1\right)\frac{\dot{\bar{\tau}}_{31}\,\mathrm{sign}\,\bar{\tau}_{31}}{\tau^{c}} \right] \times$$

$$\times \left(1 + \alpha^{c}\frac{\beta}{\beta^{c}}\right) H(1 - \pi^{c}) + \alpha^{c} H(\beta - \beta^{c}) H(\bar{\tau}_{31}\,\mathrm{sign}\,\bar{\tau}_{31} - \tau^{c})\frac{\dot{\beta}}{\beta^{c}}; \qquad (13)$$

$$\int_{0}^{t_p} \tau_{0}^{-1} \exp\left[-\frac{u_p - \gamma_p \hat{\tau}_{33}}{kT} \right] dt = 1; \; \dot{\pi}^{t} = \delta\left(\frac{t}{t_p} - 1\right)\left(\frac{1}{t_p}\right)(1 + a^{t}\beta) H(1 - \pi^{t}) + a^{t} H\left(\frac{t}{t_p} - 1\right)\dot{\beta}; \qquad (14)$$

$$\hat{\pi}_{i} = \left[\hat{\tau}_{31}\delta_{i1} + \hat{\tau}_{33} H(\hat{\tau}_{33})\delta_{i3}\right](\pi^{0} + \pi^{t})^{n_r}; \; \bar{\pi}_{i} = \left[\bar{\tau}_{31}\delta_{i1} + \bar{\tau}_{33} H(\bar{\tau}_{33})\delta_{i3}\right](\pi^{c})^{n_r}; \qquad (15)$$

$$\dot{\beta} = \dot{\beta}_{31}^{H}\,\mathrm{sign}\,\dot{\beta}_{31}^{H}; \; \beta = \int_{0}^{t} \dot{\beta}_{31}^{H}\,\mathrm{sign}\,\dot{\beta}_{31}^{H} ds; \qquad (16)$$

$$\hat{\tau}_{ik} = \eta_{pi}^{0}\eta_{qk}^{0}\overset{**}{\tau}_{pq}; \; \bar{\tau}_{ik} = \eta_{pi}^{c}\eta_{qk}^{c}\overset{**}{\tau}_{pq}; \; \overset{**}{\tau}_{ik} = \tau'_{ik} + \alpha_{\rho}\psi_{ik}^{(\rho)} + \alpha_{r}\psi_{ik}^{(r)} + \alpha_{\Phi}\psi_{ik}^{(\Phi)} + \nu_{ik}^{\Sigma}. \qquad (17)$$

In relations (4) – (17) and further, the point above the symbol means the time derivative. Tensors of elastic deformation $^{(m)}\beta_{ik}^{y}$, elastic coefficients $^{(m)}C_{ikpq}$, thermal expansion deformations $^{(m)}\beta_{ik}^{T}$ (4) and thermal expansion coefficients $^{(m)}\gamma_{ik}$ are assigned to the crystallographic basis. The tensor of the total inelastic deformation rate $^{(m)}\dot{\beta}_{ik}^{H}$ (5) consists of thermally activated $^{(m)}\dot{\beta}_{ik}^{t}$ (6) and athermic $^{(m)}\dot{\beta}_{ik}^{a}$ (7) components.

The evolutionary equations for the initial $^{(m)}\tau_{0}^{s}$ (9) and current $^{(m)}\tau^{s}$ (8) stresses of the flow of crystallographic shear take into account strain hardening (the first term in (8)), the temperature factor (T), and the ability to age or return (the second and third terms in (8)); the symbol (m) takes the values m = M for martensite, m = A for austenite. The effective stress τ'_{ik} contains, along with the components of the stresses from external loads τ_{ik}, oriented $\psi_{ik}^{(\rho)}$, $\psi_{ik}^{(r)}$, $\psi_{ik}^{(\Phi)}$, and non-oriented ν_{ik}^{Σ} structural stresses of various nature [1, 2]. Among non-oriented microstresses, stresses due to the anisotropy of the thermal expansion coefficients and the anisotropy of the elastic compliance of the crystals are particularly distinguished. A detailed derivation of evolutionary equations for calculating non-oriented stresses of various nature is contained in [1,2]. Note that the fields ν_{ik}^{Σ} are mutually balanced, they are short-range, and the wavelength of the fluctuating stress field is comparable with the characteristic size of the representative volume V_0. Oriented stresses $\psi_{ik}^{(\rho)}$, $\psi_{ik}^{(r)}$ and $\psi_{ik}^{(\Phi)}$ are long-range, reflect the effect on the stress state of crystals of spatial non-uniformity of nucleation and development of plastic deformation ($\psi_{ik}^{(\rho)}$), the appearance of cracks at meso and macroscale levels ($\psi_{ik}^{(r)}$) and uneven deformation of martensitic nature ($\psi_{ik}^{(\Phi)}$). The introduction of oriented structural stresses takes into account the stress relaxation factor in local volumes, where plastic shear ($\psi_{ik}^{(\rho)}$) occurs, the relaxation effect of structural concentrators in the places where the opening of cracks ($\psi_{ik}^{(r)}$) and the relaxation effect, where martensitic crystals ($\psi_{ik}^{(\Phi)}$) arise. The considered structural stresses $\psi_{ik}^{(\rho)}$, $\psi_{ik}^{(r)}$ and $\psi_{ik}^{(\Phi)}$ are formed at the meso and macroscale levels, therefore, a detailed derivation of the evolutionary equations for their calculation will be presented below when analyzing the corresponding defining relations. Equations (12) – (17) simulate the processes of cracking by a separation (12), a slice (13), opening damage of a thermal-fluctuation nature (14), (15), as well as their growth in the process of inelastic deformation.

Note that for convenience, the indicated relations are formulated in the corresponding bases: separation ($\vec{a},\vec{b},\vec{c}$) (12), (14), (15) and slice ($\vec{p},\vec{r},\vec{s}$ (13). Equations (12) and (13) take into account that microcracks arise only when two conditions are mandatory fulfilled. Firstly, when the path of microdeformations β (16) reaches a critical value equal to tearing cracks β^0 and shear cracks β^C. Secondly, when the pullout voltage $\hat{\tau}_{33}$or shear stress $\bar{\tau}_{31} sign\, \bar{\tau}_{31}$ reaches the corresponding critical voltage pull-off τ^0 or critical voltage cut τ^c. The damage parameter Π_Σ in (10) takes into account the influence on τ_{ik} of those damages that are revealed at the micro- and mesostructural levels.

The calculation of the parameter Π_Σ is discussed in [6], which is devoted to the formulation of the fracture criterion at the macro level. The notation α_{ik} characterizes the direction cosines of the relative orientation of the local (l, m, n) and laboratory (x, y, z) bases; η_{ik}^{o}, η_{ik}^{c} are direction cosines between the local basis for the implementation of plastic shear (l, m, n), separation bases ($\vec{a},\vec{b},\vec{c}$) and cut ($\vec{p},\vec{r},\vec{s}$); δ is Dirac delta function; δ_{ik} is the Kronecker symbol, H (x) is the Heaviside function; α_ρ, α_r and α_Φ are the concentration coefficients of structural stresses; Γ_{min}^{τ}, Γ_{max}^{τ} are boundary values of a random field, which are estimated from the analysis of the physical content of the problem.

We note that the mere appearance of microcracks of separation or shear does not mean that they will immediately develop, it is necessary to fulfill additional criteria. In accordance with the ideas of fracture mechanics, for the growth of cracks that arise, it is necessary to fulfill the force (or energy) conditions of their propagation. In this regard, microdamage vectors are introduced [6] for tear cracks and shear cracks in the form of relations (15), where n_r is a constant. Formulas (15) take into account the degree of opening of microcracks in the stress field. When the values $\hat{\pi}_i\hat{\pi}_i$ and $\bar{\pi}_i\bar{\pi}_i$ exceed the critical level $\pi_{\text{кр}}^2$, the crack will be able to propagate in the crystal. The analysis and physical meaning of the constants A_a, A_s, æ, u_s, γ_s, τ_o, m are given in [1]. Relations (4) - (17) describe the evolution of the loss of shear stability of the loaded material at the microscale level, they allow us to calculate the damageability and predict the microstrains of the corresponding nature under arbitrary modes of thermomechanical action.

Let us consider the question of finding the deformation caused by martensitic inelasticity channel. If the stresses are insignificant and there are no plastic shifts in the material, then the calculation of the deformation effects associated with the martensitic mass transfer channel can be successfully carried out using well-tested relationships [1, 3–5] for environments with phase reactions that take into account the specifics of the transformation (first or second kind), statistical properties of the structure and other features. The laws of mechanical properties are practically not investigated for the cases when plastic shears, microcracks and martensitic reactions are simultaneously initiated in the material. In this case, as already noted in the analysis of plasticity, the entire volume of the crystal is divided by the nature of the action of oriented stresses ρ_{ik} and r_{ik} into two areas. Developing the above considerations, in order to take into account the effect of plastic shears and microcracks arising on martensitic reactions, we assume that the entire representative volume where martensitic transformations are realized can be divided into two sub areas with specific volumes *a* and *b* containing oriented microstresses ${}^{+}\sigma_{ik}^{\Phi}$ and ${}^{-}\sigma_{ik}^{\Phi}$. Then, in the presence of stresses σ_{ik}, the effective stresses σ_{ik}^{Φ} acting in different areas during phase transformations can be represented in the first approximation in the form of relations (3). The introduction of effective stresses (3) allows us to formulate, following the method of [3], analytical relations for the calculation of deformations taking into account plastic shears, microdamages and martensitic transformations in the form

$$\dot{\beta}_{ik}^{+} = A_0 D_{ik} \dot{\Phi}^{+}, \dot{\beta}_{ik}^{-} = A_0 D_{ik} \dot{\Phi}^{-}, D_{ik} = D_{31}(\delta_{i3}\delta_{k1} + \delta_{k3}\delta_{i1}); \tag{18}$$

$$\dot{\Phi}^{+} = -\dot{T}_{+}^{*} \left\{ \begin{matrix} H(1-\Phi_m)H(-\dot{T}_{+}^{*})H\left[M_H - \Phi^{+}(M_H - M_K) - \dot{T}_{+}^{*}\right](M_H - M_K)^{-1} + \\ H(\Phi_m)H(\dot{T}_{+}^{*})H\left[M_H - \Phi^{+}(A_K - A_H) - A_K\right](A_K - A_H)^{-1} \end{matrix} \right\}; \tag{19}$$

$$\dot{\Phi}^{-} = -\dot{T}_{-}^{*} \left\{ \begin{matrix} H(1-\Phi_m)H(-\dot{T}_{-}^{*})H\left[M_H - \Phi^{-}(M_H - M_K) - \dot{T}_{-}^{*}\right](M_H - M_K)^{-1} + \\ H(\Phi_m)H(\dot{T}_{-}^{*})H\left[\dot{T}_{-}^{*} - \Phi^{-}(A_K - A_H) - A_K\right](A_K - A_H)^{-1} \end{matrix} \right\};$$

$$\dot{T}_{+}^{*} = \dot{T} - \frac{T_0}{q_0}\alpha_{ip}\alpha_{kq}D_{pq}\,{}^{+}\sigma_{ik}^{\Phi}, \dot{T}_{-}^{*} = \dot{T} - \frac{T_0}{q_0}\alpha_{ip}\alpha_{kq}D_{pq}\,{}^{-}\sigma_{ik}^{\Phi}; \tag{20}$$

$$\Phi_m = \int_{\{S_\Gamma\}} F(S_\Gamma) \int_{\{S_D\}} F(S_D) \int_{\{\Omega\}} f(\Omega)(a\Phi^{+} + b\Phi^{-})dS_\Gamma dS_D d^3\Omega;$$

$$\int_{\{S_\Gamma\}} F(S_\Gamma)dS_\Gamma = 1, \int_{\{S_D\}} F(S_D)dS_D = 1, \int_{\{\Omega\}} f(\Omega)d^3\Omega = 1. \tag{21}$$

Here, A_0, D_{31}, T_0, q_0 are material constants, M_K, M_H, A_H, A_K are the characteristic temperatures of martensitic transformations; S_Γ and S_D – statistical parameters of phase transformations. These parameters are described in detail in [3].

The structural-analytical model of the deformation of materials with the SME (1) – (21) represents a related system of equations reflecting the properties of materials at the micro level. This model makes it possible to formulate and solve problems of determining the structural-kinetic parameters of a material, which provide it with specified mechanical properties at meso- and macroscale levels. In order to formulate the defining relations for meso- and macroscale levels, we will use the methods proposed in [1, 3–6]. Methods for deriving the defining relations for scale levels of meso-1, meso-2 and macro-1 are described in detail in [4, 6] when initiating either the deformation phase channel only [4, 5] or the dislocation mechanism of mass transfer only [6], taking into account the influence of microdestruction. Developing the mentioned above methods for modeling phase deformation processes, it is necessary to take into account that the average statistical values of martensitic microstrain $\left\langle \beta_{\text{ik}}^{\Phi} \right\rangle$ are determined taking into account the influence of both plastic deformation and the development of damage (3) – (20).

5. Meso level-1

We introduce the volume of averaging V_{m1} ($V_0 << V_{m1} << V_{m2}$), where V_{m2} is the characteristic volume of meso-2. The deformation on meso-1 L_{ik}^{Φ} during martensitic reactions is identified with the distortion of martensitic transformations B_{ik}^{Φ} [5]. The defining equations at the scale level under consideration obey the equations of conservation of hydrodynamic type:

$$^{c}B_{\text{ik}}^{\Phi} = {}^{c}B_{31}^{\Phi}\delta_{\text{i3}}\delta_{\text{k1}},\ {}^{c}\dot{B}_{\text{ik}}^{\Phi} = -\overline{\nabla}_{\Omega}\cdot{}^{c}\bar{I}_{\beta} + {}^{c}\sigma_{\beta},\ {}^{c}L_{\text{ik}}^{\Phi} = {}^{c}B_{\text{ik}}^{\Phi}\Phi_{m1}^{0},$$

$$^{c}\bar{I}_{\beta} = {}^{c}\text{A}_{1\Phi}\overline{\nabla}_{\Omega}\left\langle {}^{c}\beta_{31}^{\Phi}\right\rangle, \left\langle {}^{c}\beta_{31}^{\Phi}\right\rangle = \int_{0}^{t}\left\langle {}^{c}\dot{\beta}_{31}^{\Phi}\right\rangle \text{dS},$$

$$\sigma_{\Phi} = {}^{c}K_{1\Phi}\left\langle {}^{c}\beta_{31}^{\Phi}\right\rangle + l_1(\overline{\nabla}_{\Omega}\times{}^{c}\vec{a}_{\Phi})\cdot\vec{e}_n;$$

$$^{c}\vec{a}_{\Phi} = \left\langle {}^{c}\dot{\beta}_{31}^{\Phi}\right\rangle \vec{e}_1,\ {}^{c}l_{1\Phi} = {}^{c}\eta_{1\Phi}\cdot\left(\frac{\operatorname{mod}\nabla_{\Omega}\left\langle {}^{c}\dot{\beta}_{31}^{\Phi}\right\rangle}{\nabla_{\Omega}^{2}\left\langle {}^{c}\dot{\beta}_{31}^{\Phi}\right\rangle}\right). \quad (22)$$

In formulas (22), the symbol "*c*" takes the values "+" or "–" in accordance with formulas (3), characterizing the equations of deformation mass transfer in the sub areas of the crystal, where plastic deformation occurs "–" or in the sub areas of the crystal, where microcracks occur "+" . ${}^{c}B_{\text{ik}}^{\Phi}$ is transformation distortion at meso-1, the vector ${}^{c}\bar{I}_{\beta}$ characterizes the distortion flux density on meso-1; ${}^{c}\sigma_{\beta}$ parameter describes the production of an internal source ${}^{c}B_{\text{ik}}^{\Phi}$ due to the initiation in the V_{m1} mesovolume of a statistical ensemble of distortions of martensitic transformations $\left\langle {}^{c}\beta_{31}^{\Phi}\right\rangle$ and a circulation flow of a deformation field ($l_1(\overline{\nabla}_{\Omega}\times{}^{c}\vec{a}_{\Phi})\cdot\vec{e}_n$ component). The parameter l_1 reflects the size of the arising vortex; ${}^{c}K_{1\Phi}$, ${}^{c}\text{A}_{1\Phi}$, ${}^{c}\eta_{1\Phi}$ are constants of matter, ∇_{Ω} is the "nabla" operator for orientational coordinate system; the symbol <....> denotes the average value.

6. Meso level-2

We introduce the volume of averaging V_{m2} ($V_{m1} << V_{m2} << V$), where V is the characteristic volume of the macro level. The distortion rate $\dot{\varphi}_{\text{ik}}^{\Phi}$ at the considered scale level will be described on the basis of the theory of large-scale fluctuations by a system of integro-differential equations, taking into account the continuity of the environment. In the framework of the model being developed, the continuity condition of the environment means the need to reconcile the slip velocities of the considered band and the shifts occurring in arbitrary deformation mesos. Using the technique developed in [5, 6], the following analytical relations can be formulated for meso-2:

$$^{c}\dot{\varphi}_{\text{ik}}^{\Phi} = {}^{c}\dot{\varphi}_{31}^{\Phi}\delta_{\text{i3}}\delta_{\text{k1}},$$

$$
{}^{c}\dot{q}_{\mathrm{ik}}^{\Phi}(\Omega_{\mathrm{i}})=\int\limits_{\{\Omega'_{\mathrm{i}}\}} f(\Omega'_{\mathrm{i}})R(\Omega_{\mathrm{i}},\Omega'_{\mathrm{i}})\frac{1}{2}[\alpha_{m3}(\Omega'_{\mathrm{i}})\alpha_{n1}(\Omega'_{\mathrm{i}})+
$$

$$
+\alpha_{m1}(\Omega'_{\mathrm{i}})\alpha_{n3}(\Omega'_{\mathrm{i}})]\overline{D}({}^{c}\dot{\varepsilon}_{mn}^{\Phi})d\Omega'_{\mathrm{i}};
$$

$$
{}^{c}\dot{\varphi}_{31}^{\Phi}(\Omega_{\mathrm{i}})=\int\limits_{\{\Omega'_{am}\}} f(\Omega'_{am})\mathrm{A}_{m}(\Omega_{am},\Omega'_{am})\frac{1}{2}[\alpha_{m3}(\Omega'_{am})\alpha_{n1}(\Omega'_{am})+
$$

$$
+\alpha_{m1}(\Omega'_{am})\alpha_{n3}(\Omega'_{am})]\overline{D}({}^{c}\overset{*}{\sigma}_{mn})D'_{am}+
$$

$$
+\int\limits_{\{\Omega'_{ma}\}} f(\Omega'_{ma})\mathrm{M}_{a}(\Omega_{ma},\Omega'_{ma})\frac{1}{2}[\alpha_{m3}(\Omega'_{ma})\alpha_{n1}(\Omega'_{ma})+
$$

$$
+\alpha_{m1}(\Omega'_{ma})\alpha_{n3}(\Omega'_{ma})]^{c}\,\overline{D}(\sigma_{mn}^{\Phi})d\Omega'_{ma};
$$

$$
{}^{c}\overline{D}(\overset{*}{\sigma}_{\mathrm{ik}})={}^{c}\overset{*}{\sigma}_{\mathrm{ik}}/{}^{c}\,\overset{*}{\sigma}_{\mathrm{i}},\ {}^{c}\overset{*}{\sigma}_{\mathrm{i}}=\frac{1}{\sqrt{2}}({}^{c}\,\overset{*}{\sigma}_{\mathrm{ik}}\ {}^{c}\overset{*}{\sigma}_{\mathrm{i}})^{1/2};
$$

$$
{}^{c}\overline{D}(\dot{\varepsilon}_{\mathrm{ik}}^{\Phi})={}^{c}\dot{\varepsilon}_{\mathrm{ik}}^{\Phi}/{}^{c}\,\dot{\varepsilon}_{\mathrm{i}}^{\Phi},\ {}^{c}\dot{\varepsilon}_{\mathrm{ik}}^{\Phi}=\frac{1}{\sqrt{2}}({}^{c}\,\dot{\varepsilon}_{\mathrm{ik}}^{\Phi}\ {}^{c}\dot{\varepsilon}_{\mathrm{ik}}^{\Phi})^{1/2}. \tag{23}
$$

In formulas (23), ${}^{c}\overline{D}(\overset{*}{\sigma}_{\mathrm{ik}})$, ${}^{c}\overline{D}(\dot{\varepsilon}_{\mathrm{ik}}^{\Phi})$ are the directors of the effective stresses and phase strain rates of the corresponding sub area; $f(\Omega_{\mathrm{i}})$ and $f(\Omega'_{\mathrm{i}})$ – distribution density by orientations $\{\Omega_{\mathrm{i}}\}$ and $\{\Omega'_{\mathrm{i}}\}$; ${}^{c}\mathrm{A}_{m}(\Omega_{ma},\Omega'_{ma})$, ${}^{c}\mathrm{M}_{a}(\Omega_{ma},\Omega'_{ma})$ and ${}^{c}R(\Omega_{ma},\Omega'_{ma})$ – influence functions that make sense of structural compliance, structural memory of form, and structural relaxation, respectively; $\{\Omega_{\mathrm{i}}\}$ – areas of orientation space where shifts occur; $\{\Omega'_{\mathrm{i}}\}$ – areas of influence [5].

7. Macro level-1

We introduce the volume of averaging V (V_{m2}<<V<<R^3), where R is the characteristic size of the macro product. As a result of orientational averaging of parameters (23), the defining relations are obtained at the macro level in the traditional form adopted in the mechanics of a deformable solids. Below is a system of defining relations for the macroscale level. The method of deriving the above equations is described in detail in [5, 6]. For areas of the crystal, where the phase deformation mechanism is initiated, the defining relations will be:

$$
{}^{-}\dot{\varepsilon}_{\mathrm{ik}}^{\Phi}={}^{-}\mathrm{A}_{\mathrm{ikpq}}^{\Phi}\ {}^{-}\sigma_{\mathrm{pq}}+{}^{-}\mathrm{M}_{\mathrm{ikpq}}^{\Phi}\ {}^{-}\varepsilon_{\mathrm{pq}}^{\Phi},\ {}^{+}\dot{\varepsilon}_{\mathrm{ik}}^{\Phi}={}^{+}\mathrm{A}_{\mathrm{ikpq}}^{\Phi}\ {}^{+}\sigma_{\mathrm{pq}}+{}^{+}\mathrm{M}_{\mathrm{ikpq}}^{\Phi}\ {}^{+}\varepsilon_{\mathrm{pq}}^{\Phi},\ \dot{\varepsilon}_{\mathrm{ik}}^{\Phi}={}^{-}\dot{\varepsilon}_{\mathrm{ik}}^{\Phi}+{}^{+}\dot{\varepsilon}_{\mathrm{ik}}^{\Phi}, \tag{24}
$$

To calculate the elastic ε_{ik}^{Y}, temperature ε_{ik}^{T} and plastic deformations ε_{ik}^{H}, following [5,6], we get:

$$
\dot{\varepsilon}_{\mathrm{ik}}^{\mathrm{Y}}=\overline{C}_{\mathrm{ikmn}}\dot{\sigma}_{\mathrm{mn}},\ \dot{\varepsilon}_{\mathrm{ik}}^{\mathrm{T}}=\mathrm{C}\gamma_{\mathrm{ik}}\dot{\mathrm{T}},\ \dot{\varepsilon}_{\mathrm{ik}}^{\mathrm{H}}=\mathrm{A}_{\mathrm{ikmn}}\,{}^{-}\dot{\overset{*}{\sigma}}_{\mathrm{mn}}+\mathrm{B}_{\mathrm{ikmn}}\,{}^{-}\overset{*}{\sigma}_{\mathrm{mn}}; \tag{25}
$$

$$
\dot{\rho}_{\mathrm{ik}}=\mathrm{G}_{\mathrm{ikmn}}\dot{\varepsilon}_{\mathrm{mn}}^{\mathrm{H}}-\mathrm{R}_{\mathrm{ikmn}}\rho_{\mathrm{mn}},\ \dot{r}_{\mathrm{ik}}=\Pi_{\mathrm{ikmn}}\overset{**}{\sigma}_{\mathrm{mn}}-\mathrm{Z}_{\mathrm{ikmn}}r_{\mathrm{mn}},\ \dot{\rho}_{\mathrm{ik}}^{\Phi}=\mathrm{R}_{\mathrm{ikmn}}^{\Phi}\dot{\varepsilon}_{\mathrm{mn}}^{\Phi}. \tag{26}
$$

8. Macro level-2

Analytical relations (24) - (25), together with equations (4) - (17) and (1) - (3), should be considered as defining equations for a structurally inhomogeneous environment in a macropoint. To solve practical problems of engineering mechanics, it is required to formulate additional equations that take into account the spatial arrangement of volumes *V*, dynamic and static properties of the environment, etc., as well as initial and boundary conditions for the corresponding variables. The foregoing reduces to the formulation of the following equations of the boundary value problem of mechanics:

8.1. Dynamic Equilibrium Equations for Stresses

$$\Delta_i \sigma_{ik} = \rho \ddot{u}_k \tag{27}$$

where ρ is the density of the medium; u_k is the displacement vector.

8.2. Continuity condition for deformation

$$e_{ksr} e_{qmt} \nabla_s \nabla_t \varepsilon^{e}_{rm} = -e_{ksr} e_{qmt} \nabla_s \nabla_t \overset{*}{\varepsilon}_{rm}, \tag{28}$$

where e_{ikr} is the Levi-Civita tensor; $\overset{*}{\varepsilon}_{ik} = \varepsilon^{T}_{ik} + \varepsilon^{\hat{o}}_{ik} + \varepsilon^{H}_{ik}$

8.3. Balance condition for temperature

$$\dot{T} = \frac{K_{ik}}{\rho \cdot c} \nabla_i \nabla_k T - \frac{q_0}{\rho \cdot c} \dot{\Phi}_{\Sigma}, \tag{29}$$

where K_{ik} is the thermal conductivity coefficient tensor; *c* is the specific heat.

The second summand on the right-hand side (29), which contains $\dot{\Phi}_{\Sigma}$, takes into account the heat effect of the martensitic transformation reaction. In the case of isotropic heat transfer at the macro level, the tensor of thermal conductivity coefficients is equal to $K_{ik}=k_0\delta_{ik}$, where k_0 is the scalar coefficient of thermal conductivity.

For the final formulation of a specific boundary value problem, differential equations for stresses (27), deformations (28) and temperature (29) must be supplemented with appropriate boundary and initial conditions.

The following important aspects should be noted. In contrast to the traditional methodology of continuum mechanics, when the components of the strain tensor are introduced as derivatives of the field of displacements of the continuum points, in this approach they are specified through the corresponding micro- and meso-characteristics taking into account the physical laws of formation of martensitic crystals at the micro level and the processes of self-organization of martensitic structures at the mesoscale level.

Physically, the analytical relations (24) - (26) take into account not only the translational-rotational nature of mass transfer at the meso-level, but also the cross-effects of interaction between elements of the environment at various scale levels.

To reflect the complex functionally mechanical properties of materials with the SME, equations for calculating the kinetic coefficients of structural compliance A^{Φ}_{ikpq}, structural shape memory M^{Φ}_{ikpq} and structural heterogeneity of interphase interaction R^{Φ}_{ikpq} are derived. In order to take into account the effect of structural changes in the material as a result of plastic deformation and damage accumulation, equations for calculating the kinetic coefficients of structural compliance A_{ikpq}, structural flowability B_{ikpq}, structural heterogeneity G_{ikpq}, structural relaxation R_{ikpq}, structural damage Π_{ikpq} and structural relaxation of damage concentrators Z_{ikpq} are derived [6]. The above kinetic coefficients were obtained as functionals representing fourth-rank tensor objects and along with the formulas for calculating elastic compliance $\overline{C}_{ikpq}$ and thermal expansion coefficients of materials $\overline{\gamma}_{ik}$, reflect nontrivial structural changes and the formation of complex deformation properties taking into account the kinetics of martensitic phase transformations and structural evolutions on micro-, meso- and macro levels.

9. Conclusion

The proposed version of the theory eliminates the main drawbacks of the previous models based on the methods of structural and analytical strength theory for materials with shape memory effect and significantly expands the possibilities of mathematical modeling of complex functional and mechanical properties of materials with a martensitic inelasticity mechanism. This result was obtained by taking into account the multilevel processes of mutual influence of the mechanisms of plastic deformation, microdestruction, and phase transformations of a martensitic nature. The results of analytical calculations showed good qualitative and quantitative agreement with experimental data in a wide range of variations in the modes of thermomechanical action [1, 3, 5, 6].

References

[1] Malinin V G 1997 Structural and analytical model of physical mesomechanics for materials with shape effect *International seminar "Actual problems of strength" named after V A Likhacheva* (Novgorod) **2(1)** 26–32

[2] Malinin V G and Malinina N A 1997 On the calculation of interphase structural stresses arising at the front of martensitic transformations *International seminar "Actual problems of strength" named after V A Likhacheva* (Novgorod) **2(1)** 33–37

[3] Likhachev V A and Malinin V G 1993 *Structural analytic theory of strength* (Saint-Petersburg: "Nauka") p 471

[4] Malinin V G, Malinina N A and Malinin G V 2002 Mechanical properties of a material with shape memory effect under complex isothermal loading mode *Attachment to the Engineering Journal. Handbook – Exhibition to Engineering Journal: Handbook* **10/1** 27

[5] Malinin V G, Malinina N A and Malinin G V 2002 Mechanical properties of a material with shape memory effect under complex isothermal loading mode *Attachment to the Engineering Journal. Handbook – Exhibition to Engineering Journal: Handbook* **10/2** 25

[6] Malinin G V 2012 Structural analytic mesomechanics of nanostractured conditions of the medium with reversible martensitic transformations *Nanoengineering* **5** 22–29

[7] Malinin V G 2012 Structural and analytical mesomechanics of tools with translational-rotary methods of deformation and fracture *Nanoengineering* **5** 29–37

The calculation of beam-wall made of material with shape memory effect by the finite element method

G V Malinin[1,3], N A Malinina[1] and O A Malukhina[2]

[1] Oryol State Agrarian University, ul. General Rodina, 69 302019, Orel, Russia
[2] Yaroslav-the-Wise Novgorod State University, ul. B. St. Peterburgskaya, 41 173003, Veliky Novgorod, Russia

[3]E-mail: malinin2002@yandex.ru

Abstract. In present article application of a method of finite elements for calculation of beam-wall, executed from a material with effect of memory of the form (SME) is considered. The equations of structural-analytical mesomechanics were used as defining relations.

1. Introduction

The introduction of new materials with previously unknown properties, the emergence of new engineering problems that require precise prediction of mechanical properties, urgently requires the creation and development of more advanced, in comparison with the known, approaches in the theory of deformation and destruction. It seems that such new approaches should be based on a rational synthesis of constructive ideas of mechanics of a deformable solid body and physical materials science, naturally combining the main achievements and methods of modern physics of plasticity and destruction as well as mechanics of a deformable solid.

The development of modern concepts of deformation of solids is based on the understanding of the interrelated process of deformation and destruction as multilevel and hierarchical, and requires of researchers to new approaches due to the need to analyze plastic deformation and the evolution of structural damage at different scale and structural levels, taking into account the localization and heterogeneity of deformation development. Alloys with SME have unique properties that are absent in most traditional materials used in industry. However, the unconventional behavior of these materials requires new models of the material in which the use of traditional methods of mechanics of a deformable solid body is possible. The development of the theory and methods for calculating engineering structures made of materials with the SME is an important task.

In this paper, we use the defining relations of the deformation type, derived from the general equations of structural-analytical mesomechanics [1–5].

In the framework of this approach, solutions of some boundary problems that allow an analytical solution are known. This is the problem of the bending of the beam, as well as the long thick-walled pipe under internal pressure. The following boundary-value problems were solved with numerical and numerical-analytical methods: calculation of a two-rod statically indefinable system with the finite element method, calculation of the truss structure with the numerical-analytical method of boundary elements [6, 7], calculation of a thin rectangular plate, working on a bend with the method finite differences [8]. In this paper, we consider a cantilever beam-wall loaded with a uniformly distributed load of constant intensity, the material of which undergoes inelastic martensitic deformations. Only isothermal martensitic transformations are considered. The problem is solved in a physically non-linear, but geometrically linear formulation. To solve the problem, the finite element method is used in combination with the method of variable parameters of elasticity.

2. Defining relations at the stage of loading and the calculation algorithm

The defining relations for the material with the SME taking into account the simplified equations of the structural-analytical mesomechanics [4, 5] are represented as follows:

$$\varepsilon_i^{\Sigma} = \frac{\sigma_i}{E} + \left(\frac{2}{3}B_{\phi}\sigma_i \frac{k\left(\sigma_i - \sigma_{н\uparrow}^{A\to M}\right)}{M_н - M_к}\right) H\left(\sigma_i - \sigma_{н\uparrow}^{A\to M}\right) H\left(\sigma_{к\uparrow}^{A\to M} - \sigma_i\right) H(\dot{\sigma}_i)$$

where σ_i is the stress intensity; E is the modulus of elasticity; B_{Φ} is the material constant; $k = \frac{T_0}{q_0} D_i$; T_0 is the thermodynamic equilibrium temperature; q_0 is the heat effect of the reaction; D_i is the distortion of the phase transformation; $\sigma_{н\uparrow}^{A\to M} = \frac{T_D - M_н}{k}$ is the voltage of the onset of direct martensitic transformation; T_D is the temperature of the onset of deformation, $M_н$ and M_K are the temperatures of the beginning and end of the direct martensitic transformation, respectively; $\sigma_{к\uparrow}^{A\to M} = \frac{T_D - M_к}{k}$ is the voltage of the end of martensitic transformation, $\dot{\sigma}_i = \frac{d\sigma_i}{dt}$; $H(\ldots)$ is a Heaviside function.

Take the following assumptions:
1) isothermal loading from the austenitic state $T_D = A_K$;
2) the material undergoes elastic and inelastic (phase) deformations;
3) accept the bilinear model of material deformation.

For the calculation the program in the Mathsad package is compiled. The calculation was carried out by the finite element method in combination with the method of variable parameters of elasticity, which is widely used in solving problems of the theory of plasticity. The algorithm for solving the problem was as follows:

1. Solve the basic equation of the finite element method (FEM) [K]{Δ}={F} where [K] is the stiffness matrix of the structure, {Δ} is the vector of displacements of nodes, {F} is the vector of nodal loads.

2. For elements experiencing phase transformation in which $\varepsilon_1 > \varepsilon_\tau$: to find the secant module $E_c = \frac{\Phi(\varepsilon_i)}{\varepsilon_i}$, $\Phi(\varepsilon_i)$ is the equation of the strain diagram.

According to it, we determine the variable parameters of elasticity $E_{\phi} = \frac{E_c}{1+\alpha}$, $\alpha = \frac{1-2\mu}{3}\frac{E_c}{E}$, $\mu_{\phi} = \frac{1-2\alpha}{2+2\alpha}$, $G_{\phi} = \frac{E_c}{2(1+\mu_{\phi})}$

– recalculate the internal stiffness matrix taking into account these parameters

$$G = \begin{bmatrix} 1 & \mu_{\phi} & 0 \\ \mu_{\phi} & 1 & 0 \\ 0 & 0 & \frac{1-\mu_{\phi}}{2} \end{bmatrix} \frac{E}{1-\mu_{\phi}^2}$$

– recalculate the element stiffness matrix $\left[K^e\right]$

3. Recalculate the system stiffness matrix $\left[\overline{K}\right]$

4. Solve the basic FEM equation again and find the displacements of the nodes$\{\Delta\}$, stresses and strains in the elements$\{\sigma\},\{\varepsilon\}$, stress intensity and strain intensity σ_I, ε_i.

5. To compare the results of the beginning and end of the calculation. In case of dissatisfaction with a given accuracy, go to the second approximation. Repeat steps 2, 3, 4, 5. The process is carried out until the convergence criterion is met.

Figure 1 shows the block diagram of the solution of the physically nonlinear FEM problem with the method of variable elastic parameters. First, the elastic problem is solved with a single load and the F_T load is determined at which max $\sigma_i = \sigma_T$. Then it is necessary to specify the load $F_i = F_T \cdot \alpha$, where $\alpha > 1$, in this problem we take $\alpha = 2$. Under such a load, the structure is guaranteed to undergo phase transformations. Next, an elastic calculation is carried out at a given load F. From the condition $\sigma_i \geq \Phi(\varepsilon_i)$, the elements in which the phase transformations begin are determined. For these elements, the new parameters of elasticity are calculated, the element stiffness matrix is recalculated. For elements that remain elastic, the element stiffness matrix remains the same. Then the system stiffness is formed, the basic FEM equation is solved and the nodal displacements $\{\Delta\}$ are determined.

According to the calculation results, the criterion for the end of the calculation of *Eps* is determined. If *Eps* < η, the calculation is completed, if not, we go to the next iteration. Here η is the specified calculation accuracy.

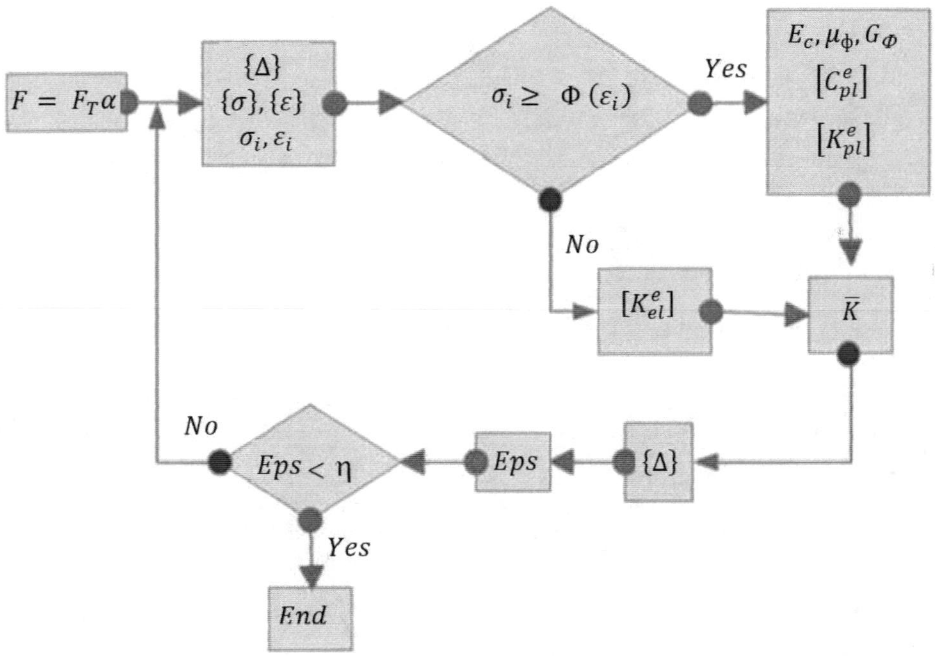

Figure 1. Block diagram of the method of variable parameters of elasticity.

3. Boundary problems for SMA

Formulation of the problem. Determine the stress-strain state of the cantilever beam-wall, loaded on the upper edge with a uniformly distributed load, at which stresses that are twice the phase yield point occur in the most loaded element. The length of the beam-wall is a = 12 mm, width b = 10 mm, thickness толщина δ= 10 mm. The plate is made of TiNi alloy, which has a shape memory effect.

IOP Conf. Series: Materials Science and Engineering **656** (2019) 012035 doi:10.1088/1757-899X/656/1/012035

Initial data are presented in table 1.

Table 1. Initial data.

Name	Formula
Characteristic temperatures	$M_н$ =330K, $M_к$ = 320K $A_н$ =370K, $A_к$ = 380K
Material constant k	k = 0,29 KMPa^{-1}
Constant $B_Ф$	$B_Ф$ = 0,06·10^{-2} MPa^{-1}
Elastic modulus E	E = 7,42·10^4 MPa
Deformation temperature T_D	T_D = $A_к$ = 380K
Tension of the beginning of straight martensitic transformations (phase yield strength) $\sigma_{H1\uparrow}^{A\to M} = \sigma_{T1}$	$\sigma_{H1\uparrow}^{A\to M} = \sigma_{T1} = 172\ MPa$
Ferroelasticity module (tangent module)	$E_{\Phi\gamma} = E_K = 414\ MPa$
Deformation at yield stress ε_T	$\varepsilon_T = \frac{\sigma_{H1\uparrow}^{A\to M}}{E} = \frac{\sigma_T}{E} = 2.324 \cdot 10^{-3}$
Calculation accuracy for displacements η	$\eta = 10^{-4}$

4. Decision

To solve the problem, we use a linear triangular element. The number of nodes of the finite element model is 90, the number of elements is 139. The node numbers of the calculated beam-wall are presented in Table 2, where zero numbers mean fixings, and loaded nodes are presented in line 1.

Table 2. The node numbers of the finite element mesh.

	1	2	3	4	5	6	7	8	9	10	11	12	13	14	15	16
1	0	6	12	18	24	30	36	42	48	54	60	66	72	78	84	90
2	0	5	11	17	23	29	35	41	47	53	59	65	71	77	83	89
3	0	4	10	16	22	28	34	40	46	52	58	64	70	76	82	88
4	0	3	9	15	21	27	33	39	45	51	57	63	69	75	81	87
5	0	2	8	14	20	26	32	38	44	50	56	62	68	74	80	86
6	0	1	7	13	19	25	31	37	43	49	55	61	67	73	79	85

We approximate the material with a linear hardening diagram shown in figure 2.

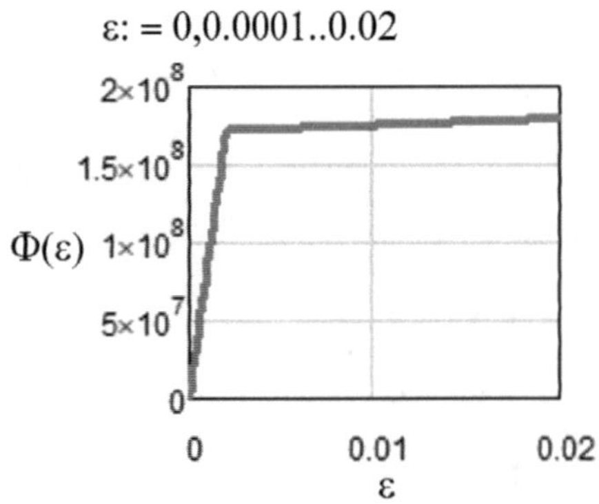

Figure 2. Diagram of the deformation of the alloy TiNi.

It took 53 iterations to obtain the required displacement accuracy $\eta = 10^{-4}$.

As a result of the calculation, we obtain the values of stress intensity σ_i, strain intensity ε_i and displacementΔ, presented in table 3, in figure 3, respectively.

Table 3. Stress intensity σ_i, MPa at the grid nodes.

2	3	4	5	6	7	8	9	10	11	12	13	14	15	16
211	195	182	174	155	133	111	90	73	59	47	39	34	32	32
194	196	182	166	146	124	105	89	74	61	50	41	34	30	31
176	180	180	163	134	112	96	83	71	60	49	40	31	25	24
177	172	172	156	127	105	89	75	64	53	43	33	25	17	15
216	190	183	157	124	101	83	67	53	41	31	22	14	9	5
239	213	194	173	137	111	89	70	53	38	26	16	8	4	0

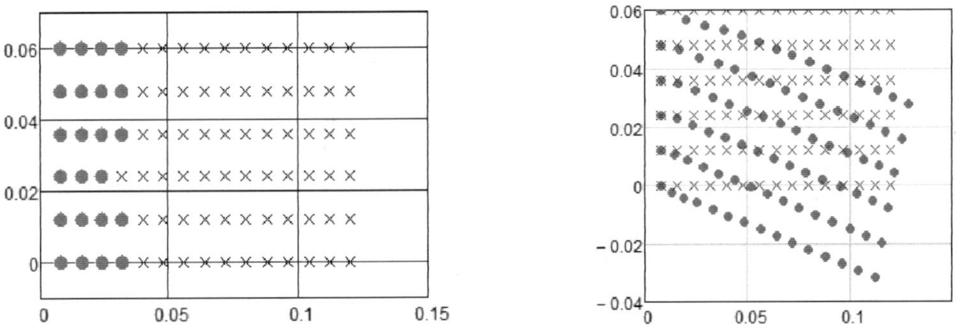

a) b)

Figure 3. The intensity of deformation and displacement: a) strain intensity ε_i and the region of origin of martensitic crystals as a result of the transformation (indicated by a circle); b) the positions of the nodes before the deformation (crosses) and after the deformation (circles) when moving nodesΔ, mm.

Table 4. The results of the calculation of the maximum values.

Maximum deflection		Maximum strain rate		Maximum stress intensity	
№ node	magnitude, Δ, mm	№ node	magnitude, ε_i, %	№ node	magnitude, σ_i, MPa
90	1.115	1	1.05	1	239

5. Conclusion

The problem of calculating the stress-strain state of a beam-wall made of a material with inelastic martensitic transformations is solved with the finite element method in combination with the method of variable elastic parameters.

References

[1] Likhachev V A and Malinin V G 1993 *Structural analytic theory of strength* (Saint-Petersburg: Nauka) p 471

[2] Malinin V G, Malinina N A and Malinin G V 2002 Mechanical properties of a material with shape memory effect under complex isothermal loading mode *Attachment to the Engineering Journal. Handbook - Exhibition to Engineering Journal: Handbook* **10/1** 27

[3] Malinin G V 2012 Structural analytic mesomechanics of nanostractured conditions of the medium with reversible martensitic transformations *Nanoengineering* **5** 22–29

[4] Malinin V G, Malinina N A, Malinin G V and Perelygin A I 2012 Method of forecasting the conversion plasticity effect in complex stress state on the basis of structural-analytical mesomechanics *Fundamental and Applied Problems of Engineering and Technology* **2** 9–18

[5] Malinin G V and Malinina N A 2013 Finding the constitutive relations for the shape memory material *Proc. 10th All-Russian Sci. & Pract. Conf. with Int. Participation "Current Issues of Mining and Metals Sector. Energy conservation. Ecology. Innovative technologies"* (Stary Oskol) pp 57–61

[6] Malinin V G 2015 Application of the finite element method for the calculation of statically indeterminate rod systems is made of a material with shape memory effect *Fundamental and Applied Problems of Engineering and Technology* **4** 3–11

[7] Malinin G V 2014 Application of boundary element method to calculate statically indeterminate farm, made from material with shape memory effect *Vestnik of Samara state University of Railways* **1** 71–77

[8] Aleksandrov A V and Potapov V D 1990 *Fundamentals of the theory of elasticity and plasticity* (Moscow: Visschaia schkola) p 400

IOP Conf. Series: Materials Science and Engineering **656** (2019) 012036 doi:10.1088/1757-899X/656/1/012036

Exergetic analysis of heat pump units for various climate conditions

A A Malyshev[1,3], O S Malinina[1], S I Arendateleva[2] and I M Zawadzki[2]

[1]Saint Petersburg National Research University of Information Technologies, Mechanics and Optics, 49, Kronverksky Av., Saint Petersburg, 197101, Russia

[2]Yaroslav-the-Wise Novgorod State University, ul. B. St. Peterburgskaya, 41 173003, Veliky Novgorod, Russia

[3]E-mail: malyshev46@list.ru

Abstract. The article deals with exergetic analysis of a heat pump for climate of Russian North-West. The dependencies between the parts of the pump and its general characteristic are shown. The influence of ambient temperature, condensation temperature, and low-grade heat source on the working efficiency of heat pump units is analyzed. The changes in thermodynamic characteristics of the heat pump at ambient temperature decreasing form 0°C to –15°C are identified. The coefficient of performance is seen to decrease two times, the coefficient of transformation – by 1.4 times; overall exergetic coefficient of performance increases by 18 %. Working efficiency and coefficient of performance of the unit in general, and the ones of compressor, compensator, evaporator and thermal expansion valve are determined. The results of the research are of use for increasing efficiency of heat pumps.

1. Introduction

It is established that heat pumps are energy-efficient thermotransformers using both traditional and renewable energy resources. The use of HP provides energy saving, economic benefits and a positive environmental effect [1, 2, 3, 4] by reducing CO_2 emissions into the atmosphere.

The advantages of heat pump installations (HPI) are the versatility of the type of energy used and a wide range of capacities, including small and medium HPP (Heating Power Plants) [2].

According to [5], heat pump installations are most effective for decentralized heat supply systems near sources of power generation and, first of all, in cottages. In these cases, electric heating is excluded (as an additional means to the HPI), in which the energy consumption is 3–4 times higher than the power consumption of HPI. Currently, in world practice, heat pump installations are among the most dynamically developing areas of energy [6, 7, 8]. According to the article [1], global sales of air-to-water heat pumps account for more than 1.8 million units. The volume of the world market for HP of all types in value terms is estimated at more than $ 10 billion. In Russia, the forecast of the market for steam-compressor heat pumps (SCHP) for 2030 is the end of the implementation of the current "Energy Strategy" – 11,000–15,000 units. (500–700 MW).

The authors [9, 10, 11] present the results of a comprehensive analysis of the HP and their comparison with boiler systems, as well as the effective combination of heat pumps with electrically heated boilers.

It is shown that, taking into account the average Russian cost of generating electricity, the annual fuel consumption of the "SCHP + electric boiler" system will be less than that of a traditional boiler house, provided that the heating coefficient of the heat pump is at least 4.0 and also at least 65 % coverage of annual heat load will be provided by HP.

In the literary source [9] exergetic approach to the analysis of real cycles of HPI is used. Based on the analysis, a method has been developed to increase the efficiency of SCHP with temperatures of water sources of low-grade heat above + 5°C.

In [12, 13], the efficiency of heat pump installations using renewable energy sources (RES) was analyzed. The energy of the sun and wind, as well as the heat of the soil is considered. The concept of application of renewable energy sources relevant for the south of Russia is developed.

In the majority of works devoted to the study of the efficiency of heat pumps, the region of positive temperatures of low-potential heat sources is considered, while data on the operation of HPI at negative temperatures is of considerable interest.

Work at negative temperatures is often associated with the use of air-source heat pumps, since in these cases it is difficult to use water as a source of low-grade heat. At the same time, air heat pumps have a seasonal loading factor on average 10–30% lower than when using water heat pumps [14], which limits their use. Air pumps are rarely seen as an alternative to geothermal heat pumps, and as an advantage of the former, there is no need to perform expensive drilling operations [15].

Presented literature review shows:

– the value of heat pumps in the energy sector is constantly increasing;

– the area of use of HPI is expanding, including in the northern regions;

– the efficiency of a HPI is most often evaluated by methods based on the analysis of integral thermodynamic and technical and economic indicators;

It is necessary to evaluate the effectiveness of individual HPI units depending on their types, external conditions and tasks for the further development of technologies associated with the use of heat pumps, in addition to analyzing the integral indicators.

It seems that for such an assessment the use of an exergetic method of analyzing thermodynamic systems is quite reasonable.

Exergetic analysis was used when considering the energy complex of HPP-HPI. A significant increase in overall efficiency has been shown through the use of a heat pump installation.

2. Results and discussion

The aim of the work is to analyze the impact on the performance of a HPI and its individual elements of the following parameters:

– ambient temperature t_a;

– condensation temperature t_c;

– low potential heat source temperature t_s.

The following initial data were taken for calculations:

– heat output $Q_к = 10$ kW;

– The water of the Gulf of Finland with the calculated temperature $t_{s1} = 5$°C is used as the low-potential source of heat;

– average room air temperature $t_r = 22$°C with the supply of heat through the fence;

– the difference between the condensation temperature and average air temperature in the room varied within $\Delta T_{cr} = T_c – T_r = (12 \div 28)$ K;

– the temperature difference between the outside air and the boiling point of the refrigerant in the evaporator was assumed to be $\Delta T_{ev} = T_a – T_0 = 7$ K; $\Delta T_{ev} = 10$ K;

– temperature of the outside air (ambient) varied in the range t_a=(–5 ÷ –25)°C

– working media R134a.

The steam compressor heat pump cycle is shown in the figure 1.

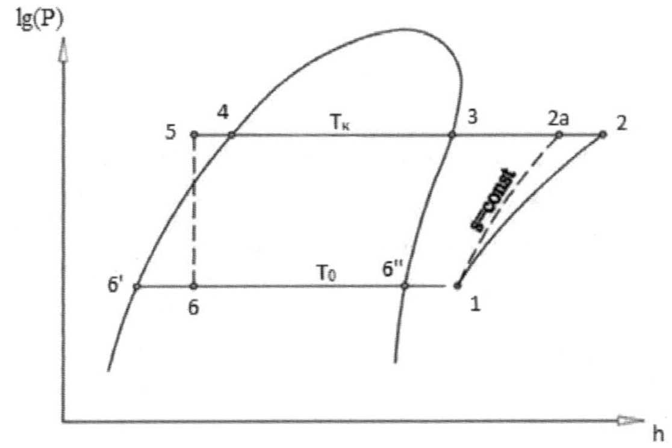

Figure 1. Steam compressor heat pump cycle.

The results of the analysis of the influence of the ambient temperature on the thermodynamic parameters of the processes of HPI and its main elements under the specified conditions are summarized in Table 1 and presented as graphical dependencies (figure 2, a, b).

Table 1. The influence of the ambient temperature on the value exergetic loss processes in the elements of HPI.

Name of elements of HPI	Ambient temperature, °C				
	–5	–10	–15	–20	–25
Compressor $DEx_{cm} = N_{el} - \Delta Ex_{cm}$, kW	0.72	0.71	0.70	0.68	0.67
Condenser $DEx_{cm} = [Ex_{cm}] - \Delta Ex_{cr}$, kW	0.79	0.78	0.76	0.75	0.73
Evaporator $DEx_{ev} = [Ex_s] - \Delta Ex_{ev}$, kW	0.13	0.13	0.12	0.12	0.12
Thermostatic valve $DEx_{tv} = [\Delta Ex_{tv}]$, kW	0.34	0.33	0.32	0.32	0.31

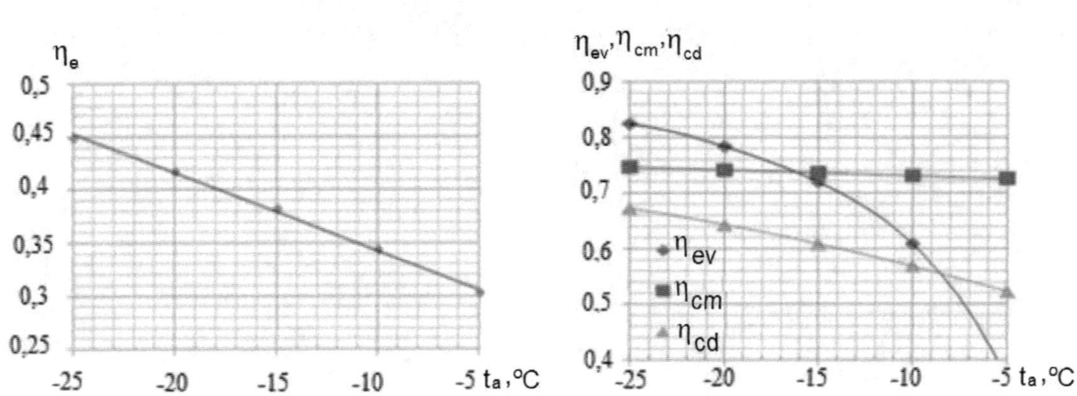

Figure 2. A graph of the dependence of the general exergetic efficiency of the heat pump (a), exergetic efficiency of the processes in the elements of the heat pump (b) on ambient temperature.

From the obtained results it can be seen that the main losses are observed in the condenser. The reason is the choice of the accepted method of heat removal of condensation into the room. For a heat pump, an increase in thermal resistance on the air side leads to an increase in the condensation temperature of the refrigerant and a deterioration of the thermodynamic characteristics of the heat pump unit. For the considered conditions, the heating coefficient was μ= 3,51.

Reducing the temperature difference between air and condensation will reduce irreversible losses and increase the efficiency of the heat pump as a whole. This can be achieved both by the development of the heat exchange surface and by intensifying heat transfer from the heated air.

It is necessary to assess the degree of influence of the temperature difference in the condenser on the thermodynamic parameters of the system to select the most appropriate direction for increasing the efficiency of a condenser. In addition, a sharp decrease in the exergetic efficiency of the evaporator with decreasing ambient temperature (figure 2, b) draws attention. This is due to the “depreciation” of exergy when the ambient temperature approaches the temperature of a low-potential source. The equalization of ambient temperature and low-potential source is the limiting case in which the thermodynamic efficiency of the evaporator equals zero.

2.1. The influence of temperature difference in the condenser

It should be noted that when using average temperatures in the room T_r, the losses in the condenser include two components:

– actual losses in the condenser,

– loss when mixing air with different temperatures in the volume of the room.

With intensive methods of heat removal in condensers (with forced air circulation) it is advisable to separate these losses. At the same time, it is necessary to perform an analysis taking into account the existing Norms and Rules for the permissible temperature difference between the air supplied and leaving the premises. This task is multiparametric and requires the specification of the object. It was decided not to separate the indicated loss components for a general assessment, and to perform an analysis of the effect on thermodynamic indicators of the difference in condensation temperatures and the average air temperature in the room at the same ambient temperature t_a = –10°C. The remaining independent parameters are taken as identical [18].

The results of the analysis of the influence of the temperature difference in the condenser on the value of the exergetic losses of the HPI are given in the table 2.

Table 2. The results of the analysis of the influence of the temperature difference in the condenser on the value of the exergetic losses of HPI.

Name	Value				
Temperature difference between condensation and average room temperature $\Delta T_{cr} = T_c – T_r$, R	12	16	20	24	28
Ambient temperature $t_{a.}$, °C			–20		
Average room temperature t_r, °C			22		
Boiling temperature t_0, °C			– 2		
Condensation temperature t_c, °C	34	38	42	46	50
Thermal load on the evaporator Q_0, kW	8.88	8.37	7.85	7.33	6.80
Heat capacity Q_{cd}, kW	10.99	10.59	10.18	9.75	9.46

Power consumed by the compressor N_{el}, kW	2.22	2.34	2.46	2.56	2.65
Heating coefficient μ	4.95	4.53	4.14	3.81	3.57
Refrigerant mass flow M_a, kg/c	0.056	0.055	0.054	0.053	0.051
Evaporator					
Evaporator inlet water temperature t_{s1}, °C	5	5	5	5	5
Evaporator outlet water temperature, t_{s2}, °C	2	2	2	2	2
Mass flow of water in the evaporator $M_s = Q_0/c_s(T_{s1} - T_{s2})$, kg/c	0.704	0.664	0.623	0.581	0.539
Average heat capacity of the coolant c_s, kJ/(kg·K)	4.200	4.200	4.200	4,200	4.200
Change in the specific exergy of water $\Delta ex_s = c_s \cdot (T_{s2} - T_{s1}) - T_a \cdot \ln(T_{s2}/T_{s1})$, kJ/kg	–0.615	–0.615	–0.615	–0.615	–0.615
Change in exergy of water flow , $Ex_s = M_s \cdot \Delta ex_s$, kW	–0.433	–0.408	–0.382	–0.357	–0.331
Change in the specific exergy of the refrigerant flow $\Delta Ex_{ev} = (h_1 - h_6) - T_a \cdot (S_1 - S_6)$, kJ/kg	4.627	4.455	4.281	4.104	3.924
Change in exergy of refrigerant flow $Ex_{ev} = M_a \cdot \Delta ex_{ev}$, kW	0.263	0.248	0.232	0.217	0.201
Loss of exergy in the evaporator $DEx_{ev} = \Delta Ex_s - \Delta Ex_s$, kW	0.17	0.16	0.15	0.14	0.13
Exergetic evaporator efficiency coefficient $\eta_{ev} = \Delta Ex_{ev} / \Delta Ex_s$	0.60712	0.60740	0.60765	0.60778	0.60797
Compressor					
Change in the specific exergy of the refrigerant flow $\Delta ex_{cm} = (h_2 - h_1) - T_{ev} \cdot (S_2 - S_1)$, kJ/kg	27.171	29.844	32.480	35.091	37.670
Change in exergy of refrigerant flow $\Delta Ex_{cm} = M_a \cdot \Delta ex_{cm}$, kW	1.544	1.661	1.766	1.858	1.936
Loss of exergy in the compressor $Ex_{cm} = N_{el} - \Delta Ex_{cm}$, kW	0.676	0.679	0.694	0.702	0.714
Exergetic compressor efficiency coefficient $\eta_{cm} = \Delta Ex_{cm} / N_{el}$,	0.696	0.710	0.718	0.726	0.731
Thermostatic valve					
Change in the specific exergy of the refrigerant flow $\Delta ex_{rv} = (-T_{ev} \cdot (S_6 - S_5))$, kJ/kg	–3.058	–3,796	–4.615	–5.518	–6.507

Loss of exergy in a thermostatic valve $DEx_{rv}= \Delta Ex_{rv}$, kW	0.174	0.211	0.251	0.292	0.334
Exergetic efficiency coefficient of thermostatic valve			0		
	Condenser				
Change in the specific exergy of the refrigerant flow $\Delta ex_{cd} = (h_5 - h_2) - T_{ev} \cdot (S_5 - S_2)$, kJ/kg	–28.741	–30.503	–32.147	–33.677	–35.087
Change in exergy of refrigerant flow $\Delta Ex_{cd}=M_a \cdot \Delta ex_{cd}$, kW	–1.634	–1.698	–1.748	–1.784	–1.803
Indoor exergy flow $\Delta Ex_{cr}=Q_{cd} \cdot (1-T_{ev}/T_r)$, kW	1.192	1.148	1.104	1.057	1.026
Loss of exergy in the condenser $DEx_{cd}= \Delta Ex_{cd}- \Delta Ex_{cr}$, kW	0.44	0.55	0.64	0.73	0.78
Exergy efficiency of the condenser $\eta_{cd}= \Delta Ex_{cr} / \Delta Ex_{cd}$,	0.729	0.676	0.632	0.593	0.569
	General indicators				
Exergy flow to the room ΔEx_{cr}, kW	1.192	1.148	1.104	1.057	1.026
Total summed up exergy $\Delta Ex_{gen}= N_{el} + \Delta Ex_s$, kW	2.653	2.748	2.843	2.918	2.982
Exergetic efficiency coefficient of the system $\eta_{gen}= \Delta Ex_{cr} / \Delta Ex_{gen}$	0.449	0.418	0.388	0.362	0.344
General thermodynamic losses $DEx_{gen}= \Delta Ex_{gen}- \Delta Ex_{cr}$, kW	1.462	1.600	1.739	1.861	1.956
	Relative components of thermodynamic losses				
In evaporator $\Omega_{ev}= DEx_{ev} / DEx_{gen}$	0.116	0.100	0.086	0.075	0.066
In compressor $\Omega_{cm}= DEx_{cm} / DEx_{gen}$	0.462	0.424	0.399	0.377	0.365
In thermostatic valve $\Omega_{rv}= DEx_{rv} / DEx_{gen}$	0.119	0.132	0.144	0.157	0.171
In condenser $\Omega_{cd}= DEx_{cd} / DEx_{gen}$	0.302	0.344	0.370	0.390	0.398

From the calculation, it follows that reducing the temperature difference ΔT_{ave} from 28 K to 12 K leads:

– to an increase in the performance of the heat pump by 16% while reducing power consumption by 16%;

– to an increase in the heating coefficient by 39%;

– to an increase of the total exergetic efficiency coefficient of the heat pump on.

2.2. *Influence of low-potential source temperature on the efficiency of HPI*

External air is taken as a source of low potential heat. Currently, many manufacturers produce units for air conditioning of residential, administrative and industrial premises with units that provide work on

IOP Conf. Series: Materials Science and Engineering **656** (2019) 012036 doi:10.1088/1757-899X/656/1/012036

the cycles of both the refrigerating machine and the heat pump with air condensers and evaporators. To reduce the well-known deficiencies in the operation of such units during periods of a year with a negative temperature, manufacturers use, depending on the conditions of use, various technical solutions (heat recovery of ventilation emissions, air circulation, combined heat supply, etc.).

The variant of the heat pump units without the use of these solutions leads to the lowest thermodynamic efficiency. At the same time, such a production of units with forced air circulation in the evaporator and condenser is typical and it is advisable to perform a thermodynamic analysis for these conditions.

The influence of ΔT_{ave} on the thermodynamic efficiency coefficient of the compressor and the condenser is shown in the figures 3 and 4.

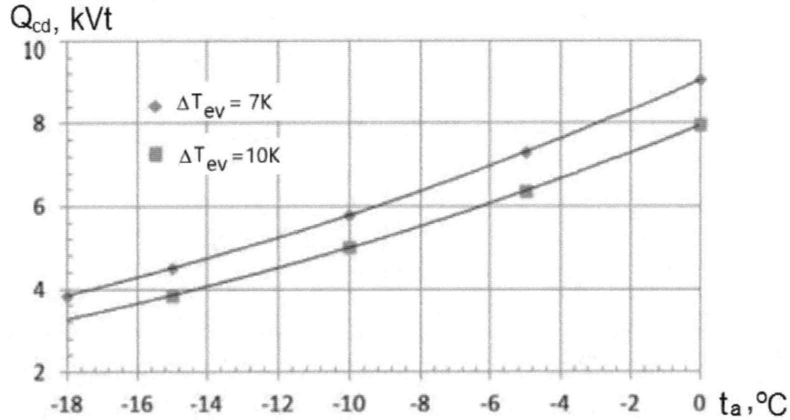

Figure 3. A graph of influence of heat pump heat capacity on ambient temperature.

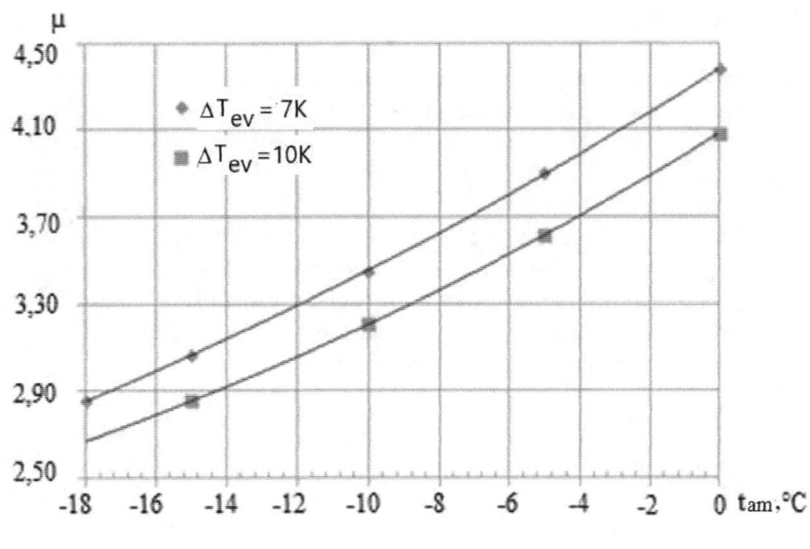

Figure 4. A graph of the dependence of the heating coefficient of the heat pump on the ambient temperature.

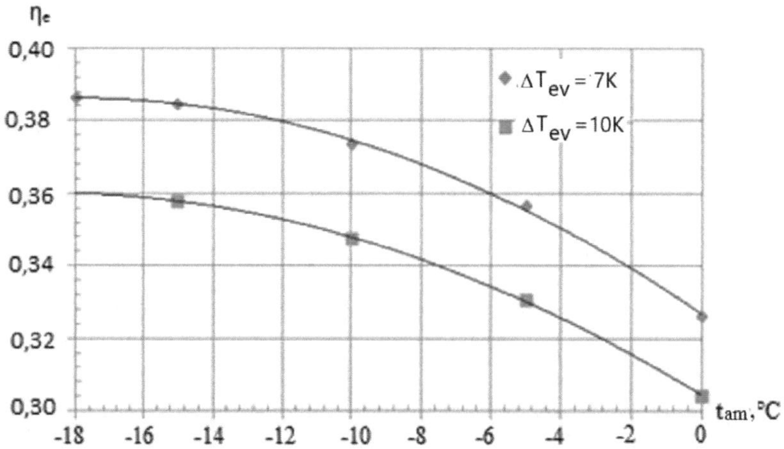

Figure 5. A graph of the dependence of the exergy efficiency coefficient of the heat pump on the ambient temperature.

The analysis performed reflects the significant influence of T_{am} on the general thermodynamic characteristics of the heat pump. When T_{am} от 0°C до –15°C, ΔT_{ev} = 7 K and ΔT_{ev} = 10 K they change accordingly:

– productivity decreases by 2 times;

– heating coefficient decreases by 1.4 times;

– common exergetic efficiency coefficient increases by 18%.

The cycles of the heat pump and the refrigerating machine are reverse, for which an increase in the difference in boiling points and condensation leads to an increase in the internal thermodynamic losses of the cycle, a decrease in power: the refrigerating of evaporator or the thermal of condenser, cooling and heating coefficients. At the same time, specific decisions of individual elements of the equipment can influence the nature of the change in exergetic efficiency, ensuring its changes in any direction, including the presence of optimum in a certain mode. Its only principal difference for the heat pump in comparison with the refrigerating machine is the following fact. For a chiller, any thermodynamic loss leads to a decrease in exergetic efficiency.

For a heat pump, internal losses lead to an increase in thermal power supplied to the room by a condenser and having an exergetic potential relative to the environment (outside air).

As we noted, the option of using an external low-potential source with a temperature above$T_{o.c}$ is more expedient. The results of its comparison with the above data when using water with a temperature of T_{s1} = 5°C and ΔT_{ev} = 7 K, of outdoor air with a temperature ofT_{am}= –10°C and ΔT_{ev}= 10 K with other things being equal:

– an increase in productivity by 77%,

– an increase of heating coefficient by 54%,

– an increase in the total exergetic efficiency by 29%.

The accepted temperature differences in the apparatus are close to those recommended in industrial heat exchangers of heat pumps.

For the option of supplying heat from the outside air in the evaporator in the temperature range from –15°C to 0°C while decreasing the temperature difference ΔT_{ev} from 10 K to 7K similar indicators will be:

– an increase in productivity by (14 ÷ 17) %,
– an increase of the heating coefficient by (7 ÷ 8) %,
– an increase of total exergetic efficiency coefficient by (7÷8) %.

3. Conclusion

1. The study established quantitative laws of the influence on the thermodynamic efficiency of a heat pump with a piston compressor and its elements, parameters of low-potential heat sources (outside air and external with a nonzero exergetic potential), temperature differences in heat exchange processes at negative ambient temperatures.

2. Increase in the total exergetic efficiency coefficient of the heat pump with a decrease in the ambient temperature is characteristic for the selected compressor in all the modes considered. The reasons are associated with an increase in the exergetic value of the external low-temperature source and the heat supplied to the room when the $T_{o.c}$ decreases.

3. The fundamental difference between the influence of internal losses on the thermodynamic efficiency of a heat pump and a refrigerating machine is shown.

4. The inexpediency of application of devices for heating air in rooms operating in natural convection mode in heat pumps is substantiated.

5. The results of the study can be used to improve the efficiency of heat pumps.

Reference

[1] Baranenko A 2018 *Vestnik Mezhdunarodnoi akademii kholoda* **2** 4–12
[2] Fumo N, Mago P and Chamra L 2009 *Applied Energy* **86(11)** 2344–2350
[3] Self S and Reddy Band Rosen M 2013 *Applied Energy* **101** 341–348
[4] Lin D, Shu H and Jiang S and Zhu Y 2007 *Renewable Energy* **32** 2603–2616
[5] Fumo Nand Chamra L 2010 *Applied Energy* **87** 2023–2030
[6] Baier Pand Pietrucha U 2006 District heating and district cooling with large centrifuge chiller-heat pumps *Proc. 10th international Symposium on District heating and cooling Hanover*
[7] Zogg M 2008 History of Heat Pumps Swiss Contributions and International Milestones Process and Energy *Engineering* **CH-3414** Oberburg 64–96
[8] Tian Y, Shi W and Wang B 2015 *Energy and Buildings* **107** 103–112
[9] Fang C, Xu Q, Wang S and Ruan Y 2018 *Energy Procedia* **152** 45–50
[10] You T, Wei B, Wenxing W and Li S 2014 *Applied Thermal Engineering* **64** 283–292
[11] Filippov S, Ionov M and Dil'man M 2012 *Termal Engineering* **59** 816–823
[12] Papin V 2013 Micro-energy complex based on a wet-steam turbine, solar collector and heat pump *PhD thesi, Novocherkassk* 8–10
[13] Heat Pumps 2005 *Technology and Environmental Impact* **Part 1 2**
[14] Byrne P and Ghoubali R 2019 *Applied Thermal Engineering* **149** 414–424
[15] Borovkov V and Al' Alavin A 2006 Izvestiya vysshikh uchebnykh zavedenii *Problemy energetiki* pp 12–20
[16] Suvorov D, Tatarinova N, Krupin D, Suvorova L and Baibakova T 2017 *Termoenergeica* **33** 66–76

Experimental research on a phase slip and pressure loss during boiling of the refrigerant R134a in a mini-channel

A A Malyshev[1], K F Kouadio[1], O V Letenkov[2,3] and I M Zawadzki[2]

[1]Saint-Petersburg National Research University of Information Technologies, Mechanics and Optics, Saint-Petersburg, Russia
[2]Yaroslav-the-Wise Novgorod State University, ul. B. St. Petersburgskaya, 41 173003, Veliky Novgorod, Russia

[3]E-mail: Oleg.Letenkov@novsu.ru

Abstract. This article is devoted to the development of a methodology for calculating the hydrodynamic characteristics of liquids boiling in mini-channels. In contrast to the well-known methods, the proposed methodology is based on the use of the true volumetric steam content and prediction of two-phase flow regimes in the field of a steady flow. The article presents experimental data on the true volumetric steam content, pressure loss, and two-phase flow pattern in mini-channels

1. Introduction

The solution to the problem of energy saving in low-temperature technology is largely due to the intensification of heat exchange in refrigeration apparatuses, optimization of heat-hydrodynamic processes and the development of new technologies for heat exchange apparatus engineering.

The study of the hydrodynamic characteristics of boiling refrigerants in pipes Dh > 6 mm is discussed in [6].

Recently, the use of boiling in pipes or channels with a diameter of Dh> 2.5 ÷ 6 mm [6] or mini-channels with Dh <1.5 mm [1, 2, 3] has become relevant.

Mini-channel technologies in heat exchange apparatus engineering have a number of advantages over known types of devices:

– high heat transfer rate. The heat transfer coefficient in mini-channels is 2 ÷ 3 times higher than the similar coefficient in the slot channels of plate evaporators [2];

– a decrease in 3 ÷ 4 times of the level of filling with the working substance;

– improved weight and size characteristics. Devices are made of aluminum casting;

– high durability at high pressures.

Up till now, mini-channel technologies have been mainly used in electronics and automotive air conditioning. Since 2012–2014, the research on the prospects of using mini-channel evaporators and condensers in refrigeration has appeared.

The monograph by P. Hrnyak [2], as well as the works by A. V. Baranenko and D. Khovalyg [1] are among the most comprehensive studies in this area. The authors [3, 4, 5] obtained a very interesting experimental material on the study of heat-hydrodynamic processes with the flow of various media in mini-channels Dh = 0.25 ÷ 0.5 mm. At the same time, the question of modeling and summarizing the accumulated experimental material remains open.

A.A. Malyshev [6] formulated the main provisions of the integrated method for the analysis of thermo-hydrodynamic processes during boiling of liquids in pipes, including:

– study of the phase slip and the calculation of the true parameters of two-phase flows;
– methodology for predicting flow patterns;
– methodology for calculating local heat transfer based on true parameters;
– methodology for calculating pressure losses in various two-phase flow regimes.

The research shows that the value of true volumetric steam content (phase slip) is the basis of a comprehensive method that provides the most correct modeling of heat-hydrodynamic processes in a wide range of mode, geometric and physical parameters.

2. Results and Discussion

In order to implement the integrated method in relation to boiling in mini-channels, experimental studies were conducted on the stand shown in figure 1.

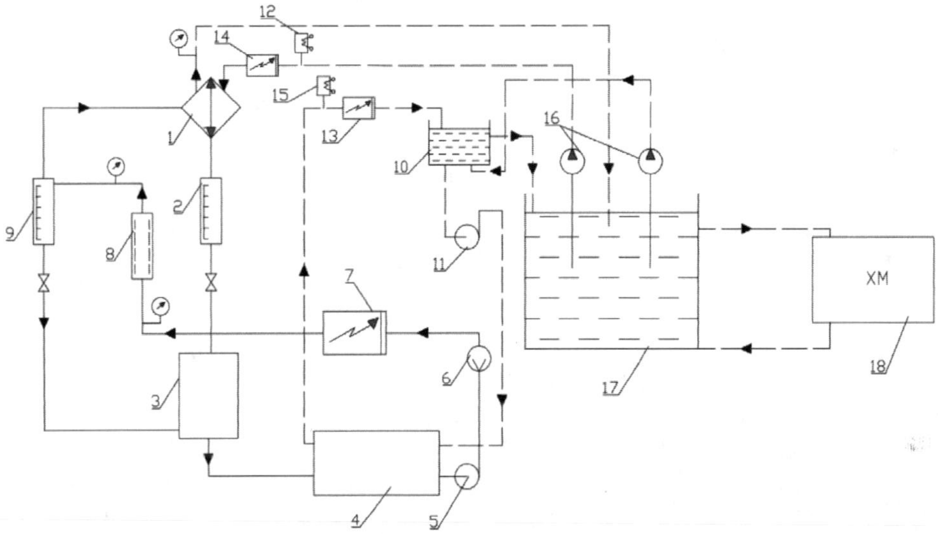

Figure 1. Scheme of the experimental stand: 1 – condenser; 2 – measuring container; 3 – receiver; 4 – subcooler; 5 – freon pump; 6 – flow meter; 7 – steam generator; 8 – experimental section; 9 – measuring container; 10 – thermostat; 11 – brine pump; 12, 15 – contact thermometers; 13, 14 – electric heaters; 16 – submersible pumps; 17 – storage tank; 18 – refrigeration machine.

In front of the experimental section 8 there is a steam generator 7, in which the initial steam content Xin is set with electric heating. Having passed the experimental section, the two-phase flow is sent to the measuring container 9, in which the phases are separated: the steam is sent to the condenser 1, and the liquid flows to the receiver 3. In the measuring container 9, the flow rate of the liquid phase is measured by the volumetric method. In the condenser 1 there is the heater 14 included in the circuit of the contact thermometer 12. The condensation temperature (equal to the boiling point) is regulated in steps using electric heating and cold refrigerant supplied from the storage tank 17. The amount of condensate is determined by the volumetric method using the measuring container 2. The condensate flows into the receiver 3, and then with the pump 5 through the flow meter 6 is supplied to the steam generator 7. In order to prevent boiling up in the pipelines and the pump, the liquid is slightly cooled down in the subcooler 4.

Figure 2 shows an outline of an experimental unit consisting of an inner steel pipe 2 and an outer transparent pipe 1, a mini-channel with a gap of $Dh = 0.5 \div 1$ mm is formed between the centers. Section B is a section of hydrodynamic stabilization, and the part of pipe A isolated from section B is a measuring section. At the measuring section, due to electric heating, the density of the heat flow is set and the wall temperature is measured, on the basis of which the heat transfer coefficients and steam content are determined. Visualization is accompanied by high-speed filming.

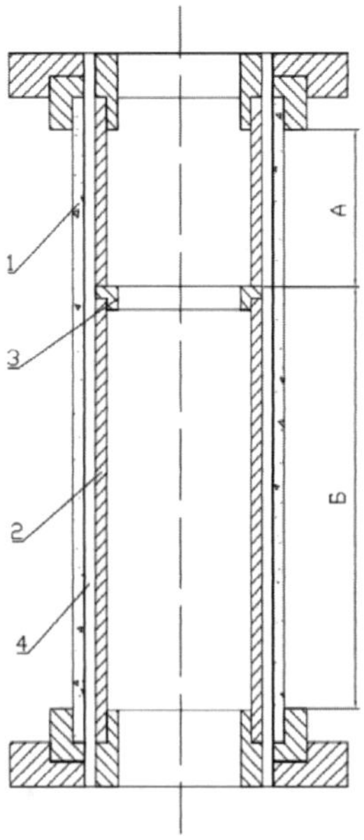

Figure 2. Outline of the experimental block 1 – transparent tube, 2 – steel tube, 3 – gasket, 4 – mini-channel, A – heat section, B – section of hydrodynamic stabilization.

The experiments were conducted on the refrigerant R134a under the following conditions:

- saturation temperature –10 ÷ +20 ° C;
- mass velocity $\omega\rho$ 100 ÷ 500 kg / s. m^2;
- mass steam content x 0.1 ÷ 0.9.

The error in determining the mass velocity based on the experimental determination of the flow rate was 5–10%. The relative error in determining the true volumetric steam content was within 1–3%. The relative error in calculating the mass consumable steam content was in the range: at x = 0.1 – 5%, at x = 0.9 – 15%.

The results of a true volumetric steam content study are presented in figure 3.

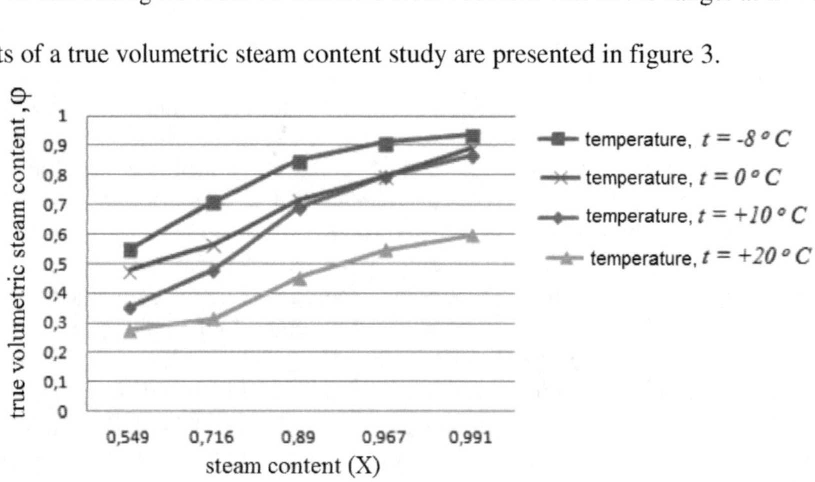

Figure 3. Dependence of true volumetric steam content (φ) on mass steam content (X) at boiling of refrigerant R134a in a mini-channel with $D_э$ = 0.5 mm and at G = *450.12* kg / s m^2.

The presented data were obtained at a mass velocity G = 450.12 kg/s m^2, which, according to [1], corresponds to the section of steady flow.

As the figure shows, with a decrease in temperature, the true volumetric steam content significantly increases and, accordingly, increases the slip coefficient S.

This is primarily due to the increase in the specific volume of steam with a decrease in temperature. So, when the saturation temperature falls from 0 to –8 ° C, the specific volume of steam increases from 0.069 to 0.092 m^3 / kg (in 1.3 times). At the same time, φ increases by an average of 33%, depending on the mass vapor content X. It is important that this flow range is characterized by a circular flow pattern and the growth of true steam content is mainly due to an increase in steam volumes with decreasing temperatures.

In the range from 0 to +20 ° C, the value sharply decreases. This is explained by the fact that with increasing temperatures in the positive range, not only steam volumes decrease and, accordingly, true steam velocities, but also the mode from the annular one goes into the projectile mode, at which the hydrodynamic mechanism changes qualitatively and the slip decreases.

Figure 4 shows the pressure drop values at the experimental site.

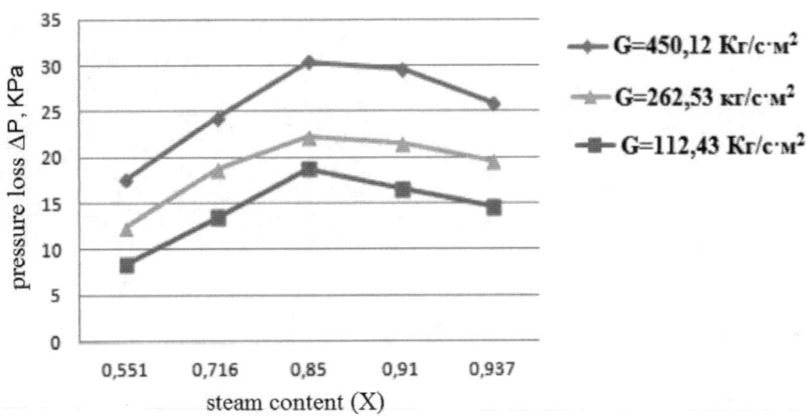

Figure 4. Dependence of pressure loss from steam content during boiling of refrigerant R134a in a mini-channel with D_e = 0.5mm and at $t = -8\ ^oC$.

The increase in pressure drop with the increasing speed is quite understandable. It is important to note that for all values of the mass velocity at X = 0.85, an extremum of the function *ΔP* is observed, followed by a decrease in pressure drops.

As the observations of the flow regimes prove, at these values of the mass consumable steam content, a transition from the annular flow mode to the emulsion one is observed. The emulsion flow is characterized by the liquid film carried away from the channel surface to the center of the flow. The absence of a liquid film on the wall leads to a decrease in friction forces in the boundary layer, which causes a certain decrease in pressure loss.

In general, the transition to the emulsion flow (dry wall mode) is of a crisis nature, since it is accompanied by a sharp decrease in heat emission in these sections of the channel.

It should be noted that the "dry wall" mode is most typical for boiling in mini-channels, because in them (unlike pipes) on a significant part of the heat transfer surface, steam generation processes occur at high values of mass steam content (X> 0.5). Therefore, one of the tasks is to develop a methodology for predicting the crisis regime of a dry wall.

As noted above, one of the tasks of the complex analysis of heat-hydrodynamic processes is the study of boiling modes.

To predict the flow modes of refrigerants, boiling in mini-channels, a modified diagram (map) of modes was developed. The coordinates of φ-Fr_{cm} were used in this diagram. The flow modes were identified on the basis of visual observations and photo-filming of the stream.

The characteristic of the observed modes (figure 5) is in qualitative agreement with the results of [7].

One of the main tasks of the analysis is the calculation of the true volumetric steam content φ , on which all the stages of the integrated approach are based.

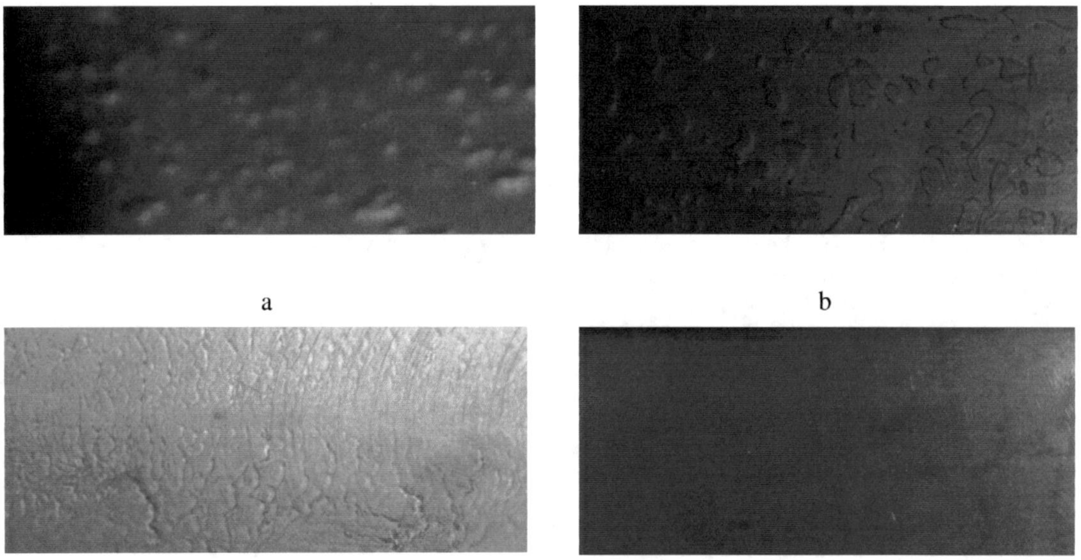

a b

c d

Figure 5. Shooting of modes of two-phase flows: a – bubble; b – projectile; c – transitional; d – annular.

In contrast to the method obtained for pipes in which the inertia forces mainly determine the phase slip (true steam content), when calculating the true parameters of two-phase flows in channels with an equivalent diameter of less than 2 mm, the surface tension forces were taken into account through the Weber criterion [6].

$$W_e = \frac{\rho' \cdot D_h \cdot \omega_{cm}^2}{\sigma} \quad (1)$$

As a result, the general form of the modified dependence in dimensionless parameters acquired the form:

$$\varphi = \beta - 0{,}06 \cdot \beta \cdot (1-\beta)^{0{,}5} \cdot \left(\frac{\frac{\sigma \cdot w_e}{g \cdot d_n^2 \cdot \rho'}}{Re_{cm}} \right)^{-0{,}23} \cdot \left(\frac{P_0}{P_{кр}} \right)^{-0{,}15} \quad (2)$$

In this equation, β-volumetric steam content, calculated from the heat balance equation

The results of the calculation according to equation (2) indicate satisfactory agreement with experiments on channels with Dh≥1.6 mm [2]. For mini-channels with an equivalent diameter of less than 1 mm, which are quite promising, additional experimental study is required, which is the focus of this research.

Using equation (2), a modified mode diagram was developed. The graph with satisfactory accuracy summarizes the boiling data of R134a, R12, R22, NH3 in pipes with Dh greater than 5 mm, as well as the authors' data for R134a boiling in a mini-channel with Dh = 0.5 mm at mass velocities in the range of 100–500 kg / cm.

It should be noted that, according to [1], a steady flow in mini-channels occurs at $\omega\rho$ more than 160 kg / cm^2, and at lower values of the mass velocity, flow reversal, pulsation and hydrodynamic instability are observed. This indicates that the selected mass velocity range is the most promising.

The new diagram of two-phase flows in the mini-channels is shown in figure 6.

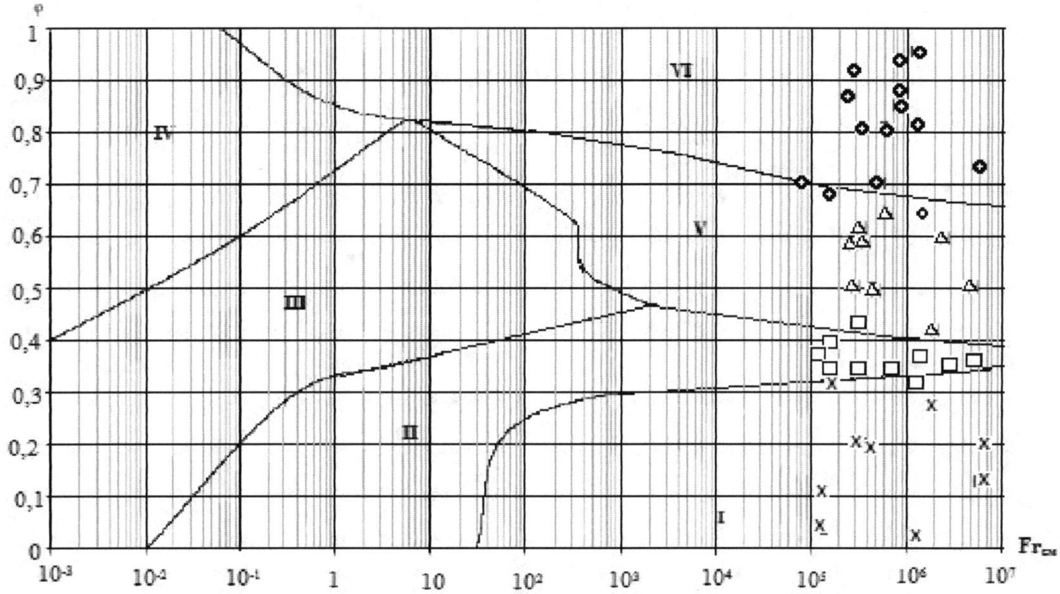

Figure 6. Modified map of boiling modes of refrigerants R134a, R22, R12, NH3 with Dh = 0.5–1.6 mm and wρ – 50–500 kg / (s m^2). Modes: I – bubble (X); II – projectile (□), III – wave, IV – stratified; V – transitional (Δ); VI – annular (◇).

In the diagram presented, in comparison with the well-known diagram [6], the range of values of the criteria Fr_{CM} = 10^4–10^7 is extended, which is most typical for the section of stable flow in mini-channels. The modified diagram describes the experimental results quite satisfactorily.

3. Conclusion

As a result of the research, the following results were obtained:

– new experimental data on true volumetric steam content in mini-channels;

– new experimental data on pressure loss in mini-channels;

– the boundary of the crisis mode of flow in the mini-channel with Dh = 0.5 mm under certain conditions was determined;

– a modified map of boiling regimes of liquids in mini-channels was obtained;

The obtained results allow us to continue the development of a general model for the flow of boiling liquids in a confined space.

References

[1] Khovalyg D M and Baranenko A V 2015 Two Phase Flow Dynamics During Boiling of R134a Refrigerant in Mini-Channels *Technical Physics* **85** **3** 34–41

[2] Niño V G, Hrnjak P S and Newell T A 2002 Characterization of Two-Phase Flow in Microchannels *ACRC TR* 202

[3] Kozulin I A, Kuznetsov V V and Shamirzayev A S 2011 The Modes of a Current and a Thermolysis when Boiling in Micro-channels of Various Orientation *Interekspo Geo – Sibir* [InterExpo Geo – Siberia] **2 (5)**
[4] Kuznetsov V V and Shamirzaev A S 2007 The Modes of a Current and Heat Exchange When Boiling Moving R318C Freon in the Annular Mini-Channel *Teplofizika i aeromekhanika* [Thermophysics and Aeromechanics]] **14 (1)** 57–65
[5] Smiths V V and Shamirzayev A S 2007 The Modes of a Current and Heat Exchange When Boiling Moving R318c Freon in the Annular Mini-Channel *Teplofizika i aeromekhanika* [Thermophysics and Aeromechanics] **4 (1)** 57–65
[6] Malyshev A A, Kisser K V and Zaytsev A V 2017 True Parameters of the Boiling Refrigerants in Pipes and Channels *Vestnik Mezhdunarodnoy akademii kholoda* [Bulletin of the International Academy of Cold] **2** 53-56
[7] Ullmann A and Brauner N 2007 The Prediction of Flow Pattern Maps in Mini-Channels *Multiphase Science and Technology* **19 (1)** 49–73

Basic factors of modeling and optimization of technological processes of construction production

O S Naumova

Yaroslav-the-Wise Novgorod State University, ul. B. St. Peterburgskaya, 41 173003, Veliky Novgorod, Russia

E-mail:osn5@yandex.ru

Abstract. The article analyzes the main goals, objectives, directions of modeling and optimization of technological processes of construction production. The basic tasks and factors of organizational and technological modeling of construction industry, as well as their influence on the change in production volumes, labor productivity growth and the efficiency of construction enterprises have been determined. The main directions of optimization of technological processes of construction production are formulated based on the modernization and development of production and technological capabilities of construction enterprises.

1. Introduction

Construction can rightfully be considered one of the most important sectors of the national economy, since it creates not only buildings and structures for housing and civil purposes, but also the main production assets, in terms of their passive component, for all industries.

Consequently, it is necessary to intensify attention to the increase in production volumes, growth of labor productivity in enterprises of the construction industry. In the conditions of development of crisis phenomena, both in the national and global economy, there is an increase in competition in the market, which is explained by the increasing demands on the quality, safety and environmental friendliness of construction products, especially in housing construction.

Determination of basic tasks and factors of organizational and technological modeling of construction production will allow evaluating their impact on the change in production volumes, labor productivity growth and the efficiency of construction enterprises.

2. Study of factors of modeling and optimization of technological processes of construction production

The main goal of the organization of construction industry is the practical implementation of targeted organizational, technical and technological solutions aimed at ensuring that the construction of the facilities of the required quality has been commissioned on schedule. In addition, the construction company should strive to minimize production costs, incl. unproductive costs at all stages of construction, starting with pre-project preparation.

Achieving these goals is possible on the basis of modeling and optimization of construction production by improving the organization of production, since it represents the interconnection of all technological processes of production of the final construction products in space and time, which, ultimately, allows optimizing types of resources and the duration of the construction of facilities and their complexes significantly.

The problems of compatibility (logical, methodological, informational, mathematical, etc.) subsystems of organizational and technological design with adjacent subsystems of architectural and construction design, construction of facilities and design of management decisions are actual in modern conditions. A distinctive feature of the construction process, in addition to a large number of parameters affecting its final cost, is the use of a large number of different types of resources and the ability to create various combinations of these resources in the form of rigid clusters of sequences. The totality of construction resources is presented in figure 1.

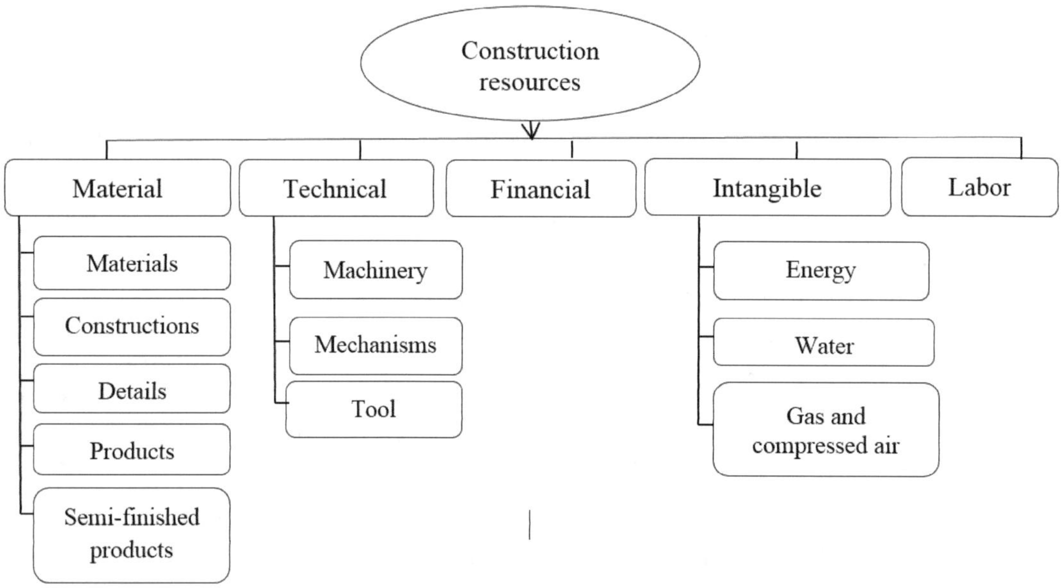

Figure 1. The complex of construction resources.

Construction is one of the most resource-intensive sectors of material production, which is confirmed by figure 1. Resources in construction production include both material resources and resources in the form of services, as well as animate resources in the form of labor force. The latter implies some degree of uncertainty in the meaning of key parameters affecting the cost of production.

There is a need to address a wide range of issues related to the automation of pre-production, which arise during the processing of design and estimate documentation (DED), the formation of promising and annual plans for construction and installation works (CIW), organization of logistics and equipment. The organization of interrelations of crews, mechanisms, vehicles and services ensuring the supply of structures, materials and semi-finished products is far from perfect.

Practice shows that modeling the work of construction enterprises in the use of resources is the most important factor influencing the increase in the pace of construction, productivity growth, commissioning of construction projects in a timely manner.

Objectively, the work of enterprises of the construction industry is significantly affected by the creation of an optimally constructed structure of interaction of resources that can ensure the flow of work. At the same time, the main negative factors counteracting the maintenance of the flow of construction, installation and special works should include simple resources with a continuing need for these resources in the future, as well as an open front of work with a lack of resources, that is, to continue the production cycle which is currently unavailable for consumption.

Therefore, it is necessary to highlight the following factors when setting the basic tasks of organizational and technological modeling of construction industry:

– reduction of terms for implementation of a complex of works (or an indicator of deviation from the established deadlines for the objects to be delivered) under given resource constraints;

– optimization of indicators (in terms of quantity and quality, delivery, the size of production stock, etc.) of the use of resources for a given period of project implementation.

When modeling the construction industry, the following basic elements should be considered first:

– amount of work to be performed;

– terms of work;

– resource endowment.

In the process of optimizing schedules, it is necessary to remember all the elements of this triad and that in case of a change in one of its elements, the remaining two are transformed, either positively or negatively, depending on the specific project or construction conditions. In order to make sure that the identified changes will not become critical for the implementation of the project, the other two elements must also be assessed.

Properly developed work plan will ensure the growth of labor productivity by several times.

Research on this issue revealed the absence of a systematic approach to the integrated assessment of the level of organization of construction process, based on such parameters as continuity, uniformity, combination, rhythm and intensity. The results of the analysis create the prerequisites for automation and modeling of the work of enterprises for the construction of objects [4].

3. Directions of modeling and optimization of technological processes of construction production

The main goal is to minimize costs, shorten the construction of facilities, to provide consistent high-quality work, and, ultimately, maximize profits with any method of modeling in the preparation of construction plans.

Construction companies need to overcome technological and technical backwardness from their Western counterparts, which today makes usually 10–15 years in order to be competitive. So, according to the vice-president of the Association of Russian Builders, Dr. Sc. Vladimir Utkin, equipment depreciation in the building materials industry reaches 60–70%. And the further development of the industry, in his opinion, is impossible without large-scale modernization.

Most of the issues that are directly related, in particular, with new construction, ranging from the allocation and design of the land, obtaining engineering support and finishing the commissioning of the final product, are extremely and artificially difficult in modern conditions. This is another factor that significantly inhibits the development of not only the industry, but also the economy as a whole.

A serious problem is the non-observance of technology and construction rules due to low organization of construction processes.

Studies have shown that the majority of enterprises in the construction industry do not observe incentives, as well as the conditions and grounds for making any significant investments in research and development, in modernization, due to the lack of confidence in making a profit. Even if we assume that the costs will pay off in five years, then it is necessary to be sure that all five years will be provided with full capacity utilization. But today, in particular, in the conditions of Novgorod region there is no large-scale construction, not so many integrated development projects, and, as a result, there are no major contracts for such a long period and the real prospect is a maximum of 1.5 years.

Thus, all of the above factors cause a decrease not only in production volumes, but also in the quality of products.

In particular, current situation in housing construction and provision of citizens with housing is highlighted in the Summary Report on the results of monitoring the effectiveness of local government bodies in the city district and municipal districts of Novgorod region in 2017. Thus, in 2017, in the territory of the region, 1151 houses (3360 apartments) with a total area of 231.6 thousand square meters were built from all sources of financing, including 1,128 houses (1128 apartments) with individual developers with a total area of 123.1 thousand m^2. Thus, the introduction of housing in the whole region amounted to 64.0% comparing with the year 2016 and to 85,4% for individual developers.

In 2017, the volume of work performed by the company "Stroitelstvo" in Veliky Novgorod was 8.9 billion rubles or 78.8% compared with 2016. In the reporting year, 34 residential buildings with a total area of 104.7 thousand square meters of residential space were commissioned, including 13 apartment buildings with an area of 100.3 thousand square meters, 21 individual houses with a total area of 4.3 thousand square meters. Introduction of housing per inhabitant for the year made 0.47 m^2.

Consequently, the primary task of optimization of technological processes and production on the basis of modernization, automation and development of the production potential of a construction enterprise operating under conditions of external instability can be called as one of the most significant problems of modern science.

The solution of the tasks can be carried out by forming a variety of strategies, which should be understood as a future-oriented system of measures, which, in turn, is aimed at ensuring the achievement of the identified specific strategic goals to achieve the required state of the enterprise's production potential. This means that it is vital to form and implement a strategy aimed at choosing the desired direction of activity from the variety of options for the development of production potential of a construction enterprise in the intended direction [2].

The followinf factors should be considered as the strategy of modernization and development of production and technological capabilities of construction enterprises:

– mobilization of resources to achieve strategic goals;

– definition of responsibility centers in the implementation of each goal;

– stimulation of the activities of the management and direct implementers on the basis of a constant analysis of the results of implementation of strategic goals.

The following can be singled out as the basic strategic goals of modernization and development of production and technological capabilities of construction enterprises:

1. Technical re-equipment and reconstruction of fixed production assets in terms of their active component, which will make it possible in the framework of simple reproduction to significantly increase the economic efficiency of functioning without expanding the staff, i.e. additional manpower, making the most of the available human resources.

2. Modernization of buildings and structures, which may be limited by redevelopment of internal premises, as well as by improving existing engineering systems, which will improve the consumer and operational qualities of buildings without rebuilding them and eliminate their functional wear.

3. Improvement of the organizational and management system of the construction enterprise, by optimizing the organizational structure, which will allow to balance the ratio of the managerial and the linear staff, eliminate problems in the relationship of links with each other, as well as reduce labor costs [3].

4. Increased attention to the development of personnel intelligence, improvement of professional and qualification level, guaranteeing the competitiveness demanded by markets, both of employees and the results of their labor (processes, technologies, products, systems), which will ensure the acceleration of the processes of modernization, automation and development of production and technological potential construction enterprise and increase their efficiency. The important principles for the formation of a new innovative approach to improving production management and personnel development should be effective measures to create: permanently trained organizations, sites specializing in innovative

production; healthy socio-psychological climate in groups; teams (specialists) working on the basis of personal plans for business and creative growth.

5. Formation of a system of criteria designed to assess the process of managing the modernization of the construction industry, which will allow for a continuous analysis of the results achieved and to determine further areas of activity, and, in addition, to identify and eliminate shortcomings.

Under existing conditions, the replacement of obsolete capacities and maximum reduction of the influence of the human factor are considered as the most effective way to increase the level of competitiveness of an enterprise in the domestic market. Accurate, comprehensive or complete modernization of production of a building enterprise is accompanied by indisputable advantages that shape the relevance of introduction of high-tech equipment and automated systems into technological processes:

– Improving the performance of existing capacity;

– Improving the quality control of construction, installation and special works, as well as the quality of final products of the construction industry;

– Phased control of technological processes in automatic, manual, adjustment, or semi-automatic modes;

– Economical use of electricity and consumables;

– High precision dosing of structural components;

– Reducing the cost of the enterprise, calculated on the payment of human resources;

– Improving working conditions.

Since the modernization and development of production and technological capabilities of construction enterprises should be carried out on the basis of an innovative approach with maximum use of the achievements of STP, it is quite obvious that the role of the individual should be increased, both of individual performers - ordinary employees of the enterprise and managers [4].

As the most significant reserve of managers of construction enterprises, measures can be identified to improve the methods of implementing personnel management, both in group and individual form, requiring them to more fully utilize all types of resources and create new working conditions and incentives [5].

As the analysis showed, the majority of enterprises in the construction industry do not pay enough attention to the problem of insufficient use of the potential of their labor resources. To change the situation it is necessary:

– to form a continuous education system at the enterprise;

– to ensure that targeted information is available to all interested parties on achievements in the development of modern techniques, technology and organization of construction, application of best practices at similar enterprises;

– to organize professional skill contests at the enterprise;

– to improve, or develop a system of qualification certification of personnel;

– to improve job descriptions for employees of the enterprise, taking into account the constantly changing conditions of production activities;

– to ensure the availability of visual information on the strategic goals of modernization and development of production and technological capabilities and the results of achieving these goals;

– to fix the procedure for the adoption, study, evaluation and implementation of scientific and technical ideas;

– to increase attention to interpersonal relations in the team and to clarify the need to help employees of the enterprise in improving working conditions, study, life, etc., which will increase their motivation to more efficient use of their labor and intellectual potential;

– to develop and implement an organization standard on enhancing the scientific, technical and socio-economic corporate culture;

– to establish constant monitoring of the implementation of measures and business plans for innovative modernization of production.

Implementation of the above activities will allow a more complete and effective involvement in the modernization and development of production and technological capabilities of construction enterprises of such an important resource as human capital. It should be noted that this direction of modernization of the enterprise does not require any significant financial costs.

4. Impact of the planned measures for modeling and optimization of technological processes of construction production on the economic efficiency of the construction company

The growth of technical equipment and the improvement of working conditions that accompany the proposed form of renovation of construction production, contributes to the improvement of personnel skills, makes working professions more attractive, since the technical re-equipment of the production potential has not only technical but also social significance.

The effectiveness of the planned measures for the modeling and optimization of technological processes and production of a construction enterprise based on the modernization and development of production and technological capabilities can manifest itself in several forms:

– reducing the duration of construction,
– reducing labor costs or increasing productivity,
– saving material and energy costs, including saving in physical terms of metals, timber, cement.

In addition, this may include:

– cost reduction of construction and installation works,
– increase in capital productivity, acceleration of working capital turnover and other results
– increase in the quality of the end product of the construction industry.

The annual economic effect is calculated based on the difference in reduced costs for events whose implementation requires capital expenditures. Indicators of economic efficiency are determined in accordance with the instructions for determining the economic efficiency of use in the construction of new equipment, inventions and innovation proposals.

5. Conclusion

Modeling and optimization of technological processes of construction production allows not only to evaluate and analyze the future nature of the construction of the facility, to identify those features that may arise, but also to have a direct impact on the construction process of facilities or their complexes.

It is proposed to focus on modernization, automation and development of the production potential of a construction enterprise as the main direction of modeling and optimization of technological processes of construction production due to:

– technical re-equipment and reconstruction of fixed production assets in terms of their active component;

– modernization of buildings and structures, which may be limited by redevelopment of internal premises, as well as by improving existing engineering systems;

– improving the organizational and management system of the construction company, by optimizing the organizational structure;

– increasing attention to the development of personnel intelligence, improving professional and qualification level;

– formation of a system of criteria designed to assess the process of managing the modernization of the construction industry.

In particular, when modeling the organization of construction, it becomes possible to obtain information about the behavior of the proposed organizational system of construction in real conditions based on which apply the most optimal solutions at the design stage to minimize production costs, reduce construction time and, as a result, increase the economic efficiency of the construction company.

References

[1] Kovalenko M and Gonchar A 2014 Formation of management decisions in the field of production modernization at a building enterprise *Economics and management of innovative technologies* **9** 5720
[2] Melekhin V and Saidov M 2014 Methodology for managing the modernization and development of the production potential of a construction enterprise in an unstable environment *Naukovedenie* **4** 70EVN414
[3] Lewandowicz E and Lisowski P 2018 Methodology to generate navigation models in building *Journal of Civil Engineering and Management* **24** **(8)** 619–629
[4] Ostroukh A 2004 Automation of the management of labor cost reduction in construction. *Vestnik of Russian New University* **4** 117–120
[5] Yang Z, Wang Y and Sun C 2018 Emerging information technology acceptance model for the development of smart construction system *Journal of Civil Engineering and Management* **24** **(6)** 457–468

IOP Conf. Series: Materials Science and Engineering **656** (2019) 012039 doi:10.1088/1757-899X/656/1/012039

The results of simulation modeling of the operation of the regenerative fifth wheel hitch of a timber trailer

V O Nikonov[1,2], V I Posmetev[1] and V V Posmetev[1]

[1]Voronezh State University of Forestry and Technologies named after G. F. Morozov, 8, Timiryazev street, Voronezh, 394087, Russia

[2]E-mail: 8888nike8888@mail.ru

Abstract. A three-dimensional mathematical model of the movement of a timber hauler with a semi-trailer along an uneven supporting surface and a computer program developed on its basis are presented that made it possible to evaluate the possibility of equipping the fifth-wheel coupling with an energy recovery system in various driving modes and to study the effect of recovery system parameters on its efficiency.

1. Introduction

An important role in increasing the turnover of timber is played by semi-trailers, the effectiveness of which is to a certain extent hampered by the design flaws of some components and assemblies of timber hauling tractors and, in particular, by the structural imperfection of fifth-wheel coupling devices. In the process of pulling away a hauling truck with a semi-trailer, accelerating, braking and driving in difficult road conditions of forest objects in the fifth wheel coupling, increased dynamic loads occur, which lead to a reduction in the speed of the hauling truck with the semi-trailer, its fuel consumption increase, and consequently performance. Increased loads in the fifth wheel coupling increase the wear of its mating parts, impair the stability of the hauling truck with the semi-trailer when braking.

The improvement of the fifth wheel coupling of a timber hauling truck with a semi-trailer, aimed at reducing the fuel consumption of a timber hauling tractor, requires in-depth study and generalization of the actual material available on the fifth wheel coupling design, accumulated both in our country and abroad. The results of numerous studies confirm the promise of reducing the fuel consumption of specialized cars by using various types of energy recovery mechanisms in their structures, which allow energy that is dissipated into the environment to be accumulated and reused in its work [1–6].

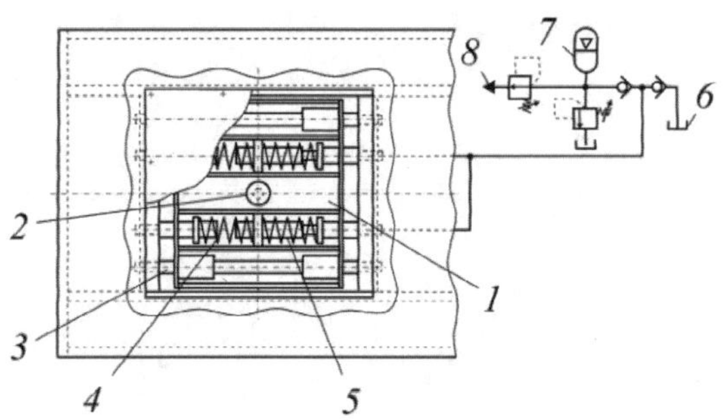

Figure 1. Scheme of recuperative spring-hydraulic fifth wheel coupling of a hauling truck with a semi-trailer: *1* – base plate; *2* – king pin; *3* – guides; *4* – springs; *5* – hydraulic cylinders; *6* – hydraulic tank; *7* – pneumohydraulic accumulator; *8* – port of the user recoverable working fluid.

IOP Conf. Series: Materials Science and Engineering **656** (2019) 012039 doi:10.1088/1757-899X/656/1/012039

Based on research conducted at the department of production, repair and operation of machines FSBEI of VGLTU named after GF Morozov, the authors propose a perspective design of a recuperative spring-hydraulic fifth-wheel hitch of a hauling truck with a semi-trailer, the scheme of which is shown in figure 1.

The work of the regenerative fifth wheel coupling of a timber hauling truck with a semi-trailer is based on the use of kinetic energy, arising from the mass inertia force of the semi-trailer when braking, accelerating, turning, changing gears and hitting obstacles in the process. Recuperative fifth wheel coupling allows accumulating and reusing hydraulic energy in the technological process of loading and unloading the assortments by a hydraulic manipulator mounted on a hauling truck with a semi-trailer. To assess the feasibility and feasibility of using a fifth wheel hitch with hydraulic energy recovery system and determining its optimal parameters, the task was set to develop and investigate a mathematical model of the movement of a hauling truck with a semi-trailer equipped with an energy recovery system in a fifth wheel hitch. Modeling is based on the methods of classical mechanics.

2. Materials and methods

The timber tractor and semi-trailer are represented in the model as two absolutely rigid bodies that perform translational and rotational movement in three-dimensional space and interact with the supporting surface in an elastically viscous manner at ten points (the number of wheels and twin wheels) figure 2. The bodies of the timber tractor and semi-trailer interact with each other through a fifth wheel coupling: between the points A and B, which is proposed to be equipped with recuperative elements. Timber tractor and semi-trailer have mass m_T and m_N and moments of inertia J_T and J_N, which are calculated for the current time relative to the current axis of rotation of the body. The position of the timber tractor and semi-trailer in space is characterized as the coordinates of their centers of gravity (x_T, y_T, z_T), (x_N, y_N, z_N), and angles of deflection (φ_{xT}, φ_{yT}, φ_{zT}), (φ_{xN}, φ_{yN}, φ_{zN}) local coordinate system of bodies from the base coordinate system.

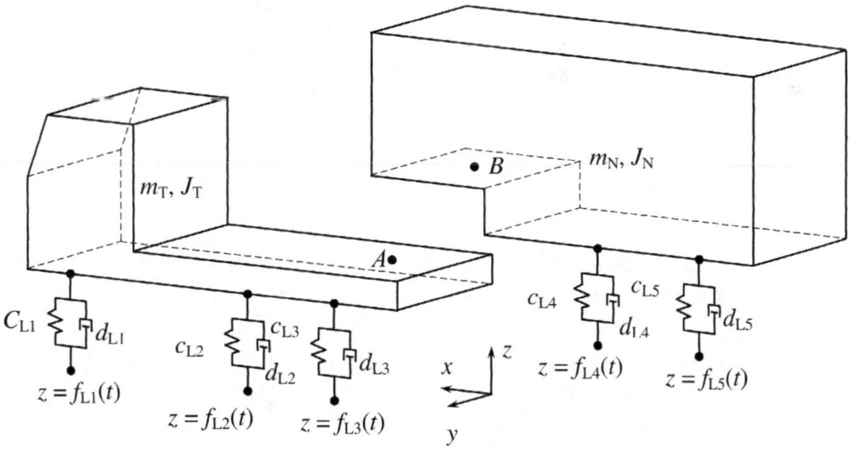

Figure 2. Design scheme for building a dynamic model of a timber hauling tractor with a semi-trailer (only the left-side wheels are shown).

The movement of a road train in a model is described by a system of differential equations based on the basic laws of the dynamics of translational and rotational motion:

for the truck tractor equation (1):

$$\begin{cases} m_T \dfrac{\partial^2 x_T}{\partial t^2} = \sum\limits_{i=1}^{3} F_{\text{Li}}^{\text{x}} + \sum\limits_{i=1}^{3} F_{\text{Ni}}^{\text{x}} + F_{Ax} \\ m_T \dfrac{\partial^2 y_T}{\partial t^2} = \sum\limits_{i=1}^{3} F_{\text{Li}}^{\text{y}} + \sum\limits_{i=1}^{3} F_{\text{Ni}}^{\text{y}} + F_{Ay}; \\ m_T \dfrac{\partial^2 z_T}{\partial t^2} = -m_T \cdot g + \sum\limits_{i=1}^{3} F_{\text{Li}}^{\text{z}} + \sum\limits_{i=1}^{3} F_{\text{Ni}}^{\text{z}} + F_{Az}; \\ J_{\text{Tx}} \dfrac{\partial^2 \phi_{\text{xT}}}{\partial t^2} = \sum\limits_{i=1}^{3} M^{x}(F_{\text{Li}}) + \sum\limits_{i=1}^{3} M^{x}(F_{\text{Ni}}) + M^{x}(F_A); \\ J_{\text{Ty}} \dfrac{\partial^2 \phi_{\text{yT}}}{\partial t^2} = \sum\limits_{i=1}^{3} M^{y}(F_{\text{Li}}) + \sum\limits_{i=1}^{3} M^{y}(F_{\text{Ni}}) + M^{y}(F_A); \\ J_{\text{Tz}} \dfrac{\partial^2 \phi_{\text{zT}}}{\partial t^2} = \sum\limits_{i=1}^{3} M^{z}(F_{\text{Li}}) + \sum\limits_{i=1}^{3} M^{z}(F_{\text{Ni}}) + M^{z}(F_A), \end{cases} \tag{1}$$

for semi-trailer equation (2):

$$\begin{cases} m_N \dfrac{\partial^2 x_N}{\partial t^2} = \sum\limits_{i=4}^{5} F_{\text{Li}}^{\text{x}} + \sum\limits_{i=4}^{5} F_{\text{Ni}}^{\text{x}} + F_{Bx} \\ m_N \dfrac{\partial^2 y_N}{\partial t^2} = \sum\limits_{i=4}^{5} F_{\text{Li}}^{\text{y}} + \sum\limits_{i=4}^{5} F_{\text{Ni}}^{\text{y}} + F_{By}; \\ m_N \dfrac{\partial^2 z_N}{\partial t^2} = -m_N \cdot g + \sum\limits_{i=4}^{5} F_{\text{Li}}^{\text{z}} + \sum\limits_{i=4}^{5} F_{\text{Ni}}^{\text{z}} + F_{Bz}; \\ J_{\text{Nx}} \dfrac{\partial^2 \phi_{\text{xN}}}{\partial t^2} = \sum\limits_{i=4}^{5} M^{x}(F_{\text{Li}}) + \sum\limits_{i=4}^{5} M^{x}(F_{\text{Ni}}) + M^{x}(F_B); \\ J_{\text{Ny}} \dfrac{\partial^2 \phi_{\text{yN}}}{\partial t^2} = \sum\limits_{i=4}^{5} M^{y}(F_{\text{Li}}) + \sum\limits_{i=4}^{5} M^{y}(F_{\text{Ni}}) + M^{y}(F_B); \\ J_{\text{Nz}} \dfrac{\partial^2 \phi_{\text{zN}}}{\partial t^2} = \sum\limits_{i=4}^{5} M^{z}(F_{\text{Li}}) + \sum\limits_{i=4}^{5} M^{z}(F_{\text{Ni}}) + M^{z}(F_B), \end{cases} \tag{2}$$

where t – time; F_{Li} и F_{Ni} – forces acting on the hull of a hauling truck or semi-trailer from the wheels of the i-th axis through the suspension on the left (index "L") and right (index "N") bead; F_{Ax}, F_{Ay}, F_{Az} – cartesian components of the force acting on the truck tractor from the fifth wheel coupling; F_{Bx}, F_{By}, F_{Bz} – cartesian components of forces acting on a semi-trailer from the side of the fifth wheel coupling; M^i – moments of the indicated forces about the axis i.

To set the perturbing effect of the bearing surface on the bodies of the timber hauling and semi-trailer, a simplified elastic-viscous wheel model was used, simplifying the "wheel" - "suspension" system to a simpler system characterized by two coefficients: stiffness and damping. The force from the side of the wheel on the hull of the hauling truck was calculated using equation 3:

$$F_{\text{i}}^{\text{z}} = c_{\text{i}}\left(z_{\text{Ni}}\left(x_i, y_i\right) + R_{\text{K}} - z_{\text{Ki}}\right) - d_{\text{i}}\left(\frac{\partial z_{\text{Ni}}\left(x_i, y_i\right)}{\partial t} - \frac{\partial z_{\text{Ki}}}{\partial t}\right), \tag{3}$$

where i – wheel index; $z_{\text{Ni}}(x, y)$ – vertical coordinate of the surface under the wheel, equal to the coordinate of the lower point of the wheel; z_{Ki} – the vertical coordinate of the point of attachment of the wheel to the body; R_{K} – wheel radius; c_{i}, d_{i} – stiffness coefficients and damping of visco-elastic interaction.

The system of equations of movement of the timber tractor and semi-trailer is a system of second-order differential equations. The numerical solution of differential equations consists in discretization of time t into equal steps of Δt, which are numbered by variable τ. At each integration step, it is necessary to calculate the forces and moments acting on the hulls of the hauling truck and semi-trailer. After that, the coordinates and velocities of the bodies for the next integration step are calculated from the known coordinates and velocities of the bodies for the current integration step. In particular, for the Cartesian component x, the numerical integration of the equations of motion of a certain body in the framework of the method under consideration is performed according to «equation 4»:

$$\begin{cases} x_{\tau+1} = x_\tau + v_{x\tau} \cdot \Delta t + \dfrac{F_{x\tau}}{m} \cdot \dfrac{(\Delta t)^2}{2}; \\ v_{x\tau+1} = v_{x\tau} + \dfrac{F_{x\tau}}{m} \cdot \Delta t, \end{cases} \tag{4}$$

Where x_τ and $v_{x\tau}$ – the coordinate and velocity of a body of mass m along the cartesian direction x at the previous integration step over time τ; x_τ and $v_{x\tau}$ – the same at the next time integration step $\tau + 1$. Similarly, numerical integration is performed for the remaining cartesian components y and z, at the same time, integration is performed for both bodies: the body of the hauling truck and the semi-trailer body.

As a result of solving the differential equations by a numerical method, table-defined functions of the dependence on the time coordinates and the relative orientation angles of the timber carrying tractor and semi-trailer are obtained. These functions are analyzed further to assess the effectiveness of the recovery system. To reproduce the operation of the fifth-wheel hitch recovery system, it is necessary to create an intensive relative movement of the tractor and semi-trailer in the model. In the developed model, this is done in two ways: by moving on a supporting surface with random irregularities, by moving on a flat horizontal surface with variable speed. Random irregularities in the model are set on the basis of the actual operating conditions of the forest truck. A significant share of the path of a timber truck accounts for low-quality access asphalted roads, unpaved forest roads, winter roads. For transmission in the model of a complex random surface relief, it was considered that the perturbing function of the wheels of each bead consists of a set of protrusions of gaussian shape. The height and length of the Gaussian peaks are set randomly and vary widely: from "obstacles" (imitating stones, stumps, protruding roots, etc.) having a small length (about 0.2 ... 0.5 m) to "hills" having a greater length (about 2 ... 5 m) figure 3.

Figure 3. An example of a support surface for the study of loads on a fifth wheel coupling of a timber hauling tractor with a semi-trailer: fragments of functions $z(x)$ for the wheels of the left (blue line) and right (red line) sides (fragments of length 20 m).

The reference surface defined the function of the surface height from the coordinates of the point of contact $z(x, y)$, as a superposition of gaussian peaks with the parameters (x_i, y_i) (protrusion position), H_i (height of the protrusion) and σ_i (standard deviation setting the protrusion width) equation 5:

$$z(x) = \sum_{i=1}^{S_X} H_{Xi} \exp\left(-\frac{(x - x_{Xi})^2}{\sigma_{Xi}^2} \right) + \sum_{i=1}^{S_N} H_{Ni} \exp\left(-\frac{(x - x_{Ni})^2}{\sigma_{Ni}^2} \right), \tag{5}$$

where S_X and S_N – number of hills and obstacles; H_{Xi} and H_{Ni} – heights of hills and obstacles; x_{Xi} and x_{Ni} – coordinates of the center of the hills and obstacles; σ_{Xi} and σ_{Ni} – характерная полуширина холмов или препятствий.

Gaussian peaks were distributed along the length of the control section of 500 m randomly according to a uniform law. In this case, the parameters H_i and σ_i were also chosen randomly according to a uniform law from intervals: from 0 to 0,6 m for H_i and from 0,2 to 4,0 m for σ_i. The number of gaussian peaks that mimic hills and obstacles was calculated in accordance with the parameter tables for various types of road and ground conditions.

The considered recovery system converts adverse fluctuations in the distance between the tractor and the semi-trailer at the junction (fifth wheel coupling) into useful energy stored by the pneumohydraulic accumulator, and used further, for example, during operation of the hydraulic manipulator. In one of the variants, the model makes it possible to estimate the average value of the recovered power N_p without taking into account the design of the recovery system and the hydraulic subsystem underlying it. In this case, N_p is estimated by the power dissipation in the conditional damper of the fifth wheel coupling: through the coefficient of viscous friction in the approximation of the elastic-viscous contact of points A and B.

In the process of modeling, characteristics are calculated that allow one to determine both the regenerative effect and the adverse effects of fitting a fifth wheel coupling system with a recovery system. The main indicators of the efficiency of the recovery system are the recovered power (instant and average) and the average acceleration of the semi-trailer (instant and average). At each step of the numerical integration, the instantaneous value of the recovered power N_p was determined from the change in distance between points A and B of the fifth wheel coupling in the longitudinal direction equation 6:

$$N_p(\tau) = d_0 \left(\frac{D}{D_0}\right)^2 \left(\frac{\left|x_A^\tau - x_A^{\tau-1}\right|}{\Delta t} - \frac{\left|x_B^\tau - x_B^{\tau-1}\right|}{\Delta t}\right)^2, \quad (6)$$

where d_0 – effective damping coefficient from recuperative hydraulic cylinders; D_0 – base diameter of recuperative hydraulic cylinders (the main calculations used the value of 50 mm); D – diameter of recuperative hydraulic cylinders used; x_A and x_B – the longitudinal coordinate of the fifth wheel coupling point of the tractor and semi-trailer respectively; indexes τ and $\tau - 1$ denote the current and previous steps of integrating the differential equations of motion over time.

Equipping a fifth wheel coupling system with a heat recovery system can degrade the damping properties of the fifth wheel coupling and lead to unfavorable fluctuations of the semi-trailer with respect to the truck hauler. Therefore, in the developed model, along with the recoverable power, the acceleration of the semi-trailer in the longitudinal direction is estimated. The instantaneous longitudinal acceleration of the semi-trailer at the integration step τ was calculated using equation 7:

$$a_N(\tau) = \frac{x_N^{\tau+1} + x_N^{\tau-1} - 2x_N^\tau}{(\Delta t)^2}, \quad (7)$$

where $x_N^{\tau-1}$, x_N^{τ}, $x_N^{\tau+1}$ – coordinates of the center of gravity of the semitrailer along the longitudinal cartesian axis OX to the previous $\tau - 1$, the current τ and the subsequent $\tau + 1$ steps of integrating the time differential equations.

After each computer experiment, the $N_p(t)$ and $a_N(t)$ functions were used to determine two indicators N_{pc} and a_{Nc}, averaged over a sufficiently long time interval of the equation 8, 9:

$$N_{pc} = \frac{1}{\tau_\kappa - \tau_v} \sum_{i=\tau_v}^{\tau_\kappa} N_p(\tau), \quad (8)$$

$$a_{Nc} = \frac{1}{\tau_{\kappa} - \tau_{v}} \sum_{i=\tau_{v}}^{\tau_{\kappa}} \left| a_N(\tau) \right|, \tag{9}$$

where τ_v and τ_κ – time integration steps at which averaging begins and ends (averaging is not done from the beginning of the computer experiment, since at the first moments of time the mechanical system comes to equilibrium and not until the end of the computer experiment, since at the end the car can leave the area with a random relief) .

3. Results

For the study of the developed mathematical model, a computer program called "Program for simulating the movement of a timber truck with a semi-trailer with a recuperative semi-coupling device" was developed in Object Pascal in the Borland Delphi 7 programming environment.

The process of investigating the recovery system of a timber hauler with a semi-trailer was to conduct multiple computer experiments with changing the set of input parameters relative to the parameters of the basic computer experiment. The basic computer experiment was carried out with the most typical parameters of a hauling truck with a semi-trailer and the conditions of its movement. After the launch of the developed computer program, the preparation of the relief of the bearing surface for the left and right side wheels was carried out. In this study, the geometric parameters of the terrain corresponded to a significantly uneven logging road. At the beginning of the simulation, the forest tractor and semi-trailer were placed near the origin, at a slight height above the supporting surface. With the beginning of the numerical integration of the equations of motion, the forest truck and the semi-trailer spontaneously lowered to the surface under the action of gravity, connected with each other at the point of the fifth wheel coupling and came to a stable position, which took about 5 seconds of modeling time.

The speed of movement of the tractor in the direction of the axis *OX* was constant in the process of modeling and was 20 km/h. This is a characteristic value of speed when driving on access roads to cuttings with asphalt pavement of poor quality, dirt roads, winter roads. In the process of driving on an uneven supporting surface, the truck tractor and the semi-trailer made a complex translational-rotational motion, remaining connected to each other through a fifth wheel coupling with model recuperative elements. The fluctuations of point *A* relative to *B* in the fifth wheel coupling caused the conditional pumping of the working fluid into the regenerative hydraulic system, and the program calculated the instantaneous recuperative power for each integration step figure 4, *a*. At certain points in time, the instantaneous values of the recovered power may exceed 40 kW, however, the average recoverable power for the entire basic experiment was 6,31 kW. To calculate the average recovered power, it was assumed that the forest truck with a semi-trailer moved on a supporting surface of 100 m.

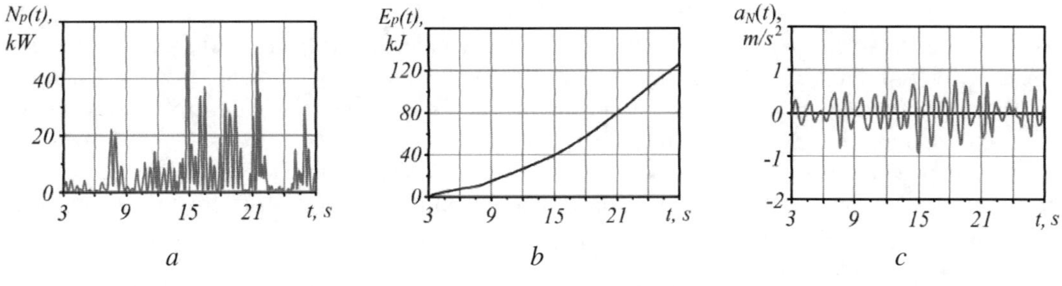

Figure 4. The results of the basic experiment: time dependencies *t* of the instantaneous recovered power N_p (*a*) accumulated by the energy recovery system E_p (*b*) and the instantaneous longitudinal acceleration of the semi-trailer a_N (*c*).

As the truck trailer moves, energy is accumulated in the recovery system, which can later be used, for example, in a hydraulic manipulator in the process of loading or unloading assortments figure 4, *b*. Energy accumulation occurs approximately linearly; deviations from linearity are associated with fluctuations in the density of the distribution of irregularities of the support surface. To assess the smoothness of the semi-trailer, the instantaneous a_N and average a_{Nc} of the «figure 4, c» longitudinal accelerations of the semi-trailer were calculated. In the basic computer experiment, the instantaneous value of acceleration could reach about 0,9 m/s^2 with the most unfavorable random combinations of irregularities of the supporting surface. However, the longitudinal acceleration averaged over the 100 m path amounted to a significantly lower value of 0,261 m/s^2, which is a rather smooth movement. In a further theoretical study, they were changed in relation to the basic values: the speed of movement of a forest truck and the diameter of the recuperative hydraulic cylinder. In this case, an analysis of changes in the performance of the recovery system N_{pc} and a_{Nc}.

From the speed of movement depends on the intensity of oscillations of the semi-trailer relative to the forest truck, and, consequently, the intensity of the accumulation of energy by the recovery system. To study the effect on the efficiency of the recovery system of the speed of movement v of a forest truck, a series of five computer experiments was carried out in which they varied v from 10 to 50 km/h in 10 km/h increments. This longitudinal speed was maintained constant during the computer experiment for the center of gravity of the tractor, while the longitudinal speed of the semi-trailer was calculated using the equations of the mechanical system, taking into account the mechanical coupling between the bodies and the rotational motion of the bodies. With an increase in speed, the average recoverable power N_{pc} increases according to a law close to a quadratic figure 5, *a*. At the same time, as the speed of movement along an uneven surface increases, the average longitudinal acceleration of the semi-trailer Figure 5, *b* increases approximately linearly. Thus, with an increase in the speed of a timber truck from 10 to 50 km/h on a road with roughness of an average height of 0,4 m, the recovered power increases according to a square law from 5,5 to 12,2 kW.

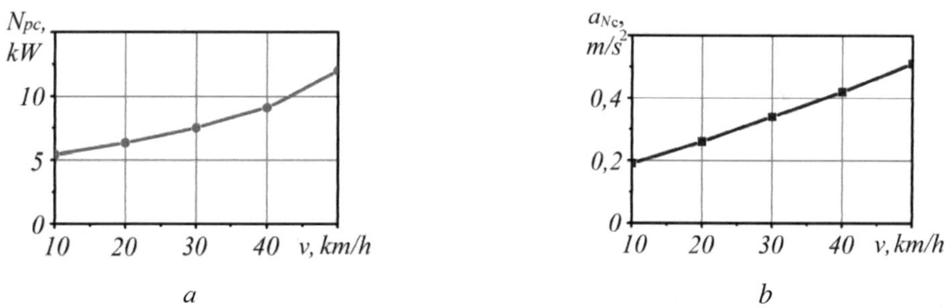

Figure 5. Influence of the speed v of movement of a timber hauling tractor with a semi-trailer on the average recoverable power N_{pc} and the average longitudinal acceleration of the semi-trailer a_{Nc}.

The geometrical parameters of the recuperative hydraulic cylinders in the fifth wheel coupling determine the intensity of pumping the working fluid into the recuperative hydraulic system. However, the same effect in the fifth wheel coupling on the efficiency of damping adverse oscillations of the semi-trailer relative to the tractor. In order to study the effect of the diameter of the recuperative hydraulic cylinders *D*, a series of nine computer experiments were carried out in which the parameter *D* was changed from 20 to 100 mm in 10 mm increments. It was found that the optimal diameter of the hydraulic cylinder is 40 ... 50 mm figure 6. At the same time, the recoverable power is quite high – about 6,5 kW, – and a rather low average longitudinal acceleration is about 0,27 m/s^2. If the diameter of the hydraulic cylinders (*D* is less than 30 mm) decreases too much, the volume of fluid pumped into the pneumatic-hydraulic accumulator with each oscillation of the semi-trailer relative to the tractor decreases, so the recovered power decreases figure 6, *a*. Also, due to insufficient effective damping, the amplitude and sharpness of unfavorable oscillations of the semi-trailer increase, which is reflected in the growth of the average acceleration figure 6, *b*. With an excessive increase in the diameter of the hydraulic cylinders (*D* over 60 mm), the fifth wheel coupling has a very high effective damping, which leads to the fact that the tractor and the semi-trailer are too tightly coupled. Because of this, the amplitude of movement of the

IOP Conf. Series: Materials Science and Engineering **656** (2019) 012039 doi:10.1088/1757-899X/656/1/012039

pistons of the hydraulic cylinders is reduced, and the flow of fluid from the hydraulic cylinder to the pneumohydraulic accumulator also decreases, figure 6, *a*. In addition, due to too high stiffness of the fifth wheel coupling, the movements of the mechanical system become sharper and the average acceleration of the semi-trailer figure 6, *b* increases.

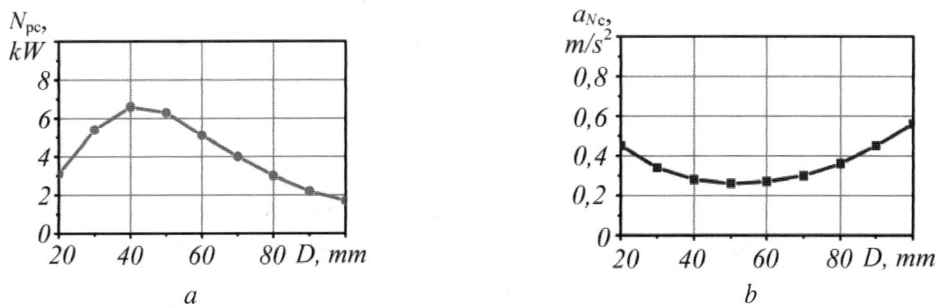

Figure 6. Influence of the diameter of *D* recuperative hydraulic cylinders in the fifth wheel coupling on the average recoverable power N_{pc} and the average longitudinal acceleration of the semi-trailer a_{Nc}.

4. Conclusion

Thus, a mathematical model has been developed, implemented as a computer program, which makes it possible to evaluate the possibility of equipping the fifth wheel coupling system with energy recovery in various driving modes and to study the effect of the parameters of the recovery system on its efficiency.

References

[1] Avaritsioti E 2016 *Transportation Research Procedia* **14** 1003–1012
[2] Jia Y, Li S and Shi Y 2018 *Energies* **11** (11) 3126
[3] Wang R, Jiang Q, Ye Q, Chen L and Meng X 2017 Nongye Jixie Xuebao *Transactions of the Chinese Society for Agricultural Machinery* **48** (8) 350–357
[4] Chen L, Zhang C, Wang R, Ye Q and Niu L 2017 Nongye Jixie Xuebao *Transactions of the Chinese Society for Agricultural Machinery* **48** (1) 303–308
[5] Hui J-Z, Yang Y-K, Zheng H-Y, Zhang Z-Y and Han W-P 2016 Zhongguo Gonglu Xuebao *China Journal of Highway and Transport* **29** (7) 143–151
[6] Wang F and Wu Y-H 2013 *Advanced Materials Research* **605–607** 384–387

Improving the efficiency of a vapor refrigerating machine with a screw compressor while adjusting the performance

A N Noskov[1,3], D S Tarasenkov[1,3], S I Arendateleva[2,4] and I M Zawadzki[2]

[1]ITMO University. 191002, Russia, St. Petersburg, Lomonosov str., 9
[2]Yaroslav-the-Wise Novgorod State University, ul. B. St. Petersburgskaya, 41 173003 Veliky Novgorod, Russia

[3]E-mail: dn.noskoff@rambler.ru
[4]E-mail: Svetlana.Arendateleva@novsu.ru

Abstract. When using an air condenser in a refrigeration machine, the condensation temperature significantly depends on the parameters of the outside air. To ensure stable operation of the throttling device, the condensation temperature must be maintained and remain no lower than a certain value.This leads to unnecessary energy consumption for steam compression in the compressor. The article presents a scheme of a refrigerating machine with a screw compressor which allows to reduce energy consumption for cold production by reducing the condensation pressure when the ambient temperature drops and installing a pump in front of a throttle device. Operation of the refrigerating machine with a screw oil-filled compressor with full capacity on different modes has peculiarities. The article describes the regulator of the screw compressor, which consists of a spool and two butterfly valves. This design allows for adjustment of the geometric degree of compression with full performance and allows to get arbitrary laws of change with decreasing performance. The article presents the dependences of cooling capacity, effective power and coefficient of performance at full and 50% performance of a screw compressor with a decrease in condensation temperature. Experimental and calculated characteristics of a refrigeration oil-filled screw compressor operating on Freon *R*22 were used. Using a scheme of a vapor refrigerating machine with a screw compressor pump in front of the throttle valve can significantly increase the energy efficiency of its work.

1. Introduction

Nowadays, oil-filled screw compressors (OSC) are widely used in vapor refrigerating machines. When the refrigerating machine with an air condenser is operating, the condensation temperature and pressure is determined by the outside air temperature and, as it is lowered, the machine's operating efficiency rises. Under reduced condensation pressure, unstable operation of thermostatic valves (TV) occurs, which leads to an uneven supply of liquid refrigerant to the evaporator. In order to ensure the stable operation of thermostatic valves, the temperature and pressure of condensation is maintained not lower than a certain value. Danfoss recommends maintaining the condensation temperature in a medium-temperature refrigerating machine at a temperature not lower than 30°C, and in a low-temperature refrigerating machine not lower than 20°C. Thus, when compressing steam in the compressor, extra work is expended.

In work [1] operation of the refrigerating machine is shown, the scheme of which is shown in figure 1. A pump is installed between the condenser and the control valve, it increases the pressure of the liquid refrigerant to a value corresponding to the saturation pressure of the refrigerant at a temperature of 30°C, which ensures stable operation of the control valve. When the condensing temperature decreases, the pressure in the condenser decreases, which leads to an increase in the efficiency of the refrigerating machine. The work expended by the pump on increasing the pressure of the liquid refrigerant is much less than the additional work expended on compressing the steam in the compressor. Work under this scheme allows for maintaining the pressure necessary for normal operation of the thermostatic valves when the outside air temperature decreases. The article provides an assessment of the effectiveness of the operation of the refrigeration machine according to this scheme with a scroll and a piston compressor.

2. Materials and methods

Operation of the refrigerating machine with a screw oil-filled compressor with full capacity on different modes has peculiarities. A significant advantage of the OSC is the ability to control the performance using internal devices. When adjusting the performance of one spool together with a decrease in performance, the geometrical degree of compression ε_Γ decreases, which leads to an increase in work losses associated with the lack of refrigerant vapor [2].

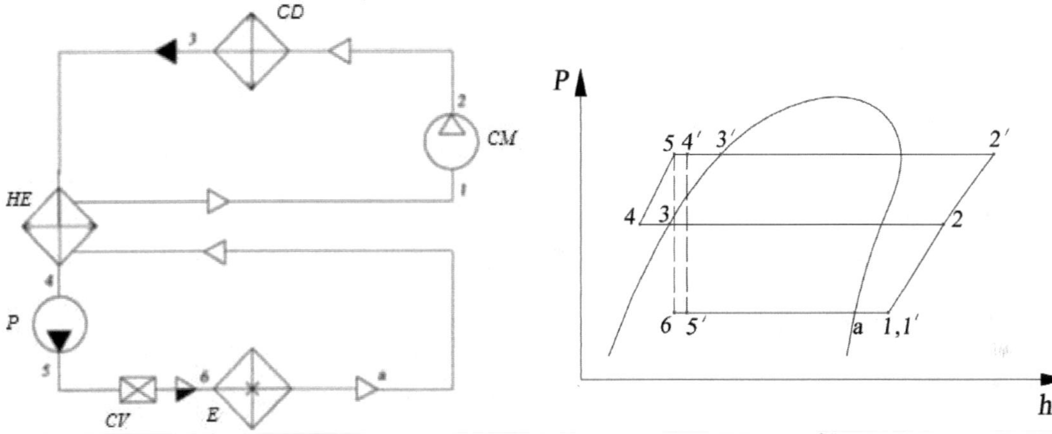

a) b)

Figure 1. Scheme (a) and the operation cycle (b) of a vapor refrigerating machine *CD* – condenser; *HE* – heat exchanger; *P* – pump; *CV* – control valve; E – evaporator; CM – compressor; the operation cycle of the refrigerating machine without a pump $1'-2'-3'-4'-5'$; cycle of operation of the refrigerating machine with a pump $1-2-3-4-5-6$.

In another work [3], a regulator consisting of a spool and two butterfly valves was suggested. This design of the regulator allows for changing the geometric degree of compression ε_Γ with full performance and get arbitrary laws of change of ε_Γ with decreasing performance.

To assess the performance of a vapor refrigerating machine with a pump in front of a control valve, it is necessary to calculate the characteristics of a OSC with a full and 50% capacity.

For this, experimental and calculated dependencies of cooling capacity, effective power and coefficient of performance of compressors operating on Freon *R*22 were determined. Experimental characteristics were obtained at the Department of Low-Temperature Engineering and Renewable Energy of the ITMO University [3].

The experimental compressor has the following geometrical characteristics: the number of settings of the leading screw (LS) and the driven screw (DS) is 4/6; outer diameters of the LS and DS screws is 160 mm; screw length is 144 mm; stroke of the screw is 144 mm; screw rotation speed is 2940 r/m;

theoretical volume capacity is 5,1 m^3/min; profiles of the teeth of the screws are of the standard range. Performance regulation was carried out by a spool and butterfly valves [3].

Calculations were made for a screw compressor having the following geometrical characteristics: the number of settings of LS and DS is 5/6; outer diameter of the LS is 137.5 mm; outer diameter of the DS is 107.6 mm; stroke of the LS is 215 mm; screw length is 200 mm; screw rotation speed is 2940 r/m; theoretical volumetric capacity is 4.8 m^3/min; the profiles of the teeth of the screws are made according to [4]. In calculating the effective efficiency, the dependencies given in [2, 3, 4, 5, 6] were used.

In Figure 2 the dependences of the effective efficiency η_e of compressors on the external degree of pressure increase π_H when operating at full capacity and relative performance $V_{Ti}/V_T = 0,5$ at the boiling temperature of the refrigerant $t_0 = -20°C$ are shown.

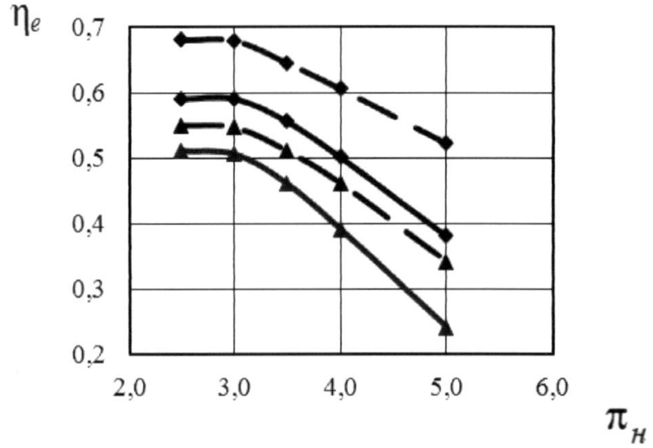

Figure 2. The dependence of the effective efficiency η_e of the screw compressor on the external compression ratio π_H.
– $V_{Ti}/V_T = 1,0$; – $V_{Ti}/V_T = 0,5$;
—— OSC by size range (experiment)
– – – OSC with the ratio of the number of teeth 5/6 R 22 (calculation)

To determine the law of change in the current value of the geometric compression ratio εGi from the relative volumetric capacity of the compressor V_{Ti}/V_T, at which the values of the effective efficiency η_e have the greatest values, an experimental study of the OSC with different end discharge windows that were changed using interchangeable inserts was conducted.

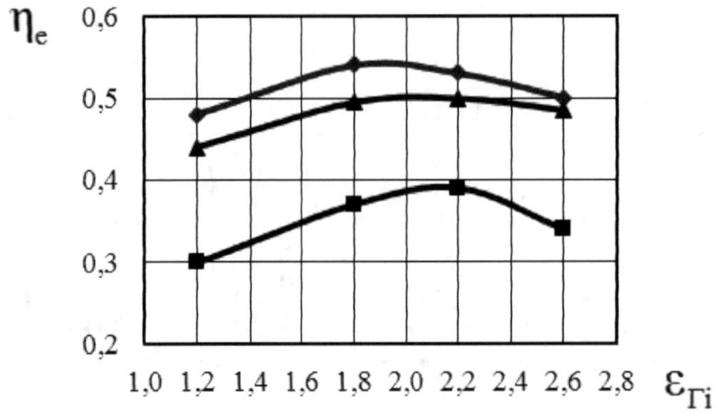

Figure 3. Dependence of the effective efficiency η_e of an experimental screw compressor on the geometric compression ratio ε_{Gi}. $-\pi_N = 2,5$; $-\pi_N = 3,0$; $-\pi_N = 4,0$.

Figure 3 shows the dependences of η_e on $\varepsilon_{\Gamma i}$ with relative performance $V_{Ti}/V_T = 0,5$ for various external degrees of compression and boiling point $t_0 = -20°C$. From the examination of the figures it can be seen that the curves shown have maxima, the position and absolute value of which depend on the external compression ratio π_H. When π_H =2,5…4 , the largest values of η_{Se} occur at $\varepsilon_{\Gamma i}$ = 2,0 ... 2,2. The dependence $\eta_e = f(\varepsilon_{\Gamma i})$ is shown in Figure 1 in [10] and is determined by the line $f - q$.

In figure 4, 5, 6 shows the dependences of the cooling capacity Q_0, the effective power N_e and the cooling coefficient of the cooling machine with OSC on the condensation temperature t_b with full and partial performance. Boiling point t_0 = –20°C.

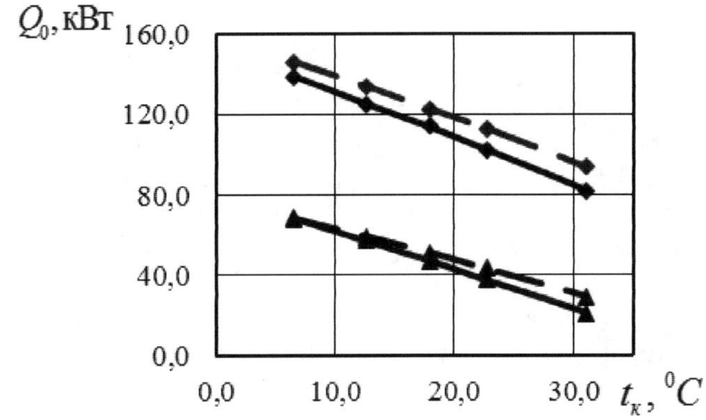

Figure 4. Dependency of refrigerating capacity of compressor Q_0 on the boiling temperature t_b. Designations are the same as in figure 2.

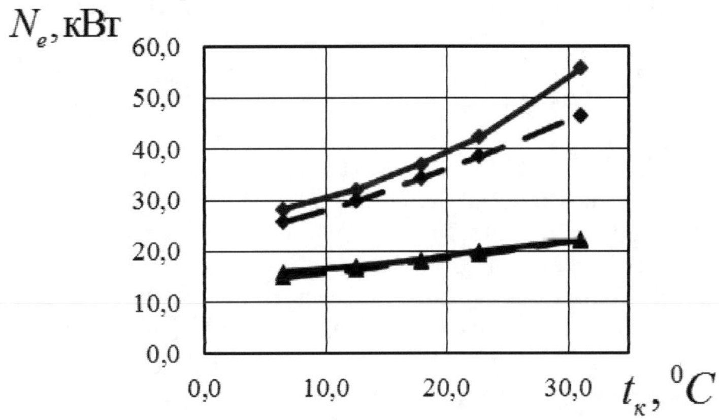

Figure 5. Dependency of effective power of compressor N_e on the boiling temperature t_b. Designations are the same as in figure 2.

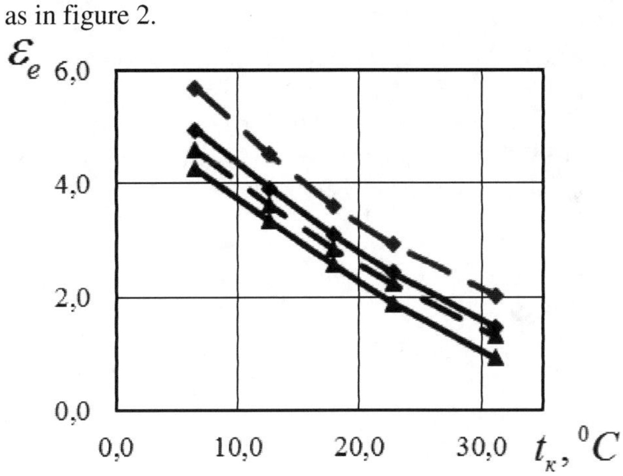

Figure 6. Dependency of refrigerating coefficient of compressor ε_e on the boiling temperature t_b. Designations are the same as in figure 2.

3. Discussion of the research results

From the presented dependences it is clear that the refrigeration coefficient of a vapor refrigerating machine with a screw compressor in full and partial capacity increases while maintaining the condensation temperature in accordance with the outdoor air temperature.

When the condensation temperature decreases from t_c = 31°C to t_c = 12,6°C, the refrigeration coefficient increases in full capacity from ε_e =1,48 to ε_e = 3,9 for the experimental compressor and from ε_e = 2,02 to ε_e = 4,5 according to the calculation results, and at 50% capacity from ε_e = 0,93 to ε_e = 3,34 for the experimental compressor and from ε_e = 1,32 to ε_e = 3,62 as calculated.

The increase in the refrigeration coefficient of a vapor refrigerating machine occurs due to a decrease in the operation of the compressor, since the outlay of work for increasing the fluid pressure by the pump is significantly less than the outlay of compressing steam. The magnitude of the pump does not exceed 2.5% of the magnitude of the compressor.

Thus, the use of a circuit with an oil-filled screw compressor and pump leads to an increase in the operating efficiency of the vapor refrigerating machine with full and partial performance while reducing the outside air temperature.

References

[1] Zherdev A A, Kolesnikov A S and Frolov Y D 2010 Cycle of vapor compression but refrigeration machines with floating pressure condensation *Vestnik Bauman MSTU* “Mashinostroenie” Iss. **1–2** p 137–146

[2] Sakun I A 1970 *Screw compressors* (M.–L.: Mechanical Engineering) p 400

[3] Stosic N, Kovacevic A and Smith I K 1998 Modelling of Screw Compressor Capacity Control *International Engineering Conference* 1307 (http://docs.lib.purdue.edu/icec/1307)

[4] Noskov A N 1998 Pat. 2109170 Rossii, MKI F 04 S 18/16. *Zubchatoe zaceplenie vintovogo kompressora* [Gearing screw compressor]

[5] Cavallini A 1995 Working Fluids for Mechanical Refrigeration *Proc. 19 th Int. Congress of Refrig* (The Hague, The Netherlands) pp 25–42

[6] Noskov A N and Zimkov A A 2013 Raschet processa vsasyvaniya maslozapolnennogo holodil'nogo vintovogo kompressora [Calculation of the process of absorption of oil-filled refrigeration screw compressor] *Holodil'naya tekhnika i kondicioniro-vanie* **1**

[7] Lazarev V L 2005 Entropy approach to the organization of monitoring and management *WPI. WOUND. Theory and control systems* **6** 61–68

[8] Stosic 2004 Screw Compressors in Refrigeration and Air Conditioning *Int Journal of HVACR Research* **10** (3) 233–263

[9] Bubeck S 2015 Convex optimization: algorithms and complexity *In Foundations and Trends in Machine Learning* Vol 8 **3–4** 231–357

[10] Tang Y, Fleming J S and Anderson H 1994 Optimisation Techniques Applied to the Design of a Refrigeration TiN Screw Compressor *International Compressor Engineering Conference* 1049 (http://docs.lib.purdue.edu/icec/1049)

Estimation of energy secrecy of low-frequency satellite communication system with arbitrary number of spatial diversity reception of signals

V P Pashintsev[1,2], A F Chipiga[1,3], D V Kostyuk[1,4] and R M Mikhajlichenko[1,5]

[1]North Caucasus Federal University, Stavropol, Russian Federation

[2]E-mail: pashitsevp@mail.ru
[3]E-mail: chipiga.alexander@gmail.com
[4]E-mail: dkostiuk@ncfu.ru
[5]E-mail: furiousfaind@gmail.com

Abstract. The imitational model for the low-frequency satellite communication system has been developed using Simulink Ver. 8.2 (MATLAB) environment. The imitational modeling has been carried out for the signal reception with carrier frequencies in $f_0 \approx 30$ MHz...1 GHz range and with both single antenna (*n=1*) and diversified (*n=2, 4, 8* antennas) signal reception. On this basis, the dependency of energetic concealment coefficient on the carrier frequency selection has been determined. The analysis of the obtained dependencies shows that in the lowered frequency range ($f_0 \approx$ 30 MHz ... 100 MHz) it is possible to increase the energetic concealment coefficient by $\gamma_{e(2)} \approx 24$ dB when using the dual antenna reception system and up to $\gamma_{e(8)} \approx 42$ dB when the signal is received by 8 antennas.

1. Introduction

There is a known method [1, 2] to achieve a high energetic concealment coefficient ($\gamma_e \approx 22...34$ dB) for the satellite communication systems (SCS) with the radio intercepting receiver in close proximity to the ground-based SCS receiver. This method implies the application of lowered carrier frequency ($f_0 = 30$ ... 100 MHz) and diversified signal reception with several ($n = 2...4$) antennas. However, such high energetic concealment coefficients (ECc) can only be achieved with the carrier frequency value of $f_0 = 30$ MHz under the conditions of the deepest (Rayleigh) fading of the received signals (when the Rice parameter $\gamma^2 = 0$) and when the fading is uncorrelated (when the correlation coefficient $R = 0$) on the outputs of the diversified antennas.

In works [3, 4] for the case when diversified reception with dual antennas ($n = 2$) is applied in the low-frequency SCS, the analytical dependency $\gamma_{e(2)} = \Psi(P_{er\,alwd}, \gamma^2, R, n = 2)$ has been obtained based on the following dependencies for the ECc: on the allowed signal reception error rate ($P_{er\,alwd}$), on the Rician fading parameter of the received signal ($0 \le \gamma^2 < \infty$) and on the fading correlation coefficient ($0 \le R \le \infty$) on the outputs of the diversified antennas. Aside from that, the dependencies of the Rice parameter $\gamma^2 = \Psi(f_0)$ and the fading correlation coefficient $R = \Psi(f_0)$ on the carrier frequency selection in the transionospheric (satellite) communication link has been obtained. Through these dependencies, another dependency $\gamma_{e(2)} = \Psi(f_0, P_{er\,alwd}, n = 2)$ has been determined. It is the dependency of the ECc on the carrier frequency selection with the determined $P_{er\ alwd}$ value and the diversified signal reception with $n=2$ antennas.

It is obvious that the application of spatially diversified signal reception with an increased number ($n>2$) of antennas allows to significantly boost the ECc ($\gamma_{e(n)}$). However, when the multiplier of diversification is more than two ($n>2$), it is impossible to obtain the analytical dependency $\gamma_{e(n)} = \Psi(f_0, P_{er\ alwd}, n > 2)$ due to the following reasons.

Known analytical dependencies $P_{er(n)} = \Psi(h^0, \gamma^2, R, n)$ are limited by the cases where only singular ($n=1$) or dual ($n=2$) reception is applied. There are no known analytical dependencies $P_{er(n)} = \Psi(h^2, \gamma^2, R, n)$ for the cases of increased diversification multiplicity ($n>2$). Hence, a conclusion follows that it is useful to utilize the capabilities of imitational modeling to compose the dependencies $P_{er(n)} = \Psi(h^2, \gamma^2, R, n)$ that can lead to the obtainment of the desired dependency $\gamma_{e(n)} = \Psi(f_0, P_{er\ alwd}, n > 2)$.

The purpose of this research is to develop the imitational model for the low-frequency SCS to obtain the dependencies $P_{er(n)} = \Psi(h^2, \gamma^2, R, n \geq 1)$ of the erroneous reception of the signals with the Rician ($0 \leq \gamma^2 < \infty$) correlated ($0 \leq R \leq \infty$) fading with an arbitrary number ($n \geq 2$) of the receiving antennas, and, based on the dependency $\gamma_{e(n)} = \Psi(f_0, P_{er\ alwd}, n > 2)$, to calculate of the SCS' energetic concealment coefficient on the carrier frequency selection f_0 when the antenna diversification multiplicity is more than two ($n > 2$).

2. The imitational model for the low-frequency satellite communication system with the spatially diversified signal reception system

The energetic gain when utilizing the spatially diverse reception system depends on the addition scheme of the diversification branches. Judging by the efficiency and the simplicity of realization, the average position is taken by the scheme of coherent addition of the diversification branches with incoherent processing of the resulting signal [5]. The imitational model for the low-frequency SCS with diversified signal reception with two ($n=2$) diversified antennas with specified addition scheme can be realized in Simulink Ver 8.2 (MATLAB) environment, as shown on figure 1 [6, 7]. The imitational model for the low-frequency SCS consists of the transmitter, transionospheric communication link, branch addition scheme and the receiver with the error rate calculator (ER).

In the transmitter, the pseudorandom sequence generator (The Random Integer block) forms the initial sequence of the information symbols $x(t)$ with the alphabet volume (code base) $M=2$, which is transformed into the orthogonal binary frequency-manipulated signal $s(t)$ by the modulator (the 2-FSK block).

Further, the signal $s(t)$ passes through the communication link, where the Rayleigh SISO subsystem (Rayleigh Single Input Single Output) models the Rayleigh fading (when the Rice parameter $\gamma^2=0$) and where the signal splits to two ($n=2$) channel signals $s_i(t)$ ($i=1, 2$) without considering the fading correlation ($R=0$) [8]. Then AWGN blocks add an additive white Gaussian noise (AWGN) with adjustable spectral density of the power N_{0i} to the signals $s_i(t)$.

Afterwards, the fluctuations $y_i(t) = s_i(t) + \xi_i(t)$ proceed to the branch coherent addition scheme. In the given scheme, the process of "weighting" (multiplying) of the received fluctuations $y_i(t)$ proportionally to the expected signal amplitude $s_i(t)$, which is taken into account by the transmission coefficient (TC) $\mu_i(t) = s_i(t)/s(t)$, takes place [5]. Further, the signals formed on the outputs of the multipliers $v_i(t) = y_i(t) \cdot \mu_i(t)$ are transmitted to the adder and the received result $\sum_{i=1}^{n} v_i(t)$ arrives to the input of the demodulator (the 2-FSK block).

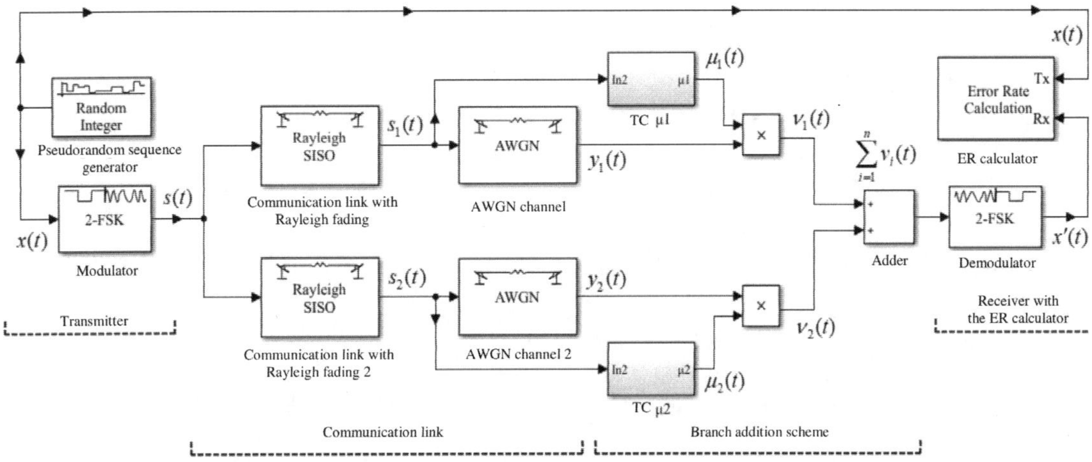

Figure 1. The imitational model for the low-frequency SCS with the diversified reception system utilizing two $(n=2)$ antennas, Rayleigh $(\gamma^2=0)$ uncorrelated $(R=0)$ fading, coherent branch addition and incoherent processing of the resulting signal.

In the receiver the demodulator produces an incoherent detection of the signal sum $\sum_{i=1}^{n} v_i(t)$ and forms the information symbol sequence $x'(t)$ on the output. The error rate calculator (the Error Rate Calculation block) calculates the number of erroneously transmitted symbols based on the comparison of the received $x'(t)$ and the transmitted $x(t)$ data flow and outputs the error rate value P_{er}.

The imitational modeling is carried out using the built-in MATLAB application – BER Error Rate Analysis Tool, which allows to obtain dependencies $P_{er}=\psi(h^2,n=2)$ [9] with Rayleigh $(\gamma^2=0)$ uncorrelated $(R=0)$ fading.

However, specified imitational model for the low-frequency SCS does not allow to obtain a dependency $P_{\mathrm{er}(n)}=\psi(h^2,\gamma^2,R,n\geq 2)$ of the erroneous reception of the signals with the Rician $(0\leq\gamma^2<\infty)$ correlated $(0\leq R\leq 1)$ fading as the model uses two Rayleigh SISO Blocks representing communication links with one input and one output, where Rayleigh $(\gamma^2=0)$ uncorrelated $(R=0)$ fading takes place [8].

To overcome this limitation, the authors propose to use the approach [10, 11], which is applied in the imitational model (figure 2) for the SCS with using adaptive coding method called OSTBC (Orthogonal Space-Time Block Coding).

Specified imitational model includes the transmitter and the receiver with the ER calculator, which operate similarly to figure 1. The difference is the use of another type of digital modulation – QPSK (Quadrature Phase Shift Keying).

The transmitter and receiver include the OSTBC Encoder and OSTBC Decoder blocks, which simulate the OSTBC encoding algorithms. This encoding is intended for the SCS using MIMO (Multiple Input Multiple Output). Principle of OSTBC [12] is to divide the sequence of information symbols into blocks and transmit each block at set time intervals over spatially diversified antennas in accordance with the coding matrix. Reception is performed on spatially diversified antennas, followed by the addition of all signals. Decoding is provided through the use of a special structure of the transmitted units. The simplest and most common example is the OSTBC Alamouti circuit which is intended for using two transmitting and one receiving antennas.

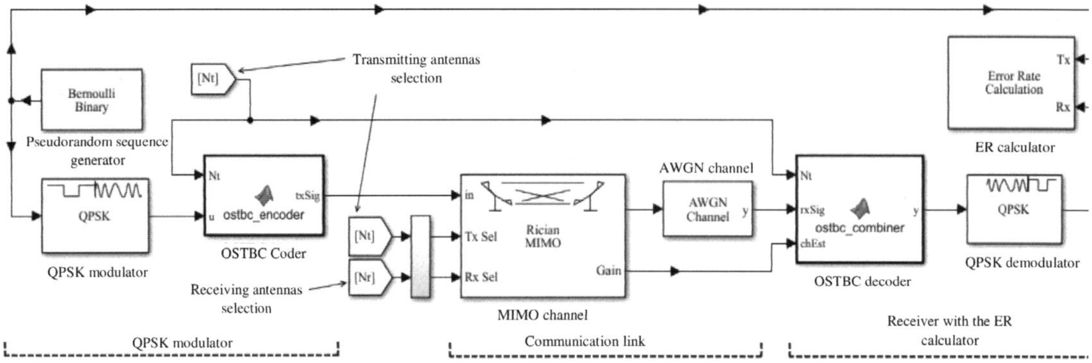

Figure 2. The imitational model for the SCS with using OSTBC.

In the communication channel segment the Rician MIMO block is used (figure 2), which is included in the class of MIMO Fading Channel (Multiple Input Multiple Output Fading Channel) blocks, which allows modeling the communication channel with the required Rician $(0 \le \gamma^2 < \infty)$ correlated $(0 \le R \le 1)$ fading and with diversity for arbitrary number of transmitting and receiving antennas $(\mathrm{Nt} \ge 2, \mathrm{Nr} = n \ge 2)$ [13].

Figure 3 shows the imitational model for the low-frequency SCS, which includes the transmitter, communication link, branch addition scheme and the receiver with the ER calculator. Specified model has been developed using Simulink Ver. 8.2 (MATLAB) environment.

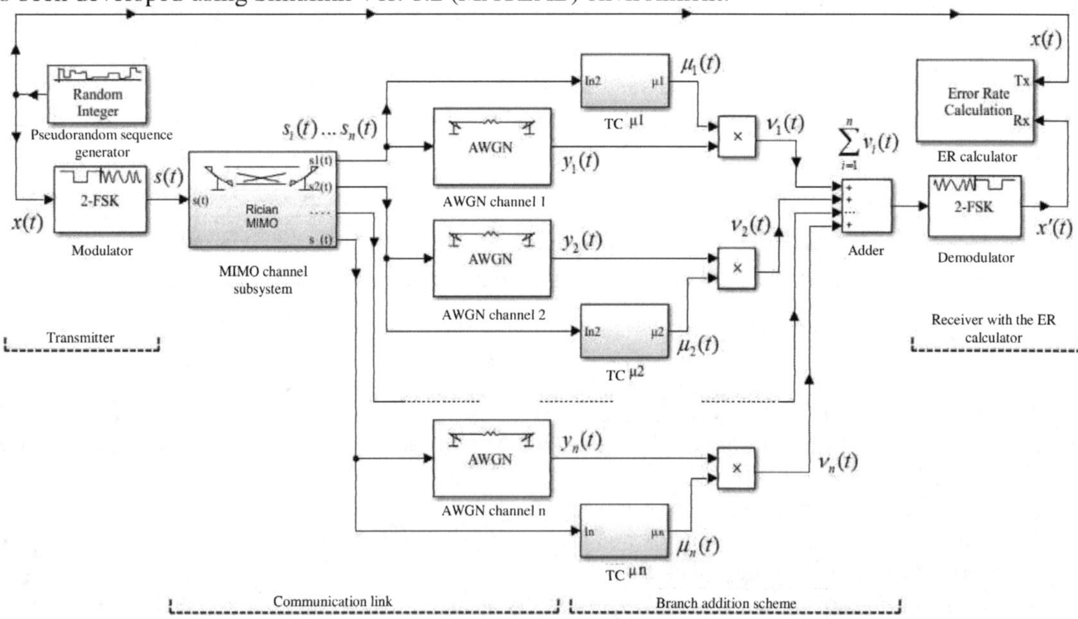

Figure 3. The imitational model for the low-frequency SCS with the diversified reception system utilizing an arbitrary number $(n \ge 2)$ of the antennas, Rician $(0 \le \gamma^2 < \infty)$ correlated $(0 \le R \le 1)$ fading, coherent branch addition scheme and incoherent processing of the resulting signal.

The segments of the transmitter, the scheme of addition of branches and of the receiver with the with the ER calculator are similar to figure 1. The novelty of the imitational model is the algorithm of the communication link segment functioning in conjunction with the other segments, which is described below.

Thus, combining the approach in the modeling of communication channels with fading and diversity (figure 2) with coherent branch addition scheme and incoherent processing of the resulting signal (figure 1), it becomes possible to get a dependency $P_{er(n)} = \Psi(h^2, \gamma^2, R, n\geq 2)$ of the erroneous reception of the signals with the Rician $(0 \leq \gamma^2 < \infty)$ correlated $(0 \leq R \leq 1)$ fading.

To confirm the functioning authenticity of the SCS imitational model, the error rate calculation P_{er} has been carried out with the spatially diversified reception based on one $(n=1)$ and four $(n=4)$ receiving antennas without the correlative fading $(R=0)$ for the two values of the Rice parameter: $\gamma^2 = 0$ and $\gamma^2 = 2$.

For statistically homogenous parallel communication links with the Rician $(0 \leq \gamma^2 < \infty)$ uncorrelated $(R=0)$ fading with the singular $(n=1)$ and the spatially diversified $(n \geq 2)$ reception of the binary equiprobable orthogonal signals and the diversification branch coherent addition scheme with incoherent processing of the resulting signal, the equation for the error rate P_{er} is determined as [5].

$$P_{er}(h^2,\gamma^2,n) = \frac{1}{2}\prod_{i=1}^{n}\left[\frac{2(1+\gamma^2)}{2(1+\gamma^2)+h^2}\exp\left(-\frac{\gamma^2 h^2}{2(1+\gamma^2)+h^2}\right)\right]. \quad (1)$$

The results of theoretical calculation (solid lines) and imitational modeling (dots) based on the formulas (1) with the Rayleigh $(\gamma^2 = 0)$ and the Rician $(\gamma^2 = 2)$ fading are presented on figures 4a and 4b respectively.

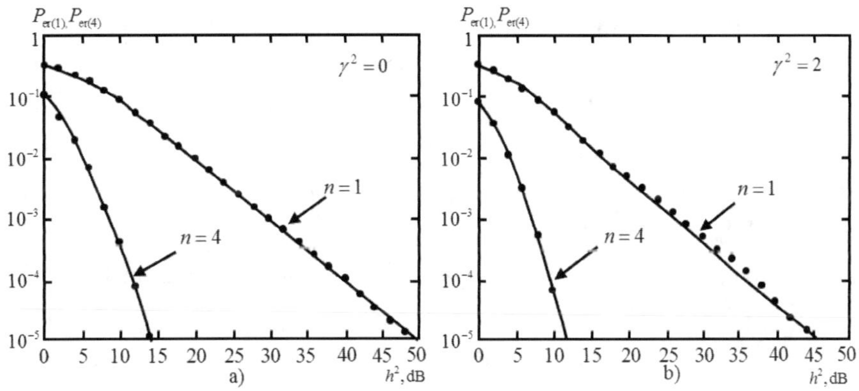

Figure 4. The results of theoretical calculation (solid lines) and imitational modeling (dots).

As observed on figure 4, it is evident that the results of imitational modeling match the theoretical calculation results fairly accurately, which tells about the reliability of the developed imitational model for the low-frequency SCS (figure 3).

According to figure 4a, to achieve the desired error rate $P_{er\ alwd} = 10^{-5}$ with the Rayleigh fading $(\gamma^2 = 0)$, it is required to provide the acceptable S/N ratio: $h^2_{alwd\ (1)} \approx 50$ dB for the singular $(n=1)$ reception and just $h^2_{alwd\ (4)} \approx 14$ dB for the quadruple $(n=4)$ reception without the correlation of fading in the antennas $(R=0)$. According to figure 4b, to achieve $P_{er\ alwd} = 10^{-5}$, it is required to provide the acceptable S/N ratio of $h^2_{alwd\ (1)} \approx 46$ dB and $h^2_{alwd\ (4)} \approx 12$ dB respectively.

3. The baseline parameters of imitational modeling for the low-frequency satellite communication system

Using the developed imitational model (figure 3), it is possible to build the graphs for desired dependencies $P_{er} = \Psi(h_2, \gamma_2, R, n\geq 1)$. Based on these graphs, it is possible to find the acceptable achieve $P_{er} = P_{er\,alwd} = 10^{-10}$) S/N ratio on the receiver's input with the singular ($n = 1$) reception $h^2_{alwd\,(1)}$ and with the diversified reception $h^2_{alwd\,(n)}$ of the signal with an arbitrary number ($n \geq 2$) of the antennas. These S/N ratios are the baseline to obtain the energetic concealment coefficient $\gamma_e = h^2_{alwd\,(1)}/h^2_{alwd\,(n)}$.

However, to obtain the desired dependency $\gamma_e = \Psi(f_0, P_{er\,alwd}, n > 2)$ of the SCS' ECc on the carrier frequency f_0 selection with the antenna diversification multiplicity $n > 2$, it is required to consider the known [14–16] frequency dependencies of the Rice parameter $\gamma^2 = \Psi(f_0)$ and of the fading spatial correlation coefficient $R = \Psi(f_0)$ in the transionospheric communication link (CL).

It is known [14] that the Rician fading parameter, which characterizes the ratio of the regular component P_r of the received signal's power to the fluctuating P_{fl} component's power, depends only on the dispersion of phasic wave front's fluctuations σ^2_φ on the exit of the ionosphere in the transionospheric CL as

$$\gamma^2 = P_r/P_{fl} = \left[\exp(\sigma^2_\varphi) - 1\right]^{-1}. \quad (2)$$

The standard deviation (SD) of phasic wave front's fluctuations for the emerging wave is described by the expression

$$\sigma_\varphi = \sqrt{2}(80{,}8\,\pi/c)\sqrt{l_s z_e \sec\alpha}\,(\sigma_{\Delta N}/f_0)\ \text{[rad]}, \quad (3)$$

where 80,8 is a coefficient with the dimension [m^3/s^2]; $l_s \approx 390\,m$ is a characteristic (average) size of the small-scale ionospheric inhomogeneities; $z_e = 5\cdot 10^5\,m$ is an equivalent depth of the homogenous ionospheric layer with the electron concentration (EC) equal to its average value on the height h_m of maximum ionospheric ionization $\overline{N}(h_m) = \overline{N}_m$; f_0 is a carrier frequency in [Hz]; α is a zenith angle.

Since in the mid-latitude ionosphere the typical values for maximum average EC are $\overline{N}_m \approx 6\cdot 10^{11}$ el/m^3 and for intensity of the small-scale ionospheric inhomogeneities are $\beta = \sigma_{\Delta N}/\overline{N}_m \approx 3\cdot 10^{-3}$, the SD of EC fluctuations in the small-scale ionospheric inhomogeneities on the altitude of maximum ionospheric ionization is approximately equal to $\sigma_{\Delta N} \equiv \sigma_{\Delta N}(h_m) = \beta\overline{N}_m \approx 2\cdot 10^9$ el/m^3.

The fading spatial correlation coefficient in the antennas diversified with the distance of $\Delta\rho_a$ is also determined by the SD of phasic wave front fluctuations σ_φ (3) on the ionosphere's exit [15, 16]

$$R = \exp\left[-(\Delta\rho_a/\Delta\rho_k)^2\right] = \exp\left[-(\Delta\rho_a\sigma_\varphi/l_s)^2\right] \quad (4)$$

where $\Delta\rho_k = l_s/\sigma_\varphi$ - signal fading spatial correlation interval.

Let us consider that the receiving antennas (from $n = 1$ to $n = 8$) are positioned on a single straight line with the constant diversification interval $\Delta\rho_a = const$. Let us select the value of latter to be typical for the shortwave communication system with spatially diversified signal reception [17]: $\Delta\rho_a = 300\,m$. It is evident that the correlation of signal fading on the outputs of the 1st and the 2nd (adjacent) antennas $R_{12}(\Delta\rho_a)$ shall be the most pronounced, on the outputs of the 1st and the 3rd antennas it shall be reduced $R_{13}(2\Delta\rho_a) < R_{12}(\Delta\rho_a)$, on the outputs of the 1st and the 4th antennas it shall be even less $R_{14}(3\Delta\rho_a) < R_{13}(2\Delta\rho_a)$, and so on until the outputs of the opposite (1st and n) receiving antennas: $R_{1n}((n-1)\Delta\rho_a) < R_{1(n-1)}((n-2)\Delta\rho_a) < \ldots < R_{12}(\Delta\rho_a)$

In accordance with the typical parameters for mid-latitude ionosphere ($\sigma_{\Delta N} \approx 2\cdot 10^9$ el/m^3; $l_s \approx 390\,m$; $z_e = 5\cdot 10^5\,m$) and with the analytical dependencies (2–4) with transionospheric propagation of the radiowaves with the carrier frequency $f_0 = 30\ldots 1000$ MHz and the zenith angle

$\alpha = 65°$, values presented in table 1 are the values for the SD of phasic wave front fluctuations σ_φ on the exit of the inhomogenous ionosphere, the values for the Rice parameter γ^2, the values for the fading spatial correlation interval $\Delta\rho_k$ and for the fading correlation coefficient R_{1n} on the outputs $n = 1, ..., 8$ of the receiving antennas with $\Delta\rho_a = 300\,m$.

Table 1. The dependency of SCS' fading parameters on the carrier frequency selection.

	1	**2**	**3**	**4**	**5**	**6**	**7**	**8**	**9**
f_0, MHz	30	50	80	100	150	200	300	500	1000
σ_φ, rad	1.71	1.03	0.64	0.51	0.34	0.26	0.17	0.1	0.05
γ^2	0.056	0.53	1.96	3.3	8.02	14.64	33.56	94.1	378
$\Delta\rho_k$, m	227	379	607	758	1138	1517	2276	3793	7586
$R_{12}(\Delta\rho_a)$	0.18	0.53	0.78	0.85	0.93	0.96	0.98	0.99	0.998
$R_{13}(2\Delta\rho_a)$	0	0.08	0.38	0.535	0.76	0.85	0.93	0.98	0.99
$R_{14}(3\Delta\rho_a)$	0	0.003	0.11	0.25	0.54	0.7	0.85	0.95	0.98
$R_{15}(4\Delta\rho_a)$	0	0	0.02	0.08	0.33	0.54	0.76	0.9	0.975
$R_{16}(5\Delta\rho_a)$	0	0	0	0.02	0.18	0.38	0.65	0.86	0.961
$R_{17}(6\Delta\rho_a)$	0	0	0	0	0.08	0.24	0.54	0.8	0.945
$R_{18}(7\Delta\rho_a)$	0	0	0	0	0.03	0.15	0.43	0.74	0.926

The analysis of the obtained values (table 1) for the Rice parameter $\gamma^2 = \psi(f_0)$ and for the fading correlation coefficient $R = \psi(f_0)$ in the transionospheric CL indicates that as the carrier frequency in the SCS decreases from the traditional values $f_0 = 1$ GHz to the smallest $f_0 = 30$ MHz, the Rice parameter decreases from $\gamma^2 = P_r / P_{fl}$ 380 (i.e. there is practically no fading) to $\gamma^2 = P_r / P_{fl}$ 0,05 (the regular component of the fading is practically absent, i.e. the fading is nearly Rayleigh), the fading spatial correlation coefficient decreases from $R \approx 1$ (almost total correlation) to $R \approx 0$ (decorrelation).

4. The imitational modeling and the evaluation of energetic concealment coefficient for the low-frequency satellite communication system

Based on the imitational model (figure 3) and the data from table 1 (γ^2, n, R_{1n}), the imitational modeling has been carried out for the low-frequency SCS' functioning with the transmission of signals via the transionospheric CL with statically homogenous correlated $(0 \le R \le 1)$ Rician $(0 \le \gamma^2 < \infty)$ fading and with the singular $(n = 1)$ and the diversified signal reception with $n = 2, 4, 8$ antennas, the dependency curves have been obtained $P_{er(n)} = \psi(h^2, \gamma^2, R, n \ge 1)$.

As an example, on figure 5 there are dependency graphs $P_{er(8)} = \psi(h^2, \gamma^2, R, n = 8)$ for the evaluation of SCS' noise immunity with the diversified signal reception with $n = 8$ antennas. These graphs are built with various carrier frequencies $f_0 = 30 ... 1000$ MHz according to the data from table 1.

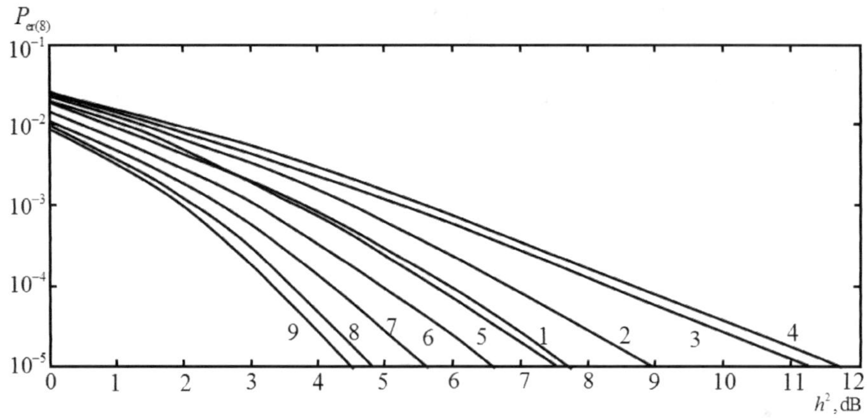

Figure 5. The dependency graphs $P_{er} = \psi(h^2, \gamma^2, R, n = 8)$ for the evaluation of SCS' noise immunity with various carrier frequency values f_0: 1) 30 MHz; 2) 50 MHz; 3) 80 MHz; 4) 100 MHz; 5) 150 MHz; 6) 200 MHz; 7) 300 MHz; 8) 500 MHz; 9) 1000 MHz.

The analysis of figure 5 indicates that with $f_0 = 30$ MHz, the value to achieve the desired $P_{er} = P_{er\ alwd} = 10^{-5}$ S/N ratio is $h^2_{alwd(8)} = 7,7$ dB. As the carrier frequency increases, the noise immunity curves initially shift towards the right up to the maximum value $h^2_{alwd(8)\max} = 11,7$ dB at $f_0 = 100$ MHz, but afterwards shift towards the left until the minimum value $h^2_{alwd(8)\min} = 4,5$ dB at the carrier frequency $f_0 = 1000$ MHz. These changes are described by the fact that, according to table 1, at the frequency of $f_0 = 30$ MHz the fading of received signals in the SCS has a nearly Rayleigh character $(\gamma^2 = P_r / P_{fl} \approx 0,056)$ and is weakly correlated $(R_{12} \approx 0,18)$. At the frequency of $f_0 = 100$ MHz the fading of received signals shall have lesser depth (Rician character $(\gamma^2 = P_r / P_{fl} \approx 3,3)$ and is strongly correlated $(R_{12} \approx 0,85)$). At the frequency of $f_0 = 1000$ MHz the fading of received signals is nearly absent $(\gamma^2 = P_r / P_{fl} \approx 378)$ and is totally correlated $(R_{12} \approx 0,998)$.

Based on these curves (figure 5) the formulation (figure 6) has been found that describes the dependency of acceptable S/N ratio $h^2_{alwd(8)}$ on the SCS receiver's input with the application of $n = 8$ diversified antennas on the Rice parameter γ^2 and the carrier frequency f_0 selection. Similarly, on figure 4 there are functional dependencies built $h^2_{alwd(n)} = \psi(\gamma^2, f_0)$ for the cases when signals are received using $n = 1$ antenna and when the reception is diversified with $n = 2$ and $n = 4$ antennas. The acceptable S/N values $h^2_{alwd(n)}$ depending on the selection of carrier frequency f_0 in the SCS and on the Rice parameter γ^2 of received signals are provided in table 2.

The analysis of figure 6 and of the data from tables 1 and 2 indicates that in the lowered carrier frequency range $f_0 = 80 \ldots 100$ MHz the acceptable S/N values $h^2_{alwd(n)}$ on the receiver's input are maximized with the application of diversified reception with *n = 2, 4, 8* antennas. It is described by the fact that in this range a fairly deep fading of the received signals $(\gamma^2 \approx 2 \ldots 3)$ is observed with a relatively insignificant spatial correlation $(R_{12} \approx 0,8)$.

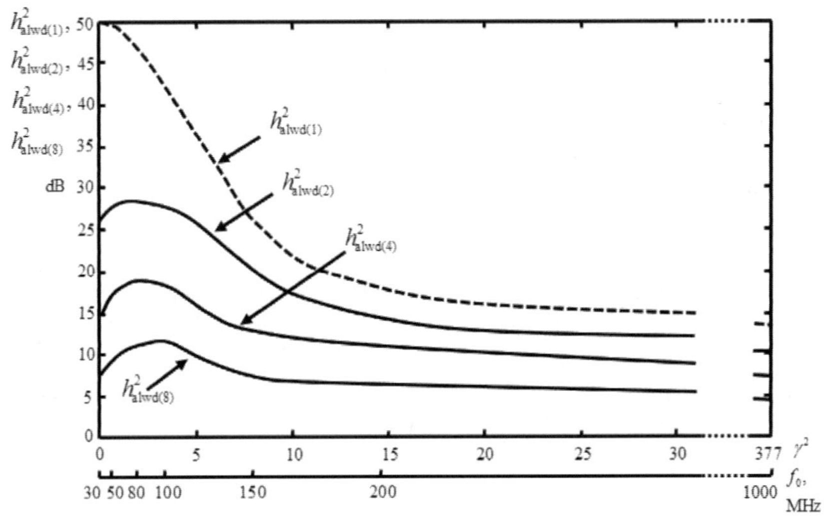

Figure 6. The dependency graphs for the acceptable signal/noise ratios $h^2_{alwd(n)}$ on the SCS receiver's input with the reception based on $n = 1$ antenna and $n = 2, 4, 8$ antennas on the carrier frequency f_0 selection.

Table 2. The dependencies of acceptable S/N ratios on the SCS' carrier frequency selection.

	1	**2**	**3**	**4**	**5**	**6**	**7**	**8**	**9**
f_0 , MHz	30	50	80	100	150	200	300	500	1000
γ^2	0.056	0.53	1.96	3.3	8	14.6	33.6	94.1	378
$h^2_{доп(1)}$, dB	49.99	49.5	46.2	42	26	18	15	13.9	13.5
$h^2_{доп(2)}$, dB	26.2	27.5	28.4	27.7	21	14.7	11.9	10.8	10.4
$h^2_{доп(4)}$, dB	14.5	16.5	18.8	18.3	13	11.2	8.7	7.9	7.5
$h^2_{доп(8)}$, dB	7.7	8.9	11.2	11.7	7.5	6.6	5.6	4.8	4.5
$\gamma_{э(2)}$, dB	23.8	22	17.8	14.3	5	3.3	3.05	3.07	3.08
$\gamma_{э(4)}$, dB	35.5	33	27.4	23.7	13	6.8	6.3	5.97	5.98
$\gamma_{э(8)}$, dB	42.3	40.6	35	30.3	18.5	11.4	9.4	9.1	9

As the carrier frequency of SCS grows up to the traditional values $f_0 = 1000$ MHz, it is possible to observe the decrease in acceptable S/N ratios on the receiver's input down to $h^2_{alwd(n)} \approx 10 \ldots 4$ dB with $n = 2 \ldots 8$ due to the practical absence of fading ($\gamma^2 \approx 378$) and its correlation ($R_{12} \approx 1$) in the diversified antennas. This corresponds to the known [5] conclusion stating that the gain from diversified reception grows with the increase of received signal's fading depth and with the decrease of their spatial correlation, as well as conforms with the conclusion about the unpracticality of applying spatially diversified reception in the communication channels without fading.

Thereby, the obtained (figure 6, table 1) dependencies $P_{\text{er}(n)} = \psi(h^2, \gamma^2, R, n)$ allow to calculate (table 2) the ECc $\gamma_{\text{e}(n)}$, expressed in decibels, as the difference of acceptable S/N ratios: $\gamma_{e(n)}(\gamma^2)_{\text{dB}} = h^2_{\text{alwd}\,(1)}(\gamma^2)_{\text{dB}} - h^2_{\text{alwd}\,(n)}(\gamma^2)_{\text{dB}}$

Based on the results in table 2, presented on figure 7 are the dependency graphs for ECc $\gamma_{e(n)} = \psi(\gamma^2, f_0)$ in the SCS with diversified reception based on $n = 2, 4, 8$ antennas on the Rice parameter γ^2 and on the carrier frequency f_0 selection.

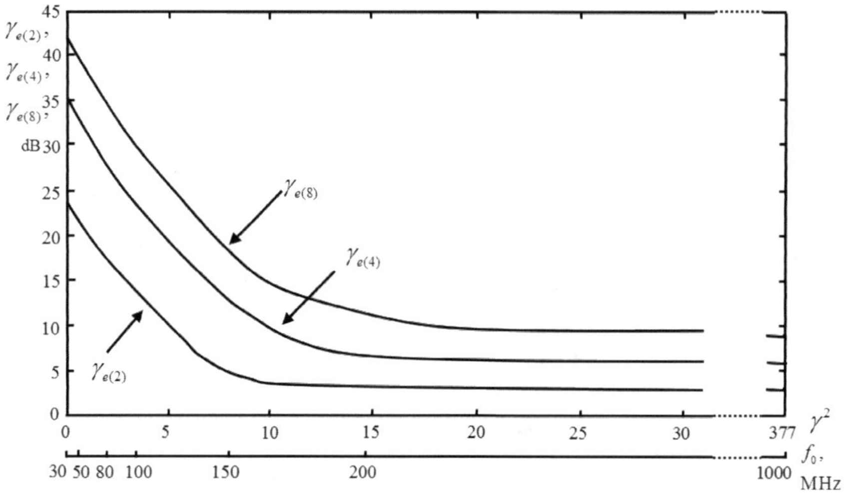

Figure 7. The dependency of SCS' energetic concealment coefficient on the carrier frequency f_0 selection with the diversified signal reception based on $n = 2, 4, 8$ antennas.

5. Conclusion

The imitational model for the low-frequency SCS (fig. 3) with spatially diversified signal reception system with the general Rician fading has been developed in Simulink Ver. 8.2 (MATLAB) graphical modeling environment.

This imitational model allows to obtain the dependency graphs $P_{er(n)} = \psi(h^2, \gamma^2, R, n \geq 1)$ of erroneous signal reception rate with Rician $(0 \leq \gamma^2 < \infty)$ correlated $(0 \leq R \leq 1)$ fading with an arbitrary number $(n \geq 1)$ of the receiving antennas. As an example, presented on figure 5 is a dependency graph $P_{\text{er}(8)} = \psi(h^2, \gamma^2, R, n = 8)$ with $n = 8$ receiving antennas.

Based on the dependencies $P_{er} = \psi(h^2, \gamma^2, R, n)$ and the expressions (2–4), provided on figure 6 are the graphs for the dependency of acceptable signal/noise ratios $h^2_{\text{alwd}(n)}$ on the SCS receiver's input with the singular $(n = 1)$ and the diversified signal reception based on $n = 2, 4, 8$ antennas on the carrier frequency f_0 selection.

Based on the results of figure 6 and table 2, the desired dependencies $\gamma_e = \psi(f_0, P_{\text{er alwd}}, n)$ are obtained $\gamma_e = \psi(f_0, P_{\text{er alwd}}, n)$ and presented on figure 7. Specified dependencies show the dependency of SCS' energetic concealment coefficient on the carrier frequency f_0 selection with the diversified reception based on $n = 2, 4, 8$ antennas.

The analysis of these graphs indicates about the possibility of increasing the energetic concealment coefficient in the low-frequency SCS from $\gamma_{e(2)} \approx 24$ dB when the dual antenna signal reception is applied to $\gamma_{e(4)} \approx 35$ dB with the quad antenna signal reception system and up to $\gamma_{e(8)} \approx 42$ dB with the reception system based on 8 antennas.

References

[1] Chipiga A F and Senokosova A V 2007 Information protection in space communication systems using changes in radio wave propagation conditions *Cosmic Research* Vol 45 **1** 52–59

[2] Chipiga A F and Senokosova A V 2009 A method to ensure energy security of satellite communication systems *Cosmic Research* Vol 47 **5** 393–398

[3] Tsimbal V A, Pashintsev V P and Chipiga A F 2012 Analiticheskaya zavisimost' ehnergeticheskoj skrytnosti sputnikovoj svyazi ot vybora nesushhej chastoty radiovoln *Trudy 18 Mezhdunarodnoj nauchno-tekhnicheskoj konferentsii "Radiolokatsiya, navigatsiya, svyaz'"* (Voronezh) pp 2113–2120

[4] Chipiga A F, Pashintsev V P, Tsymbal V A and Shimanov S N 2016 Procedure for calculating the dependence of the energy concealment factor on carrier frequency selection for low-frequency satellite communications system *Automatic Control and Computer Sciences* **50 (6)** 408–414

[5] Andronov I S and Fink L M 1971 *Peredacha diskretnykh soobshhenij po parallel'nym kanalam* (Moscow: Sovetskoje radio Publ) p 408

[6] Malinkina O V, Natal'in A B and Sergienko A B 2009 *Tsifrovaya svyaz': Metodicheskie ukazaniya k vypolneniyu laboratornykh rabot* (SPb.: SPbGEHTU «LEHTI» Publ) p 44

[7] Giordano A A and Levesque A H 2015 *Modeling of Digital Communications Systems using Simulink* (Hoboken, New Jersey: John Wiley & Sons Publ) p 377

[8] *MathWorks Documentation. SISO Fading Channel* [Electronic resource] 2019 Available at: https://www.mathworks.com/help/comm/ref/sisofadingchannel.html

[9] *MathWorks Documentation. Bit Error Rate (BER)* [Electronic resource] 2019 Available at: https://www.mathworks.com/help/comm/ug/bit-error-rate-ber.html

[10] *MathWorks Documentation. Adaptive MIMO System with OSTBC* [Electronic resource] 2019 Available at: https://www.mathworks.com/help/comm/examples/adaptive-mimo-system-with-ostbc.html

[11] Jeruchim M C, Balaban Ph and Shanmugan K S 2000 *Simulation of Communication Systems. Second Edition* (Boston, New York: Kluwer Academic Publishers Publ) p 908

[12] Banket V L 2009 *Signal'no-kodovye konstrukcii v telekommunikacionnyh sistemah* (Odessa: Feniks Publ) p 180

[13] *MathWorks Documentation. MIMO Fading Channel* [Electronic resource] 2019 Available at: https://www.mathworks.com/help/comm/ref/mimofadingchannel.html

[14] Maslov O N and Pashintsev V P 2006 *Modeli transionosfernykh radiokanalov i pomekhoustojchivost' sistem kosmicheskoj svyazi* (Samara: PGATI Publ) p 357

[15] Pashintsev V P, Koval' S A and Senokosova A V 2009 *Porsev. Prostranstvennaya korrelyatsionnaya funktsiya korotkovolnovogo kanala svyazi Izvestiya instituta inzhenernoj fiziki* **2 (12)** p 81–84

[16] Pashintsev V P, Solchatov M Eh, Gakhov R P and Eremin A M 2003 Model' prostranstvenno-vremennogo kanala kosmicheskoj svyazi *Fizika volnovykh protsessov i radiotekhnicheskie sistemy* Vol 6 **5** 63–69

[17] Serkov V P 1981 *Rasprostranenie radiovoln i antennye ustrojstva* (L.: VAS Publ) p 468

Optimal experimental design for multi-response modeling of technological processes

S A Popov[1]

[1]Yaroslav-the-Wise Novgorod State University, ul. B. St. Peterburgskaya, 41 173004, Veliky Novgorod, Russia

E-mail: Stanislav.Popov@novsu.ru

Abstract. The problem of optimal experimental design for the construction of multi-response models of the technological process is considered. Optimal experimental design can significantly improve the accuracy of model building and their predictive ability.

1. Introduction

The location of observation points of input parameters $\mathbf{X}_i$, $i=\overline{1,n}$, n is the number of observations in the region of experiment planning of the technological process in the space of input parameters, determines the accuracy of estimating the coefficients of the model, which connect input and output parameters, and the accuracy of prediction of response by the model type [1]

$$Y = F(B, X) + E, \tag{1}$$

where **Y** is the vector of output parameters, **F** is the multi-response function of the vector of input parameters **X** and the vector of coefficients **B**, **E** is the vector of observation errors of the output parameters.

Practically all production tasks of the modeling of technological processes are of multi-response nature and the experimental designs are built for the multi-response space of output parameters. If the observation points are arranged in the space of input parameters in accordance with the optimal experiment design, then the prediction accuracy according to model (1) will be the maximal with the same number of observations. The design of experiments Δ, thus, is a set of observation points $\Delta = \{X_1, X_2, \dots, X_n\}$ in the space of the input parameters of the technological process.

2. Prediction by the model

Prediction of the future values of output parameters (responses) is the main goal of building a mathematical model. These values provide estimates of the coefficients of the model (1), which determine the prediction equations. If the model used is correct (adequate), then there are two possible sources of inaccuracy in the predictions: errors in the estimates of the **B** coefficients and errors in the observations of the response vector. These two sources contribute to the differences between the predicted and actually observed response values.

It is usually assumed that the errors generated by these sources are statistically independent. Then it is possible to calculate the covariance matrix of prediction errors. The observed value of the vector **Y** will be determined by the formula

$$Y = F(X, B + \delta B) + E,$$

where δ**B** is the estimation errors of the model coefficients.

Decomposition of a Taylor series, limited to only linear terms, leads to equality

$$\delta Y = \frac{\partial F}{\partial B}\delta B + E.$$

Then the covariance matrix of response estimates is represented by expression (2)

$$V_y = P(\hat{B}, X)^T V_B P(\hat{B}, X) + V_E, \quad (2)$$

where $P(B,X) = \{\frac{\partial f_1(B,X)}{\partial B}, \frac{\partial f_2(B,X)}{\partial B}, \ldots, \frac{\partial f_m(B,X)}{\partial B}\}$, $\hat{\mathbf{B}}$ is the model coefficient estimates (1), $\mathbf{V_B}$ is the coefficients covariance matrix and $\mathbf{V_E}$ is the errors covariance matrix.

Optimal experimental design allows you to increase the accuracy of estimation of the coefficients of the model. The covariance matrix of observation errors $\mathbf{V_E}$ is assumed constant, therefore the quality of the model prediction is estimated by the value of the covariance matrix of the average prediction error $\mathbf{V_Y}$ in the center of the design area without taking into account matrix $\mathbf{V_E}$ as

$$V_Y = P(\hat{B}, X)^T V_B P(\hat{B}, X)\,. \quad (3)$$

Thus, as a criterion for the optimality of the design in this case, you can take the determinant of the covariance matrix $\det \mathbf{V_Y}$. To calculate this optimality criterion, it is necessary for a given experimental design to calculate the covariance matrix of the coefficient estimates, and then, using formula (3), calculate the covariance matrix of the response estimates in the center of the design area and its determinant. The optimal designs obtained for this criterion of optimality are called *G*-optimal. It is known, however, that to obtain the same exact *G*-optimal designs, instead of the optimality criterion in the form of $\det \mathbf{V_Y}$, one can use the optimality criterion in the form of $\det \mathbf{V_B}$. This greatly increases the accuracy of the coefficients calculation and speeds up the design process, as it simplifies the calculation of the optimality criterion.

Optimal designs that use the value of $\det \mathbf{V_B}$ as a criterion for optimality are called *D*-optimal.

3. *D*-optimal design of experiments

The criterion of *D*-optimality requires such an arrangement of experimental points in the design area of the experiment Σ, in which the determinant $\det \mathbf{V_B}$ of matrix $\mathbf{V_B}$ is minimal. The *D*-optimal design minimizes the amount of ellipsoid of scattering coefficient estimates in the form of [2]

$$D = det\, V_B. \quad (4)$$

Thus, the *D*-optimal design is defined as follows:

$$detV_B(\Delta^\circ) = \min_{\Delta\in\Sigma} det\{\sum_{j=1}^{n} P(\hat{B}, X_j)V_E^{-1}[P(\hat{B}, X_j)]^T\}^{-1} = \max_{\Delta\in\Sigma} det\{\sum_{j=1}^{n} (\hat{B}, X_j)V_E^{-1}[P(\hat{B}, X_j)]^T\}, \quad (5)$$

where $\Delta^o = (X_1^o, X_2^o, \ldots, X_i^o, \ldots, X_n^o)$ is the optimal design in the sense of the *D*-optimality criterion.

There are discrete *D*-optimal designs, which are built for a predetermined number of experiment points, and continuous *D*-optimal designs, the number of points of which significantly exceeds the number of model coefficients. Continuous designs are characterized by a finite spectrum and frequency of experiment points at each point of the design spectrum.

The number of points *r* of the spectrum of continuous *D*-optimal designs satisfy the following inequality [3]

$$l + 1 \le r \le \frac{(l+1)(l+2)}{2},$$

where *l* is the number of estimated coefficients of model (1).

It is known that the continuous *D*-optimal designs coincide with the *G*-optimal designs. In this case, it is easier to build the *D*-optimal designs, while the resulting design will also be *G*-optimal. The analytical way of constructing continuous *D*-optimal designs and their subsequent rounding to discrete designs is used to obtain designs only in the simplest cases. In practice, there are a large number of tasks in which the design area Σ_X does not have a regular geometric shape, and the model is presented in a rather complicated way, for example, in the form of a computer program. In these cases, to build optimal designs, it is necessary to use numerical methods associated with minimizing the determinant det $\mathbf{V_B}$ of the covariance matrix of coefficient estimates [3].

Extremum problem (5) has the dimension of *mn*, where *n* is the number of points of the experiment design, *m* is the number of observed output parameters and is complex in the sense of convergence of the computational procedure and computer time spent using standard extremum search procedures.

4. Building multi-response sequential designs

A priori experimental designs, which are constructed prior to the experiment, can only be constructed for linear single-response models. For nonlinear models, the use of a priori experimental designs is based on an average result and usually does not provide the necessary accuracy of estimating the output parameters for a particular product situation. Often, as a condition to stop the construction of an optimal design, the accuracy of estimating the coefficient estimates is set. This condition determines the number of points in the design. However, even for a large number of observation points of an a priori design for a specific task, the accuracy of estimation is much worse than required. This is due to the fact that the optimal design is built for the average values of the parameters, and the actual values of the output parameters can often differ significantly. The use of such design is justified when the design needs to be built prior to the experiment, for example, due to technical limitations imposed on the technological process.

More efficient is the sequential planning procedure. In this case, some initial design of the experiment is set up with a relatively small number of observation points *n*. As such a design, you can use, for example, a uniform experimental design that uniformly covers the entire design area.

Using this design, experiments are performed and estimates of the coefficients of the model are calculated. Based on the estimates obtained, the next $(n + 1)$ *-th* observation point is found, adding which to the experimental design ensures the best estimation accuracy. The number of observations increases until the required actual accuracy of estimating the output parameters for this model is achieved.

The algorithm for building sequential design can be represented by the following expression [4].

$$detV_B(\Delta + X^o, B) = \min_{X\in\sum} detV_B(\Delta + X, B) \qquad (6)$$

where $\mathbf{X}^{\circ}$ is the point added to design Δ.

The accuracy of estimating the output parameters in the center of the design area is calculated as follows.

$$V_y = P(\hat{B}, X)^T V_B P(\hat{B}, X), \qquad (7)$$

where $\mathbf{X}$ is the input parameter vector in the center of the experiment design.

To perform the search for extremum, in accordance with the expression (6), the initial design of the experiment is set. The initial design is generated using a random number generator for a given number of experiment points and for a given area of the independent input parameters.

The search for the optimal design is carried out several times for various generated initial design generated by computer. The design that provides the smallest determinant of the covariance matrix for estimating the coefficients of the model (5) is taken as the optimal experimental design. Usually it takes no more than 20-50 iterations to find the global extremum using formulas (6,7).

5. Algorithm for building *D*-optimal design using the half-division method

After specifying the initial design, the optimal design is searched for in accordance with the above mentioned algorithm using the half-division method. A penalty function is also introduced to ensure the search for an extremum in a given area of variation of the parameters. The search algorithm is as follows [5].

1. The number of control points *n* is set and an arbitrary initial experiment design is determined. $\Delta^1 = \{X_1^1, X_2^1 \dots, X_n^1\}$.

2. The first control point is selected to search for a particular extreme $\mathbf{X}_i$.

3. Separately, for each parameter, the partial search method solves a particular extremal problem for one variable of the control point $\mathbf{X}_i$ and for the rest of the fixed points $\mathbf{X}_1, \mathbf{X}_2, \dots, \mathbf{X}_i, \dots, \mathbf{X}_n$. Trial steps

are performed and a step is taken in the direction of decreasing the objective function. If the specified parameter is outside the specified limits of its variation, then the value of the objective function is given by the introduction of a penalty.

If in both directions the objective function increases, then the length of the trial step is reduced by half. As the stop criterion the search step length, which determines the accuracy of the input parameters, is taken.

The search is repeated for all parameters and this point of the plan. The resulting point is added to the design $\Delta^1 = (X_1^1, X_2^1, \ldots, X_i^1, \ldots, X_n^1)$ instead of the starting point. Thus, a design $\Delta^2 = \Delta^1 - X_i^1 + X_i^2$ is defined in the form $\Delta^2 = (X_1^1, X_2^1, \ldots, X_i^1, \ldots, X_n^1)$, where the point is defined as follows:

$$\det\{V_B(\Delta^1 - X_i^1 + X_i^2, B)\} = \min_{X \in \Sigma} \det\{V_B(\Delta^1 - x_i^1 + X, B)\} \quad (8)$$

1. The search is repeated for all points of the initial design.
2. The criterion for stopping the entire procedure is the relative variation in the objective function. If this criterion is not met, the search procedure 2–4 is repeated.

For example, an optimal design is obtained for 14 points with a single covariance matrix of observation errors and a two-response quadratic model with two responses (output parameters) with 12 coefficients, the form of which is represented as follows:

$$y_1 = b_1 + b_2 x_1 + b_3 x_2 + b_4 x_1 x_2 + b_5 x_1 x_1 + b_6 x_2 x_2$$
$$y_2 = b_7 + b_8 x_1 + b_9 x_2 + b_{10} x_1 x_2 + b_{11} x_1 x_1 + b_{12} x_2 x_2 \quad (9)$$

Figure 1 shows the initial design for 14 points of the experiment. As an optimality, the expression is taken $D' = D^{-1} = -Log(\det \mathbf{V_B})$ for which increase means decrease in the covariance matrix $(\det \mathbf{V_B})$.

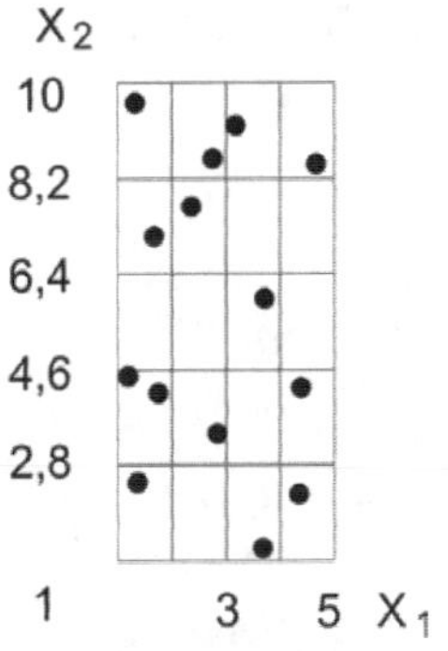

Figure 1. The experimental design for 14 points, obtained using a random number generator and used as an initial approximation for model (9). The marked points show the experimental design. D' =22.789.

Figure 2 shows the optimal design, obtained on the basis of algorithm (8).

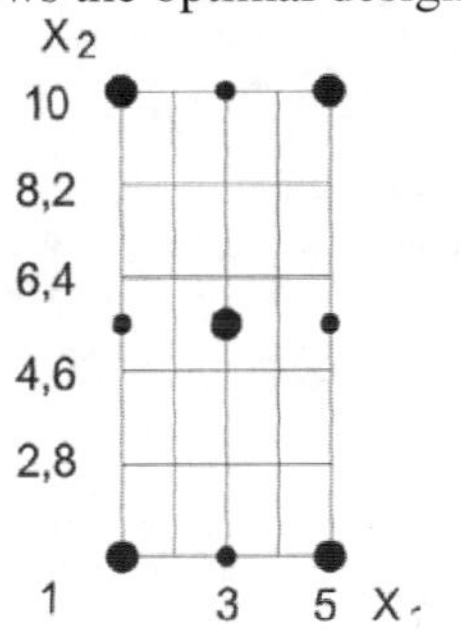

Figure 2. The experimental design obtained after optimization by the search method. Larger circles marked points of the design, including two observations. D' =25.109. Such design exactly corresponds to the theoretical one.

For model (9) for 12 design points, an optimal experiment design was constructed in the sense of criterion (6) with uniform probability distributions of coefficients. Thus, as an optimality criterion in this case, for a model that is nonlinear in terms of coefficients, the average determinant of the covariance matrix of coefficient estimates was taken as

IOP Conf. Series: Materials Science and Engineering **656** (2019) 012042 doi:10.1088/1757-899X/656/1/012042

$$\overline{det}V_B(\Delta^o) = \min_{\Pi\in\Sigma} \overline{det}\, V_B(\Delta). \qquad (10)$$

When calculating, the admissible ranges of values for each of the model coefficients were specified in the form of a lower bound and the length of the coefficient variation with a diagonal unity covariance matrix of observation errors.

The number of statistical tests was 10,000. As an initial design, uniform design for 12 points was used. For this design and non-linear model, the average determinant $\overline{det}V_B(\Delta)$ of the covariance matrix $\mathbf{V_B}$ of the coefficients is = 28.607.

Figure 3 shows the experimental design obtained by the half division method for uniform a priori probability distributions of the coefficients.

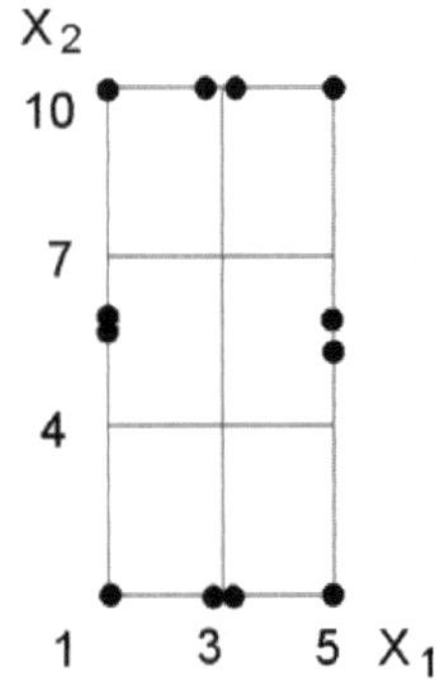

Figure 3. The optimal experiment design obtained by the search method for 12 points. D' =28.785.

Figure 4 shows experimental design built for the case when all coefficients are known. D' =28.785.

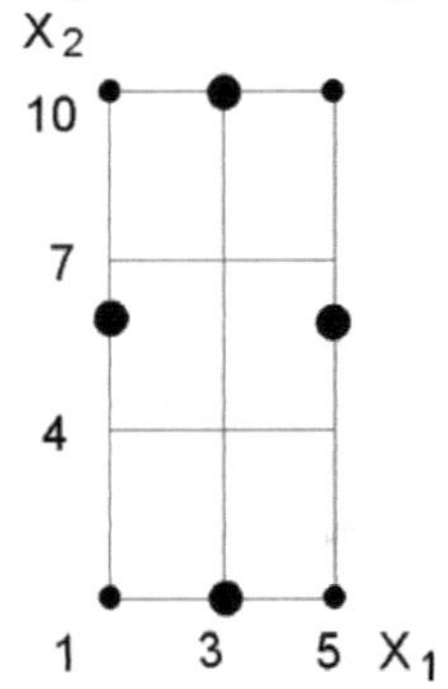

Figure 4. The experimental design for 12 points, obtained with the known coefficients of the model. Large circles marked points of the design, including two observations. D' = 31.800.

6. The sequential algorithm of design building

The sequential design program is implemented in accordance with expression (6) and is based on all the programs described earlier: obtaining an initial uniform design, calculating estimates of the coefficients of a non-linear model with an unknown observational error covariance matrix, calculating the coefficient of the covariance matrix of estimating coefficients, and finding the optimal design point using the half division.

1. The initial design of the experiment is set Π^s (s = 0) such and with so many points n that the information matrix is sufficiently well defined.
2. In accordance with this design, measurements of technical parameters are performed, estimates of the coefficients of the model and the criterion of optimality of the design (10) are calculated.
3. To the design is added another point in the middle of the area of variation of the measured parameters.

4. The search method searches for the $(n + s)$ *-th* point of the experiment $(s = s + 1)$ X_{n+s} such that adding this point to the existing experiment design $\Pi^{s+1} = \Pi^{s} + X_{n+s}$ provides the maximum increase in the estimation accuracy, i.e.

$$\det\{V_B(\Pi^s + X^o_{n+s}, B)\} = \min_{U \in \sum_U} \det\{V_B(\Pi^s + X_{n+s}, B)\}. \quad (11)$$

5. Observation is performed at the obtained point of the design. The stop criterion is the specified limit value of the determinant of the covariance matrix.

6. If this criterion is not met, the next point is added to the design.

The procedure of sequential designing allows to obtain the accuracy of determining the coefficients of the model not worse than the specified one.

Table 1 shows the *D*-optimality criteria in the form of a quantity D' calculated for various types of experimental designs for the model in the form of two-dimensional polynomial (9).

Table 1. The value of D' - criterion for different experimental designs and number of points.

Number of points \ Form of the design	Uniform with given coefficients	Optimal for given coefficients	Uniform for given probability distribution of coefficients	Optimal for given probability distribution of coefficients	Sequential experimental design
12	31.031	30.338	28.607	28.785	31.162
20	32.583	33.553	29.246	29.493	33.268
50	35.274	34.671	41.959	32.241	36.128
100	37.05	39.120	34.587	35.744	38.757

7. Conclusion

The modeling of technological processes in the form of multi-response models is a necessary condition for obtaining adequate models. The use of optimal multi-variant experimental designs by several orders of magnitude improves the accuracy of estimating the coefficients of a multi-response model. The specific examples show how to construct a priori and sequential experiment design.

The use of the procedure of sequential design reduces the determinant of the covariance matrix of the coefficient estimates in comparison with the a priori optimal design. built for a given probability distribution of coefficients by 2–4 times.

References

[1] Rencher Alvin C 2002 *Methods of multivariate Analysis* (John Wiley & Sons. Inc. Publication Brigham Young University) 339–345

[2] Popov S A and Klykov N N 2015 Comparison of playback errors during video compression *Vestn. Novgorod. Gos Univ.* **8(91)** 96–98

[3] Popov S A 2016 Multiresponse optimisation in the discretized parameter space *Vestn.estn. Novgorod. Gos. Univ.* **4(95)** 53–57

[4] Popov S A and Abramov A M 2017 Multiresponse model based design parameters optimization: engineering approach *EAI Endorsed Transactions on Energy Web* **17(12)**: e2 Available at: http://eudl.eu/issue/ew/4/12 (accessed: 20.05.2019)

[5] Popov S A 2018 Method of indirect measurements on the basis of vector-observation *The International Scientific and Practical Conference on Innovations in Engineering and Technology* (Veliky Novgorod) Available at: https://iopscience.iop.org/article/10.1088/1757-899X/441/1/012039 (accessed: 20.05.2019)

IOP Conf. Series: Materials Science and Engineering **656** (2019) 012043 doi:10.1088/1757-899X/656/1/012043

Determination of the admissible intervals of the input technological parameters that provide the necessary values of the parameters – acceptance criteria

S A Popov

Yaroslav-the-Wise Novgorod State University, ul. B. St. Peterburgskaya, 41 173004, Veliky Novgorod, Russia

E-mail: Stanislav.Popov@novsu.ru

Abstract. An inverse optimization problem is considered to determine the optimal intervals of the input technological parameters. The solution is based on multi-response technological process models for the multi-dimensional area of input and output parameters. Expressions for calculating the coefficients of the multi-response models are given. It is shown how the intervals of admissible values of the input technological parameters are built. As an optimality criterion, the percentage of yield of products is used, parameters – acceptance criteria.

1. Parameters - acceptance criteria

The solution of the problem of determining the admissible areas of independent technological (control) parameters x_i, $i = 1,2,\ldots k$ allows rejecting admittedly unsuitable products at the early stages of the technological process and increasing the yield of good products. In practice, there is a area of changes of these parameters, determined by physical causes, which is defined by the system of inequalities in the form $x_{i1} \le x_i \le x_{i2}$. Belonging to the class of acceptability is determined on the basis of output parameters – acceptance criteria, which are measured for the finished products.

The task of determining the permissible areas of the input process parameters is to determine such intervals for the values of the input process parameters that ensure that this product belongs to the class of acceptability at an early stage of the process.

Such a problem is usually solved for the case of a single parameter, the criterion of acceptability [1], which, obviously, limits the applicability of the obtained results so much that they have no practical value. In this paper, the solution of the problem of determining the admissible areas of technological parameters (tolerances) is considered for the case of several output parameters – acceptance criteria, taking into account the multi-response of output parameters.

2. Intervals of the admissible values of output parameters – acceptance criteria

The area of admissible values of output parameters – acceptance criteria $\mathbf{Y} = \{y_1, y_2, \ldots, y_m\}^{\mathrm{T}}$ is represented as a m-dimensional parallelepiped Ω in the space of these parameters. The point **Y**, representing the finished product in the output parameter area, belongs to this space if its coordinates satisfy the system of inequalities

$$\begin{cases} y_{11} \le y_1 \le y_{12} \\ y_{21} \le y_2 \le y_{22} \\ \dots \\ y_{m1} \le y_m \le y_{m2} \end{cases}.$$

where some interval boundaries can be assigned to infinity. Vector $\mathbf{X} = \{x_1, x_2, \ldots, x_k\}^{\mathrm{T}}$ is the input controllable technological parameters area which defines a point in the output parameters space, parameters – acceptance criteria Ω corresponding to this product.

If the transformation is known $\mathbf{X} \Rightarrow \mathbf{Y}$, then for a given point in the space of input technological parameters it is possible to determine the corresponding point in the space of output parameters – acceptance criteria.

If point $\mathbf{X}$ corresponds to point $\mathbf{Y}$ inside the space Ω ($\mathbf{Y} \notin \Omega$), then this product meets the requirements of technical conditions (TC) for these parameters. Thus, a given m -dimensional area Ω is displayed in the space of input technological parameters in the form of a k -dimensional space Γ ($k \ge m$).

If image Γ of area Ω is defined, then it gives the possibility to set the technological parameters $\mathbf{X} \notin \Gamma$ that provide a priori good products. If the set of parameters – acceptance criteria is determined by the TC, then the composition of the technological parameters $\mathbf{X} = \{x_1, x_2, \ldots, x_k\}^{\mathrm{T}}$ is determined using the methods of selection of informative parameters.

Image Γ usually has a complex shape. In order to set the input technological parameters independently of each other, you need to set in the area Γ a k -dimensional parallelepiped Π, which will determine the independent tolerances on the input technological parameters.

The criterion of optimization D to determine the optimal parallelepiped Π may be the yield of good products by technological parameters. This criterion can be expressed as [2]

$$D = \int_{x_{11}}^{x_{12}} \int_{x_{21}}^{x_{22}} \cdots \int_{x_{k1}}^{x_{k2}} \mathrm{f}(x_1, x_2, \ldots, x_k) dx_1 dx_2 \ldots dx_k, \tag{1}$$

where $\Delta x_i = x_{iH} - x_{iL}$ is the maximum length of the interval of variation of the input parameter x_i, determined by the TC.

If all input parameters are controllable, then to determine the tolerances of the input parameters, it is necessary to introduce additional restrictions. You can, for example, build area Π according to the criterion of the maximum relative width of the intervals of these parameters for a given area of their variations in the form

$$D = \prod_{i=1}^{k} \frac{x_{iH} - x_{iL}}{\Delta x_i}, \tag{2}$$

where $\mathrm{f}(x_1, x_2, \ldots, x_k)$ is probability density function of the error of technological parameters reproduction, and the point inside k -dimensional parallelepiped Π is determined by the system of inequalities

$$\begin{cases} x_{11} \le x_1 \le x_{12} \\ x_{21} \le x_2 \le x_{22} \\ \dots \\ x_{k1} \le x_k \le x_{k2} \end{cases}.$$

Figure 1 shows an example of constructing admissible intervals for three input controllable parameters by criterion (2).

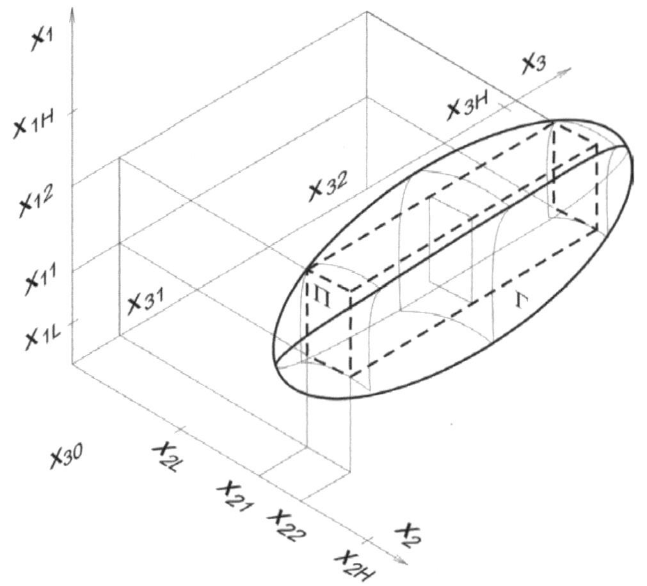

Figure 1. Building of admissible intervals of values of input parameters in the form of parallelepiped Π inscribed in the area of admissible values Γ for three input controllable parameters and two output parameters.

3. Multi-response model of the area of admissible values of output parameters

Transformation $\mathbf{X} \Rightarrow \mathbf{Y}$ is represented as a multi-response model.

$$\mathbf{Y} = \mathbf{F}(\mathbf{X}, \mathbf{B}) + \mathbf{E}, \tag{3}$$

where $\mathbf{F}(\mathbf{X}, \mathbf{B}) = \{\mathrm{f}_1(\mathbf{X}, \mathbf{B}), \mathrm{f}_2(\mathbf{X}, \mathbf{B}), \ldots, \mathrm{f}_m(\mathbf{X}, \mathbf{B})\}^{\mathrm{T}}$ is the m-dimensional vector of functions, $\mathbf{X} = \{x_1, x_2, \ldots, x_k\}^{\mathrm{T}}$ is the vector of input technological parameters, $\mathbf{Y} = \{y_1, y_2, \ldots, y_m\}^{\mathrm{T}}$ is the vector of output parameters – acceptance criteria, $\mathbf{B} = \{b_1, b_2, \ldots, b_l\}^{\mathrm{T}}$ is the l-dimensional vector of coefficients, the values of which are to be estimated, $\mathbf{E} = \{e_1, e_2, \ldots, e_m\}^{\mathrm{T}}$ is the observation error of the output parameters (3).

In case of a normal distribution of observational errors $\mathbf{E}$ with zero expectancy and with covariance matrix $\mathbf{V}_{\mathbf{E}}$ for problems of maximum likelihood and pseudo maximum likelihood methods, the objective function for obtaining estimates of coefficients in expression (3) depends on the coefficients through the likelihood function logarithm in [2,3]

$$\varphi(\mathbf{B}) = \sum_{i=1}^{n} [\mathbf{Y}_i - \mathbf{F}(\mathbf{X}_i, \mathbf{B})] \mathbf{V}_{\mathbf{E}}^{-1} [\mathbf{Y}_i - \mathbf{F}(\mathbf{X}_i, \mathbf{B})]^{\mathrm{T}} . \tag{4}$$

where n is the number of vector $\mathbf{Y}$ observations.

As estimates of coefficients $\hat{\mathbf{B}}$, the values minimizing function (4) are taken. If the model's equations $\mathbf{Y} = \mathbf{F}(\mathbf{B}, \mathbf{X})$ in the neighborhood of point $\hat{\mathbf{B}}$ can be approximated by equations linear with respect to coefficients in the form

$$\mathbf{Y} = \mathbf{Y}^* + \left(\mathbf{P}^{\mathrm{T}}(\mathbf{X}) | \hat{\mathbf{B}}\right)\left(\mathbf{B} - \hat{\mathbf{B}}\right), \tag{5}$$

where $\mathbf{P}(\mathbf{X}) = \left\{ \frac{\partial \mathrm{f}_1(\mathbf{B}, \mathbf{X})}{\partial \mathbf{B}} | \hat{\mathbf{B}}, \frac{\partial \mathrm{f}_2(\mathbf{B}, \mathbf{X})}{\partial \mathbf{B}} | \hat{\mathbf{B}}, \cdots, \frac{\partial \mathrm{f}_m(\mathbf{B}, \mathbf{X})}{\partial \mathbf{B}} | \hat{\mathbf{B}} \right\}$, then the estimates of the coefficients with an unknown covariance matrix $\mathbf{V_E}$ can be calculated using the system of equations [4, 5]

$$\begin{cases} \hat{\mathbf{B}}^{s+1} = \hat{\mathbf{B}}^{s} + \mathbf{K}\sum_{i=1}^{n}\mathbf{P}(\mathbf{X}_i)\mathbf{S}_\mathbf{E}^{-1}\left[\mathbf{Y}_i - \mathbf{F}(\mathbf{X}_i, \hat{\mathbf{B}}^{s})\right] \\ \mathbf{S}_\mathbf{E} = \frac{1}{n}\sum_{i=1}^{n}\left[\mathbf{Y}_i - \mathbf{F}(\mathbf{X}_i, \hat{\mathbf{B}}^{s})\right]\left[\mathbf{Y}_i - \mathbf{F}(\mathbf{X}_i, \hat{\mathbf{B}}^{s})\right]^{\mathrm{T}} \end{cases}, \qquad (6)$$

where $\mathbf{S_E}$ is estimation of covariance matrix $\mathbf{V_E}$.

The covariance matrix of coefficient estimates (6) is equal to

$$\mathbf{K} = \left[\sum_{i=1}^{n}\mathbf{P}(\mathbf{X}_i)\mathbf{S}_\mathbf{E}^{-1}\mathbf{P}^{\mathrm{T}}(\mathbf{X}_i)\right]^{-1}. \qquad (7)$$

As functions $\mathbf{F}(\mathbf{X},\mathbf{B}) = \{\mathrm{f}_1(\mathbf{X},\mathbf{B}), \mathrm{f}_2(\mathbf{X},\mathbf{B}), \ldots, \mathrm{f}_m(\mathbf{X},\mathbf{B})\}^{\mathrm{T}}$ for modeling technological processes, quadratic polynomials with all interactions are usually used. In this case, function $\mathbf{P}(\mathbf{X})$ in the expression (5) takes the form

$$\mathbf{P}^{\mathrm{T}}(\mathbf{X}) = \begin{Bmatrix} 1, x_1, x_2, \ldots, x_k, x_1^2, x_2^2, \ldots, x_k^2, x_1x_2, x_1x_3, \ldots, x_{k-1}x_k \\ 1, x_1, x_2, \ldots, x_k, x_1^2, x_2^2, \ldots, x_k^2, x_1x_2, x_1x_3, \ldots, x_{k-1}x_k \\ \ldots \\ 1, x_1, x_2, \ldots, x_k, x_1^2, x_2^2, \ldots, x_k^2, x_1x_2, x_1x_3, \ldots, x_{k-1}x_k \end{Bmatrix}.$$

Then, step regression procedure is applied to this model with decrease in the number of members. In the step regression procedure, the coefficients are found for which the inequality holds

$$\frac{\hat{b}_i}{\sqrt{k_i}} \le t_{1-\frac{\alpha}{2}}, \qquad (8)$$

where k_i is the diagonal element of matrix $\mathbf{K}$ for the i-th coefficient, $t_{1-\frac{\alpha}{2}}$ is the quantile of the Student's distribution at the significance level α.

Members that include coefficients that satisfy inequality (8) are excluded from the corresponding functions, and the estimates of the coefficients are recalculated and the procedure for reducing the number of members of the model is repeated.

4. Determining the image of the area of acceptable values of output parameters

After performing the step regression, clarifying the type of functions $\mathrm{f}_1(\mathbf{X},\mathbf{B}), \mathrm{f}_2(\mathbf{X},\mathbf{B}), \ldots, \mathrm{f}_m(\mathbf{X},\mathbf{B})$ and determining the estimates of all significant coefficients of model (2), expression $\mathbf{Y} = \mathbf{F}(\hat{\mathbf{B}},\mathbf{X})$ allows us to build area Γ. In some rare practical situations this area can be built analytically. Generally, the principle of building of this area consists in determining the projections of this area on separate planes of the projections of the space of technological parameters and building for each of these projections an inscribed rectangle that represents the projection of area of admissible values Π on this plane.

Consider the process of building area Γ on the illustrative example of a two-response function with two technological parameters in the form

$$\mathbf{F}(\mathbf{X},\mathbf{B}) = \begin{cases} y_1 = b_1 + b_2x_1 + b_3x_2 + b_4x_1^2 + b_5x_1x_2 + b_6x_2^2 \\ y_2 = b_7 + b_8x_1 + b_9x_2 + +b_{10}x_1^2 + b_{11}x_1x_2 + b_{12}x_2^2 \end{cases}. \qquad (9).$$

For specified intervals of valid values of parameters – acceptance criteria in the form

$$\begin{cases} y_{11} \le y_1 \le y_{12} \\ y_{21} \le y_2 \le y_{22} \end{cases}.$$

In the space of technological parameters x_1, x_2, it is possible to obtain two areas of admissible values: for parameter y_1 – area Γ_1, and for parameter y_2 – area Γ_2. The joint area of admissible values of technological parameters Γ is defined as $\Gamma = \Gamma_1 \cup \Gamma_2$ (figure 2).

To build area Γ with a confidence probability P, as model $\mathbf{Y} = \mathbf{F}(\hat{\mathbf{B}}, \mathbf{X})$ coefficient values the corresponding one-sided boundaries of their confidence intervals are used as the values of the coefficients of the model (to obtain the minimum area) in the form

$$b_i \le \hat{b}_i \pm \sqrt{k_i} t_{1-\alpha},$$

where $P = 1 - \alpha$, $i = 1,2,\ldots l$.

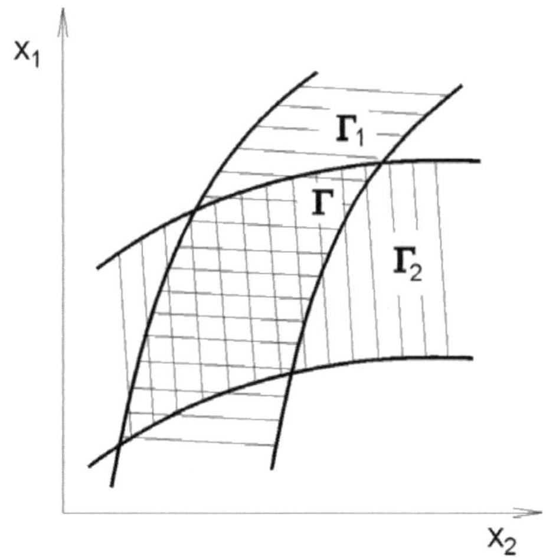

Figure 2. Obtaining the joint area of admissible values of technological parameters for a two-response quadratic function: Γ_1 is the area of admissible values for parameter y_1, Γ_2 is the area of admissible values for parameter y_2, $\Gamma = \Gamma_1 \cup \Gamma_2$

5. Calculation of intervals of admissible values of technological parameters

To obtain independent intervals of admissible values of technological parameters, a *k*-dimensional parallelepiped must be inserted in area Γ in accordance with the optimality criterion (1), which represents the yield of the product under given technological tolerances. Figure 3 shows an illustration of a two-dimensional problem (9) with quadratic functions $\mathbf{Y} = \mathbf{F}(\hat{\mathbf{B}}, \mathbf{X})$.

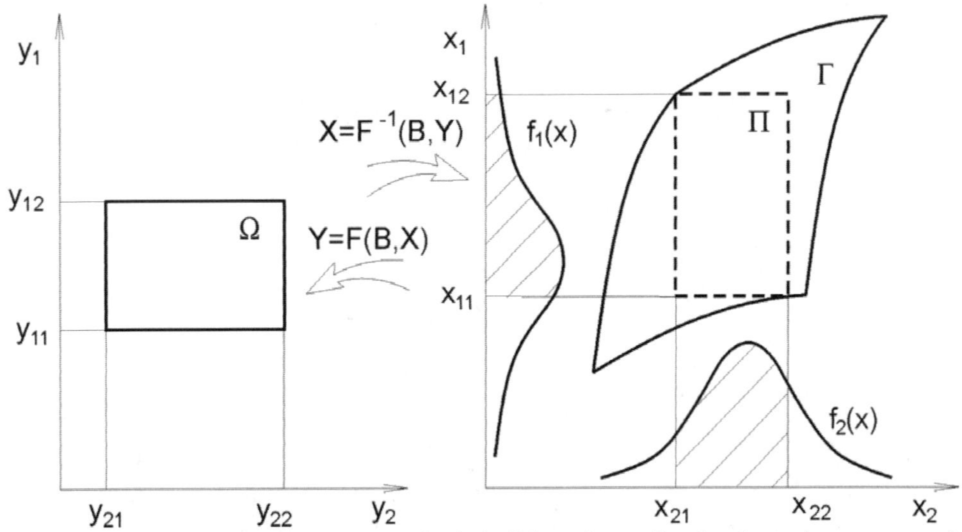

Figure 3. Building of rectangular area Π of admissible values of technological parameters for two-dimensional quadratic function $\mathbf{F}(\mathbf{B}, \mathbf{X})$, $f_1(x)$ and $f_2(x)$ are distribution density of technological parameters x_1 and x_1 respectively.

If, along with uncontrolled input parameters, there are also controlled technical parameters, then optimization by criterion (3) makes it possible to determine the values of these parameters that provide the highest yield. Figure 4 shows the example of determining the optimal tolerances for input uncontrolled parameters x_1 and x_1 with distribution densities φ_1 и φ_2, and the optimal value x_{30} of the controlled input parameter x_3 providing the highest yield by criterion (3).

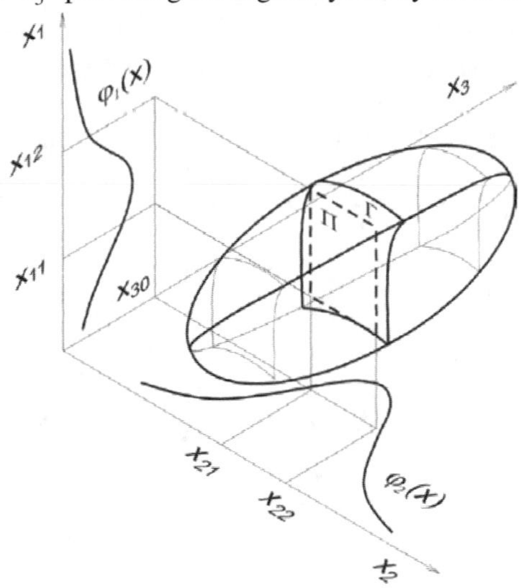

Figure. 4. Determination of the admissible intervals for two uncontrolled input parameters and the optimal value of one controlled input parameter for the two-response model.

The use of the optimality criterion (2) makes it possible to build admissible intervals of input parameters, which provide the most convenient control of technological process, when there is no need to ensure the exact values of input parameters.

Due to the large dimension of the input parameters space and the complex shape of the area of admissible values of the parameters – acceptance criteria in the multicriteria case, the solution of optimal problems by criteria (2) and (3) is performed by the Monte Carlo method.

Figure 3 shows that with proper adjustment of the technological process by shifting the centers of the distribution of the input technological parameters, even with a constant variation of these parameters, it is possible to increase the yield. This allows, also, determining the maximum value of the variation of the input technological parameters to ensure the required percentage of yield. When statistically independent errors of observation of technological parameters and optimization criterion (1) takes the following form:

$$D = \int_{x_{11}}^{x_{12}} f_1(x)dx \int_{x_{21}}^{x_{22}} f_2(x)dx .$$

The building of area Π is carried out by statistical modeling method.

6. Conclusion

The proposed method of determining technological tolerances allows determining independent intervals of admissible values of technological parameters, which makes it possible to reject unsuitable products in the technological process to ensure the necessary values of parameters – acceptance criteria of finished products. This method, in addition, makes it possible to work out recommendations on how to adjust the technological process in order to increase the yield, both by optimizing the position of the distribution centers of the technological parameters and in relation to their spread.

References

[1] Maindonald J H 1984 *Statistical Computation* (New York: John Wiley and Sons) 311–364

[2] Rencher and Alvin C 2002 *Methods of multivariate Analysis* (A John Wiley & Sons Inc. Publication Brigham Young University) 339–345

[3] Bard Y 1974 *Nonlinear Parameter Estimation* (New York: Acad. Press) 294–303

[4] Popov S A 2016 Multi-criteria optimization in discretized parameter space *Vestnik NovSU* **4 (95)** 53–57

[5] Popov S A and Abramov A M 2017 Multiresponse model based design parameters optimization: engineering approach *EAI Endorsed Transactions on Energy Web* **17(12)** *e2* http://eudl.eu/issue/ew/4/12. (accessed 20.05.2019)

IOP Conf. Series: Materials Science and Engineering **656** (2019) 012044 doi:10.1088/1757-899X/656/1/012044

Simulation investigation of a heat pump, operating on a low-temperature source of heat and a constant condensation pressure

N O Rachkovskiy[1], Iu V Tatarenko[1], V V Mitropov[1], O S Naumova[2] and K A Bordashev[2,3]

[1]Saint-Petersburg National Research University of Information Technologies, Mechanics and Optics, pr. Kronverksky, 49 197101, St. Petersburg, Russia
[2]Yaroslav-the-Wise Novgorod State University, ul. B. St. Peterburgskaya, 41 173003, Veliky Novgorod, Russia

[3]E-mail: Kirill.Bordashev@novsu.ru

Abstract. The simulation investigation was carried out at the “Vapor Compression Heat Pump” stand, which allows simulating the annual cycle of operation of the heat pump. Values and further calculations were determined by the operating characteristics of the cycle. Calculations were made according to generally accepted dependencies. As a result of simulating the annual cycle of the heat pump operation, the operational and energy characteristics of the heat pump were determined under the specified conditions: maintaining the specified condensing pressure p_c=1,31 MPa and the changing temperature of the low-potential heat source – water t_{w4}, which varied from from – 0,3°C to + 10,1°C. Based on the obtained data, it was estimated the change in the heat pump coefficient of performance (COP) as a function of the low–grade heat source temperature, as well as the application potential of renewable energy sources.

1. Introduction

The source of heat for heating, ventilation and air conditioning systems is usually fossil fuel (gas, coal, peat, shale, oil), which is burned at power plants (GRES, TPP) to produce electricity and heat. The result is a deterioration of the environmental situation, both in the production area and in the combustion area. According to studies, life support systems consume about 50% of the total energy consumption in buildings [1]. These facts are the basis for the development of renewable energy sources.

Soil, rock, water and air can be used as a source of low-grade heat for the heat pump. These sources of energy are considered renewable.

Life-support systems that use groundwater or water bodies as a source of low-potential heat are advantageous in terms of energy efficiency [2].

The temperature of the source of low-potential heat and its thermal power are important parameters affecting the energy efficiency and productivity of the heat pump [3]. A higher boiling point, which corresponds to the temperature of the source of low-grade heat, has a positive effect on the heating coefficient μ (hereinafter referred to as the international term – coefficient of performance or COP), provided that the condensation temperature is kept constant [4]. The thermal power of the low-potential heat source affects the amount of heat output and the power consumption of the heat pump compressor drive. The amount of energy extracted from the source of low-grade heat affects its efficiency and payback period [5, 6].

In recent decades, the world has seen a trend of increasing the use of heat pumps, both in developed countries such as the United States, Japan, some countries of the European Union, and in developing countries. The people's Republic of China is a prime example of countries that pay attention to the alternative energy.

Recently, the People's Republic of China (PRC) has experienced a high level of urbanization, resulting in increased demand for thermal energy for heating, ventilation and air conditioning. 22% of all energy consumed in China falls on the operation of buildings, of which 60% is spent on life support systems. The coal-fired boilers for heating are gradually prohibited. Based on the above circumstances, heat pumps are an obvious alternative, and also have a high potential for development. The average cost of a heat pump installing is 50 – 65 us dollars per $1m^2$ of the building area [7].

Over the past few years, China has paid attention to energy saving and greenhouse gas emissions. In this regard, a number of laws and regulations of the government were adopted:

The Renewable Energy Law of the People's Republic of China, 2005;

The Energy Conservation Law of the People's Republic of China, 2007;

Energy Conservation Regulations of Civil Buildings, 2008;

The Action Plan for China's Energy Strategy (2014–2020), 2014;

The Thirteenth Five-Year Plan for Development and Utilization of Geothermal Energy, 2017.

According to the national statistics of the PRC, by the end of 2016, more than 22,000 heat pumps with a serviced area of more than 487 million m $м^2$ were installed.

It is necessary to note that over the past 10 years, the average annual sales of heat pumps in Italy were about 1300000 units (as of 2018). The heat pump market was stimulated by the introduction of a special tariff in 2014, that was available only to private consumers using the heat pump as the main heating system.

The review shows that in the Russian Federation (RF) it is necessary to pay more attention to this topic in connection with its active development in the countries close to the borders of the Russian Federation, and not only to them.

2. Results and discussion

Stand "Vapor compression heat pump" allows you to simulate the annual cycle of operation of the heat pump both in cooling mode and in heating mode, which makes it possible to obtain operating characteristics close to the actual.

The schematic diagram of the stand of the vapor compression heat pump is shown in figure 1 and includes the following main elements: a spiral compressor, a plate condenser, an evaporator, a source of low-grade heat – an isolated tank with cold water, a heat storage device – an isolated tank with hot water, a fan coil unit, an automatic control Board, shut-off and connecting valves, a set of control and measuring devices. The working substance circulating in the circuit is R134a. The source of low-grade heat – insulated tank with cold water is equipped with electric heaters installed electric power of 6 kW and filled with 15% ethylene glycol solution – these measures allow you to maintain the desired temperature of the source of low-grade heat from negative to positive values of the liquid temperature. Heat storage is an insulated tank with hot water is filled with prepared tap water, purified from insoluble substances, it does not contain specific impurities. The cycle of the experimental stand in the $h – lgp$ diagram with the main nodal points is shown in figure 2.

Registration of values was made at the steady-state operation mode: at constant value of $p_к$, and also at constant value of temperature t_s. It should be noted that the operating mode was determined by the temperature value of the low-potential heat source t_s.

The aim of the study is to obtain and accumulate experimental data on the operation of the vapor compression heat pump cycle, as well as the values of the heating coefficient (transformation coefficient or conversion coefficient) of the steam compression heat pump in the conditions of maintaining a constant condensation pressure and changing temperature values of the low-potential heat source, simulating the annual course of the temperature of the natural source of low-potential heat.

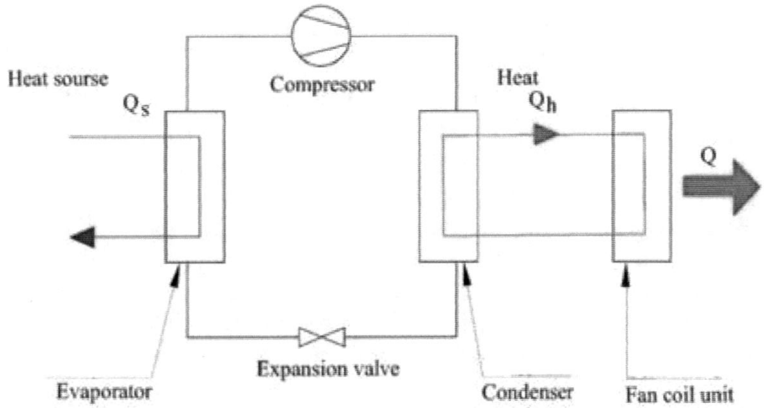

Figure 1. Schematic diagram of experimental stand.

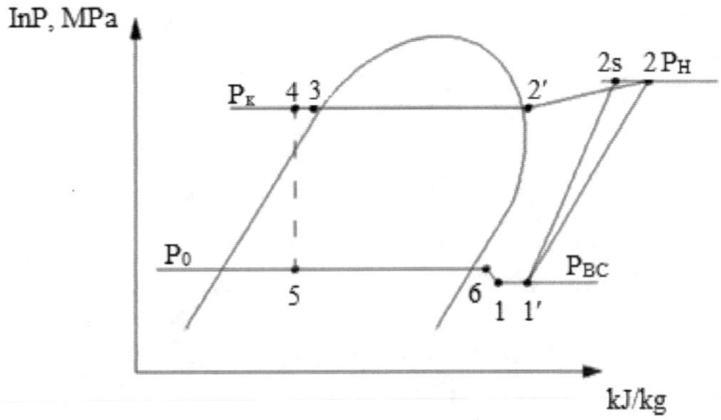

Figure 2. The cycle of the experimental stand in *h* – *lgp* diagram.

Based on the values obtained as a result of the physical modeling the parameters of the cycle at the nodal points were determined, and the necessary calculations were made to estimate the performance of the heat pump: the values of the heat transferred from the heat pump to the consumer Q and the values of the conversion coefficient of the heat pump *COP* were calculated.

The heat transferred from the heat pump to the consumer by means of a fan coil unit is determined by the following dependence:

$$Q = \frac{G \cdot c_a \cdot (t_{a2} - t_{a1})}{3600}, \quad (1)$$

where G – mass air flow through the fan coil unit, kg/h;

c_a – the heat capacity of air, J/(kg·K);

t_{a1} – the temperature of the air at the inlet of the fan coil unit , °C;

t_{a2} – the air temperature at the outlet from the fan coil unit, °C.

The conversion factor of the actual heat pump cycle depends only on the heat Q and the electrical energy expended on the compressor drive W:

$$COP = \frac{Q}{W}. \tag{2}$$

It should be noted that the conversion factor is the most important parameter in assessing the efficiency of the heat pump.

As a result of the experimental study data were obtained on the basis of which the performance data of the heat pump were built. The choice of a low-potential heat source temperature was based on the accumulated long-term climate data in different regions of Russia.

Figure 3 shows the dependence of the temperature change of the working substance in the first circulation circuit t_1, t_2, t_3, t_4, t_5 on the temperature of the source of low-potential heat t_s. Analyzing the presented figure, it can be noted that the increase in the temperature of the source of low-potential heat t_s leads to a reduction in the temperature difference between t_2 and t_5, which positively affects the performance of the heat pump – this leads to an increase in the value of the heating coefficient. Temperature t_3, t_4 remain virtually unchanged, which confirms the maintenance of the given values of the condensation pressure during all the physics experiment.

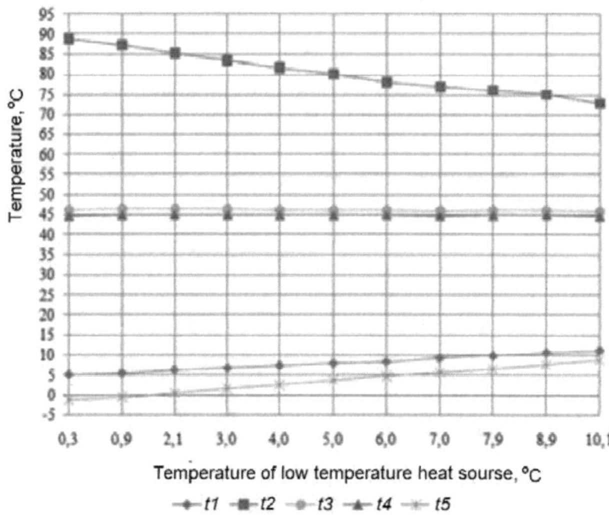

Figure 3. Graph of the controlled temperatures of the working substance depending on the temperature of the low-grade heat source t_s.

From figure 4 one can notice that with increasing value of temperature of low-grade heat source t_s the controlled values of the temperature values at the inlet to the t_{w2} capacitor are linearly reduced. The controlled values of the temperature values at the inlet d of the condenser t_{w1} and the temperature at the outlet of the air heater t_{w3} increase in proportion to the increase in the temperature of the source of low-grade heat t_s. The values of the above parameters when the temperature of the source of low-potential heat t_s is 5.0°C and 6.0°C are out of the General series (linear approximations t_{w1}, t_{w2}, t_{w3}), this is due to the influence of external factors on the system.

The data presented in figure 5 show that as the temperature of the low-potential heat source t_s increases, the amount of heat transferred from the heat pump to the consumer increases. Linear approximation of the calculated data reflects the absence of significant deviations.

Analyzing the data presented in figure 6, we can make the following conclusion: an increase in the temperature of the source of low-potential heat t_s, leads to an increase in the value of the heating coefficient (conversion factor). The difference between the calculated values and the values of the linear approximation does not exceed 3%, which suggests a high accuracy of the values and calculations. There is a direct correlation between the values of the transmitted heat Q and the value of the conversion coefficient *COP*.

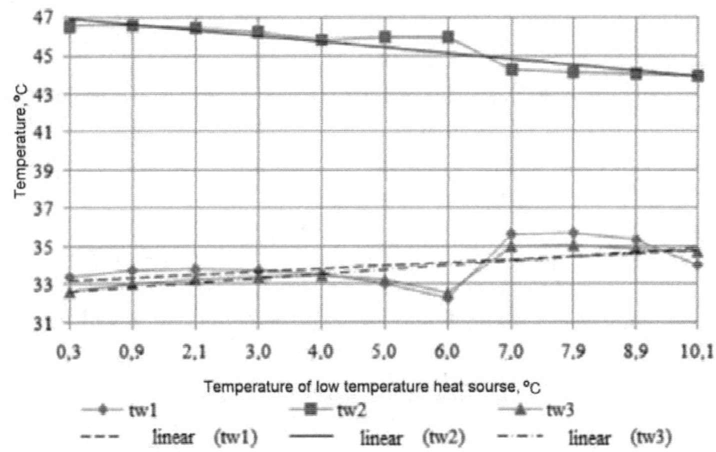

Figure 4. Graph of the controlled parameters of the working fluid on the warm side of the heat pump depending on the temperature of the low-grade heat source t_s.

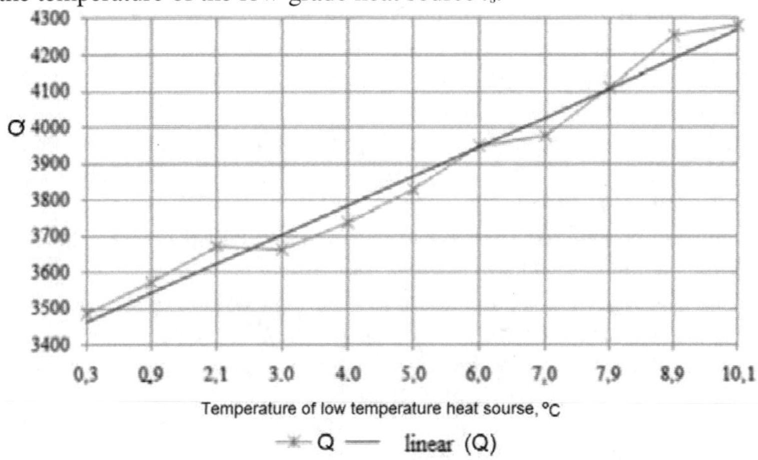

Figure 5. Characteristic of the dependence of heat transmitted to the consumer from the temperature of the source of low-potential heat t_s.

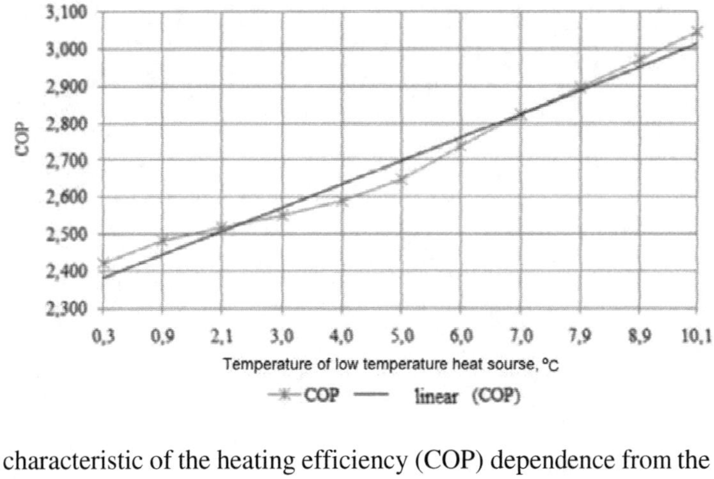

Figure 6. The characteristic of the heating efficiency (COP) dependence from the temperature of the low-grade heat source t_s.

3. Conclusion

As a result of the study, the values of the conversion coefficient of the Vapor compression heat pump in the conditions of maintaining a constant condensation pressure $p_к$ =1.31 MPa and when the temperature of the low-potential heat source changes in the range from 0.3°C to 10.1°C, simulating the annual temperature of the natural source of low-potential heat.

The European Parliament adopted the "Directive on the use of renewable energy sources" in 2008, according to which the use of heat pumps with a conversion factor below the value of 2.875 is prohibited.

Taking into account the requirement of the Directive for European countries, the use of a heat pump on the working substance *R*134*a* and condensation pressure $p_к$ =1,31 MPa, possibly at a source of low-grade heat above 7°C.

After analyzing the values of sea water temperatures in different regions of the Russian Federation, it can be concluded that the use of a heat pump at the declared values is possible only in the coastal zone of the Black sea (Tuapse and Sochi) [8]. In the conditions of energy shortage in the southern regions of Russia as well as in order to reduce operating costs in the cold period of the year for heating in a mild winter in the design of new facilities and buildings, it is necessary to take into account the results of this study.

Reference

[1] Peréz-Lombard L, Ortiz J and Pout C 2008 A review on buildings energy consumption information *Energy Build* **40** 394–398
[2] Neuberger P and Adamovský R 2019 Analysis and Comparison of Some Low-Temperature Heat Sources for Heat Pumps *Energies* **12** 1853
[3] Sivasakthivel T, Murugesan K and Thomas H R 2014 Optimization of operating parameters of ground source heat pump system for space heating and cooling by Taguchi method and utility concept *Appl Energy* **116** 76–85
[4] Zarrella A and De Carli M 2013 Heat transfer of short helical borehole heat exchnagers *Appl Energy* **102** 1477–1491
[5] Choi W, Ooka R and Nam Y 2018 Impact of long-term operation of ground-source heat pump on subsurface thermal state in urban areas *Sustain Cities Soc* **38** 429–439
[6] Hepburn B D P, Sedighi M and Thomas H R 2016 Field-scale monitoring of a horizontal ground source heat system *Geothermics* **61** 86–103
[7] Zhaohong Fang *et al* 2019 *IOP Conf. Ser.: Earth Environ. Sci.* **249** 012028
[8] *Climate of the seas of Russia* [Portal of the Unified state system of information on the situation in the world ocean] 2019 Available at: http://portal.esimo.ru

IOP Conf. Series: Materials Science and Engineering **656** (2019) 012045 doi:10.1088/1757-899X/656/1/012045

Equivalent parameters of thermal comfort of the room

T V Ryabova[1], A B Sulin[1], A A Nikitin[1], A L Emelyanov[1] and K A Bordashev[2,3]

[1]ITMO University, Kronverksky Pr., 49 197101 Saint Petersburg, Russia
[2]Yaroslav-the-Wise Novgorod State University, ul. B. St. Petersburgskaya, 41 173003 Veliky Novgorod, Russia

[3]E-mail: Kirill.Bordashev@novsu.ru

Abstract. Modern approaches to the substantiation of comfort parameters of the microclimate and rooms are mainly focused on the requirements of State Standart ГОСТ Р ИСО 7730 and ASHRAE Standard 55. The article describes a method for calculating the dependencies of the required air temperature on mobility, mobility and radiation temperature difference at various levels of metabolism and insulation of clothing for three classes comfort.
For each class of comfort pre-built tables of iso-comfort microclimatic parameters, which are calculated for different levels of human metabolism and clothing characteristics. The obtained values from the tables of iso-comfort microclimatic parameters are substituted into the empirically derived equation for calculating the equivalent temperature. The results in the form of the equivalent comfortable temperature are approximated as a function of the level of metabolism and the characteristics of the clothes, as a result of which generalized calculated expressions of the parameter of the equivalent comfortable temperature are obtained. The results in the form of the equivalent comfortable temperature are approximated as a function of the level of metabolism and the characteristics of the clothes, as a result of which generalized calculated expressions of the parameter of the equivalent comfortable temperature are obtained. Thus, six data arrays with thermal comfort parameters for rooms of three classes are summarized by six calculated algebraic expressions. The obtained calculated expressions of equivalent comfortable temperature can be used in the design of air conditioning systems and in programming automatic climate control systems.

1. Introduction

By nature of its impact on human heat balance, the microclimate is divided into moderate (neutral), cooling and heating [1]. For a moderate (neutral) microclimate, such a combination of its parameters is characteristic, it provides the heat balance of a person with the environment in the range of ± 2 W, with a fraction of heat transfer from evaporation of sweat not exceeding 30%. If there is negative heat balance below –2 W, the microclimate is considered to be cooling, and with a heat balance above 2 W and if the proportion of heat loss from evaporation of sweat is more than 30%, it is considered to be heating. A comfortable climate is usually defined as "a state of mind that brings satisfaction with the thermal environment" [2]. Human thermal sensations are the result of the interaction of six factors: air temperature, average radiation temperature, relative air mobility, humidity, activity level and thermal characteristics of clothes.

The human thermoregulation system allows for physiological regulation of the heat balance and ensure the thermal comfort of the body in various conditions. Heat is primarily generated by metabolism, due to digestion and muscular activity. At rest, about 56% of the total heat production occurs in the internal organs and about 18% occurs in muscles and skin. During physical exertion, heat production by muscular work increases significantly, the percentage of heat released by muscles can increase up to 90% [3].

Under normal conditions, these processes lead to an average temperature inside the body of about 37°C [4, 5]. The human thermoregulation system tends to maintain this value when disturbances from internal or external conditions increase. The resulting effects of changing climatic conditions produce a response in the form of autonomous physiological responses, as well as activation of behavioral regulation. Autonomic regulation is controlled by hypothalamus, which controls heat flow through dermal blood flow, sweating, and aspiration. Behavioral regulation is manifested in changes in activity and clothing [6, 7]. Figure 1 shows the heat transfer components of the human body for a two-core “core-shell” model [2].

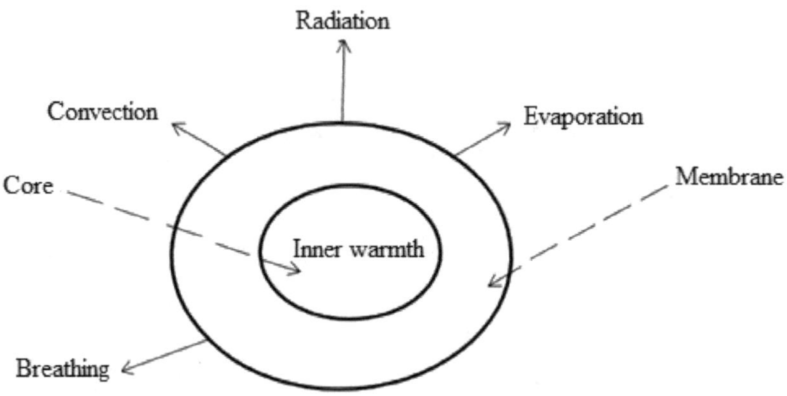

Figure 1. Components of the human body’s heat transfer.

2. Results and discussion

To estimate the thermal state of a person in the internal and external environment, multiparameter biometeorological indices are widely used.

Effective temperature (ET) is a combination of relative air humidity and temperature at which the effect of thermal sensation and heat transfer will be the same. The values of this parameter are expressed in degrees of the effective temperature, that is, in degrees of the temperature of still air saturated with water vapor. When calculating the effective temperature, the approximate dependence proposed by F.A. Missenard is used [6]:

$$\mathrm{ET} = e - 0{,}4(t - 10)\left(1 - \frac{f}{100}\right),$$

where f is the average daily relative air humidity, %; t is the average daily air temperature, °C.

The effective temperature index is used in still air conditions. To account for the effect of air movement, B.A. Aĭzenshtat introduced the concept of equivalent-effective temperature [2].

Equivalent-effective temperature (EET or ET) is determined by the relation:

$$\mathrm{ET} = 37 - \frac{37 - t}{0{,}68 - 0{,}0014f + \frac{1}{1{,}76 + 1{,}4v^{0{,}75}}} - 0{,}29t\left(1 - \frac{f}{100}\right),$$

where t is the air temperature, °C; f is the relative humidity, %; v is the average wind speed.

The index of the normal equivalent-effective temperature (NEET) is known, it is used for the analytical assessment of the thermal sensation of a dressed person. This index was proposed by I.V.

Butieva, it indicates sensitivity, taking into account the influence of wind and is determined by the formula [2].

$$\mathrm{NEET} = 0{,}8 * \mathrm{ET} + 7°\mathrm{C}$$

In 2005 S.S. Andreev proposed a new index — the biologically active temperature (BAT), which makes it possible to determine the complex effect the temperature, air humidity, wind speed, and total solar radiation have on humans. The values of this indicator can be calculated by the formula [2]

$$\mathrm{BAT} = 0{,}8 * \mathrm{NEET} + 9°\mathrm{C}$$

E.G. Golovina introduced another indicator — the radiation-equivalent-effective temperature (REET), which is calculated according to the following dependencies [2]:

$$\mathrm{NEET} = 12\lg[1 + 0{,}025T + 0{,}001(T-8)(f-60) - 0{,}45(33-T)\sqrt{v} + 185B],$$

where T is the air temperature, °C; f is the relative humidity, %; v is the wind speed, m/s; B is the solar radiation absorbed by the body surface, kW/m^2.

In this study, the task was to generalize the iso-comfort microclimate parameters to the given level of comfort for each class of premises.

The methodology for the integrated assessment of the comfort level of a moderate microclimate according to State Standart ГОСТ Р ИСО 7730 [1] is based on the definition of PMV and PPD thermal comfort indices. The predicted average PMV rating is an index that is used to determine the average sensitivity to the microclimate parameters on a 7-point scale. PPD is an indicator that sets the predicted percentage of people dissatisfied with the microclimate who are either too warm or too cold. The functional dependence PPD = f(PMV) is obtained on the basis of the mathematical processing of the results of the subjective assessment of microclimatic conditions for a large sample of subjects. In Figure 2 this functional dependence is given in graphical interpretation for the premises of three comfort classes A, B, and C [1].

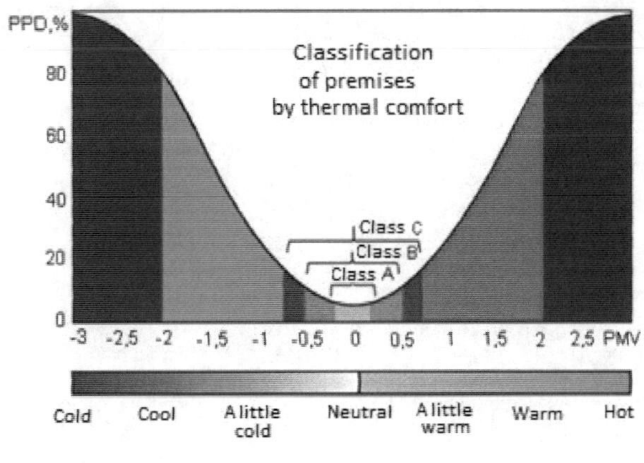

Figure 2. Functional dependence PPD = f(PMV).

This function, in fact, is a regression equation of the form:

$$PPD = 100 - 95\exp(-0{,}03353 * PMV4 - 0{,}2179 * PMV2),$$

The equation relates the thermal sensations of a group of people in a particular environment to the results of the calculation using the heat balance equation [6]. This semi-empirical equation is used to predict the average rating on the ordinary rating scale of thermal comfort for a group of people. The standard provides for three classes of rooms in terms of thermal comfort.

The premises of comfort class A: –0,2 < PMV <+0,2; (PPD < 6 %).

The premises of comfort class B: –0,5 < PMV < +0,5; (PPD < 10 %).

The premises of comfort class C: –0,7 < PMV < +0,7; (PPD < 15 %).

In [3, 4], the solution of the inverse problem with respect to the method for determining the level of comfort is described [5]. The method of successive approximations is used to calculate the totality of microclimate parameters that provide a given level of thermal comfort (iso-comfort parameters). The target functions of the calculations were to determine the required air temperature depending on mobility and relative humidity in the absence of radiation heat exposure (the first series of calculations) and to determine the required air temperature depending on mobility and radiation temperature difference at a constant (50%) relative air humidity (second series of calculations). Radiation temperature difference here means the difference between the air temperature and the average radiation temperature of the enclosing surfaces. Table 1 and figure 3 present the calculations of iso-comfortable air temperature as a function of mobility and radiation temperature difference for specific characteristics of clothing and metabolic rate for a given comfort class B.

Table 1. The dependence of air temperature on mobility and radiation temperature difference for PPD conditions = 10%, thermal resistance of clothing is 1.0 clo. $t_r = t_a$, °C, work category 1a (68 W/m^2; 1.1 met).

		PMV = + 0.49		
Δt_r,°C	v = 0.1, m/c	v = 0.3, m/c	v = 0.5, m/c	v = 0.7, m/c
–8	27.60	27.63	29.27	27.65
–4	25.81	26.29	28.10	26.63
0	23.98	24.92	26.90	25.59
4	22.12	23.52	25.67	24.51
8	20.356	22.08	22.901	23.41
		PMV = – 0.49		
Δt_r,°C	v = 0.1, m/c	v = 0.3, m/c	v = 0.5, m/c	v = 0.7, m/c
–8	23.23	23.72	23.94	24.066
–4	21.45	22.40	22.81	23.06
0	19.63	21.04	21.65	22.03
4	17.89	19.65	20.46	20.97
8	16.252	18.227	19.248	19.877

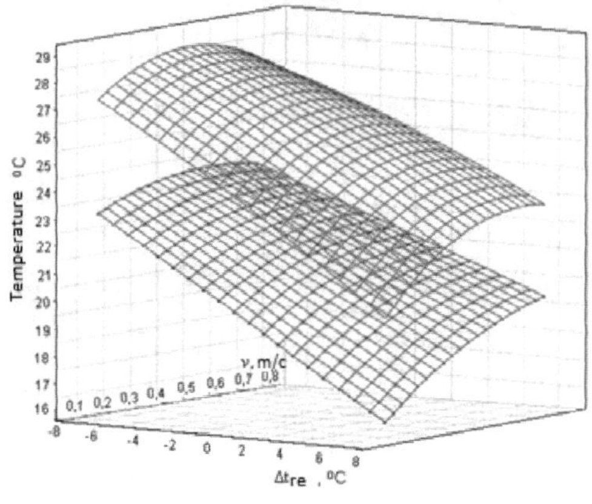

Figure 3. Surfaces of iso-comfort temperatures in coordinates "mobility – radiative temperature difference".

Thus, for each class of premises a database was obtained for a wide range of clothing characteristics (clo) and metabolic rate (met), consisting of tables of iso-comfortable parameters.

A fragment of the database of iso-comfort parameters for premises of class B is shown in figure 4.

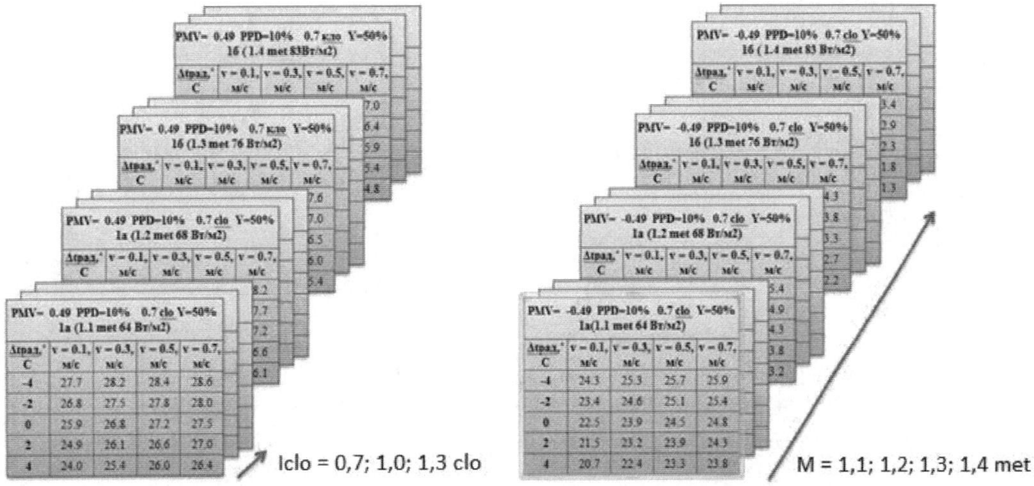

Figure 4. Fragment of the database of iso-comfort parameters.

Based on the described database, we obtained a generalizing criterion of thermal comfort, the use of which greatly simplifies the calculation of iso-comfort parameters. The derivation of this criterion is based on the empirically derived equation for the equivalent temperature proposed in [3] (a parameter corresponding to the same thermal sensations). This equation is a function of air temperature, average radiation temperature, mobility, and clothing characteristics:

$$t_{eq} = 0{,}55t_a + 0{,}45\bar{t_r} + \frac{0{,}24-0{,}75\sqrt{v_a}}{1+I_{cl}}(36{,}5 - t_a),$$

where t_{eq} is the equivalent temperature, °C; t_a is the air temperature, °C; t_r is the average radiation temperature, °C, v_a is the air velocity, m/s; I_{cl} is the thermal resistance of clothing, clo.

When substituting into this equation the values of the parameters corresponding to a given level of thermal comfort from the database of iso-comfort parameters, the equivalent comfort temperature is calculated. This parameter corresponds not only to the same (equivalent) thermal sensations, but also satisfies the required level of thermal comfort.

Let us consider an example of calculating the equivalent comfortable temperature. Table 2 shows iso-comfort parameters for premises of class B for a given metabolism value of 1.1 met and thermal resistance of clothing 1 clo.

Table 2. Dependence of air temperature on mobility and radiation temperature difference for the following conditions: thermal resistance of clothing is 1.0 clo, metabolic rate is 1.1 met.

	PMV = + 0.49			
Δt_r,°C	v = 0.05, m/c	v = 0.1, m/c	v = 0.15, m/c	v = 0.2, m/c
–4	26.03	26.26	26.43	26.54
–2	25.12	25.35	25.60	25.78
0	24.22	24.43	24.76	25.00
2	23.35	23.49	23.92	24.22
4	22.48	22.55	23.06	23.42

PMV = – 0.49				
Δt_r,°C	v = 0.05, m/c	v = 0.1, m/c	v = 0.15, m/c	v = 0.2, m/c
–4	21.98	22.20	22.53	22.76
–2	21.12	21.29	21.71	22.00
0	20.27	20.37	20.87	21.23
2	19.44	19.44	20.03	20.45
4	18.60	18.60	19.19	19.66

The values of iso-comfort parameters (air temperature, air mobility, average radiation temperature) and thermal resistance of clothes are substituted into the formula of equivalent temperature and then the equivalent comfortable temperature is calculated.

$$t_{eqc} = 0{,}55t_{ac} + 0{,}45\overline{t_{rc}} + \frac{0{,}24-0{,}75\sqrt{v_{ac}}}{1+I_{cl}}(36{,}5 - t_{ac}), \quad (1)$$

where t_{ac} is the comfortable air temperature, °C; $\overline{t_{rc}}$ is the comfortable average radiation temperature, °C; v_{ac} is the comfortable air velocity, m/s; I_{cl} is the thermal resistance of clothing, clo.

As a result, based on the data of Table 2, we determine the values of the equivalent comfortable temperature if the thermal resistance of clothes is 1.0 clo and the metabolic rate is 1.1 met (table 3).

Table 3. Equivalent comfortable temperature for the following conditions: thermal resistance of clothing is 1.0 clo, metabolic rate is 1.1 met.

PMV = +0.49				
Δt_r,°C	v = 0.05, m/c	v = 0.1, m/c	v = 0.15, m/c	v = 0.2, m/c
–4	24.61	24.47	24.38	24.26
–2	24.63	24.47	24.42	24.37
0	24.66	24.45	24.46	24.45
2	24.73	24.41	24.50	24.53
4	24.79	24.37	24.52	24.60
Equivalent comfortable temperature is 24,61 °C				
PMV = – 0.49				
Δt_r,°C	v = 0.05, m/c	v = 0.1, m/c	v = 0.15, m/c	v = 0.2, m/c
–4	20.70	20.42	20.38	20.30
–2	20.78	20.41	20.44	20.41
0	20.86	20.39	20.48	20.50
2	20.96	20.36	20.51	20.58
4	21.05	20.43	20.55	20.66
Equivalent comfortable temperature is 20,65 °C				

Consequently, the values of the equivalent comfortable temperature for the conditions of thermal resistance of clothing of 1.0 clo and metabolic rate of 1.1 met for premises of comfort class B are:

t_{eqc} = 24.64 °C if PMV = +0.49;

t_{eqc} = 20.65 °C if PMV = –0.49.

Thus, the iso-comfort parameters given in Table 2 are summarized by a single value equivalent to a comfortable temperature.

Table 4 shows the results of calculations of the equivalent comfortable temperature for premises of comfort class B.

Table 4. Equivalent comfortable temperatures for premises of comfort class B.

	PMV = + 0.49		
	0.7 clo	1.0 clo	1.3 clo
1.1 met	26.66	25.28	23.93
1.2 met	26.07	24.61	23.19
1.3 met	25.69	24.18	22.71
1.4 met	24.92	23.31	21.75
	PMV = – 0,49		
	0.7 clo	1.0 clo	1.3 clo
1.1 met	23.73	21.77	19.85
1.2 met	22.76	20.65	18.58
1.3 met	22.13	19.91	17.75
1.4 met	20.86	18.45	16.09

For convenience of practical application, the calculation results are approximated using the built-in subroutines of the Excel spreadsheet editor in the form of exponential functions:

for PMV = +0.49, $t_{eqc} = 40{,}23 * 0{,}77\text{met} * 0{,}82\ \text{clo}$;

for PMV = –0.49, $t_{eqc} = 53{,}60 * 0{,}77\text{met} * 0{,}70\ \text{clo}$.

The data generalizing the calculated expressions of the parameter of equivalent comfort temperature correspond to the required level of thermal comfort in the premises of comfort class B. Thus, a single generalizing criterion for determining the parameters of thermal comfort is obtained.

Table 5 shows the calculated expressions for the equivalent comfortable temperature for premises of all comfort classes according to the standard [1].

Table 5. Parameter of equivalent comfort temperature.

Class of premises according to State Standard ГОСТ Р ИСО 7730	Predicted average PMV score, points	Equivalent comfortable temperature, °C
A	+0.19	$t_{eqc} = 43{,}04 * 0{,}72^{met} * 0{,}79^{clo}$
	–0.19	$t_{eqc} = 48{,}32 * 0{,}64^{met} * 0{,}73^{clo}$
B	+0.49	$t_{eqc} = 40{,}23 * 0{,}77^{met} * 0{,}82^{clo}$
	–0.49	$t_{eqc} = 53{,}60 * 0{,}58^{met} * 0{,}70^{clo}$
C	+0.69	$t_{eqc} = 38{,}63 * 0{,}80^{met} * 0{,}84^{clo}$
	–0.69	$t_{eqc} = 44{,}25 * 0{,}65^{met} * 0{,}72^{clo}$

The calculated expressions of the comfort parameters of the microclimate using the equivalent comfort temperature, obtained from formula (1), have the form:

– comfortable air temperature

$$t_{ac} = \frac{t_{eqc} - 0{,}45\Delta\overline{t}_{rc} - \dfrac{0{,}24 - 0{,}75\sqrt{v_{ac}}}{1 + I_{cl}} \cdot 36{,}5}{1 - \dfrac{0{,}24 - 0{,}75\sqrt{v_{ac}}}{1 + I_{cl}}}; \tag{2}$$

– comfortable average radiation temperature

$$\Delta\bar{t}_{rc} = \frac{t_{eqc} - t_{ac} - \dfrac{0,24 - 0,75\sqrt{v_{ac}}}{1 + I_{cl}}(36,5 - t_{ac})}{0,45}; \quad (3)$$

– comfortable air velocity

$$v_{ac} = \left[\frac{0,24 - \dfrac{t_{eqc} - t_{ac} - 0,45\Delta\bar{t}_{rc}}{36,5 - t_{ac}}(1 + I_{clo})}{0,75} \right]^2. \quad (4)$$

As the approbation, the parameter of equivalent comfort temperature proposed in this work, we will determine the necessary average radiation temperature in the premise of comfort class B.

The life support system maintains an air temperature of 25 °C in the room, the air mobility is 0.3 m/s, staff's clothing has a characteristic of 1.1 clo, the level of metabolism during work is 1.15 met.

From the expressions for the equivalent comfortable temperature for the premises comfort class B (table 5), we calculate the equivalent comfortable temperature:

for PMV = +0.49, t_{eqc}= 23.83

for PMV = –0.49, t_{eqc} = 19.39.

As a result of the calculation using formula (3), the following values were obtained for the desired average radiation temperature:

for PMV= + 0.49, $\bar{t}_{rc}$ = 24.49 °C;

for PMV= –0.49, $\bar{t}_{rc}$ = 14.60 °C.

Thus, to ensure the state of comfort of the corresponding premise of class comfort B, given the initial data, the average radiation temperature should be in the range of 14.6 °C to 24.5 °C.

3. Conclusion

The method of determining the necessary combinations of microclimate parameters and personal parameters that provide the required level of thermal comfort in accordance with the State Standart ГОСТ Р ИСО 7730 is considered. It is proposed to use an empirically derived calculation expression for the equivalent temperature, on the basis of which a generalizing parameter of the equivalent comfort temperature is obtained, which corresponds to the required level of thermal comfort in the room. Six calculated data arrays with thermal comfort parameters for premises of comfort classes A, B and C are summarized by six algebraic expressions. These calculated expressions of equivalent comfortable temperature can be used in the design of air conditioning systems and in programming of automatic climate control systems.

References

[1] ANSI/ASHRAE Standard 55 2017 *Thermal Environmental Conditions for Human Occupancy* 41p

[2] Holopainen R, Tuomaala P, Hernandez P, Häkkinen T, Piira K and Piippo J 2014 Comfort assessment in the context of sustainable buildings: Comparison of simplified and detailed human thermal sensation Methods *Building and Environment* Vol 71 60–70

[3] Halawa E and Van Hoof J 2012 The adaptive approach to thermal comfort: a critical overview *Energy and Buildings* Vol 51 101–110

[4] State Standard GOST R ISO 7730-2009 2011 *Analytical determination and interpretation of thermal comfort using calculation of the PMV and PPD indices and local thermal criteria of comfort* (Moscow: Standartinform) p 38

[5] Syed I-ul-H G, Muhammad H K, Muzaffar A 2016 Revisiting Fanger's thermal comfort model using mean blood pressure as a bio-marker: An experimental investigation *Applied Thermal Engineering* Vol 109 Part A 35–43
[6] Poddubnyi R A, Ryabova T V and Sulin A B 2016 Construction of matrices from comfortable values of microclimate parameters *Vestnik DGTU Tekhnicheskie nauki* **41(2)** 68–74
[7] Sulin A B, Ryabova T V, Ivanov S V and Poddubnyi R A 2017 Calculation substantiation of microclimate parameters with a given level of thermal comfort *Kholodil'naya tekhnika* **4** 37–41

IOP Conf. Series: Materials Science and Engineering **656** (2019) 012046 doi:10.1088/1757-899X/656/1/012046

Determination of the allowable minimum thickness of protective coating for welding in carbon dioxide

S B Sapozhkov[1,2], S I Arendateleva[1] and O V Ushakova[1]

[1]Yaroslav-the-Wise Novgorod State University, ul. B. St. Peterburgskaya, 41 173003, Veliky Novgorod, Russia

[2]E-mail: ssb@novsu.ru

Abstract. Among industrial technologies, carbon dioxide welding occupies a leading position. However, this welding method is characterized by increased spattering of metal and splashing it on the welded piece of work. A protective coating layer applied to metal is effective means of preventing the consequences of this disadvantage. The paper developed the method for determining the minimum allowable thickness of the coating, sufficient to prevent the interaction of molten metal drops with the surface of the welded pieces of work.

1. Introduction

As is known, protective coatings are used to protect the surface of welded products and welding equipment from splashes of molten metal during welding in CO_2 [1–5].

However, the thickness of required coating layer is assigned only on the basis of experience. In this case, a small thickness of the coating leads to poor quality protection of the surface of the piece of work, a large layer thickness to excessive consumption of the coating [1].

Currently, there are a number of methods for determining the thickness of the protective coating [1, 6], but the disadvantages of these methods are the lack of specific recommendations for calculating the thickness of the coating depending on the thickness of the metal being welded.

2. Main part

The task of the article was to determine the allowable minimum thickness of protective coating layer depending on welding mode, thermophysical characteristics of the coating, the temperature of drop at the moment of contact, the contact diameter of the interaction of drop with the surface of the piece of work to be welded and the thickness of the metal to be welded

In solving the problem, it was assumed that:

1. At the moment of flying a drop from the arc gap, the amount of heat contained in it is determined by the equation:

$$q = IU\eta, \quad (1)$$

where q is the amount of heat introduced by the arc into the metal of the electrode per unit of time, J/s; I is welding current, A; U is arc voltage, V; η is effective efficiency of heating of the electrode metal by electric arc.

2. During the flying until the moment of contact with the surface of the metal being welded, the drop loses some of the heat to radiant and convective heat transfer. The specific flux of total heat transfer by drop to radiative and convective heat exchange with the environment is determined by the equation:

$$q_{To} = \alpha(T_k - T_o), \tag{2}$$

where α is total surface heat transfer coefficient which is the sum of the coefficients of convective heat transfer and radiant heat transfer; $T_к$ and T_o are drop temperature at the moment of flying from the electrode and the environment, respectively, °C.

3. It follows from the boundary adiabatic condition of the 3-rd kind that drops during flying cool insignificantly, i.e. we have: a → 0.

4. The amount of heat contained in drop at the moment of contact with the surface of the welded pieces of work is determined by the equation:

$$q_k = q - q_{To}. \tag{3}$$

5. Energy heat content of drop should be less than the amount of heat required for decomposition and evaporation of the protective coating layer δ.

To determine the thickness of the protective coating layer, we considered the process of temperature increment on the plate surface at the moment of introduction of heat from normally circular source with the condition that it spreads in depth in time t_0 in the direction of axis OZ (figure 1).

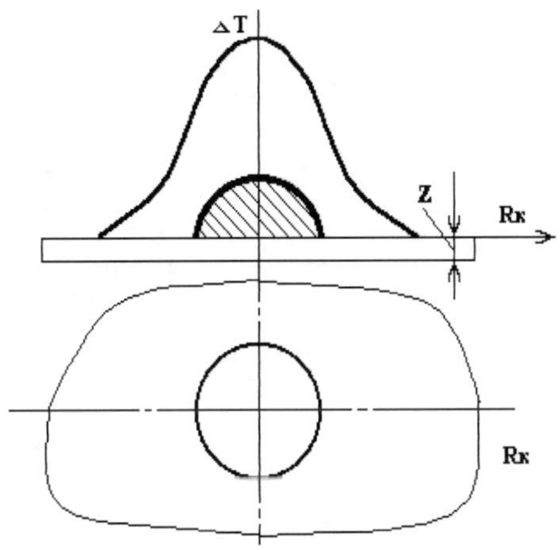

Figure 1. The scheme of heating the surface of the welded metal at the point of contact with drop of molten metal.

Such a process is expressed by the equation:

$$\Delta T = \frac{2Q}{cp} \frac{e^{-R_k^2}/[4a(t+t_0)]}{4\pi a(t+t_0)} \frac{e^{-z^2/(4at)}}{(4\pi at)^{1/2}}, \tag{4}$$

where ΔT is the temperature increment at the considered point, °C; Q is the heat contained in drop at the point of contact with the surface of the metal being welded (enthalpy ΔH), J; cp is the product of the heat capacity by the density of the droplet material, J/deg cm^3; $R_к$ is the radial distance from the considered point to the axis of the imprint of drop (radius of the imprint of the drop), cm; a is the coefficient of thermal diffusivity, cm^2 / s; t is the time counted from the moment of introduction of heat, s; z is the depth to which heat is distributed (thickness of the metal being welded), cm; k is the concentration coefficient of the heat flux of drop.

Time t_0, from which the heat propagates in depth along OZ axis, is determined by the equation [6–9]:

$$t_0 = 1/(4ak), \tag{5}$$

where a is thermal diffusivity of the protective coating, k is concentration coefficient of the heat flux of drop and is determined by the equation:

$$k = (3.46 / dk)^2, \tag{6}$$

where *dк* is contact diameter of interaction of drop with surface, см.

If you specify the condition that q is the amount of heat contained in drop at the moment of its contact with the surface of the welded piece of work; cp is the volumetric heat capacity of the protective coating; r is the radius of the drop; t is the time elapsed since the contact of the droplet with the protective coating, and z is the depth to which the heat transferred by the drop to the protective coating layer extends, then it can be assumed that z will be the thickness of the layer of protective coating necessary to prevent splashing on the surface of the welded part (hereinafter δ).

By mathematical transformations, the equation was obtained:

$$\delta = \sqrt{\left[\ln\left(\frac{(IU\eta)tk}{cp\pi T\delta_1}\right)\right]/k} \tag{7}$$

The complexity of further determining the thickness of the protective coating layer lies in the absence of data on the volumetric heat capacity and thermal diffusivity of the coatings used to protect the surface of the welded products. The coefficient of thermal diffusivity is determined by the equation:

$$\alpha = \lambda / cp. \tag{8}$$

The density of protective coatings p was determined experimentally by equation:

$$p = \frac{m}{Q}, \tag{9}$$

where m is the mass of the protective coating, kg, Q is the mass volume of the protective coating, m^3.

The specific heat capacity of the protective coatings was determined by calorimetric measurement (through the heat of combustion of the material under study).

To determine the thermal conductivity coefficient λ, the plate method was used [8]. The principle diagram of the device is presented in figure 2.

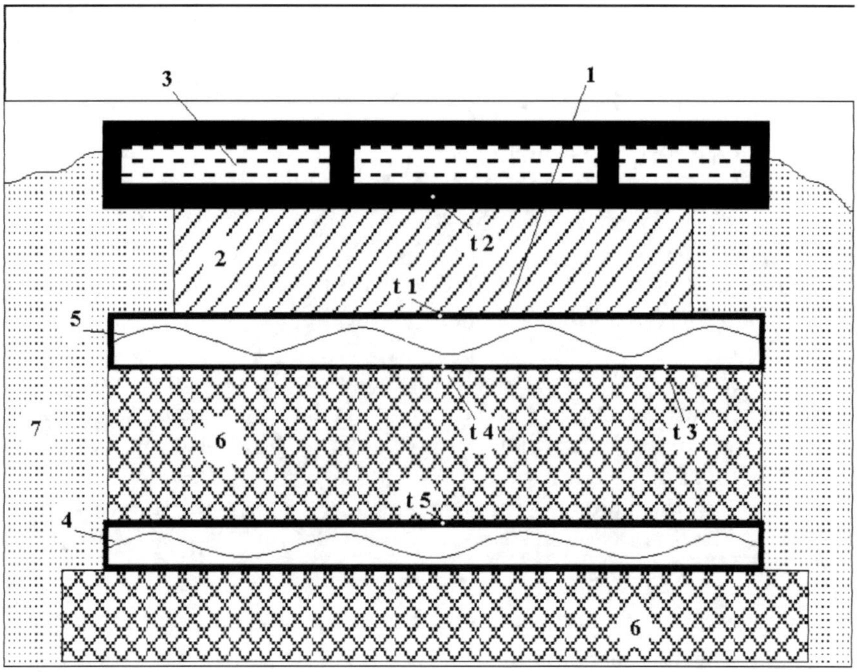

Figure 2. Scheme of device for determining the coefficient of thermal conductivity of protective coatings on the plate principle [9].

Here, in the middle, there is flat electric heater 1, test sample 2 and refrigerator 3 above it. To compensate for heat leakage, additional electric heaters 4–5 are installed from below and from the sides. The device is carefully isolated from heat loss to the external environment, t_1–t_5 – thermoelements for temperature measurement. At steady state of the system, all the heat released in the main heater 1 will pass through the sample and will be absorbed by the refrigerator. Since the amount of heat released is H and the temperatures t_1 and t_2 are known, the thermal conductivity coefficient is determined from the following relationship:

$$\lambda = \frac{H\delta_0}{F(t_1 - t_2)}, \qquad (10)$$

where F is the surface area of the main heater, mm^2; H is the thermal power of the heater; δ_o is the thickness of the layer of the protective coating under investigation, mm.

The following equation was used to determine H:

$$H = I^2 R, \qquad (11)$$

where I is the electric current value, A; $R = U/I$ is the electrical resistance of the heater, V/A.

Measurements of I and U, t_1 and t_2 should be made only after the temperature t_3 is equal to t_1, and the temperature t_4 is equal to t_5. Only in this case, all the heat released in the main heater 1 really passes through test sample 2 [8].

According to the method described above, the thermophysical properties of a number of protective coatings were determined, the data of which are summarized in table 1.

Table 1. Volumetric heat capacity of protective coatings.

S No.	Coating composition	cp, J/deg cm^3
1	Sulfite-alcohol bard (GOST 8578) – 140g, soap – 30g, soda ash – 25g, water – 1000g	928.1
2	Sulfite-alcohol bard (GOST 8578)-60 – 100g, soap – 25–45g, soda ash – 15–25g, kaolin – 25–50g, water – 1000g	985.5
3	50-80g of sulphite-alcohol bards (GOST 8578), 30-40g of soap, 15–20g of soda ash and 30–60 g of abrasive waste (Patent No. 2134186), water –1000 g.	1254.5
4	Zircon (TsMTU 4469-54) – 40%, polyvinyl butyral (GOST 9439) – 3%, solvent – 646 (GOST 18188) – 57%	1101.1

Using thermal diffusivity and volumetric heat capacity cp of protective coatings according to equation (7), calculations were carried out to determine the minimum layer of protective coatings. These calculations are summarized in table 2.

Таблица 2. Minimum layer of protective coatings.

S No.	Coating thickness, mm		Error in %
	Theoretic	Experimental	
1	0.0348	0.0366	5.2
2	0.024076	0.0273	13.4
3	0.02418	0.0252	4.04
4	0.05548	0.0614	10.7

Based on calculated data, nomograms of the dependence of the thickness of the protective layer of coating on the thickness of the product being welded (figure 3, a) and welding current (figure 3, b) were constructed.

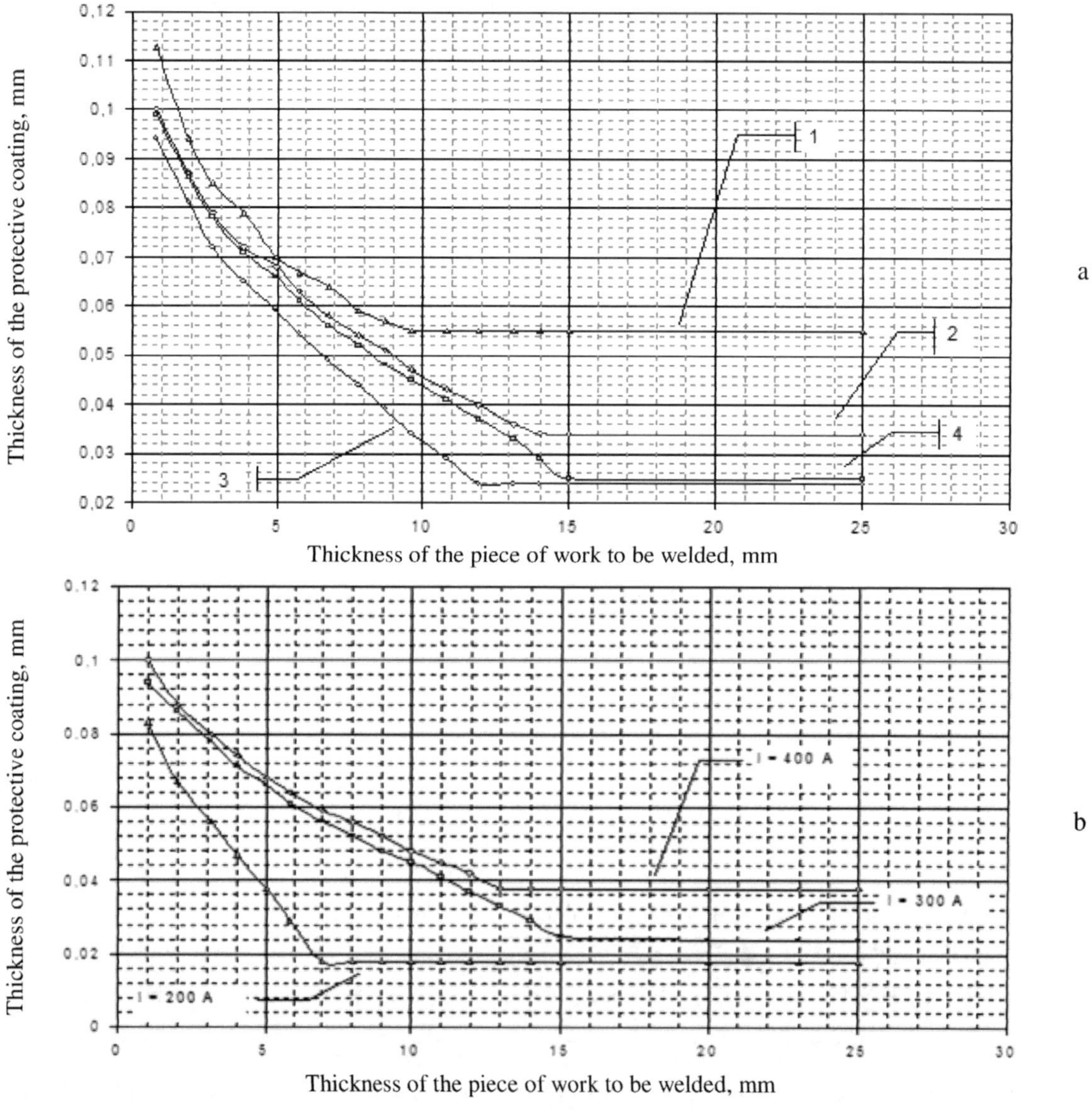

Figure 3. Nomograms of dependence of the thickness of the protective coating on thickness of the piece of work to be welded (a) (for designations, see table 1) and for coating No. 4 on the welding current (b)

To determine the error of the measurement results and the data obtained by calculation, the results were processed according to the method proposed in [7–8]. Such processing of the results consists in estimating the value of the sample mean (arithmetic average) and determining its accuracy.

Analysis of theoretical and experimental data showed that the allowable minimum coating thickness, sufficient to prevent strong adhesion of drops to the surface, for water-based coatings reaches 200 or more microns, for solvent-based coatings –646 ranges from a few microns to several tens of microns , and coatings based on liquid glass – from 20 to 30 microns.

3. Conclusions

1. Method has been developed for determining the allowable minimum layer thickness of protective coating depending on the welding mode, thermal characteristics of the coating, drop temperature at the moment of contact, the contact diameter of the interaction between drop and surface, and the thickness of the piece of work being welded. It has been established that the allowable minimum coating thickness, sufficient to prevent durable adhesion of drops to the surface, for water-based coatings reaches 200 and more microns, for solvent-based coatings –646 ranges from a few microns to several tens of microns, and for coatings on based on liquid glass – from 20 to 30 microns.

2. It has been established that the strength of adhesion of splashes with the surface of piece of work being welded and the nature of their interaction in the contact zone are determined by the following factors: the nature of the contacting materials, the energy heat content of drop, which is sufficient to destroy the oxide films, the surface roughness and contact diameter of interaction.

References

[1] Fedko V T, Sapozhkov S B, Sokolov P D and Yastrebov A P 2004 *Elements of theory and technology of anti-spatter surface shielding when welding in carbon dioxide* Monograph (Tomsk: TSU Publishing) p 138

[2] Solonenko O P, Kudinov V V, Smirnov A V, Cherepanov A N, Popov V N, Mikhalchenko A A and Kartaev E V 2005 Micro-Metallurgy of Splats: Theory, Computer Simulation and Experiment *JSME Inter. J. Ser.* B Vol. 48 **3** pp 366–380 DOI: 10.1299/jsmeb.48.366

[3] Knotek O and Elsing R 1987 Monte Carlo Simulation of Lamellar Structure of Thermally Sprayed Coatings *Surf. Coat. Technol* **32** p 261–271

[4] Ghafouri-Azar R, Mostaghimi J, Chandra S and Chamchi M A 2003 Stochastic Model to Simulation of a Thermal Spray Coating *J. of Thermal Spray Techn.* Vol 12 **(1)** p 53–69

[5] Steinke T B, Ker M 2006 Monte Carlo Simulation of Thermal Sprayed Coatings *Proc. ITSC'06 Seattle* (USA, May 15–18) p 329–334

[6] Fedko V T and Sapozhkov S B 1997 Processing the results of a passive experiment by studying the dependence of the thickness of the protective coating layer on the varying value of thermal conductivity of this coating *Proc. of the 10th scientific. conf.* (Urga: Published TPU) p 81–83

[7] Fedko V T, Tomas K I and Sapozhkov S B 1998 Protecting the surfaces of welded components against molten metal splashes in CO2 welding *Welding International* **1** p 58–62

[8] Sapozhkov S B and Fedko V T 2002 Determination of coating thickness to protect the surface from molten metal splashes when welding in carbon dioxide *Metal technology* **5** p 12–16

[9] Mikheev M A, Mikheeva I M 1961 *Short course of heat transfer: Textbook for non-energy specialties* (M.–L.: Gosenergoizdat publishing) p 180–182

Technological method of assessing the quality of adhesion of the coating to the basis of the tool material during indentation with the exposure time of loading the sample

I V Shvetsov[1,4], B Y Mokritsky[2] and D V Zhelnov[3]

[1] Yaroslav-the-Wise Novgorod State University, ul. B. St. Peterburgskaya, 41 173003, Veliky Novgorod, Russia
[2] Komsomolsk-on-Amur State Technical University, Russia
[3] Department of Expertise and Certification of the Union "St. Petersburg Chamber of Commerce and Industry", Russia

[4]E-mail: Igor.Shvetsov@novsu.ru

Abstract. Techniques of assessing the quality of adhesion of the coating to the base of a tool material during penetration of the indenter with the sample loading time exposure have been considered. Data of efficient tool hardening have been given. Comparative tests of the tool subjected to various kinds of hardening thermoplastic processing show relative increase in its performance. Within the given work, we understand the quality of tools as the correspondence of controlled parameters of physical mechanical characteristics of tools to those values which define the performance properties of a metal cutting tool, or, comparing them, it is possible to prefer those which have performance advantages.

1. Introduction

Within the given work, we understand the quality of tools as the correspondence of controlled parameters of physical mechanical characteristics of tools to those values which define the performance properties of a metal cutting tool, or, comparing them, it is possible to prefer those which have performance advantages.

Below, the information about individual techniques (processes, ways) of assessing the quality of tool materials is given.

According to the task set, a solution for expressive operative assessment of the quality of adhesion of the coating to the base in different tool materials was developed. The solution allows selecting that material which has the best prognosis of adhesion quality according to parameters of acoustic emission. The solution was developed because often (when developing new materials, analyzing data, etc.) it is important to know both the hardening adhesion value itself and the comparative result (better or worse than in any other material) of the tool quality assessment.

2. Results and Discussions

The solution provides several loading phases, which allow relaxing stresses, when introducing an indenter. The supposed loading pattern is illustrated in figure 1. In the first phase shown in figure 1*b*, the indenter penetrates the sample to the depth exceeding the coating thickness. As a result of penetrating, there occurs a plastic flow shear of the coating, deformation of the base, formation of cracks in the coating and interstices between the coating and the base, resulting in the initial act of the coating exfoliation. All these processes generate signals of AE which are registered (interval $t_0...t_1$) by the diagram record «total count — time». When the indenter stops penetrating, the AE signals practically cease. For some time (interval $t_1...t_2$) the indenter is exposed under the load without displacing, then the load is removed and the indenter is taken out. In this case, under the action of inner stresses the coating is deformed by the way of peeling from the base, i.e., the interstices between the coating and the base merge. This is also accompanied by AE signals generation (an interval $t_2...t_3$).

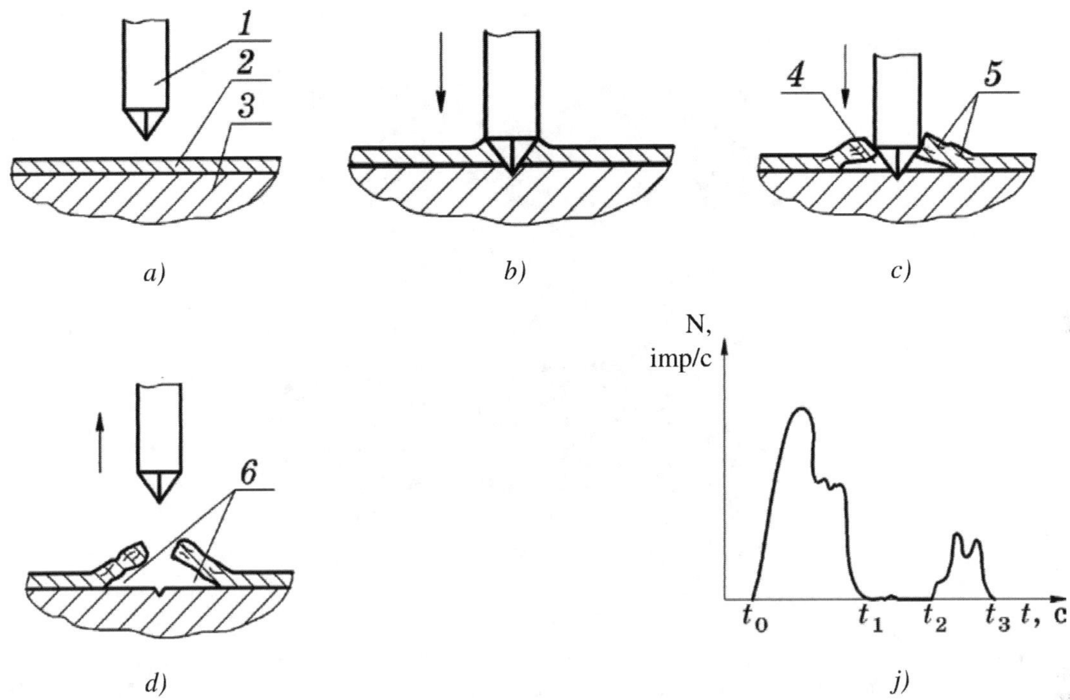

Figure 1. Diagram of implementing of quality assessment of adhesion of the coating to the base by indenting acoustic method: *a*– initial phase; *b*– implementation phase; *c*– dwell phase; *d*– unloading phase; *j*– diagram «total count of AE – time»; *1* –indenter; 2 – coating; *3*– base; *4*– interstices; *5*– cracks; *6*– fused interstices.

Total count of AE pulses for the whole testing period (interval $t_0...t_3$) is designated by N_1, total count for the loading time ($t_2...t_3$) is designated by N_2. They are used to calculate the parameter $K = (N_1 - N_2)/N_1$ which characterizes the quality of adhesion of the coating to the base. It is also considered that the closer is the parameter *K* to unit, the higher is the quality (strength) of adhesion to the base.

Trials of the given technique compared to other similar techniques have shown its acceptability, reliability, simplicity. Implementing this technique, it is possible to reach the absolute parameter of the quality of adhesion, i.e., the strength of adhesion of the coating to the base, but under the given conditions it is advisable to speak about relative trials due to the fact that if this sample has the parameter *K* closer to unit than in the other sample, then the quality of adhesion of the coating to the base is higher.

An example of registration of total count of AE signals is shown in figure 2, where various phases of testing the sample are expressed.

It shows that the intensity of AE signals increased during the indenting (the left part on the record of the signals), then it decreased and stabilized during the dwell of the indenter (the central part on the record of the AE signals), then increased dramatically and destabilized during the withdrawal of the indenter out of the sample (the right part on the record of the AE signals).

The use of a relative, rather than an absolute value as an evaluative parameter eliminates the influence of a number of random factors on the result, and, particularly, it eliminates the influence of the state of the indenter, preparation of sample surfaces, etc.

Thanks to this technique, it is possible to arrange operatively the random measurable row (as the example in section 2 shows) of a group of the examined tool samples according to the expected quality of adhesion of the coating to the base and select that sample which has the value *K* closer to unit and, correspondingly, the best strength of adhesion of the coating to the base, and, consequently, the best quality of the tool (with other equal parameters).

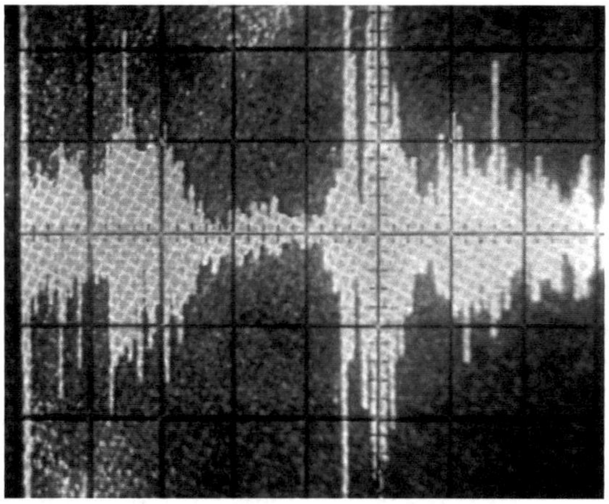

Figure 2. Recording of AE signals during different phases of penetration of the indenter.

Technological methods of evaluating the quality of tools according to their crack resistance during penetration of the indenter. We agree to understand crack resistance as the ability of a material to resist the formation and growth of cracks when a load is applied. The later a crack (cracks) is formed, the higher the tool performance can be expected.

In some cases, it is convenient to operate not with time of formation of cracks or rate of their growth, but with the number of cracks formed [1, 2, 3, 4]. The simplest assessment of crack resistance is provided when a standard (sample) is used, by comparing the tested sample with the standard.

So, in the solution, the tool with known efficiency was taken as a standard, and its crack resistance was conditionally taken as a base value, e.g., as unit. Then, the examined tool was taken to assess its crack resistance (and, consequently, its efficiency) in comparison with the standard. The standard was loaded with a hardometer indenter, and total count N_1 of acoustic emission signals was recorded. Then, under similar conditions, the examined tool was loaded, and the corresponding number N_2 of signals was recorded.

The ratio N_2 to N_1 was calculated, and, by the value of this ratio, the crack resistance of the examined tool was judged from the statement: the smaller the ratio is, the higher the resistance of the tool material to the formation and growth of cracks is. The physical meaning of this is in the following premise: the smaller total count N_2 of the examined sample is, the fewer events in it occurred during the loading time. The most probable events are formation of cracks, or growth phases of cracks. If they are fewer in the tested sample than in the standard, consequently, the sample is expected to have higher performance due to better crack resistance. This solution cannot provide high accuracy of quality control of a tool, but it is convenient for sorting and rejecting of tools.

Accuracy of quality assessment of a tool can be enhanced due to the possibility of statistic processing of results and the separation of AE signals into amplitude spectrum. The controlled body is loaded with an indenter, the signals of acoustic emission arising during the indenting are taken, and their amplitude analysis is made with an amplitude analyzer, i.e., the spectrum of acoustic emission signals is formed.

Crack resistance is defined by dimensionless value:

$$T = \left(1 - \frac{n}{N}\right) \cdot \sum_{i=1}^{N} N_1/(N \cdot N_1),$$

where N is a number of channels of the amplitude analyzer, items; n is the number of the channel with the maximum number of acoustic emission signals; N_i is a number of signals in i-channel of the analyzer, items; N_n is a number of signals in n-channel of the analyzer, i.e., the maximum number of signals in the channels of the analyzer, items.

The closer is the value T to unit, the higher is the crack resistance of the controlled product.

Results of accuracy assessment for quality control of the tool by the indicated techniques are given in table 1.

Table 1. Comparative results of tool quality control when assessing its crack resistance by acoustic emission parameters.

Controlled tool	Tool ordinal number in a series of periods of durability in specific operating conditions	Tool ordinal number in a series by parameter N_2/N_1 in solution	Tool ordinal number in a series by parameter T in solution
T15K6	7	6	7
T15K6+TiN	6	7	6
VK6	5	5	5
VK6+TiN	4	4	4
VK6+Ti+TiN+Ti	3	3	3
VK6+TiN+TiCN++Al₂O₃	2	1	2
VK6+TiC+Ti+TiN+Zr+ZrN	1	2	1

Note. Results of 3 trials are given for each case.

The data in the table show that for two cases the technique of express evaluation gave inaccurate results in relation to nature trials in cutting. The technique of signal accounting for each threshold signal showed the complete identity with the experiment in cutting.

Technological methods of evaluating the quality of tool materials by their crack resistance in scribing. Those techniques which allow examining the parameters of acoustic emission in kinetics of destruction (microdestruction) of the tested material with or without a coating possess the great possibilities of evaluating the tool quality. Destruction kinetics takes place in any kind of loading a sample, but indenting, using a typical pattern, allows reading information about the local volume of tool material (figure 3). Loading, using a scribing pattern (scratching), has larger possibilities.

In the solution, the loading has been performed with relative displacement (scribing) of a pyramidal or cone indenter, and the sample to assess the crack resistance of coatings, including multi-layer ones, and the quality of adhesion of layers of the coating between themselves and to the base. Informativity of the given technique has been enhanced by tracking the cracking kinetics with the displacement of the sample for some length in relation to the indenter. The technique is implemented as follows (figure 3).

IOP Conf. Series: Materials Science and Engineering **656** (2019) 012047 doi:10.1088/1757-899X/656/1/012047

The tested and standard samples are put on a two-coordinate table. A standard sample was made of the same material as the tested one, but it does not have a coating. The indenting is made into the sample coating for the depth *h* (figure 7.3*a*), smaller than the thickness *t* of the coating. The displacement of the table with the sample is made for some fixed length, and total count N_1 of acoustic emission signals for the displacement time is recorded. Next, the indenter is introduced into the sample (figure 3*b*) to the depth exceeding thickness of the coating by a certain value h_1. Again, the sample is displaced for the same length, and the total count N_1 of AE signals is recorded. The indenter is removed out of the tested sample, and the standard sample is loaded (figure 3*c*), by indenting for the same value h_1. The displacement is made for the same length, and the total count N_3 of AE signals is recorded.

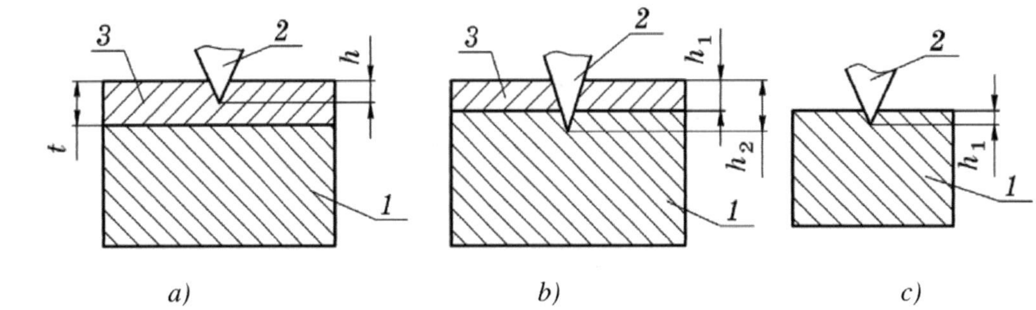

Figure 3. Sample loading circuit: *a, в, c*– loading phases; *1* – base; *2*–indenter; *3*– coating; *t*– coating thickness; *h*, h_1 and h_2– depths of indenter penetration.

Relative crack resistance can be defined, using the ratio: $T = \frac{N_2-N_3}{N_1}$ at that, the closer is value *T* to 1, the higher is crack resistance of the coating.

The quality of adhesion of the coating to the base can be evaluated, using the ratio $K=(N_2-N_1)/N_3$, at that, the closer is it to 1, the higher is the adhesion quality (figure 4).

To evaluate the crack resistance of multi-layer coatings, the examined sample is loaded, using the other circuit. Thus, in figure 4, total coating thickness is designated by *t* (t_1, t_2, t_3 are thicknesses of layers of the coating). To evaluate the crack resistance of an upper layer of the coating, everything is made as described above, but, as a standard, the sample site with removed (or absent) upper layer is used, and the total count N_3 is recorded. For the other layers all actions are similar, but it should be borne in mind that subsequent loadings are performed with periodic displacements of the indenter by a certain value *a* exceeding 3…5 times the width *l* of indenting, to prevent chipping of the coating layers between the penetrations of the indenter.

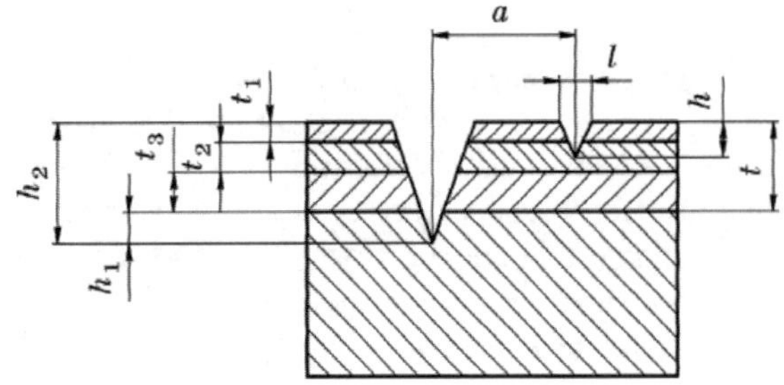

Figure 4. Circuit of loading the sample with a multi-layer coating.

Technological methods for evaluating the quality of tool material under dynamic loading of the sample. The extended possibilities of evaluating the crack resistance of tool materials can be achieved due to the loading process transformation, namely, progressive displacement of the indenter along the sample body. The essence of the technique is as follows. A wedge-shaped (cone-shaped) indenter is introduced into the sample, and displaced. Then it is shifted in relation to the path of primary displacement, loaded similarly, and moved to the same path. The quality of adhesion of the coating to the base is judged by the growth of total count of AE signals during the repeated loading related to the displacement value. This was made by the next way. Flat coated samples were taken. The indenter was introduced into the coating (figure5), and the total count of AE signals was recorded. The sample was displaced (speed 15 mm/min) for some (about 20 mm) distance. The displacement was ceased, and the value N_1 of total count was recorded.

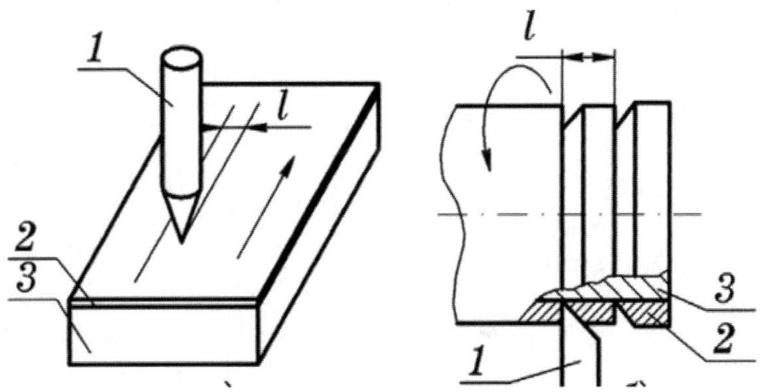

Figure 5. Circuit of loading samples: *a* – of flat shape; *b* – of cylindrical shape; *1* – indenter; *2* – coating; *3* – base.

The indenter was taken out, shifted to the side for some (0,5; 1; 1,5; 2 mm) distance *l*. Again, the indenter penetrated the same depth and was shifted for the same distance with the same speed. Then the shift was ceased, and the value N_2 of total count of AE signals was recorded. The value N_2 proved to be larger than the N_1 because of shift acts and chips of the coating between the traces of scribing of the indenter resulted from shear stresses. The larger is the growth value, the smaller is the value *l* of displacement during the repeated loading (and, consequently, the more are chips of the coating). Crack resistance was evaluated by the ratio

$$T = \frac{\Delta N}{l} = \frac{N_2 - N_1}{l}.$$

Various samples were compared by the value *T*, and the crack resistance was judged, considering that the larger is the value *T*, the lower is the crack resistance, since, with the constant value *l* of shear, larger value of the growth Δ*N* of total count of AE signals is the evidence of occurrence of more cracks and chips.

In some cases, samples can be made in shape of a cylinder and have different coatings at different sites. In these cases, the loading after pressure piling of the indenter to the sample was performed by rotating the sample. Concerning the rest, the technique was the same as described above.

In the solution, the circuit of loading by the indenter was used when the sample was displaced with plane-parallel movement (figure 6). Actually, this is the circuit of loading by the way of wearing out the coating (till its exfoliation from the base by scouring). When setting the reciprocating motion of the sample, the acoustic emission signals were recorded.

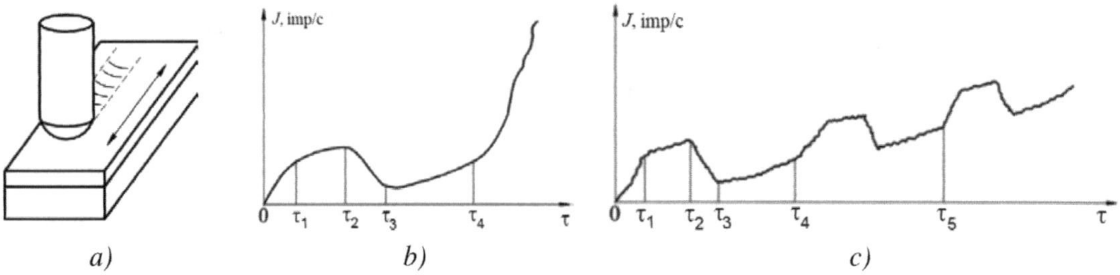

Figure 6. Circuit of loading the sample and examples of AE recording: *a* – circuit of loading the sample; *b*– recording of intensiveness of AE over time when testing the sample with a one-layer coating; *c*– the same with a multi-layer coating.

Interaction of the indenter and the sample under a certain loading caused the recording of AE signals of definite intensity over time (site $0 - \tau$ in figure 6*b*). Then the sample was displaced relative to the indenter with some speed for some distance. This loading period corresponds to the site $\tau_1 - \tau_2$ in figure 6*b*. Then, without changing the loading and speed, the sample was displaced backwards for the same distance (site $\tau_2 - \tau_3$). AE intensity decrease in this period is associated with the fact that the sample surface destructed with the first shift is not able to collapse with the same intensity, nevertheless the destruction goes on. With successive multiple shifts, the destruction goes on with slow increase of the cracking process (site $\tau_3 - \tau_4$). At some moment (τ_4), the cracking (and acoustic emission) intensity increases dramatically due to the accumulation of fatigue failures. As the examinations of the samples and their microsections have shown, these destructions occur at the border of separation coating-base. Consequently, the point of time τ_4 indicates the start of the coating exfoliation process and the time period $(0 - \tau_4)$ indicates the time of accumulating deformations under cyclic loadings till the start of destroying the sample at the border base-coating.

When testing the sample with a multi-layer coating, Figure 6*c*, after reaching the moment $\tau_4$of the start of sharp growth of intensity of AE signals, the cyclic loading of the sample was continued, followed by the loading of the second ($\tau_4 - \tau_5$) and successive layers (after some bursts on the chart caused by the following destruction of one layer of the coating and the start of introducing the indenter into the following layer of the coating).

Under such loading conditions, the criterion of evaluating the quality of the tool is considered to be the time till the moment of sharp growth of intensity of acoustic emission signals (interval $0 - \tau_4$or $\tau_1 - \tau_2$).

The results of evaluation of the tool quality were checked in comparison with the results of evaluation, using the classical method (the coating shift). For this, a number of samples were taken, numbered, and tested, using the classical method and the supposed technique. According to the testing results, the samples were arranged into random measurable rows, and the error of their arrangement in the row was evaluated, compared to the classical method (table 2). The data given in the Table differ by the loading options, namely: option 1 – loading on the indenter 40 kgf, speed of the sample displacement 30 mm/min for distance 20 mm; option 2 – loading on the indenter 40 kgf, speed of the sample displacement 30 mm/min for distance 40 mm; option 3 – loading on the indenter 40 kgf, speed of the sample displacement 30 mm/min for distance 20 mm; option 2 – loading on the indenter 28 kgf, speed of the sample displacement 10 mm/min for distance 20 mm.

IOP Conf. Series: Materials Science and Engineering **656** (2019) 012047 doi:10.1088/1757-899X/656/1/012047

Sequence of arranging the samples in the order (column 1, table 2) of increasing their quality, obtained by evaluating with the classical method was taken as a basis. The rest of the obtained results were compared with this basis option, the number of the obtained errors were defined.

Table 2. Sequence of arranging the numbers of the samples by increased quality of adhesion of the coating to the base.

Evaluated by the classical method	Evaluated by the supposed technique		
	Option 1	Option 2	Option 3
2	2	2	2
8	8	8	8
4	4	6	4
6	10	4	6
10	6	10	9
9	9	9	10
12	12	12	12
11	1	11	11
1	11	1	1
7	7	7	7
5	5	5	5
3	3	3	3

From the data of table 2 it follows:

1. the modes of loading samples do not influence significantly the testing results;
2. the first option gave two errors (the disorder of arranging samples № 10 and 6 and samples № 1 and 11);
3. the second option gave one error (the disorder of arranging samples № 6 and 4);
4. the third option gave one error (the disorder of arranging samples № 9 and 10).

Errors of arranging samples are not principal, the number of errors is not great, this allows considering the supposed technique to be quite informative and reliable.

In the solution the cyclic loading of the sample has been implemented by rotating the samples. For this the surveillance samples were manufactured in shape of a cylinder. A set of samples made with different coatings on one mandrel allows, when rotating the samples and reciprocating the indenter, to test at once a series of products and identify those coatings that have the best quality of adhesion to the base (figure 7).

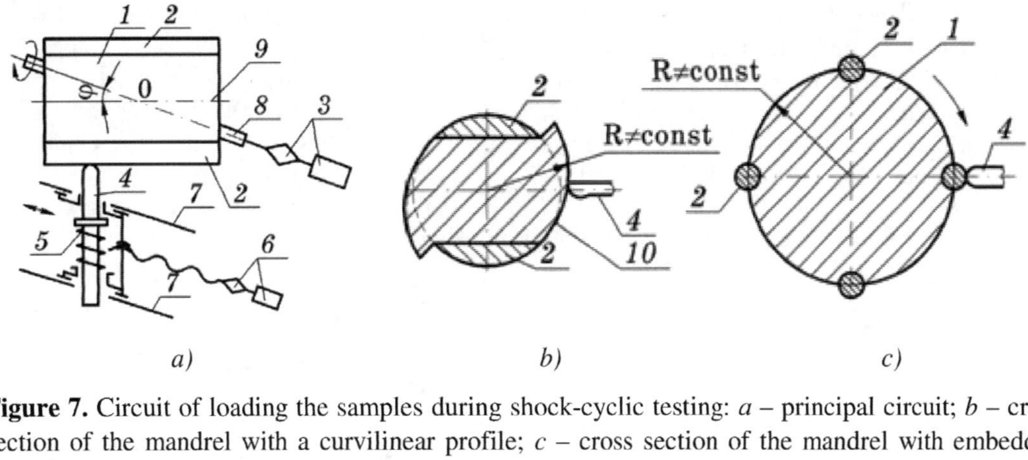

Figure 7. Circuit of loading the samples during shock-cyclic testing: *a* – principal circuit; *b* – cross section of the mandrel with a curvilinear profile; *c* – cross section of the mandrel with embedded samples.

3. Conclusion

The closest compliance with the conditions of intermittent cutting is achieved by implementing the technique of shock-cyclic loading of samples, figure 7. Mandrel *1* with tested surveillance samples *2* is rotated by actuator *3*. The indenter *4*, spring – loaded to the mandrel by a spring *5*, receives the progressive displacement from the actuator *6* along the guides *7*. The axis of rotation *8* of the mandrel *1* and the axis *9* of its symmetry intersect at some angle φ. Cyclic nature of the loading is provided by rotation of the mandrel. Nature of the loading varied by effort is provided by intersection of axes *8* and *9*. Shock nature of the loading is provided by projection *10* on the mandrel, Figure 7*b*, or embedded form of samples *2* in mandrel *1*, figure 7*c*. Impact effort variability of the indenter on the samples can also be provided with a conical shape of the mandrel. Similar phenomena can be observed, when considering other physical chemical processes. This can occur during welding [5, 6, 7, 8].

References

[1] Mokritsky B Y 2010 *Povyshenie rabotosposobnosti metallorezhushchego instrumenta putyom upravleniya svojstvami instrumental'nogo materiala* [Hardening of composite coatings of inserts for machining stainless steel 09H17N7U] (Vladivostok: Dal'nauka Publ) p 232 (in Russian)

[2] Mokritsky B Y, Usov T I and Konyukhov Y V 2017 *Upravlenie effektivnostyu primeneniya metallorezhushhego instrumenta* [Management of efficiency of cutting tools: monograph*]* (Komsomolsk-on-Amur: KnAGTU Publ) p 14 (in Russian)

[3] Shvetsov I V, Belyakov V N, Zhelnov D V, Letenkov O V and Nikulenkov O V 2018 Ocenka rabotosposobnosti abrazivnogo instrumenta [Diagnostics of Wheel Life in Grinding] *Vestnik Mashinostroeniya* **5** pp 78–81 (in Russian)

[4] Shvetsov I V 2017 *Physical-chemical-mechanical manufacturing processes: monograph* (Veliky Novgorod: Yaroslav the Wise Novgorod State University Publ) p 100 (in Russian)

[5] Makarov S V and Boots S B 2014 Fiziko-himicheskie svojstva zhidkogo stekla primenyaemogo pri izgotovlenii svarochnyh elektrodov (Elektronnyj resurs) [Physical and chemical properties of liquid glass used in the manufacture of welding electrodes (Electronic resource)] *Current state and problems of natural sciences: proceedings of the all-Russian scientific and practical conference of young scientists, graduate students and students* (Tomsk: TPU Publishing House) pp 156–158

[6] Chernykh E I, Gatowski Yu M and Boots S B 2014 Strukturno-funkcional'nye principy proektirovaniya v elektroplazmennyh tekhnologiyah (Elektronnyj resurs) [Structural and functional principles of design in electroplasmic technologies (Electronic resource)] *Current state and problems of natural Sciences: proceedings of the all-Russian scientific and practical conference of young scientists, postgraduates and students (Electronic resource)* (Tomsk: TPU Publishing House) pp 271–273

[7] Boots A S and Boots C B 2015 Mekhanizm raspredeleniya teploty v svarnyh soedineniyah, vypolnennyh s primeneniem zashchitnyh pokrytij v uglekislom gaze (Elektronnyj resurs) [The mechanism of heat distribution in welded joints made with the use of protective coatings in carbon dioxide (Electronic resource)] *Current state and problems of natural Sciences: proceedings of the II all-Russian scientific-practical conference of young scientists, graduate students and students* (Tomsk: TPU Publishing House) pp 204–205

On dynamics and control when finishing body parts

I V Shvetsov[1,4], D V Zhelnov[2] and Y M Zubarev[3]

[1] Yaroslav-the-Wise Novgorod State University, ul. B. St. Petersburgskaya, 41 173003 Veliky Novgorod, Russia
[2]Department of Expertise and Certification of the Union "St. Petersburg Chamber of Commerce and Industry", Saint Petersburg, Russia
[3] Peter the Great Saint-Petersburg State Polytechnic University, Saint Petersburg, Russia

[4]E-mail: Igor.Shvetsov@novsu.ru

Abstract. For designing machines, the optimal treatment regimens need to perform dynamic quality of the machine elastic system on the basis of imitating modeling of technological systems during the cutting process. When finishing body parts in some cases, problems arise with a high dynamic quality of the machine elastic system, as the workpiece has, as a rule, sufficiently high rigidity. The simulation modeling allows to obtain reliable qualitative and quantitative characteristics of the process being modeled. In the article the mathematical description of the dynamic model of milling machine is given taking into account the main mechanisms of the excitation process in multi-processing.

1. Introduction

When finishing the body parts there are problems associated with insufficiently high dynamic quality of the elastic system of the machine due to the high rigidity of the workpiece [1, 2, 3, 4, 5]. The quality of the material can be estimated when there is a dependence of the concentration level of a gas on the diagnostic parameter. But this dependence is ambiguous, and it is impossible to directly assess the quality of the material being processed and to make a direct diagnostic conclusion. A change in, for example, the concentration of gaseous compounds formed during mechanical processing is a consequence of variation in physicochemical parameters of the preform, i.e. presence of anomalies. The histograms taken during machining on a planer, indicate the presence of surface hardness anomalies, which indicate the place where it is necessary to check quality of the workpiece, i.e. in this case, the measurement of hardness. Taking into account dependence of the intensity of carbon monoxide formation on hardness, we can assume that its values are the lowest at the minimum values of the measured concentration of carbon oxide and the highest at the maximum values of the concentration in the treatment zone.

In this case, when diagnosing the state of the material, there is no direct conclusion about the parameter being diagnosed, but specific places are indicated where it is necessary to carry out control measurements of the parameter by technological measuring means, thereby significantly reducing the number of such measurements. This is especially true with limited possibilities of the number of technological measurements, when the reliability of the result increases, and in the case of large areas of products, the possibility of volumetric control of its physicochemical properties in principle opens up.

Particularly noteworthy is the continuous monitoring of chemical composition of the material, which is difficult or impossible by other means. Determining the presence of abnormal areas or deviations of the chemical composition of chips is problematic, since in the first case considerable time is required under extreme conditions of the process of plastic deformation and material destruction. The chemical composition of chips is changing, and in the second case it is limited by the number of possible samples for chemical analysis.

Therefore, for a more complete continuous diagnosis of the chemical composition of the material being processed, it is necessary:

– optimally select the gas analyzer that registers the selected and controlled gas, taking into account the chemical composition of the material and the instrument;

– select the optimal modes of machining products;

–use the extensive information obtained from the information output of the gas analyzer, including static, analysis of variance and the calculation of the correlation function.

At constant cutting speed and constant feed, the concentration level of a number of gases in the treatment area is a function of the allowance, hardness, condition of the cutting tool and other factors. Considering that the cutting depth and hardness of the workpiece material require identical and unidirectional actions to optimize the cutting process, you can use information about the intensity of the gases formation as equivalent information about the level of loads on the tool and carry out optimization aimed at reducing the concentration of the formed gases, similar to optimization on the selection of the force (moment) of cutting according to the appropriate algorithm. It should be borne in mind that the effect of tool wear, with the exception of chipping and catastrophic wear, also increases the concentration level of the gases produced and also requires actions of the same direction and character as the effect of changes in the allowance and hardness.

Optimum design of machines and rational technological modes choice can be performed while solving the problem of dynamic simulation of the machine technological system during the cutting process.

2. Results and discussion

Everything that human activity is directed at is called an object (Latin objection is an object). The development of the methodology is aimed at streamlining the receipt and processing of information about objects that exist outside of our consciousness and interact with each other and the external environment.

Replacing one object with another in order to obtain information about the most important properties of the original object with the help of a model object is called modeling. Thus, modeling can be defined as the representation of an object by a model for obtaining information about this object by conducting experiments with its model. The theory of replacing some objects (originals) with other objects (models) and studying the properties of objects on their models is called modeling theory.

It should be noted that from the point of view of philosophy, modeling is an effective means of knowing nature. The modeling process involves the presence of the object of study; the researcher who has been assigned a specific task; model created to obtain information about the object and necessary to solve the problem. Moreover, in relation to the model, the researcher is, in fact, an experimenter, only in this case, the experiment is conducted not with a real object, but with its model. Such an experiment for an engineer is a tool for directly solving organizational and technical problems.

The modeling is based on information processes, since the creation of model M itself is based on information about a real object. In the process of implementing the model, information about this object is obtained, at the same time, during the experiment with the model, control information is entered, the processing of the obtained results, i.e. Information underlies the entire modeling process.

One of the most important aspects of building modeling systems is the problem of goal. Any model is built depending on the goal that the researcher sets for it, therefore one of the main problems in modeling is the problem of target use. The similarity of the process taking place in the model to the real process is not the goal, but the condition for the model to function properly, and therefore the goal should

IOP Conf. Series: Materials Science and Engineering **656** (2019) 012048 doi:10.1088/1757-899X/656/1/012048

be to study any aspect of the object's functioning. To develop a simulation model and carry out its simulation of a queuing system with one source of requests, one machine and a queue for the problem of mechanical engineering.

An important point in the study of dynamic cutting processes is their dynamic imitation using specially designed stands and modern computer technology. When bringing the original multidimensional model of a technological system with a blank to a set of simple models, models are formed with the following types of dynamic structure - single or double. Developed measuring and computing complexes that include the following elements: a metal-cutting machine, a stand for full-scale dynamic modeling, primary converters, amplifiers, matching devices, an application package, etc. Such complexes are an effective original full-scale dynamic simulator for a wide class of technological systems for mechanical processing of low-hard blanks with different elastic-dissipative and inertial characteristics. The design features of the stands allow you to simulate the necessary dynamic characteristics of the contours. For more complex machining process systems, the number of contours can be increased to four. The application package provides real-time measurements; filtering, spectral analysis and statistical processing of measurement results.

Mechanical engineering is one of the leading industries, affecting almost all spheres of human activity. The products of the engineering industry are distinguished by complexity, a variety of design solutions and the use of a wide variety of materials with different physical and mechanical properties. The design of the product and the material from which it is made have a significant impact on the choice of shaping methods and technological equipment. For the production of the same parts can be used various methods of shaping and means of technological equipment, which leads to a large variety of production process options. Under these conditions, the most rational method of choosing the optimal variant of the production process is the method of simulation modeling. On the basis of this method, a model has been developed that simulates the production process for processing a given range of parts. The determination of the optimal variant of the production process is based on the analysis of economic indicators and indicators of product quality. The simulation model was based on the modular principle of production organization. In this regard, a classification module was developed.

The simulation model was based on the modular principle of production organization. In this regard, the classification of modules was developed, which were divided into five groups: design modules, procurement modules, process modules, parameter modules and technological base modules. The developed simulation model consists of five stages.

At the first stage, modeling of technological routes for processing individual parts is performed. The objective of this simulation stage is the formation of technological modules of the part, which contain information about the methods and strategies for processing the part, as well as the name of the technological equipment necessary to achieve the specified values of technical parameters at each stage of processing.

At the second stage, an analysis of the machining accuracy is performed. The task of this modeling stage is to form intervals of machining accuracy and to determine allowable values of cutting conditions and machine processing time for each interval of accuracy. Analysis of the machining accuracy is based on the calculation of the total error, which is determined by the formula

$$\Delta = f\times(\epsilon, \Delta_Y, \Delta_U, \Delta_N, \Delta_T, \Sigma\Delta_\Phi),$$

where ϵ is the error associated with the installation of the workpiece on a cutting machine or tool, μm;

Δ_Y is the error caused by elastic deformations of the technological system, μm;

Δ_U is the error resulting from dimensional wear of the cutting tool, μm;

Δ_N is the error associated with setting the cutting tool, μm;

Δ_T is the error caused by thermal deformations of the technological system, μm;

$\Sigma\Delta_\Phi$ is the error associated with the geometrical deviations of the equipment, μm.

At the third stage, the formation of all possible variants of the technological process (group technological modules) is carried out.

At the fourth stage, the production process is modeled on the basis of queuing theory. The task of this stage is to determine the parameters of the production system for various variants of the production process. At this stage a closed queuing network will act as a model.

At the fifth stage, the choice of the optimal variant of the production process is made according to three groups of indicators: the machining accuracy, the cost of the investment project and the parameters of the production system.

The definition of the optimal variant can be divided into two stages. At the first stage, a set of optimal solutions is determined based on the Pareto method - Pareto-optimal solutions. At the second stage using the ideal point method, from the set of Pareto-optimal solutions, the choice of the variant of the production process, which has the optimal combination of economic and quality parameters, is made. To calculate the distance to the ideal point, we chose the weighted Euclidean distance formula:

$$d_{bE}(x_i, x_j) = \sqrt{\sum_{k=1}^{p} W_k (x_i^{(k)} - x_j^{(k)})^2}$$

where d_{bE} is the measured Euclidean distance,

$x_i^{(k)}$ is the value of the k-th property of the object i,

$x_j^{(k)}$ is the value of the k-th property of the object j,

W_k is the "weight" of the k-th criterion,

p is the number of criteria.

The optimal variant of the production process will be considered the variant with the minimum value of the weighted Euclidean distance.

The full dynamic model of a metal-cutting machine tool in simulation tasks is described by the following system.

By cyclic coordinates

$$J_{jnp}\ddot{\varphi}_j = M_{gj} - M_{cj}; j=1,s \quad (1)$$

where $\ddot{\varphi}_j$ is the rotation angle of the motor shaft of the j-th drive; J_{jnp} is the reduced moment of inertia of the j-th drive; M_{gj} is the driving moment of the j-th engine; M_{cj} is reduced moment of resistance on the j-th drive; S is the number of concurrent drives.

For elastic deformations of the drives:

$$\ddot{\theta}_j^* + 2n_j^*\dot{\theta}_j^* + \omega_j^{*2}\theta_j^* = k_{gj}^* M_{gj} - k_{cj}^* M_{cj}, \quad (2)$$

where θ_j^* is the column vector of the normal coordinates of the j-th drive; n_j^* is the dissipative matrix of the j-th drive, n_j^* is the diagonal matrix of squares of the natural frequencies of the j-th drive; k_{gj}^*, k_{cj}^* are Vectors-columns, characterizing the reduction of the engine torque and the moment of resistance to normal coordinates.

According to the spindle assembly:

$$\ddot{y}_{ш}^* + 2n_{ш}^*\dot{y}_{ш}^* + \omega_{ш}^{*2} y_{ш} = y_p^* P_p^* + k_{ш}^* P_{ш}^*, \quad (3)$$

where $y_{ш}^*$ is the vector of normal coordinates, characterizing the transverse oscillations of the spindle; $n_{ш}^*$ is the dissipative matrix spindle node; $\omega_{ш}^{*2}$ is the matrix of squares of the natural frequencies of the spindle assembly; P_p^* is the cutting forces vector; $P_{ш}^*$ is the vector of driving forces; k_p^* and $k_{ш}^*$ are matrices characterizing the reduction of cutting forces or forces driving the spindle assembly to normal coordinates.

By oscillations of the bearing systems:

$$\ddot{\eta}_\delta^* + 2n_\delta^*\dot{\eta}_\delta^* + \omega_\delta^{*2}\eta_\delta^* = \bar{k}_{ш}^* R_{ш}^* + \bar{k}_p^* P_p^* + \\ + \sum_{i=1}^{s}(k_{ni}^* Q_i + k_{Ri}^* \cdot R_{ni}^* + k_{g\delta}^* M_{gi}), \tag{4}$$

where η_δ^* is the vector of the normal coordinates of the carrier system; n_δ^* is the diagonal dissipative matrix, ω_δ^{*2} is the diagonal matrix of squares of natural frequencies; $k_{ш}^\delta, \bar{k}_p^*, k_{ni}^*, k_{Ri}^*, k_{g\delta}^*$ are matrices of reducing the reaction vectors of the corresponding subsystems to the normal coordinates of the carrier system, namely: $\bar{k}_{ш}^*, R_{ш}^*$ is the spindle head; k_{ni}^*, Q_i is for the effort of supplying the i-th operating drive; k_{Ri}^*, R_{ni}^* is for support reactions of the i-th operating drive; k_p^*, P_p^* is for cutting forces; $k_{g\delta}^*, M_{gi}$ is for the moment of the i-th engine.

We proceed to the description of the dynamic processes in the drives. It is convenient to use the description of structural schemes through transfer functions.

The electromagnetic moment of the j-th engine is determined by the ratio

$$M_{gj} = C_{Mj} I_{gj} \Phi_j, \tag{5}$$

where C_{Mj} is the moment constant of the j-th engine; I_{gj} is the armature current j-th engine; Φ_j is the magnetic flux of the j-th engine. For typical block diagrams of drives for heavy drives of heavy machine tools, we write the formulas for current and flow in Laplace images:

$$\bar{I}_{gj} = W_{gj}^{(з)}(P)\bar{U}_{з.я.j} + W_{ogj}^{(v)}(P)\bar{y}_{oj} + \\ + W_{ogj}^{(w)}(P)\bar{\ddot{y}}_{ogj} + f_{gj}(P, \bar{I}_{gj}, \bar{\Phi}_j, \bar{y}_{oj} \ldots). \tag{6}$$

$$\bar{\Phi}_j = W_{\Phi j}^{(з)}(P)\bar{U}_{зв j} + W_{oвj}^{(s)}(P)\bar{y}_{oвj} + W_{oвj}^{(v)}(P)\bar{\dot{y}}_{oвj} + \\ + W_{oвj}^{(w)}(P)\bar{\ddot{y}}_{oвj} + f_{вj}(P, \bar{\Phi}, \bar{y}_{oвj}, \ldots), \tag{7}$$

where $\bar{I}_{gj}$ is the image of the armature current of the j-th motor; $\bar{U}_{зяj}$ is the image of the voltage setting at the j-th drive armature; $\bar{y}_{ogs}, \bar{\dot{y}}_{ogj}, \bar{\ddot{y}}_{ogj}$ are images of feedback signals in the armature circuit of the j-th drive, respectively, in position, speed, acceleration; $W_{gj}^{(з)}(P)$ is the transfer function of the armature contour of j-th drive; $W_{ogj}^{(s)}(P), W_{ogj}^{(v)}(P), W_{ogj}^{(w)}(P)$ are transfer functions of the armature chain of the j-th drive in position, speed, acceleration, respectively; f_{gj} is the function characterizing the nonlinearity in the armature circuit of the j-th drive; $\bar{\Phi}_j$ is the image of the excitation flow of the j-th drive.

The feedback sensor signal is:

$$\dot{y}_{ogj} = \dot{\varphi}_j + \sum d_{ij}\dot{\eta}_i. \tag{8}$$

To close the mathematical description of the system, it is necessary to introduce a dynamic characteristic of cutting force into the model. Restricting ourselves to the case of fine milling and introducing a spatial nonstationary model of the milling process, we give the dynamic characteristic in the form

$$T(\omega t)\dot{P}_P^* + \left[KP_o^*(\omega t) + \sum_{i=1}^{M} \delta_i K_{pi}^*(\omega t) \right] P_p^* = -k^* E^*(\omega t)\Delta, \tag{9}$$

where ω is the angular velocity of the spindle; T is the square matrix of the third order, which elements characterize the delay of the corresponding vector components, $P_p^*; \delta_i$ is the delta function, which characterizes the impact when the tooth enters or leaves the cutting zone; M is the number of teeth cutters; Δ is the vector of deformations; k^* is the cutting coefficient characterizing the properties of the material being processed and the size of the cutting zone.

3. Conclusion

Relation (9) takes into account all the main mechanisms of excitation in the process of multi-blade processing, namely: quasi-static perturbation associated with a change in the thickness of the cut layer; shock external disturbance associated with the entrance and exit of teeth from the cutting zone; the delay of the cutting force relative to the change in the thickness of the cutting layer; trail disturbance; the impact nature of the formation conditions of the dynamic cutting force. All ratios form a closed mathematical model of the machine control system in the process of cutting, and their analysis at the design stage allows us to predict bringing the object under real working conditions.

The developed simulation model allows to:

1) Identify all possible options for the production process of processing parts and the structure of the production system.

2) Assess the accuracy of manufacturing parts at each stage of processing for different methods and strategies of shaping.

3) Determine the list of technological equipment necessary for parts processing and their number.

4) Determine the costs required for the implementation of each variant of the production process.

5) Assess the effectiveness of the structure of the production system.

6) Determine the optimal variant of the production process, taking into account economic indicators and indicators of product quality.

References

[1] Weits V L, Maksarov V V and Skhirtladze A G 2002 *Cutting materials* (St.-Petersburg: SZTU) p 232

[2] Weits V L and Maksarov V V 2001 *Dynamics of technological systems of machining*: Monograph in 5 parts. Part 1: Schematization of processes in technological systems of mechanical processing (St Petersburg: PIMash) p 184

[3] Shvetsov I V, Belyakov V N, Zhelnov D V, Letenkov O V, Nikulenkov O V 2018 Evaluation of abrasive tool performance *Bulletin of Mechanical Engineering* Vol 5 **(4)** 78–81 (in Russ.)

[4] Shvetsov I V, Belyakov V N, Zhelnov D V, Letenkov O V and Nikulenkov O V 2018 *Russian Engineering Research* Vol. 38 **4** 610–613 © Allerton Press, Inc., 2018. Original Russian Text © Shvetsov I V, Belyakov V N, Zhelnov D V, Letenkov O V and Nikulenkov O V 2018 *Vestnik Mashinostroyenia* Vol 5 **4** 78–81 (in Russ.)

[5] Shvetsov I V 2017 *Physical and chemical and mechanic processes in production* (Veliky Novgorod: Yaroslav-the-Wise Novgorod State University) p 100 (in Russ.)

IOP Conf. Series: Materials Science and Engineering **656** (2019) 012049 doi:10.1088/1757-899X/656/1/012049

Vibration stability criterion in assessing the dynamic quality and limiting capabilities of machine tools

I V Shvetsov[1,3], D V Vasilkov[2] and I Ya Tarikov[2]

[1]Yaroslav-the-Wise Novgorod State University, ul. B. St. Petersburgskaya, 41 173003 Veliky Novgorod, Russia
[2]Baltic State Technical University «VOENMEH» named after D F Ustinov, ul. Pervaya Krasnoarmeiskaya, 1 190005 St. Petersburg, Russia

[3]E-mail: Igor.Shvetsov@novsu.ru

Abstract. The vibration resistance of the dynamic system of a cutting machine is considered as a criterion for assessing its dynamic quality and limiting capabilities. An example of diagnosing a group of machines of the same type by the criterion of vibration resistance with an assessment of their current state is given.

1. Introduction

Important factors affecting the accuracy of processing, durability are geometric and kinematic accuracy of the machine [1]. The design of machine tools and their most critical parts and mechanisms is largely subject to the criterion of accuracy. This criterion includes manufacturing accuracy, preservation of accuracy in operation (provided by the smallness of the given elastic, temperature deformations and vibration amplitudes, tuning accuracy) and preservation of accuracy over the established repair periods.

Geometric accuracy of machines is characterized by [1]:

a) the accuracy of the support surfaces for basing the tool and the workpiece;

b) the accuracy of the working bodies movement;

c) the accuracy of the relative position of the working bodies movement rails;

d) the accuracy of the location of the rails relative to the base surfaces.

Kinematic accuracy is characterized by the consistency of mutually related relative movements of the parts carrying the instrument and the workpiece.

The strength and durability of machines are also important factors affecting the performance. Wear greatly reduces accuracy, reduces vibration resistance and, as a result, performance. Therefore, the permissible wear of carrier and rail systems parts (guides, spindle bearings, tool seats and accessories, etc.) is determined by the criteria of accuracy and vibration resistance during machining [1, 2].

Increased requirements for dimensional accuracy and parts shape, the emergence of new hard-to-machine materials, the intensification of cutting conditions, as well as the widespread introduction of automation of technological processes and the creation of machines with control and regulation systems caused a sharp increase in the role of dynamic processes in machine tools. There are a number of dynamic processes during machining that have a significant impact on the quality of processing. This is the cutting process, friction processes, processes in engines [2–7].

A set of dynamic processes indicators in machine tools that determine the listed requirements of stability, accuracy, durability, etc. are united by the general concept of the machine dynamic quality [2]. The main indicators of the machine dynamic quality are the following:

a) stock and degree of stability;

b) deviations of the parameters of the dynamic system under external influences;

c) speed.

The stability margin determines the possibility of changing one or another parameter of the system without losing stability. The loss of stability by the system appears in vibrations or "undermining" of the instrument, in the uneven jump-like movement of the nodes or in their jamming. Thoroughly, vibration resistance is the most important indicator of the machine dynamic quality. It allows to determine the range of allowable modes and identify the machine limits.

2. Material and research methods

As the main criterion characteristic, which determines the limiting possibilities of the machine in assessing the dynamic quality, the boundary of the stability area was adopted [8] (figure 1). It divides two areas – the area of permissible modes, providing quiet cutting with a low vibration level, and the area of unacceptable modes with an increased vibration level. The stability boundary is built in the plane of the cutting process parameters.

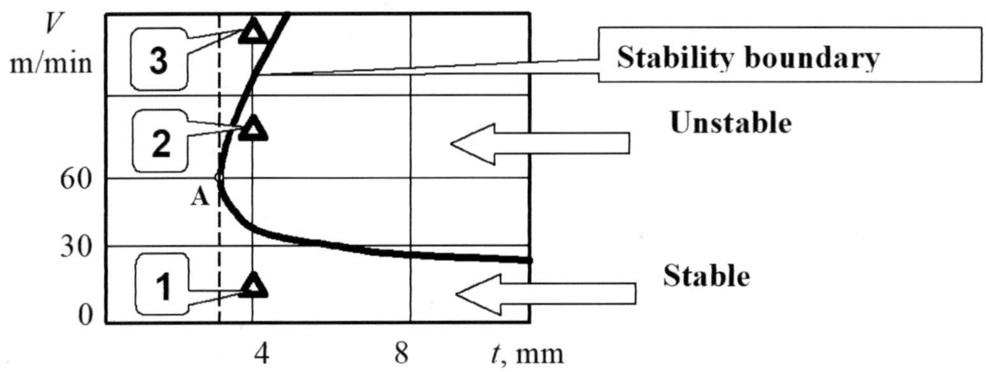

Figure 1. Stability area boundary of the machine when turning.

The range of acceptable modes can be divided into three characteristic areas:

• the area of absolute stability, bounded by the axis of ordinates (cutting speed V) and a dotted line passing through the point of absolute stability A;

• the area of stability at low cutting speeds, limited by the abscissa axis (cutting depth t) and the stability boundary below point A;

• the area of stability at high cutting speeds, limited by the axis of ordinates (cutting speed V) and the stability boundary above point A.

The change in the machine state is diagnosed by the displacement of the stability boundary, which can be characterized quantitatively. In this case, the characteristic features of each specific machine manifest themselves through the position of the stability boundary.

As an example, we can consider the dynamic manifestation of the technological system when processing by turning in the plane of parameters presented in Figure 1. We fix the depth of cut $t = 4$ mm and choose three characteristic modes with cutting speeds respectively: $V_1 = 20$ m / min, $V_2 = 75$ m /

IOP Conf. Series: Materials Science and Engineering **656** (2019) 012049 doi:10.1088/1757-899X/656/1/012049

min and V_3 = 110 m / min. Figure 2 shows diagrams of vibration displacement for the three modes. It can be seen from the diagrams that in modes 1 (figure 2, a) and 3 (figure 2, c) the average vibration amplitude level is 10 μm, and in mode 2 (figure 2, b) it is 30 μm. It turns out that when going beyond the boundary of the stability area, the vibration level increases threefold. Thus, we can conclude that the stability boundary defines the limiting machine capabilities and highlights the range of acceptable process conditions.

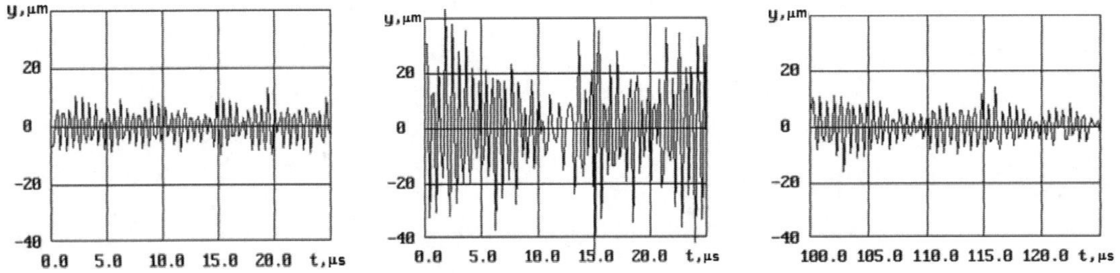

a) *b)* *c)*

Figure 2. Vibration displacement diagrams: a – mode 1; b – mode 2; c – mode.

To study stability, we consider the dynamic systems of machine tools when turning (figure 3, a, c) and milling (figure 3, b, d) [8].

The kinematics of formative movements during turning processing are presented in figure 3, a. The line of the cutting force action P_z is aligned with the cutting speed V. To the cutter in this direction, the mass m_z, rigidity coefficients c_z and damping b_z are given. The force P_y acts in the y direction. The mass m_y, the coefficients of rigidity c_y and damping b_y are given to the cutter in this direction. The slice section consists of the thickness a and the width b of the slice, which are respectively equal to $b = \frac{t}{\sin\varphi}$; $a = S \sin\varphi$; t is the depth of cut; S is feed; φ is the main angle in the plan.

As a model of minimum dimensionality, when machining by turning, a single-mass double-circuit flexural dynamic model with lumped parameters (figure 3, c) can be adopted, which is quite applicable for calculating machines stability [2].

With regard to the problem of dynamics in terms of the technological system stability, the normal contour (*OY* axis) and the tangent contour (*OZ* axis) are unequal:

- a normal contour determines the shaping accuracy, forms a vibration trace on the surface and makes a decisive contribution to the boundary formation of the stability area;
- The tangent contour improves the surface roughness, takes over a part of the energy generated during the cutting process and accordingly expands the stability area.

The exception in the model of the tangent contour *OZ* narrows the stability area to approximately 10% [9]. The transition admissibility to a single-dynamic model is confirmed by numerous dynamic tests of machine tools. But such a simplification of the dynamic system requires the periodic identification of parameters when conducting dynamic tests.

IOP Conf. Series: Materials Science and Engineering **656** (2019) 012049 doi:10.1088/1757-899X/656/1/012049

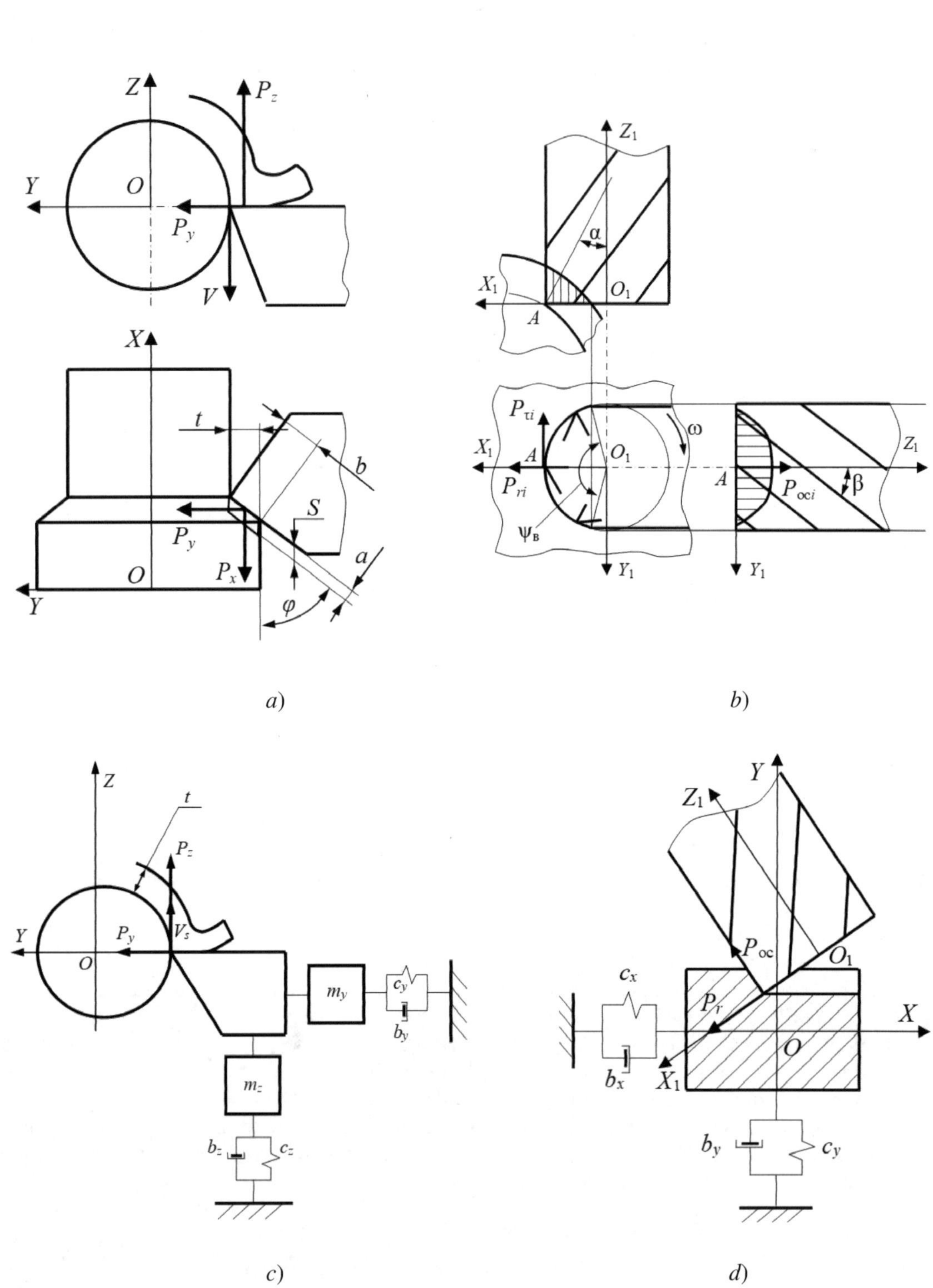

Figure 3. Kinematics of forming movements during machining by turning (*a*), by milling (*b*) and the model of a two-circuit dynamic system with machining turning (*s*), milling (*d*).

Consider the kinematics of shaping movements during contour milling (figure 3, b). Here the angle α is the angle of the cutter position measured between the normal to the forming point A of the workpiece and the axis of the cutter rotation. The instantaneous thickness a_i and the slice width b_i are determined for the i-th tooth of the cutter; s_z is the working feed per tooth cutter; r is the cutter radius; ψ_i is instantaneous angle of the cutter rotation; $\psi_в$ is the cutting angle; β is the inclination angle of the screw helix of the tooth. The tangent, the radial $P_{\tau i}$ and the axial P_{oci} components of the cutting force act on the i-th tooth of the milling cutter involved in the cutting.

Cutting force during milling is a periodic function, which complicates the solution of the stability problem and requires appropriate linearization. Analysis of the cutting forces nature change during milling showed [2, 8, 9] that while cutting with two or more milling teeth, the uneven milling with respect to the average value does not exceed 10%. Neglecting this value dramatically simplifies the calculation of cutting forces during milling. This is achieved by introducing average integral values of thickness a_m, and width b_m of the slice. The model of the technological system for contour milling is presented in Figure 3, d. The same rules for shaping the surface can be applied to it as to the model during turning (Figure 3, b), i.e. it can be mathematically represented by a single-loop dynamic model in the direction of the OY axis.

When carrying out machine diagnostics, mass m_z, rigidity coefficients c_z and damping b_z in dynamic circuits are given. To highlight the dominant dynamic contours of the machine dynamic system, we introduce into consideration the quality factor D of the oscillating system [10], which is determined by the formula

$$D = \sqrt{\frac{m_y c_y}{b_y}} \tag{1}$$

We will search for high-quality dynamic circuits by the amplitude spectrum of vibration displacements or vibration velocities in the frequency range 0 ... 1000 Hz based on the condition evaluation

$$1 - 1/(2D_s) \le f/\omega_s \le 1 + 1/(2D_s), \tag{2}$$

where ω_s is the s-th natural frequency of the dynamic loop; f is the current frequency; D_s is the good quality of the s-th own form of oscillations.

For a high-quality contour, the reduced parameters are determined and a single-dynamic model is built [8]

$$m_y \ddot{y} + b_y \dot{y} + c_y y = P_y\,; \tag{3}$$

$$T_p \dot{P}_y + P_y = -k_y^* y\,, \tag{4}$$

where T_p is the lag time constant, $T_p = l_p / V_s$; l_p is the path of chip formation equal to the length of the chip contact line with the front surface of the instrument; k_y^* is the contour transfer coefficient $k_y^* = k_y \sin\varphi$, k_y is the coefficient of reducing the force P_y to the shear thickness a.

The stability criterion of the dynamic machine system is calculated on the basis of the algebraic Hurwitz criterion [11] for the system of differential equations (3) - (4) defined by the inequality

$$(T_p c_y + b_y)(T_p b_y + m_y) > T_p m_y (c_y + k_y^*) \tag{5}$$

The construction of the stability boundary is carried out by step-by-step verification of the criterion (5) based on the D-partitioning method.

Table 1 shows the results of determining the for 7 machines of the same model.

IOP Conf. Series: Materials Science and Engineering **656** (2019) 012049 doi:10.1088/1757-899X/656/1/012049

Table 1. The results of determining the machines stability boundary.

No.	Diagnostic results	Stability boundary
1	Dominant node Spindle Frequency of dominant node, Hz 131 Damping parameter 56 Rigidity, kN / mm 55.5 Analysis of test results: The machine has high rigidity. The range of allowable cutting conditions is wide. The machine is in satisfactory technical condition.	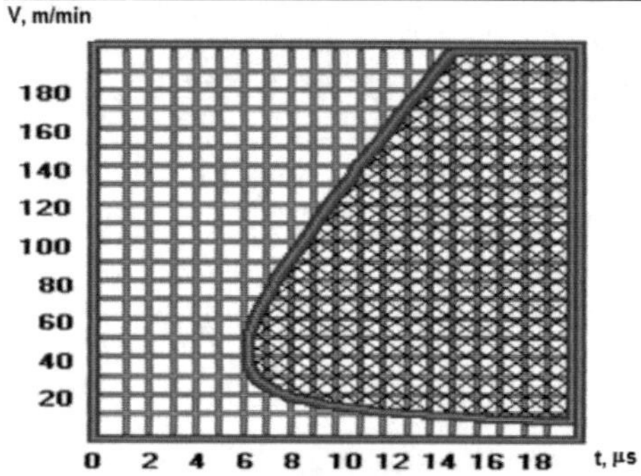
2	Dominant node Spindle Frequency of dominant node, Hz 136 Damping parameter 32 Rigidity, kN / mm 25.0 Analysis of test results: The range of allowable cutting conditions is narrow. The machine requires inspection and determination of the type of repair.	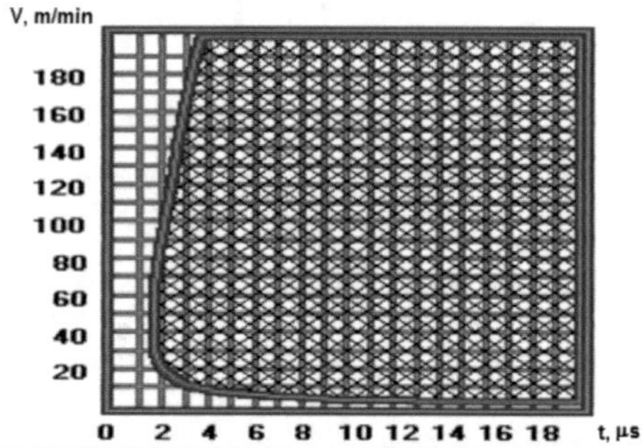
3	Dominant node Tool holder Frequency of dominant node, Hz 78 Damping parameter 35 Rigidity, kN / mm 47.0 Analysis of test results: The machine has high rigidity. The range of allowable cutting conditions is wide. The machine is in satisfactory technical condition.	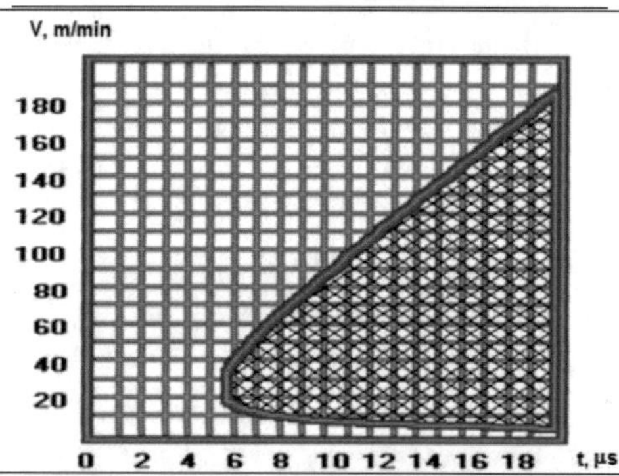
4	Dominant Node Support Frequency of dominant node, Hz 43 Damping parameter 25 Rigidity, kN / mm 25.5 Analysis of test results: *The range of allowable cutting conditions is average. The machine requires preventive maintenance.*	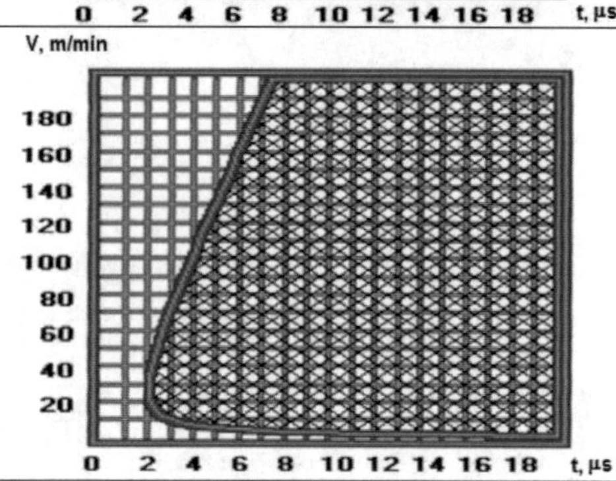

5	Dominant Node Support Frequency of dominant node, Hz 30 Damping parameter 35 Rigidity, kN / mm 24.0 Analysis of test results: The machine has high rigidity. The range of allowable cutting conditions is wide. The machine is in satisfactory technical condition.	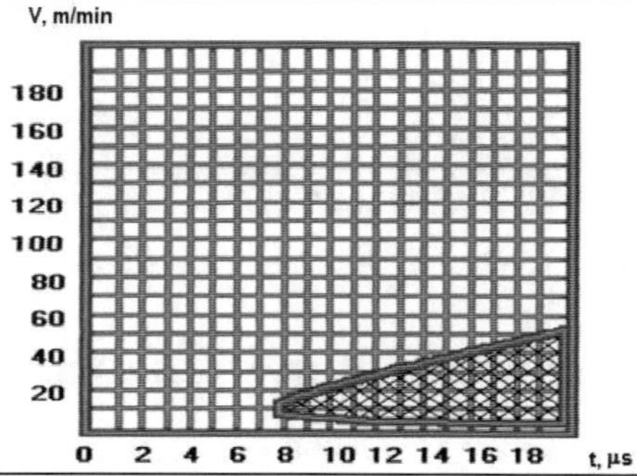

3. Discussion of research results

The studies of the machine dynamic system stability when machining with a blade tool on turning and milling operations showed the applicability of the general approach to determining the stability boundary area based on simplified models. The solutions obtained allow us to meet the requirements for the quality of manufacturing precision parts, taking into account the current state of the metal-cutting equipment.

The criterion of vibration resistance is an indicative tool for assessing the current state of the metal-cutting machine and its ultimate capabilities. From the test results shown in Table 1, it can be seen that different machines have different dynamic contours. In this case, the boundary line of the stability area depends directly on the state of the machine. In examples 1, 3, 5 the machines are in a satisfactory technical condition. In example 2 the machine has a narrow stability boundary, requires inspection and determination of the repair type. In example 4, the machine has a valid range of cutting conditions and requires preventive maintenance.

The results obtained indicate the practical applicability of the presented model solutions.

References

[1] Reshetov D H and Portman B T 1986 *Accuracy of machine tools* (Moscow: Mechanical Engineering) p 336

[2] Kudinov V A 1967 *Dynamics of machines* (Moscow: Mechanical Engineering) p 359

[3] Maksarov V 2015 Improving the accuracy of hydraulic power cylinders manufacturing using vibration-proof cutting tool *Agronomy Research* 13(3) **9** 671–679

[4] Olt J, Liyvapuu A, Madissoo M and Maksarov V 2016 Dynamic simulation of chip formation in the process of cutting *International Journal of Materials and Product Technology* 53(1) **15** 1–14

[5] Olt J and Maksarov V V 2015 Cutting process simulation on the basis of rheological properties of metals *Annals of DAAAM and Proceedings of the International DAAAM Symposium* **39** 229–237

[6] Maksarov V V, Khalimonenko A D and Matrenichev K G 2017 Stability analysis of multipoint tool equipped with metal cutting ceramics *IOP Conference Series: Earth and Environmental Science* **7(8)** 082030

[7] Olt J, Liivapuu O, Maksarov V, Liyvapuu A and Tärgla T 2016 Mathematical modelling of cutting process system. *Springer Proceedings in Mathematics and Statistics* **14** 173–186

[8] Vasilkov D V, Nikitin A V and Cherdakova V S 2018 Stability of the dynamic system of the machine tool when machining with a blade tool *IOP Conf. Series: Earth and Environmental Science* **194** (2) 022045

[9] Wulfson I I 2019 *Dynamics of machines. Oscillation* (Moscow: Yurait Publishing House) p 275

[10] Elyasberg M E 1993 *Self-oscillations of metal-cutting machines: Theory and practice* (St. Petersburg: OKBS Publ) p 180

[11] Sue D and Meyer A 1972 *Modern theory of automatic control and its application* (Moscow: Mechanical Engineering Publ) p 544

Electrochemical metallization of molybdenum alloys

A A Tikhonov[1,3], S I Arendateleva[1] and T I Perepelitsa[2]

[1]Yaroslav-the-Wise Novgorod State University, ul. B. St. Peterburgskaya, 41 173003, Veliky Novgorod, Russia
[2] GK Rosatom, JSC ATOMPROEKT, 197183, St.-Petersburg, ul. Savushkina 82

[3]E-mail: atnu1@yandex.ru

Abstract. The experimental data about influence of chemical and electrochemical operations of surface treatment of alloys of molybdenum on adhesion and solderability of electroplatings are provided. Influence of structures and the modes of electrolytes on quality of coverings, adhesion and solderability of the coverings applied on molybdenum alloys is studied. It is proved that use of combination in one electrolyte of two operations: electrochemical etching and at once the carried-out operation of copper plating, leads to increase in adhesion of coverings to molybdenum alloys. It occurs because the oxide layer is rather fully pitted during anodic dissolution, and the most important, it doesn't manage to be formed before deposition again as it occurs in many other known technological processes. The developed technology of galvanic metallization of molybdenic alloys allows: to reduce the number of technological operations, to increase efficiency of process, to use cheaper inventory, to increase adhesion of electroplatings to alloys of molybdenum and to improve their solderability.

1. Introduction

Molybdenum possesses a number of valuable properties: high melting temperature 2620°C; corrosion and chemical resistance in various environments, including sulfuric, hydrochloric and orthophosphoric acids; high modulus of rigidity (Young's modulus); low coefficient of thermal linear expansion; density 10,2г/см^3, i.e. it is almost two times lighter than tungsten. That is why molybdenum has higher specific strength at temperatures below 1370°C. Molybdenum has such limitation as low scale resistance, besides it forms brittle welding compounds and can hardly be wrought, soldered and metalized. Some limitations of molybdenum can be corrected by alloying [1]. Consideration should be given to relatively new, so-called pseudo-alloys of molybdenum with copper of brands MD40, MD50. In the given work, exactly these alloys were used for galvanic metallization; they are called like this, because they are made by method of powder metallurgy, not by alloying. Due to copper alloying of molybdenum, these alloys have higher heat and electrical conduction than pure molybdenum, they are more technological, and also have rather high heat resistance [2]. Thanks to its properties, molybdenum and its alloys can be used in electronics, instrument making, aerospace, atomic and other industries. However, molybdenum alloying does not improve much its poor ability to be soldered and metalized. First of all, this is associated with its great affinity to oxygen, as an oxide film interferes with above-mentioned processes. While soldering, it is recommended to remove an oxide film from the surface of molybdenum and molybdenum-base alloys by means of dipping into the melt consisting of 70% caustic soda and 30% sodium carbonate at temperature not higher than 400°C, or with the help of electrolytic etching in 80% aqueous solution of sulfuric acid at temperature 50 – 60°C. However, this does not allow to solve completely the problem of performing high quality soldering of molybdenum and molybdenum alloys in the conditions of increased demands to this process [3]. To metalize molybdenum and its alloys, various ways have been suggested: galvanic method, chemical method, vacuum condensation, magnetron sputtering, and others. For example, there is a well known way including the following operations: anodic stripping of oxides in 10% solution of etching acid at current density 5 A/dm^2 for 5 minutes, thermal processing in vacuum at temperature 700 – 900°C. The next way consists of the operations: electrochemical degreasing in 10 – 20% solution of caustic soda at current density 5–8 A/dm^2, activation in hydrochloric acid, chrome plating in standard sulfuric electrolyte, etching in hydrochloric acid, nickel plating in acid electrolyte with pH less than 1, etching in hydrochloric acid, and nickel plating in standard sulfuric electrolyte.

Much consideration has been given to metalize molybdenum and molybdenum alloys; multi-operative technologies have been developed, in which ecologically harmful substances are often applied, and expensive equipment (vacuum furnaces, furnaces with hydrogen environment, etc.) is used. However, these technological processes have low productivity, and they do not solve to the end the problem of increasing adhesion of coatings and improving their solderability [4, 5].

This work is devoted to further improvement of electroplating technology of molybdenum alloys, increased adhesion of applied coatings, and improvement of their solderability.

2. Materials and Methods

Plates of copper-containing molybdenum preudo-alloys MD40 and MD50 with 40% – 50% of copper, depending on the brand, were used as samples. Percentages in these alloys of molybdenum are given by weight as all other percentages given in this work. The samples had size 61x42x1mm. The samples of alloy MD40 were used in all experiments. The samples of alloy MD50 were used only in the last series of experiments (III-5). Distilled water and chemicals of brand "chp" (chemically pure) were used to prepare electrolytes. lectrochemical operations were performed in 1 liter bath equipped with thermostating system, a propeller blade mixer, and rectifier VTKG-3 with current stabilization system. Adhesion strength of coatings with molybdenum alloys was checked by methods of scratch-brushing, filing, undercutting, detachment. Solderability of the coating was checked by solder flowing area POS61 (tin-lead solder containing about 61% tin) on the tin-bismuth coating and by contact angle determination with the help of a unit for programmable heating and welding VSU200. Samples with dosed test charge 0.5 g of solder POS61 were placed in the unit VSU200, vacuumized and preserved in nitrogen. Vapors of formic acid derived at gas bubbling through liquid formic acid were used as a fluxing agent. Fluxing was performed at temperature 170°C during 6 minutes. Soldering was performed at temperature of isothermal hold time 220°C during 1.5 minutes.

3. Examination of impact of preparing molybdenum alloy surfaces on adhesion and soldering of galvanic coatings

Adhesive strength of coatings on molybdenum alloys is much less, if compared with that of galvanic coatings with copper, brasses, plain steels, and other well-metalized metals and alloys. This is associated with the fact that a molybdenum surface is coated with an oxide film which is badly etched, and, after removing, it can be formed again very soon. Due to this, much consideration has been given to composition selection and etching regimes of molybdenum alloys.

The first variant of technological scheme consisted of the operations: degreasing at temperature 60 – 80°C, washing in hot tap water at the same temperature, chemical etching, washing in cold tap water, washing in cold distilled water, electrolyte 1 tin-bismuth galvanic alloying (presented in table 2), washing in reclaim rinse, washing in hot water 60 – 70°C, drying, and quality control. The first variant has been presented by experiments denoted with I-1, I-2, I-3, I-4, where Roman numeral denotes the variant number, and Arabic numeral matches the number of chemical etching solution presented in table 1. The second variant of technological scheme differs from the first one in that, before tin-bismuth alloying, the so-called "flash" nickel plating in electrolyte 2 presented in table 2 was used, and chemical etching was performed in solution 3 (table 1), correspondingly, the experiment is denoted as II-3. In the third variant of technological scheme, electrochemical etching was used instead chemical one, with electrolytes given in table 4 (experiments III-1, III-2, III-3, III-4), and, before tin-bismuth alloying, the operation of galvanic coppering was performed in sulfuric acid electrolyte 3 (table 2) .

Samples for all experiments were degreased in the following solution (g/l): caustic soda 10–20, sodium carbonate 40–50, tri-sodium phosphate 40–50, sodium metasilicate 3–5, synthanol (DS-10 (ДС-10)) 1–2. Temperature of this electrolyte was maintained from 60 to 80°C. At first, in this solution chemical degreasing has been performed for 5 min, then cathode degreasing has been done at current density 5–10 A/dm^2 for 3 min, and then anode degreasing has been done at current density 3–8 A/dm^2 for half a minute. Degreasing quality was controlled by sample surface wettability with water.

Table 1. Solution compositions and chemical etching regimes.

Components and regimes	Solution 1	Solution 2	Solution 3	Solution 4
Potassium ferricyanide, g/l	-	-	-	200
Caustic soda, g/l	-	-	-	50
Hydrofluoric acid, %	-	-	6	-
Sulfuric acid, %	-	80	18	-
Hydrochloric acid, %	39	-	-	-
Temperature, °C	20–25	20–25	20–25	58–60
Time, minutes	5	5	5	5

Table 2. Electrolyte compositions and electrolysis regimes.

Components and regimes	Electrolyte 1	Electrolyte 2	Electrolyte 3
Stannous sulphate, g/l	40–60	-	-
Bismuth sulphate, g/l	0.8–1.5	-	-
Nickelous chloride, g/l	-	100–150	-
Cuprous sulphate, g/l	-	-	200–250
Sulfuric acid, g/l	100–110	-	50–70
Hydrochloric acid, g/l	-	100–150	-
Sodium chloride, g/l	0.3–0.8	-	-
Preparation OS-20, g/l	4–5	-	-
Cathodic current density, A/dm^2.	1	10	2
Time, minutes	15–30	3–5	15–30
Temperature, °C	18–30	25–40	18–30

In experiments I-1, I-2, I-3, I-4, II-3 with samples of alloy MD40, smooth 6 to 8 mcm thick tin-bismuth coatings without visible defects were obtained. However, as table 3 shows, adhesion of these coatings is very low, especially of those obtained in experiments I-1, I-2, I-3. These samples withstood scratch-brushing trials only, the rest did not withstand the trial. Potassium ferricyanide-base solution recommended to etch molybdenum copper-containing alloys was tried [6]. At etching in this solution (experiments I-4), selective etching took place. Molybdenum was etched off faster than copper. The alloy surface became dark-red due to formation of copper matt layer on the MD40 alloy surface. However, as table 3 shows, adhesion of tin-bismuth coatings on samples in experiments I-4 did not become better, if compared with that of similar coatings on samples obtained in other experiments I-1, I-2, I-3. Thus, chemical etching only is not enough to prepare the surface of molybdenum alloys. In relation to this, experiments II-3 were carried out, in which, to increase adhesion, the sublayer of fine nickel of acid flash electrolyte with pH less than 1 was used, before alloying with tin-bismuth coatings. Such technology also did not solve completely the problem of low adhesion. Though the adhesion became a little better, as the samples could withstand the scratch-brushing and scratching trials, and other samples obtained in experiments I-1, I-2, I-3, I-4 could withstand scratch-brushing only. Nevertheless, such adhesion is not enough to practice metalized MD40 alloys.

Table 3. Adhesion of galvanic coatings of tin-bismuth alloys on MD40 alloy treated in chemical etching agents.

Experiment number	Adhesion measurement methods 1	2	3	4
I-1	+	-	-	-
I-2	+	-	-	-
I-3	+	-	-	-
I-4	+	-	-	-
II-3	+	+	-	-

In tables 3 and 5 the following adhesion measurement methods are given: 1 – the manual scratch-brushing method with a bras brush, 2 – the method of gridding with a scalpel (a scratch mark method), 3 – the method of filing the coated sample edge with a file, 4 – the method of tearing off a wire soldered (by POS61) to the coating. When tearing off, the samples have been considered to stand the trial, if the breakage occurred along the solder, i.e. the adhesion exceeded the solder strength, and, accordingly, the samples have been considered to not stand the trial, if the coating came off the surface of the molybdenum alloy.

While soldering, the solder POS61 practically did not flow on the MD40 alloy samples coated with the tin-bismuth alloy (experiments I-1, I-2, I-3, I-4), and, accordingly, the samples had very poor or poor solderability (see table 7). As all copper samples coated with the tin-bismuth alloy were soldered excellently, the above-indicated solder flowed well. Taking this into consideration, technology of coating the MD40 alloy with a two-layer coating consisting of a copper sublayer and a galvanic tin-bismuth alloy has been worked out. To practice this technology, electrolytes for electrochemical treatment presented in table 4 were used.

Table 4. Electrolyte compositions and electrochemical treatment regimes.

Components and regimes	1	2	3	4	5
Sodium carbonate, %	20				
Orthophosphoric acid, %		90			
Sulfuric acid, %			20	80	45–65
Cuprous sulphate, %					2–3
Anode current density, A/dm^2	2	3–6	1–10	3–30	10–30
Cathode current density, A/dm^2		3–10			0.5–2
Cathode treatment time, min.		2			4–6
Anode treatment time, min.	5	4	5	5	2–4
Temperature, °C.	20–30	20–30	20–30	78–80	70–80

Technology recommended for molybdenum metallization, and providing electrochemical anodic etching in 20% sodium carbonate solution has been tested. However, at anodic etching, primary dissolving of copper occurred in this electrolyte of the MD40 alloy (experiments III-1), this being testified by the electrolyte bluish color appearing during anodic treatment of the MD40 alloy. Then electrolysis in sulfuric mixed electrolyte of coppering (table 2, electrolyte 3) was performed at cathodic current density varying from 1 to 8 A/dm^2. Along margins of the sample of the MD40 alloy with dimensions 42x61x1MM, a copper coating grew up, but any copper coating was not formed in the sample center. This testifies that an oxide film from the surface of molybdenum included into the MD40 alloy was not etched off completely, or was formed again.

Before coppering and tinning, the samples of the MD40 alloy in experiments III-2 were etched in 90% orthophosphoric acid, at first on anode at current density from 3 to 10A/dm^2 for 4 minutes, then on cathode at current density from 3 to 6A/dm^2 for 2 minutes. Before coppering and tinning (experiments III-3), the MD40 alloy samples were etched on anode in 20% sulfuric acid, and to make experiments (III-4) – in 80% sulfuric acid, treatment regimes are indicated in table 4. Thickness of galvanic copper coatings and that of tin-bismuth alloy coatings were varied from 6 to 12 mcm, herewith the coating thickness did not impact on the adhesion.

Table 5. Adhesion of galvanic two-layer coatings of copper and tin-bismuth alloy on alloys prepared electrochemically.

	Adhesion measurement methods			
Experiment number	1	2	3	4
III-2	+	+	+	
III-3	+	+		
III-4	+	+		
III-5	+	+	+	+

Results of the trials for adhesion strength of coatings with the MD40 alloys obtained in experiments with the surface electrochemical treatment demonstrated that combination of anodic treatment and cathodic one (experiment III-2) gives better results, if compared with anodic treatment only (experiments III-3 and III-4). This is most likely to be associated with the fact that the sample surface during the cathodic treatment is being saturated with hydrogen which slows down formation of oxides on the surface of molybdenum. Nevertheless, in this case the adhesion is also insufficient to be used in production, as the samples did not stand all the tests. In experiments I-1, I-2, I-3, I-4, II-3, III-1, III-2, III-3, III-4 as well as in many well-known technologies of galvanic metallization of molybdenum alloys, the coating high adhesion was not obtained by means of different ways of preparing the surface. This is due to the fact that, after etching and before coating with a galvanic coating, details are being washed in water and transferred into another bath. While washing and transferring, an oxide film is partly formed again on the molybdenum surface, and this decreases adhesion strength of the coating on molybdenum alloys. In a new way suggested (experiment III-5), anodic etching of an oxide film off the molybdenum alloy surface and the following coppering were performed in one universal bath consisting of cuprous sulphate 2 – 3 % and sulfuric acid 45 – 65%, without extracting details from the electrolyte. Thus, the pause between the operation of anodic etching and that of coating is excluded practically completely. Copper plates of brand M0 were used as electrodes. In experiments III-5, anodic etching at current density 10–30 A/dm^2 had been performed for 2–4 min, then, without removing the sample from the electrolyte, the polarity was reversed at once, with electrolyte being mixed at the same time. Cathode electrodeposition of copper had been started by current rush with cathode current density 8-10 A/dm^2 with the exposure time 0,5–0,7 min, then the working cathode current density 0,5–2 A/dm^2 was set, and electrolysis was performed for 4–6 min. As a result, a fine copper coating was formed on the MD40 alloy samples. Further, copper was grown up in electrolyte 3 presented in table 2. The tin-bismuth coating from electrolyte 1 was placed on copper coating (table 2). Coatings obtained by this way (experiments III-5) had the highest adhesion to the МД40 and МД50 alloys. As table 5 shows, they could stand all tests, including tearing-off the copper wire soldered by POS61, the adhesion strength being higher than the strength of solder POS61. Besides, the technology developed allowed to improve solderability and flowability of solder POS61 on molybdenum alloys with a two-layer coating of copper and tin-bismuth. Solderability of the coating was checked by solder POS61 flowability area on the tin-bismuth coating and by contact angle determination according to formula (1) given in the work [7].

$$\theta = 7{,}12(V/\sqrt{S^3}) \qquad (1),$$

where θ – angle, V – solder volume, S – flowing area.

Herewith, solderability was evaluated like in the work [8], using the tables similar to table 6.

Table 6. Coating solderability evaluation.

Solderability	Value of contact angle determined by flowing area
Very poor	More than 50°
Poor	50–34°
Satisfactory	34–12°
Good	12–3°
Excellent	Less than 3°

Table 7. Solderability of MD40 alloy and copper alloyed with tin-bismuth by POS61 solder.

Number of sample (experiment)	Base-metal	Solderability
I-1	Alloy MD40	Very poor
I-2	Alloy MD40	Very poor
I-3	Alloy MD40	Very poor
I-4	Alloy MD40	Poor
II-3	Nickel alloy MD40	Good
III-2	Copper alloy MD40	Good
III-3	Copper alloy MD40	Satisfactory
III-5	Copper alloy MD40	Excellent
IV-1	Copper	Excellent

As table 7 shows, samples I-1, I-2, I-3 with one-layer tin-bismuth coatings, the surface of which had been etched chemically before coating with tin-bismuth alloy, had very poor solderability. When soldering these samples, the solder POS 61 practically did not flow on the tin-bismuth coating, but appeared to "fall" into the coating and was fused into it, practically without flowing. This was most likely to be conditioned by very low adhesion of the galvanic tin-bismuth alloy with the molybdenum MD40 alloy surface. In sample I-4, solderability was poor, i.e. it was improved insignificantly, if compared with that of samples I-1, I-2, I-3, this seems to be due to copper enrichment of the molybdenum MD40 alloy surface, which occurs in selective etching in solution 4, table 1. The use of two-layer coatings, a nickel one with the tin-bismuth alloy and a copper one with the tin-bismuth alloy, as well as the use of electrochemical etching methods, when preparing the surface of the molybdenum MD40 alloys, allowed to improve solderability from very poor to satisfactory (sample III-3) and even good (samples II-3 and III-2). Before coating with a galvanic tin-bismuth coating, samples III-2 and III-3 were coated with copper coatings of almost similar thickness from one electrolyte, using similar electrolysis regimes (see table 2). Similarity was also preserved, when coating with tin-bismuth coatings, however, their solderabilities were different. The main difference between samples III-2 and III-3 was in electrochemical preparation of the surface. At first, samples III-2 underwent anodic etching, then – cathodic one in the solution of orthophosphoric acid, and samples III-3 were etched only on anode in the solution of sulfuric acid (see table 4). As a result, the adhesion strength of galvanic coatings on samples III-2 became better than on samples III-3 (see table 5). This appears to have become the main cause of improving the solderability of samples III-2, if compared with that of samples III-3 (see table 7). The solderability of the molybdenum alloy samples coated with an electrochemical tin-bismuth alloy is influenced by the composition and structure of an underlying coating, their adhesion to the molybdenum alloy surface, and adhesion of the galvanic tin-bismuth alloy to these underlying coatings. Thus, at approximately similar adhesion of the coatings on samples II-3 and III-4, a nickel sublayer, as it appears to be more porousless and dense, allowed to achieve better results of soldering, other things being equal. The solderability of the samples with nickel and copper sublayers became almost similar (see table 7), when the adhesion of the copper sublayer in sample III-2 became slightly higher than that of the nickel sunlayer coating in sample II-3 (see tables 3 and 5). Attempts of further improving the solderability of samples II-3 and III-2 due to enlargement of tin-bismuth coating thickness from 6 mcm to 12 mcm, and variation of current density from 1 to 2 A/dm^2 practically did not change the ability of the samples of series II-3 and III-2 to be soldered.

The situation with solderability has been changed sharply, the solderability has significantly improved and become excellent, by using new technology implemented in experiment series III-5 (see table 4). As a new method suggests, anodic etching of an oxide film off the surface of the molybdenum alloys MD40 and MD50 followed by coating with copper is performed in one universal bath composed of cuprous sulphate 2–3 % and sulfuric acid 45–65%, without extracting details from the bath. This practically excludes a pause between the operations of anode etching and coating as well as formation of a new oxide film on the molybdenum alloy surface. Correspondingly, such technology increased significantly adhesion and solderability of the molybdenum alloys. Experimental data obtained allowed to determine the dependence between adhesion of galvanic coatings and their solderability. Despite the fact that solderability of galvanic coatings is influenced by many factors, adhesion of galvanic coatings with the surface of molybdenum alloys is one of the most significant factors to solder molybdenum alloys. The higher is adhesion strength of coatings with molybdenum alloys, the higher is solder quality, as a rule. Thus, it is possible to line up the molybdenum alloy samples coated with the tin-bismuth alloy as solderability increases, depending on adhesion. The samples of molybdenum alloys I-1, I-2, I-3, I-4 with the surfaces prepared by a chemical way of etching had very poor or poor solderability, as well as the lowest adhesion of coatings. Samples II-3, III-2, III-3 prepared by electrochemical ways of etching and coated with layers of the nickel and tin-bismuth alloy, or the copper and tin-bismuth alloy had solderability from satisfactory to good, depending mostly on adhesion as well. Among metalized samples of molybdenum alloys, samples III-5 obtained by new technology of combining anode etching and coating with copper in one universal bath possessed the best, excellent solderability. Flowing area of dosed test charge of solder POS 61 on samples III-5 was less not more than by 10%, if compared with flowing area of dosed test charge of solder POS 61 on tinned copper samples IV-1. Herewith, samples III-5 had excellent solderability by POS61 close to that of tin-bismuth coatings on copper samples IV-1 (see table 7).

4. Conclusions

Use of chemical solutions of etching only is not enough in technology of preparing surfaces of molybdenum copper-containing alloys, to achieve necessary adhesion of galvanic coatings. Successive use of anode and cathode treatments in preparing the surface increases, though slightly, adhesion of galvanic coatings to molybdenum alloys. The highest adhesion of coatings, higher than the strength of solder POS61, was obtained by combining two operations in one electrolyte № 5 (table 4): electrochemical etching and coppering operation performed at once, practically instantly. Adhesion of coatings to molybdenum alloys increases, since an oxide film was sufficiently etched off during anode dissolving, and it had not enough time to be formed again, as it is common in many other known technological processes. Impact of adhesion strength of electrochemical coatings on solderability has been studied. It has been defined that two-layer copper-alloy and tin-bismuth galvanic coatings placed on copper-containing pseudo alloys of brands MD40 and MD50 with adhesion higher than the strength of solder POS61 possess excellent solderability. Technology of electrochemical metallization of molybdenum alloys which allows to achieve rather high adhesion and excellent solderability has been developed. Priority and originality of developed technology of galvanic metallization of molybdenum alloys was confirmed by the patent [9].

Acknowledgement

The author expresses his gratitude to engineer Vasilyev V A for help in conducting experiments on determining solderability on a programmable heat and welding installation VSU200.

References

[1] Nikonov N V 2014 *Molibden. Svojstva, primenenie, proizvodstvo, produkcija* [Molybdenum. Properties, application, production, products.] (M: OOO "Metotehnika" Publ) p 34 (in Russian)

[2] Makedonskij B K, Morozova I G and Skazkin V E 2012 Psevdosplavysistemy med' – molibden [System pseudo-alloys copper – molybdenum] *Modern problems and ways of theirsolution in science, transport, production and education* Available at: http://www.sworid.com.ua/konfer29/750.pdf (accessed 18.03. 2019)

[3] Petrunin I E *and al* 2003 Soldering guide. Spravochnikpopajke [Reference book on soldering.] (M.: Mashinostroenie) p 480 (in Russian)

[4] Nishhev K N, Martynenko V A, Beglov V I, Grishanin A V, Eliseev V V, Malygin M Ju, Novopol'cev M I and Judin V A 2013 Issledovanija svojstvmnogoslojnojmetallizaciistruktur "kremnijnamolibdene", poluchennojmetodommagnetronnogoraspylenija [Researches of properties of the multilayered metallization of structures "silicon on a molybdenum" received by method of magnetron dispersion] *Izvestijavysshihuchebnyhzavedenij* (Povolzhskij region: Fiziko-matematicheskie nauki) **3 (27)** p 248–260

[5] Azhogin F F *and al* 1986 *Gal'vanotehnika. Spravochnik* [Galvanotechnics. Reference book] (M: Metallurgija) p 736 (in Russian)

[6] *Sekrety gal'vaniki ot GalinyKoroljovoj. Pokrytie molibdena pod pajku* [Galvanics secrets from Galina Korolyovoy. A molybdenum covering under soldering] URL: Available at: https://www.subscribe.ru/archive/socio.science.galvanika/201105 (accessed 21.03.2019)

[7] Lanin V L 2008 Evaluation of solderability of electronic components and details in electronics *Componenty i tekhnologii* **2** p 28–32

[8] Kireev S Yu and Perelygin Yu P 2011 Methods of determining solderability of coatings *Galvanotekhnika i obrabotka poverkhnosti* **2** Vol XIX p 52–57

[9] Tikhonov A A 2018 *Sposob gal'vanicheskoj metallizacii molibdenovyh splavov* [Way of galvanic metallization of molybdenic alloys.] Patent RF, no. 2653515 Bul **13** p 7

Technology of electrodepositing and properties of indium-containing chromium alloys

A A Tikhonov[1,3], D A Filippov[1], S I Arendateleva[1] and T I Perepelitsa[2]

[1]Yaroslav-the-Wise Novgorod State University, ul. B. St. Petersburgskaya, 41 173003 Veliky Novgorod, Russia
[2]GK Rosatom AO Atomproekt 82, Savushkina, Sankt-Peterburg, 197183, Russia

[3]E-mail: atnu1@yandex.ru

Abstract The experimental results of the influence of electrolyte composition and electrodepositing regimes on the properties of galvanic gradient indium-containing composite coatings based on chromium are presented. The dependences of micro hardness, coefficient of friction and wear resistance of coatings on electrodepositing regimes and electrolyte composition were studied. The effect of coating thickness on properties of gradient coatings was investigated. It is shown that the gradient alloy chromium – indium – molybdenum – vanadium – boron nitride (elbor), obtained by electrolytic deposition, has a high wear resistance and is far superior to the wear resistance of the coating obtained from standard chromium sulphate electrolyte.

1. Introduction

Galvanic chromium plating has been widely used while producing details and instruments, to increase micro-hardness and wear resistance. Chromium coatings possess high micro-hardness (to 13 GPa) and wear resistance, they also have good anti-frictional characteristics and high corrosion resistance. More often, chromium plating is performed from standard sulphate electrolytes based on hexavalent chromium. Main disadvantage of these methods of obtaining is their ecological danger. Besides, the process of chroming has low dispensability and current output. In chroming, hydrogen absorption by metal (hydrogen uptake) occurs resulting in decreased fatigue strength. To improve properties of chromium coatings and increase chroming rate, supplements are introduced in electrolyte composition, ultrasound is used, impulse and reversible current is used [1, 2]. Chroming process from electrolytes based on trivalent chromium is ecologically more pure than that from standard sulphate electrolytes based on hexavalent chromium. These electrolytes have higher current output and, correspondingly, chromium steels absorb less hydrogen. However, mechanical properties of coatings obtained from electrolytes based on trivalent chromium are inferior to those obtained from electrolytes based on hexavalent chromium [3, 4]. To increase wear resistance and hardness, composite electrochemical coatings (CEC) have been made, basing on chromium. However, obtaining CEC based on chromium is associated with a number of difficulties due to abundant hydrogen excretion on cathode; because of this the content of the second phase in CEC based on chromium does not exceed 2–3 %, as a rule. Nevertheless, even insignificant amount of the second phase in the coating can essentially improve its properties due to restructuring [5]. Use of chromium-based nano-composite galvanic coatings gives still more effect. Such coatings can be obtained by the way of introducing ultradisperse powder of aluminum oxide in the chroming electrolyte. Herewith, fatigue of base metal decreases, and adhesion increases, along with the increase of micro-hardness [6-8]. Diamond-cluster galvanic coatings possess wear resistance 2.5–5.7 times more, compared with coatings obtained by the routine hard chroming plating, with their micro-hardness increasing to 14 GPa. Micro-hardness of the coatings modified by nanotubes increases to 1.5–3 times. Simultaneously, porosity and unevenness of distributing the coating on the surface of a detail decrease [9–11].

The given work is devoted to methods of improving properties of chromium coatings by the way of alloying, making the second phase in them, obtaining gradient coatings with increased content of indium in the upper surface.

2. Experiment: materials and methods

Samples were made of unalloyed carbonaceous steel with carbon content 1 %. Before coating, all samples were ground, polished, degreased with Viennese lime, pickled in 20% solution of sulfuric acid, washed first in tap, then in distilled water.

Making electrolytes, distilled water and chemically pure chemicals were used. To obtain a composite electrochemical coating, cubic boron nitride (cBN) with size of grains of base fraction (95%) 1 μm and finer was introduced into the electrolyte composition. Galvanic plating was performed in baths with volume 0.5, 1 and 3 liter, provided with thermo stating system, a propeller blade mixer, and sources with current stabilization systems.

To increase wear resistance of the chromium coatings, chromium alloying with indium, as well as chromium-molybdenum and chromium-molybdenum-vanadium micro-alloys were used. To precipitate chromium-molybdenum alloys, stable sulphate electrolytes have found applications in industry. Such coatings have wear resistance about 2 times higher, compared with pure chromium ones, also their protective ability is better, inner tensions are lower, and micro-hardness is higher. Vanadic acid introduced into chroming electrolytes contributes to increased current output, finer grains of the coating, as well as eliminates negative influence of increased concentration of sulphate ions due to their binding into sulphate-vanadic complex. Further improvement of the coatings was performed by the way of additional alloying of chromium-molybdenum-vanadium coatings with indium. Chromium, molybdenum and vanadium on the one hand, and indium on the other hand, are metals possessing a number of opposite properties. If Cr-Mo-V coatings have high melting temperature and hardness, rather often they possess significant inner tensions (though lower than chromium does) and are easily hydro treated. Unlike them, indium is fusible, soft, plastic, and its coatings are not hydro treated. Besides, indium possesses the lowest friction coefficient among all metals – 0.05–0.06 [12]. Use of such materials was aimed at removing these disadvantages and obtaining anti-frictional and wear-resistant alloys with improved mechanical properties. In order to use more fully advantages of each component of the alloy, a gradient coating with increased indium content in upper layers was made on their base. In gradient coatings, the coating composition varies smoothly across the coating thickness. Correspondingly, micro-hardness of coatings and friction coefficient (unlike multi-layer coatings which have alternate layers of different materials with distinct borders) may vary smoothly by the coating' thickness. Making of gradient or thickness modulated coatings is a very actual problem, as it allows to improve essentially the performance properties of coatings (wear and corrosion resistance), and eliminate the negative effect of "thin ice", etc. [13]. On the base of the electrolytic chromium-indium-molybdenum-vanadium alloy, a composite electrochemical gradient coating was made due to introducing cubic boron nitride into its composition, in order to improve further micro-hardness and wear resistance. As the experimental results have shown, such a coating is significantly superior in wear resistance to the coatings obtained from the standard sulphate electrolyte of chroming.

All coatings obtained were examined for composition, micro-hardness, and wear resistance. The composition of the obtained electrolytic alloys was determined by the method of micro X-ray spectral analysis (MRSA) using an X-ray microanalyzer MR-7.Micro-hardness was defined on steel samples covered with studied alloys. Initial samples had shape of plates sized 20x10x1.5 mm. Microhardness was determined on steel samples coated with the studied alloys. The initial samples had the form of plates with dimensions of 20x10x1.5 mm. The microhardness of the coatings was determined on a microhardometer by the Vickers method, by measuring the print diagonals obtained from indentation of the diamond pyramid. Wear resistance of the coating was studied with a rotary-type friction machine and reciprocating stand of construction. The reciprocating unit allows wearing out two samples simultaneously. As an opposite body, the plate made of high-carbon steel U10A (Russian mark, English analog BW1B, Chinese analog T 10A), was used. The steel was hardened and tempered at low temperature. In a rotating-type friction machine, plastic strips made of glass-filled phenol-formaldehyde polymer AG-4C (Russian mark) were used as a counter body. In preliminary experiments, plastic was stuck on copper or steel pads. It was found that it was impossible to provide conditions for uniform wear process of coatings on a roller, by using a rigidly fixed opposite body. As the coatings were wearing out locally, wear process of the support plate material introduced an error in measurements of the coatings' wear amount. In order to achieve uniform wearing of the coatings, the plastic tape was laid in two layers on 6 mm thick rubber gasket and held in tight state on a cylindrical drum with a clip of the holder with which the rotary-type friction machine was equipped. Run-in was first performed with force 10 N, then the loading gradually increased to 100 N. Tests were carried out at frequency of the roller's rotation 300 rpm (sliding velocity 0.785 m/s). The coatings' wear was evaluated by measuring their mass which was defined with an analytical balance, accuracy to 0.1 mg. To evaluate the wear scar, a microscope MPB-2 (Russian mark)) was used.

3. Results and discussion

At first, the content of indium in chromium alloys obtained by electrodeposition, and their appearance depending on the concentration of indium compounds in initial electrolytes were examined. For this electrolytes for electrodepositing the alloys were prepared, their composition being presented in table 1. Electrolytic chromium-indium-molybdenum-vanadium (Cr-In-Mo-V-BN) alloy is a gradient composite coating. Cubic boron nitride was selected as the second phase to introduce into the coating, since it is inferior only to diamond in hardness and contributes to wear resistance increase [14]. Herewith, cubic boron nitride manifests hydrophilic properties in the electrolytes used, so it can be quickly and well wetted by the electrolyte. Indium was introduced in the electrolyte as indium sulphate, or in the acidified electrolyte as indium oxide. Concentration of indium-containing compounds varied in electrolytes N1-3. The content of indium in the electrolyte was calculated on the base of concentrations of indium-containing compounds in the electrolyte. In order to gradually increase the content of indium in the coating, cathode current density in electrolyte N3 has been decreasing smoothly from 70A/dm^2 to 20A/dm^2.

Table 1. Compositions of electrolytes and regimes of electrodepositing indium-containing chromium alloys.

Components and regimes	Electrolyte N 1 Cr-In	Electrolyte N 2 Cr-In-Mo	Electrolyte N 3 Cr-In-Mo-BN
Chromic anhydride, g/l	80	250	80-150
Sulfuric acid, g/l	0,8	2,5	0,8-1,5
Indium sulphate, g/l	-	-	0,5-37,5
Indium oxide, g/l	2-50	0,5-20	-
Molybdic acid, g/l	-	15	10-30
Vanadic acid, g/l	-	-	10-30
Cubic boron nitride, g/l	-	-	10-50
Cathodic current density A/dm^2	25	25	70-20
Electrolyte temperature, C	50	50-65	55-65

The concentration of indium in the electrolytic alloys based on chromium has been changed quite a lot, depending on the indium concentration in initial electrolytes. The experimental results are presented in figure. 1.

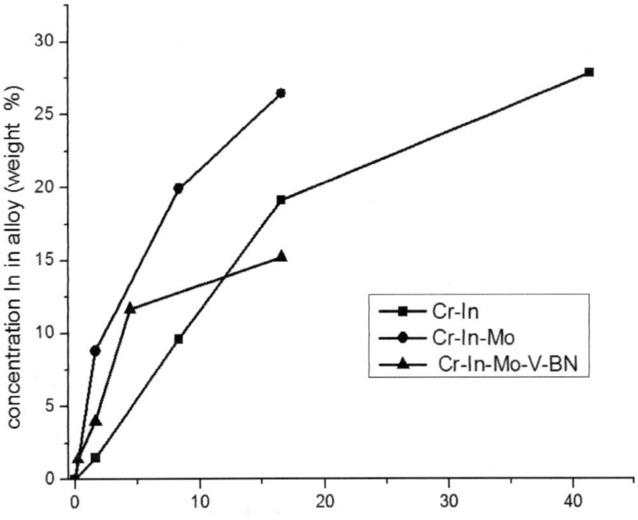

Figure 1. Dependency concentration of In in alloy versus concentration of In in electrolyte.

As can see from figure 1, at the increase indium concentration in the electrolytes, its content in the chromium-based indium-containing coatings also increases. Herewith, at first, the indium content in the coatings increases more sharply at lower concentrations of indium in the electrolytes. Then, as the double electrolytic layer was being saturated with indium ions, the concentration increased slower. As it follows from figure 1, when the concentration of indium ions in the electrolytes becomes similar, more indium is deposited in the electrolytic alloy Cr-In-Mo. It is known that introduction of molybdic acid into composition of the standard electrolyte of chroming impacts essentially on kinetics of electrodepositing chromium. The higher was the concentration of molybdic acid, the lower was the current output observed. Decreased rate of depositing chromium with increased concentration of molybdate-ions has been conditioned by decreased content of ions of hexavalent chromium in a cathode film due to increased concentration of more hardly restored molybdenum ions in it. Also, molybdate-ions contribute to formation of a less hard, looser, easily hydrogen-penetrated film. This loosing action of molybdate-ions on a cathode film is likely to contribute to deposition of indium ions into the electrolytic alloy. Despite the fact that electrolytes for electrodeposition of the alloy Cr-In-Mo-V-BN contain molybdate-ions, in this case it is necessary to consider possible impact of other electrolyte components as well. Decreased indium content in Cr-In-Mo-V-BN coatings deposited from electrolyte N2 is associated with the impact of vanadic acid on cathodic process. As vanadic acid is being introduced into the standard chromium solution to studied concentration 30 g/l, a more compact, hard and less porous film is formed, compared with a film formed in presence of molybdenum ions. This is correlated with the impact of vanadic and molybdic acids on the current output. Thus, the current output was equal to 17–18% in the standard chromium solution at current cathode density 50 A/dm^2 and temperature 50°C. As vanadic acid of concentration 30g/l was being added into this electrolyte, the current output became equal to 22–23%, and, in similar conditions of electrolysis and with the same concentration of molybdenum acid, the current output decreases to 16–17% [15]. Presence of vanadic acid in electrolytes results in decreased indium content in coatings which is aggravated at increased concentration of sulphate ions, since indium is introduced into electrolyte N3 as indium sulphate. Besides, it is necessary to consider that Cr-In-Mo-V-BN electrodeposition was performed at gradual decrease of current cathodic density from 70 A/dm^2 to 20 A/dm^2 made to obtain a gradient coating. Electroplating of alloys Cr-In-Mo and Cr-In was made at constant current cathode density 25 A/dm^2.

The experiments have proved that the appearance of a coating is greatly influenced by the indium content in alloys. Table 2 gives the dependence of metallic luster on the indium content in the coating.

Table 2. Dependence of coatings' appearance on indium content in electrolytic indium-containing alloys of chromium.

Indium content in coating, %	Cr-In	Cr-In-Mo	Cr-In-Mo-V-BN
1.4-1.5	gl/lu	-	lu/sl
4	-	-	lu/sl
8.8-11.7	lu/sl	lu/sl	sl/mt
15.2-19.9	sl/mt	sl/mt	mt/mt
26.4–27.8	mt/mt	mt/mt	-

As table 2 shows, chromium-indium coatings with low indium content (1.5% by weight) had the highest luster (lu) degree. Such, 15.8 μm thick, coatings electrodeposited as glassy (gl) ones. Then, as the coating became thicker, the degree of luster decreased a little, and 30.3 μm thick coatings became lustrous. The same effect, namely, decreased luster of the coatings with increased thickness has been observed for all studied indium-containing alloys of chromium. The electrolytic alloys Cr-In–Mo-V-BN at low indium concentration in the coating (1.4% and 4%) and the alloys Cr-In(9.6%) and Cr-In(8.8%)–Mo at the coating's thickness 15,5±1 μm were lustrous. In 30,8±1.6 μm thick coatings, these alloys became semi-lustrous (sl). Coatings Cr-In (19.1%) and Cr-In(19.9 %)-Mo deposited from semi-lustrous to mat ones, depending on thickness of the coatings. At indium content from 26,4% to 27,8% in these coatings, in the whole studied range of thicknesses, mat coatings deposited. No doubt, indium content in chromium-based coatings is the most significant factor influencing the degree of luster of coatings. Besides, the luster degree of coatings decreases if there is vanadic acid and boron nitride in the electrolyte. Thus, if the coatings Cr-In(15.2%)-Mo-V-BN deposited as mat ones in the whole studied range of thicknesses, the coatings Cr-In (19.1%) and Cr-In(19.9%)-Mo with thicknesses less than 15.5 μm deposited as semi-lustrous ones.

Composition impact of the electrolytes and regimes of electrolysis (table 3) on composition of the gradient coatings Cr-In-Mo-V-BN and rate of their deposition (table 4) have been studied. Electrolyte with minimal concentrations of components (electrolyte N3 in table 3), electrolyte with average concentrations of components (electrolyte N4 in table 4), and electrolyte with maximum concentrations of components (electrolyte N5 in table 4) were used for this purpose.

Table 3. Compositions of electrolytes and regimes of electrolysis for electrodeposition of gradient composite coatings Cr-In-Mo-V-BN.

Components and regimes	Electrolyte N4	Electrolyte N5	Electrolyte N6
Chromic anhydride, g/l	80	100	150
Sulfuric acid, g/l	0.8	1.0	1.5
Indium sulphate, g/l	0.5	5.0	10
Molybdic acid, g/l	10	20	30
Vanadic acid, g/l	10	20	30
Cubic boron nitride, g/l	10	25	50
Cathode current density, A/dm^2	70-20	70-20	70-20
Electrolyte temperature, °C	60	60	60

Table 4. Dependence of compositions of galvanic alloys and rate of plating them upon composition of electrolytes.

Alloy composition and electrodeposition rate	Electrolyte N4	Electrolyte N6
Indium, %	1.4	11.7
Molybdenum, %	0.5	0.7
Vanadium, %	0.02	0.04
Cubic boron nitride, %	0.37	0.51
Chromium	rest	rest
Electrodeposition rate, μm/hour	32.4	29.2

Table 5. Impact of thickness of gradient coatings Cr-In-Mo-V-BN electrodeposited from electrolytes N4, N5 and N6 on microhardness and static friction coefficient (of resting state).

Mechanical properties	Electrolyte N4	Electrolyte N5	Electrolyte N6
Friction coefficient, t_1/t_2	-/0.20	-/0.17	0.16/0.14
Micro-hardness, GPa, t_1/t_2	-/10.9	-/10.4	10.1/8.94

In the table 5 there are results of measurement at thickness of the coating t_1=5.5±1μm, and in the denominator – at thickness t_2=30.8±1.6 μm.

Results of measuring micro-hardness of the coatings have shown that, despite the increased content of hard particles of cubic boron nitride in the coating from 0.37% to 0.51% and increased content of molybdenum and vanadium, the indium content is a major factor determining the amount of micro-hardness. The indium concentration increased from 1.4% to 11.7%, correspondingly, despite the other factors, the micro-hardness decreased from 10.9 GPA to 8.94 GPa. As table 6 shows, the increased indium content in chromium galvanic alloys decreases the coefficient of friction. The decreased friction coefficient and micro-hardness, along with the increased thickness of coatings Cr-In-Mo-V-BN, is related to the increased indium content in upper layers of the coating caused by smooth decrease of cathodic current density.

In this work, much consideration has been given to the studies of wear resistance of indium-containing chromium alloys. Results of the testing presented in table 6 have demonstrated perceptivity of using indium-containing chromium alloys in the friction conditions examined. The given data have shown that, at first, as the indium content increased in the alloys, the wear process decreased, despite the decreased micro-hardness. The greatest wear resistance paired with steel friction belonged to sample N2 coated with the Cr-Mo-In alloy, indium content 19.9%, and sample N5 coated with the Cr-In-Mo-V-BN alloy, indium concentration 11,7%, this is likely to be due to the lower friction coefficient.

Table 6. Dependence of wear of coatings paired with hardened steel friction upon their composition and conditions of obtaining.

Composition of alloys	HV, GPa	Wear, μm/km	Conditions of obtaining
1 Cr	10.3	3.9	Electrolyte N1 without indium, Cathode current density 50 A/dm^2
2 Cr-Mo-In(19,9%)	8,57	0.41	Electrolyte N2 with indium 8,3 g/l
3 Cr-In(1,4%)-Mo-V-BN	10.9	0.82	Electrolyte N4
4 Cr-In-Mo-V-BN	10.4	0.49	Electrolyte N5
5 Cr-In(11,7%)-Mo-V-BN	8.84	0.36	Electrolyte N6
6 Cr-Mo-V	10.6	1.03	Electrolyte recommended [16]

IOP Conf. Series: Materials Science and Engineering **656** (2019) 012051 doi:10.1088/1757-899X/656/1/012051

One of main purposes of this work is to increase service of molds used to produce details from glass-filled plastics of brands AG-4S (Russian mark) due to increased wear resistance of chromium coatings. Concerning this, the wear resistance of developed coatings paired with plastic AG-4S (Russian mark) friction has been studied, using unit SMTs 2 (Russian mark) (see table 7).

Table 7. Impact of conditions of obtaining and composition of coatings on their wear paired with glass-filled plastic AG-4S (Russian mark) friction after 20000 rounds of roller.

Composition of alloys	HV, GPa	Wear, g/m $\cdot 10^{-6}$	Conditions of obtaining
1 Cr	10.3	2.8	Electrolyte N1 without indium
2 Cr-Mo-V	10.6	0.6	Electrolyte recommended [16]
3 Cr-Mo-In(8,8%)	9.92	0.33	Electrolyte N2 with indium 1,65 g/l
4 Cr-Mo-In(19,9%)	8.57	0	Electrolyte N2 with indium 8,3 g/l
5 Cr-Mo-In(26,4%)	6.62	0.76	Electrolyte N2 with indium 16,5 g/l
6 Cr-In(9,6%)	9.35	1.64	Electrolyte N1with indium 8,3 g/l
7 Cr-Mo-In(11,7%)-V-BN	8.94	0	Electrolyte N6

Despite decreased micro-hardness of the coatings, alloying with indium can lead to increased wear resistance of the coatings, compared with pure chromium coatings. This is most likely to be related to decreased friction coefficient of such coatings paired with glass-filled plastic AG-4S (Russian mark) friction. Besides, degree of hydrogen-charging coatings (which can be less for indium-containing coatings), graininess of a coating, whether there are structure defects, and other factors should be considered. Concerning the wear resistance of galvanic Cr-Mo-In alloys, you can see in table 8 that the alloys containing 19.9% of indium possess the maximum wear resistance. The roller coated with such alloy does not demonstrate the sign of losing the coating' mass even after 20000 rounds. At further increase of indium content in the Cr-Mo-In alloys to 26.4%, their wear resistance decreases due to decreased micro-hardness of the coatings (composition N5, table 7). Good results have been obtained by using a gradient composite coating Cr-Mo-In(11.7%)-V-BN, it having a zero wear after 20000 rounds of the roller with such coating.

4. Conclusion

As a result of the work done, technology of electrodepositing indium-containing chromium-based alloys with maximum indium content 27.8% has been suggested. It has been found that molybdate-ion contributes to indium electrodepositing in a chromium-based alloy, and vanadate-ion, on the contrary, hinders. It has been defined that, as indium content in the alloys and thickness of the coatings increase, luster degree and micro-hardness of the coatings decrease. However, after polishing, the alloys with higher indium concentration have lower friction coefficient. We succeeded in making a gradient chromium-based coating. In these coatings, micro-hardness of lower layers neighboring directly the metal-base (steel) has high hardness which decreases gradually as thickness of the coating increases. And upper layers possess low friction coefficient, thank to which they can work for a long time without lubricating. Galvanic alloy Cr-Mo-In (11.7%)-V-BN paired with glass-filled plastic friction as well as with steel friction has the greatest wear resistance. A gradient composite coating Cr-Mo-In (11.7%)-V-BN may be recommended to increase resistance of molds and other details working in the conditions of dry friction, especially of those contacting with glass-filled plastics.

References

[1] Solodkova L N and Rudryavtsev V N 2007*Elektroliticheskoe khromirovanie* [Electrolytic chrome plating] (Moscow: Globus Publ) p 191 (in Russian)
[2] Kolesnikov A A, Zarembo V I and Zarembo D V 2015 *Alternative Energy and Ecology* **4** 86–101
[3] Protsenko V and Danilov F 2009 *Electrochemica Acta* **54** 5667–5672

[4] Demakov A G, Pavlov L N, Reformatorskaya I I, Vinogradov S S and Kudryavtsev V N 2010 *Galvanotechnika I obrabotka poverchnosti* [Electroplating & surface treatment] **18** 27–32

[5] Borodin I N 1982 *Uprochnenie detaley kompozitsionnymi pokrytiaymi* [Hardening of parts with composite coatings] (Moscow: Mashinostroenie Publ) p 141 (in Russian)

[6] Salakhova R K 2009 *Aviatsionnye materialy I tekhnologii* [Aviation materials and technologies] **2** 20–24 (in Russian)

[7] Safonov V V, Shishilin S A, Semochkin V S and Zakharevich A M 2015 *Surface Engineering and Applied Electrochemistry* **51** 517–522

[8] Tyurikov V V 2009 *Aviatsionnye materialy I tekhnologii* [Aviation materials and technologies] **1** 13–17 (in Russian)

[9] TseluikinV N, Kanaf'eva O A 2015 *Chemical and Petroleum Engineering* **51** 54–57

[10] Grafushin R V, Vinokurov E G, Makhina V S and Burukhina T F 2018 *Galvanotechnika I obrabotka poverchnosti* [Electroplating & surface treatment] **26** 26–32 (in Russian)

[11] Gasanova O A, Tseluikin V N and Tseluikina G V 2015 *Perspektivnye Materialy* [Perspective materials] **5** 78–81

[12] Perelygin Yu P 1993 *Elektroosaxhdenie, svoijstva I oblast' primeneniyaindiya In ego dvoijnykhsplavov* [Electrodeposition, properties and field of application of indium and its double alloys] (Penza) p 84

[13] GamburgYu D 2001 *Russian Journal of Electrochemistry* **37** 585–590

[14] Tikhonov A A and Filippov D A 2018 *IOP Conf. Series: Materials Science and Engineering* **441** 012054

[15] Shluger A M, Ignat'evV I and Tok L D 1979 *Sovremenye metody naneseniya gal'vanicheskikh I khimicheskikh pokrytij Materialy seminara [Modern methods of applying electroplating and chemical coatings. Materials of seminar]* pp 124–128

[16] Azhogin F F et all 1986 *Gal'vanotekhnika. Spravochnik* [Electroplating. Handbook] (Moscow: Metallurgiya Publ) p 736

IOP Conf. Series: Materials Science and Engineering **656** (2019) 012052 doi:10.1088/1757-899X/656/1/012052

Model of stress-strain state of wooden rod under eccentric compression and transverse load

A S Varenik[1] and K A Varenik[1]

[1]Yaroslav-the-Wise Novgorod State University, ul. B. St.-Peterburgskaya 41 173003 Veliky Novgorod, Russia

E-mail: vkirillv89@mail.ru

Abstract. The analytical model of the stress-strain state of a wooden compressed-bent rod at short-term loading by forces with unequal eccentricities and at presence of transverse loading is offered. In the calculation model the real diagrams of work of wood on compression and stretching are laid. The approximation of the curved compression chart takes into account the elastic-plastic properties of the wood. The method allows to solve the problem of determination of the limiting bearing capacity both by the criterion of buckling and by criteria of durability. The theoretical results are confirmed by experimental data and can be used for practical calculations of compressed-bent elements.

1. Introduction

The share of wooden structures in the total volume of building constructions In Russia remains very insignificant. But at the same time there are some предпосылки for the development of wooden building. This is facilitated by the availability of the constantly renewable resource base, modern production technologies and some general tendencies in the building market connected with the decrease in the production cost. The current regulations in the design of wooden structures were introduced in the 1930s and have not altered significantly since that time. The computational models no longer reflect the real work and bearing capacity of structures as well as being non-economical.

Modern software systems allow to compile any engineering design based on the computational models which take into account the peculiarities of the construction material. One of the major ways to improve the methods of wooden structures calculating is accounting of the non-linear mechanical properties of wood.

In this paper, an analytical model of the stress-strain state is developed and the problems of stability of eccentrically compressed wooden rods are solved under the application of compressive forces with unequal terminal eccentricities and transverse loads. The proposed method of calculation allows more accurately than in the known solutions, to simulate the work of compressible-bending elements with virtually any, actually encountered in practice, the conditions for fixing the supporting ends and loads.

Presented in this paper, the research results are aimed at improving the methods of calculating compressible-bending wooden elements of building structures.

2. Problem statement and basic assumptions

The general theoretical approaches of analytical modeling of the stress-strain state of an eccentrically compressed rod are presented in [1] and [2].

A hingedly supported wooden rod, compressed by a longitudinal force P, applied with unequal eccentricities e_1 and e_2 and bendable by a transverse load q (figure 1) is considered. The stress-strain state of the most loaded cross section of the rod is investigated.

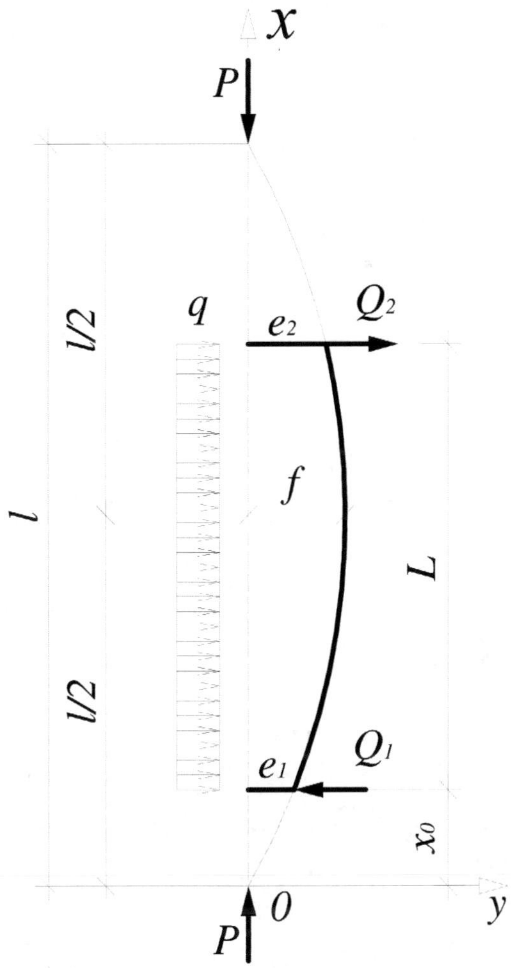

Figure 1. The design-scheme of the compressible-bending rod taking into account the lateral load.

To simplify the solution, a number of general assumptions are made.

The curvature of the element is determined by the approximate expression:

$$\frac{1}{\rho} = \frac{d^2y}{dx^2}. \quad (1)$$

It is assumed that the curved axis of the rod is a half-wave sinusoid:

$$y = fSin\frac{\pi x}{l}, \quad (2)$$

where l – estimated rod length; f – deflection (maximum displacement) when $x=l/2$.

The stress distribution over the cross section is determined by the distribution of strains in the cross section and the dependence σ–ε. It is assumed that deformations are distributed according to a linear law and the axis of zero deformations coincides with the axis of zero stresses. The dependence σ–ε for any fiber corresponds to the experimental σ–ε diagram obtained from the results of tensile and compression wood tests.

To approximate the work of wood under compression, we use a cubic parabola of the form:

$$\sigma = \mathrm{A}_1\varepsilon - \mathrm{A}_2\varepsilon^3. \tag{3}$$

The first term of this expression corresponds to Hooke's law and the coefficient A_1 is taken to be equal to the modulus of elasticity of wood under compression along the fibers. When ε=0, $\frac{d\sigma}{d\varepsilon} = A_1 = E$. The second term provides the σ–ε dependences for the decrease in the rate of increase of stresses with respect to the rate of increase in deformations and the transition beyond the tensile strength to a falling section. When $\varepsilon=\varepsilon_{пп}$, $\frac{d\sigma}{d\varepsilon} = 0$.

The relationship between stresses and strains in the stretched zone of the section corresponds to Hooke's law:

$$\sigma_p = E_p\,\varepsilon. \tag{4}$$

In the process of deformation of the rod in the cross section there are compressed and stretched zones. The distribution of stresses and strains over the cross section of the rod is shown in figure 2.

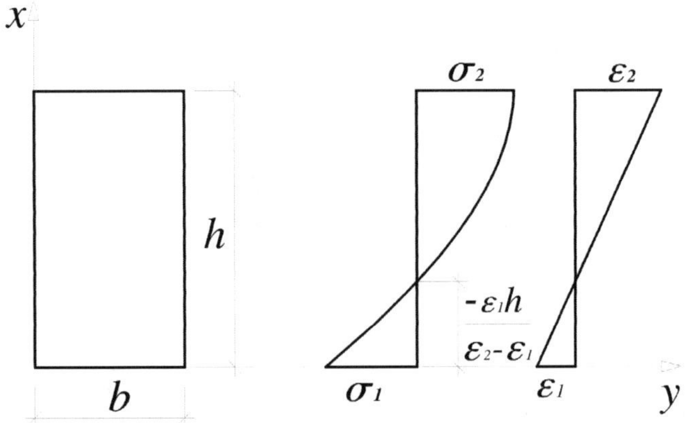

Figure 2. The design scheme of stresses and strains in the cross section of the rod.

Using an approximate expression for the curvature and the representation of the curved axis of the rod by a sinusoid, one can write:

$$\frac{1}{\rho} = \frac{\varepsilon_2-\varepsilon_1}{h} = f\frac{\pi^2}{l^2}. \tag{5}$$

From here we find the deflection of the middle section:

$$f = \frac{l^2}{\pi^2 h}(\varepsilon_2 - \varepsilon_1). \tag{6}$$

3. The determination of bearing capacity

The equilibrium equations are written as follows:

$$P_{ВН}=P; \tag{7}$$

$$\mathrm{M}_{ВН} = \mathrm{P}f - Q_2\left(x_0 + L - \frac{l}{2}\right) - \frac{q}{2}\left(x_0 + L - \frac{l}{2}\right)^2,$$

where q – a lateral distributed load;

$$Q_2 = \frac{P(e_2-e_1)-\frac{qL^2}{2}}{L}.$$

For convenience, the following ratio is introduced:

$$q = \frac{nP}{L}, \tag{8}$$

where n is lateral load intensity.

Finally, in the equilibrium equation (7) we have:

$$\mathrm{M}_{ВН} = \mathrm{P}f - P\left(x_0 + L - \frac{l}{2}\right)\left(\frac{e_2-e_1}{L} - \frac{n}{2} + \frac{n}{2L}\left(x_0 + L - \frac{l}{2}\right)\right). \tag{9}$$

The main vector $P_{вн}$ and the main moment $M_{вн}$ are diagrams of normal stress:

$$P_{\text{вн}} = b\left(\int_{-\frac{\varepsilon_1 h}{\varepsilon_2-\varepsilon_1}}^{h} \sigma_c(x)dx - \int_0^{-\frac{\varepsilon_1 h}{\varepsilon_2-\varepsilon_1}} \sigma_p(x)dx\right); \tag{10}$$

$$M_{\text{вн}} = b\left(\int_{-\frac{\varepsilon_1 h}{\varepsilon_2-\varepsilon_1}}^{h} \sigma_c(x)(x-\frac{h}{2})dx + \int_0^{-\frac{\varepsilon_1 h}{\varepsilon_2-\varepsilon_1}} \sigma_p(x)(\frac{h}{2}-x)dx\right).$$

Or, taking into account (3) and (4):

$$P_{\text{вн}} = b\left(\int_0^{\varepsilon_2}(A_1\varepsilon - A_2\varepsilon^3)\frac{h}{\varepsilon_2-\varepsilon_1}d\varepsilon - \int_{\varepsilon_1}^{0} E_p\varepsilon\frac{h}{\varepsilon_2-\varepsilon_1}d\varepsilon\right); \tag{11}$$

$$\text{M}_{\text{вн}} = b\left(\int_0^{\varepsilon_2}(A_1\varepsilon - A_2\varepsilon^3)\left(\frac{(\varepsilon-\varepsilon_1)h}{\varepsilon_2-\varepsilon_1} - \frac{h}{2}\right)d\varepsilon + \int_{\varepsilon_1}^{0} E_p\varepsilon\left(\frac{h}{2} - \frac{(\varepsilon-\varepsilon_1)h}{\varepsilon_2-\varepsilon_1}\right)\frac{h}{\varepsilon_2-\varepsilon_1}d\varepsilon\right).$$

After integration and mathematical transformations, we finally have:

$$\text{P}_{\text{вн}} = \frac{bh}{\varepsilon_2-\varepsilon_1}\left(0{,}5A_1\varepsilon_2^2 - 0{,}25A_2\varepsilon_2^4 - 0{,}5\text{E}_\text{p}\varepsilon_1^2\right); \tag{12}$$

$$\text{M}_{\text{вн}} = \frac{bh^2}{(\varepsilon_2-\varepsilon_1)^2}\left(\frac{A_1}{12}(\varepsilon_2^3 - 3\varepsilon_2^2\varepsilon_1) - \frac{A_2}{40}(3\varepsilon_2^5 - 5\varepsilon_2^5\varepsilon_1) - \frac{\text{E}_\text{p}}{12}(\varepsilon_1^3 - 3\varepsilon_1^2\varepsilon_2)\right).$$

The compressive force applied to the ends of the rod with unequal eccentricities e_1 and e_2 (picture 1):

$$e_1 = f Sin\frac{\pi x_0}{l}; \tag{13}$$

$$e_2 = f Sin\frac{\pi(x_0+L)}{l}.$$

From conditions (13) we have the equations of communication:

$$\Phi_1(\varepsilon_1, \varepsilon_2, l, x_0) = 0; \tag{14}$$

$$\Phi_2(\varepsilon_1, \varepsilon_2, l, x_0) = 0.$$

Let the load on the rod increase in time according to a linear law:

$$P = P_\text{н} + \alpha t, \tag{15}$$

where $P_\text{н}$ - initial load value; α – a linear load rise parameter; t - time.

Differentiating equilibrium equations (7) and equations (14), we obtain a system of differential equations of motion:

$$\begin{cases} \frac{\partial P_{\text{вн}}}{\partial \varepsilon_1}\dot{\varepsilon_1} + \frac{\partial P_{\text{вн}}}{\partial \varepsilon_2}\dot{\varepsilon_2} = \alpha; \\ \left(\frac{\partial M_{\text{вн}}}{\partial \varepsilon_1} + P\frac{l^2}{\pi^2 h}\right)\dot{\varepsilon_1} + \left(\frac{\partial M_{\text{вн}}}{\partial \varepsilon_2} - P\frac{l^2}{\pi^2 h}\right)\dot{\varepsilon_2} + k_1\dot{l} + k_2\dot{x_0} = k_3; \\ \frac{\partial \Phi_1}{\partial \varepsilon_1}\dot{\varepsilon_1} + \frac{\partial \Phi_1}{\partial \varepsilon_2}\dot{\varepsilon_2} + \frac{\partial \Phi_1}{\partial l}\dot{l} + \frac{\partial \Phi_1}{\partial x_0}\dot{x_0} = 0; \\ \frac{\partial \Phi_2}{\partial \varepsilon_1}\dot{\varepsilon_1} + \frac{\partial \Phi_2}{\partial \varepsilon_2}\dot{\varepsilon_2} + \frac{\partial \Phi_2}{\partial l}\dot{l} + \frac{\partial \Phi_2}{\partial x_0}\dot{x_0} = 0; \end{cases} \tag{16}$$

where $\frac{\partial P_{\text{вн}}}{\partial \varepsilon_1} = bh\left(\frac{0{,}5A_1\varepsilon_2^2 - 0{,}25A_2\varepsilon_2^4 + \text{E}_\text{p}(0{,}5\varepsilon_1^2 - \varepsilon_1\varepsilon_2)}{(\varepsilon_2-\varepsilon_1)^2}\right)$;

$$\frac{\partial P_{\text{вн}}}{\partial \varepsilon_2} = bh\left(\frac{A_1(0{,}5\varepsilon_2^2 - \varepsilon_2\varepsilon_1) - A_2(0{,}75\varepsilon_2^4 - \varepsilon_2^3\varepsilon_1) + 0{,}5\text{E}_\text{p}\varepsilon_1^2}{(\varepsilon_2-\varepsilon_1)^2}\right);$$

$$\frac{\partial \text{M}_{\text{вн}}}{\partial \varepsilon_1} = bh^2\left(\frac{\frac{\text{A}_1}{12}(-\varepsilon_2^3 - 3\varepsilon_2^2\varepsilon_1) - \frac{\text{A}_2}{40}(6\varepsilon_2^5 - 5\varepsilon_2^6 - 5\varepsilon_2^5\varepsilon_1) - \frac{\text{E}_\text{p}}{12}(3\varepsilon_1^2\varepsilon_2 - 6\varepsilon_1\varepsilon_2^2 - \varepsilon_1^3)}{(\varepsilon_2-\varepsilon_1)^3}\right);$$

$$\frac{\partial \text{M}_{\text{вн}}}{\partial \varepsilon_2} = bh^2\left(\frac{\frac{\text{A}_1}{12}(\varepsilon_2^3 - 3\varepsilon_2^2\varepsilon_1 + 6\varepsilon_2\varepsilon_1^2) - \frac{\text{A}_2}{40}(9\varepsilon_2^5 - 15\varepsilon_2^5\varepsilon_1 - 15\varepsilon_2^4\varepsilon_1 + 25\varepsilon_2^4\varepsilon_1^2) - \frac{\text{E}_\text{p}}{12}(3\varepsilon_1^2\varepsilon_2 + \varepsilon_1^3)}{(\varepsilon_2-\varepsilon_1)^3}\right);$$

$$k_1 = -P\left(\frac{2l}{\pi^2 h}(\varepsilon_2 - \varepsilon_1) + \frac{e_2-e_1}{2L}\right) - \frac{P}{2}\left(\frac{e_2-e_1}{L} - \frac{n}{2} + \frac{n}{L}\left(x_0 + L - \frac{l}{2}\right)\right);$$

$$k_2 = P\left(\frac{e_2-e_1}{L} - \frac{n}{2} + \frac{n}{L}(x_0 + L - \frac{l}{2})\right);$$

$$k_3 = \alpha f - \alpha\left(x_0 + L - \frac{l}{2}\right)\left(\frac{e_2-e_1}{L} - \frac{n}{2} + \frac{n}{2L}\left(x_0 + L - \frac{l}{2}\right)\right);$$

$$\frac{\partial \Phi_1}{\partial \varepsilon_1} = -\frac{l^2}{\pi^2 h}Sin\frac{\pi x_0}{l};$$

$$\frac{\partial\Phi_1}{\partial\varepsilon_2} = \frac{l^2}{\pi^2 h} Sin\frac{\pi x_0}{l};$$
$$\frac{\partial\Phi_1}{\partial l} = \frac{\varepsilon_2-\varepsilon_1}{h}\left(\frac{2l}{\pi^2} Sin\frac{\pi x_0}{l} - \frac{x_0}{\pi} Cos\frac{\pi x_0}{l}\right);$$
$$\frac{\partial\Phi_1}{\partial x_0} = \frac{l(\varepsilon_2-\varepsilon_1)}{\pi h} Cos\frac{\pi x_0}{l};$$
$$\frac{\partial\Phi_2}{\partial\varepsilon_1} = -\frac{l^2}{\pi^2 h} Sin\frac{\pi(x_0+L)}{l};$$
$$\frac{\partial\Phi_2}{\partial\varepsilon_2} = \frac{l^2}{\pi^2 h} Sin\frac{\pi(x_0+L)}{l};$$
$$\frac{\partial\Phi_2}{\partial l} = \frac{\varepsilon_2-\varepsilon_1}{h}\left(\frac{2l}{\pi^2} Sin\frac{\pi(x_0+L)}{l} - \frac{x_0+L}{\pi} Cos\frac{\pi(x_0+L)}{l}\right);$$
$$\frac{\partial\Phi_2}{\partial x_0} = \frac{l(\varepsilon_2-\varepsilon_1)}{\pi h} Cos\frac{\pi(x_0+L)}{l};$$
$$\frac{dP}{dt} = \alpha; \frac{df}{dt} = \frac{l^2}{\pi^2 h}(\dot{\varepsilon}_2 - \dot{\varepsilon}_1).$$

Solving the system (16) is a Cauchy problem with initial conditions obtained from the joint solution of nonlinear equations (7) and (13) for ε_1, ε_2, l, и x_0. The numerical integration of differential equations can be done using the Runge-Kutta method.

System (16) is linear with respect to derivatives $\dot{\varepsilon}_1$, $\dot{\varepsilon}_2$, $\dot{l}$, $\dot{x}_0$. In a matrix form, it looks:

$$A\,\vec{X} = \vec{B}, \quad (17)$$

where A is a coefficient matrix; $\vec{X}$- a column matrix (vector) of unknowns; $\vec{B}$- a free column matrix.

$$A = \begin{bmatrix} a_{11} & a_{12} & a_{13} & a_{14} \\ a_{21} & a_{22} & a_{23} & a_{24} \\ a_{31} & a_{32} & a_{33} & a_{34} \\ a_{41} & a_{42} & a_{43} & a_{44} \end{bmatrix}; \quad \vec{X} = \begin{bmatrix} \dot{\varepsilon}_1 \\ \dot{\varepsilon}_2 \\ \dot{l} \\ \dot{x}_0 \end{bmatrix}; \quad \vec{B} = \begin{bmatrix} b_1 \\ b_2 \\ b_3 \\ b_4 \end{bmatrix}.$$

The roots of the system are expressed by Kramer's formulas:

$$\dot{\varepsilon}_1 = \frac{\Delta\varepsilon_1}{\Delta}; \dot{\varepsilon}_2 = \frac{\Delta\varepsilon_2}{\Delta}; \dot{l} = \frac{\Delta l}{\Delta}; \dot{x}_0 = \frac{\Delta x}{\Delta}, \quad (18)$$

where Δ, $\Delta\varepsilon_1$, $\Delta\varepsilon_2$, Δl, Δx_0 are system identifiers.

To find the critical state condition, we use the criterion proposed by R.S. Sanzharovsky [3]:

$$\delta M = \delta M_{вн} = P\delta f; \quad (19)$$
$$\delta P_{вн} = 0.$$

The condition for stability loss is the determinant of the system (16) being equal to zero and composed of coefficients with variations.

The solution of the system (16) allows the evaluation of the stress-strain state of the wooden rod during any short-term loading.

4. Conclusion

The proposed analytical model provides information on the stress distribution over the cross section and on the deflection in the process of loading a wooden rod. Introduction to the calculation of the relevant criteria (limit deflection, ultimate strength) allows us to perform checks on both the first and the second group of limit states. In addition, for practical use, the buckling criterion is introduced into the design model. The reliability of the theoretical results of the proposed calculation method was confirmed by experimental studies [1, 2].

The developed technique is substantially closer to the actual work of the element than the normative method. By means of using a curvilinear stress-strain compression diagram, the physical non-linearity of the material is taken into account. Thus, the developed model of the stress-strain state of a wooden rod under off-center compression and lateral load can be proposed for introduction into computational practice of compressible-bending wooden elements.

References

[1] Varenik A S 2013 The stability of eccentrically compressed wooden elements under short-term loading *Vestnik NovSU* Series Technical sciences **75** 7–12 [In Russian]

[2] Varenik A S and Varenik K A 2014 The stability of wooden rods in compression by forces applied with unequal eccentricities *Vestnik NovSU* Series Technical sciences **81** 31–35 [In Russian]
[3] Sanzharovskij R S 1984 *Creep resistance of building structures* (Leningrad: LGU Publishing House) p 216 [In Russian]

IOP Conf. Series: Materials Science and Engineering **656** (2019) 012053 doi:10.1088/1757-899X/656/1/012053

The use of industrial waste and cullet for building foam glass

K A Varenik[1,2], A S Varenik[1], A V Kirillov[1], M V Shuvalov[1]

[1]Yaroslav-the-Wise Novgorod State University, ul. B. St.-Peterburgskaya, 41 173003 Veliky Novgorod, Russia

[2]E-mail: vkirillv89@mail.ru

Abstract. There is a huge raw material base for the production of foam glass. The aim of the work was to develop a technology for producing foam glass from cullet and industrial waste. Glass makes about 5% of municipal waste. Calcium carbonate which is the waste of ANP fertilizer production can be used as a foaming agent. The research task was to obtain effective heat-insulating and structural and heat insulating material with the corresponding physical and mechanical characteristics for low-rise and high-rise construction. The analysis of existing approaches to the production of foam glass was carried out in the course of the work, laboratory study of various mixtures composition was held and optimal mixtures of cullet and a foaming agent were proposed. Experiments showed that the density of foam glass made from different kinds of glass differs by up to 15%, which makes it possible to dispense with the expensive operation of sorting cullet.

1. Introduction

One of the rational ways to improve the energy efficiency of buildings under construction and renovation is wider use of heat insulating materials utilizing local resources. Foam glass is among such materials, which require further study and implementation.

At present, high requirements are imposed on heat insulating materials: firstly, to be high-quality heat and water insulators, which maintain their qualities constant for a long time; secondly, they must comply with all sanitary and hygienic standards and be environmentally friendly. Foam glass is one of the materials that meet these requirements. It is environmentally friendly, fireproof and does not emit toxic substances due to its inorganic composition.

The use of foam glass in construction is possible in two variants: as a heat-insulating or structural and heat-insulating material. The use of insulating foam glass with low heat conductivity (within 0.03 ... 0.07 W / m °C) can significantly increase the heat transfer resistance of the walls of the building. Structural and insulating foam glass combines both the heat engineering properties and strength required for wall material (at least 10 kgf / cm^2).

The first in the world to mention about foam glass as a building material was academician Kitaygorodsky I.I. at the All-Union Conference on Standardization and Production of New Effective Materials, held in Moscow in 1932. Theoretical foundations of the technology for producing foam glass of various compositions and purposes are most fully considered in the work of Demidovich B.K. [1]. In recent years, there has been considerable interest in the research of foam glass [2]. Still, foam glass has not yet gained popularity in Russia due to the fact that even experienced builders do not know the technology of using foam glass as a heat insulator and few people consider foam glass as a construction material. It should be noted that the introduced technologies of foam glass use are focused, as a rule, on heat-insulating material with a low coefficient of heat conductivity. However, it is safe to say that foam glass with a efficient ratio of heat conductivity and strength would also find demand in the market as, for example, a wall material for low-rise constructions.

Large-scale introduction of foam glass in construction requires further research into optimization of compositions and improvement of production technology, taking into account the specifics of local resources.

Relevance of laboratory research work is determined by the need to improve the technology of production of foamed glass of the optimal pore structure using cullet with calcium carbonate as a foaming material which is production waste of Acron PJSC (Veliky Novgorod).

The pore structure of the foam glass is formed in accordance with general laws of foaming in pyroplastic silicate systems, similar to the heat-insulating materials obtained by bulking of silicate melts. However, the foaming mechanisms of aluminosilicate molten cullet with specified pore-forming components require extensive experimental research.

The issues of the relationship between the structure and properties of foamed glass with the chemical composition and characteristics of glass preparation, as well as technological modes of heating and foaming are to be studied.

Providing the required pore structure demands an integrated approach that takes into account the influence of various factors on the formation of foam glass from an aluminosilicate melt. For example, the choice of optimal temperature regimes allows to get a foam glass with a wide range of physical and mechanical properties. The formation of foam glass occurs as a result of the gradual accumulation of gaseous products when heated. The mechanism of foam glass structure formation is described by experimental curves of the foaming process. Experimental data and theory provide scientific explanations of the properties and structure of foam glass at various stages of its making. Control of the thermophysical state of the mixture and the dynamics of its change make it possible to obtain foaming curves.

2. The study of physical and mechanical parameters of raw materials

The structure, composition and properties of foam glass with improved structural properties are directly related to the quality of glass batch preparation, chemical and phase composition of the initial components, heating regime temperatures and foaming, etc. Thus, ensuring the optimal pore structure of the foam glass is a complex task of determining the influence of various factors on the process of forming the properties of foam glass as a material obtained by bulking at high temperatures.

Cullet can be of various origins: window and bottle glass, cathode ray tubes, lamps, etc. Preliminary experimental studies conducted in the laboratory of building materials of "Building Structures" Department of the Novgorod State University named after Yaroslav the Wise, showed the possibility of obtaining foam glass of the required quality from various raw materials. Similar results were obtained by Belokopytova A.S. [3]. The density of foam glass made from different types of glass varies by up to 15%, which eliminates the costly operation of sorting the cullet.

Calcium carbonate ($CaCO_3$, chalk) taken from Acron PJSC ammonium nitrate phosphate fertilizer waste dump was used as a blowing agent:

$$CaCO_3 \rightarrow CaO + CO_2 \uparrow,$$

where CaO is lime; CO_2 is carbon dioxide.

The study of phase changes of the batch structure was carried out by various methods that allowed solving specific questions posed.

Methods of preparation of cullet to be used as a filler in glass concrete were determined in the laboratory and it was established that the average particle size of cullet should not exceed 1 mm, and the maximum size should be 2 mm.

The analysis of the chemical composition of the used raw materials, performed in the chemical laboratory of PJSC Acron, gave the following results.

Bottle cullet contains the following amount of oxides in %: SiO_2 – 65.2; Al_2O_3 – 10.7; TiO_2 – 0.8; Fe_2O_3– 1.2; CaO – 6.5; MgO – 0.7; Na_2O – 0.2; K_2O – 14.7; SO_3 – 0.

Chalk contains: SiO_2 – 1.3; Al_2O_3 – 0.7; TiO_2 – 0.1; Fe_2O_3 – 0.1; CaO – 54.7; MgO – 0.3; Na_2O – 0; K_2O – 0; SO_3 – 42.8.

The investigated batch was glass powder with an average specific surface area of S≈ 6000 cm^2/g, which corresponds to a typical technological process of foam glass production.

It was tried to bring in bound water in the form of silica gel into the raw mixture. During thermal decomposition of silica gel, water vapors are released that contribute to more intensive foaming due to the increase in the gas content, a significant decrease in viscosity, surface tension of the melt and make maximum foaming temperature lower, which will reduce the requirements for the raw glass composition.

Based on the analysis of literature [1], the following hypothesis is put forward. To improve the quality of foam glass made from cullet, it is necessary to artificially increase the content of bound water in the system at the stage of mixture preparation, which can be achieved either by hydration and hydrolysis of fine glass powder, or by directly adding an aqueous solution of sodium silicate to the system. During thermal decomposition of silica gel, as it was mentioned above, water vapors are released, which can contribute to more intensive foaming due to an increase in the gas content, a significant decrease in viscosity, melt surface tension and a decrease in the maximum foaming temperature, which obviously reduces the requirements for the composition of the raw glass.

3. Experimental studies of technological regimes

Optimal temperature regime is one of the most important stages in the rational production technology of foam glass, since, depending on the chosen mode, foam glass with a wide range of properties can be obtained. The work of R.V. Gorodova is devoted to the study of temperature conditions of sintering. [4].

Laws of granules structure formation during heat treatment were considered in the work.

Since foam glass is a heat-insulating material, it is necessary to strive to obtain granules with minimum coefficient of heat conductivity, which is achieved with a decrease in their average density. Strength of granules in this case is not a determining parameter, since during operation the material does not experience significant load, and in addition, strength of foam glass is several times greater than strength of other heat insulation materials of the same heat conductivity or average density.

Average density is controlled by changing the temperature, duration of foaming and humidity of the foaming mixture. Kinetics of granule foaming was studied to select optimal duration of the process.

The results of the experiment show that minimum values of density of the granules lies in the temperature range from 790 to 830 ° C with a foaming time of 45–50 minutes.

The results obtained indicate that the heating regimes of the batch with linear increase in temperature at the initial stage and subsequent exposure at a constant temperature close to the sintering temperature can significantly improve the quality of heating, and, accordingly, the quality of the foam glass.

The effect of the moisture content of the initial mixture on the granules foaming is also of interest. It should be noted that optimal foaming temperature range from 790 to 830 ° C is characteristic of all mixtures.

The smallest density foam glass – 286 kg / m^3 was obtained from a mixture with a moisture content of 14.4% foamed at the temperature of 790 ° C; the largest density – 461 kg / m^3 was obtained from a mixture with a moisture content of 13.0% foamed at the temperature of 750 ° C.

In addition, it should be mentioned that foam glass mixtures with the lowest moisture content are subject to the greatest fluctuations in the average density due to the change in the foaming temperature.

As a result of laboratory experiments, methods for preparing cullet as a filler in glass concrete were identified and it was established that the average size of cullet particles should not exceed 1 mm.

In the case when the glass is ground finely enough, the risk of destruction of glass concrete is significantly reduced.

It was also proved that the interrelation between the glass dispersity and carbon concentration affects on the bulk density of foam glass. Low bulk density of foam glass (160 kg / m^3) is obtained from highly dispersed mixtures; foam glass with a higher bulk density is obtained from low dispersed mixtures. It was also found that the amount of glass dispersion necessary to obtain a sample with a certain density and physical-and-mechanical properties depends on the carbon content in the mixture: in mixtures with 0.25–0.35% carbon content, a finer grinding of glass is necessary, whereas in mixtures, where the content of carbon is lower (0.15 – 0.20%) lower dispersion of the mixture is sufficient.

As a result of laboratory experiments, it was proved that the pretreatment of cullet with a solution of hydrochloric acid makes it possible to obtain reinforced foam glass concrete on its basis.

An important stage of the technology under discussion is bringing in the substances that promote hydrolysis of the glass granules surface.

When silicate glasses with a high SiO_2 content are dissolved in water, alkali metal oxides pass into the solution faster than SiO_2, as a result of which silica gel film forms on the glass surface. Chemical composition of silica gel is $nSiO_2•mH_2O$. It is inorganic polymer containing silanol groups (-Si-O-Si -) - O-H on the surface. Thus upon hydration and hydrolysis of glass particles, hydroxide groups appear on the surface.

It was concluded that silica gel containing chemically bound water in the batch for obtaining foam glass creates more favorable conditions for foaming. During thermal decomposition of silica gel, water vapors are released, which can contribute to more intensive foaming due to an increase in the gas phase content, a significant decrease in viscosity, melt surface tension and a decrease in the maximum foaming temperature, which obviously reduces the requirements for the composition of raw glass.

The use of aqueous solution of sodium hydroxide makes it possible to moisten and briquette the batch by pressing and also to foam the foam glass without bringing in additional gas forming agents.

Laboratory studies have also shown the possibility of using fly ash from the Novgorod CHP-20 which consists 40–60% of SiO_2 in raw mixture and does not require pre-crushing.

4. Obtaining foam glass samples with specified physical and mechanical parameters

The main way to reduce energy consumption for heating buildings is to increase heat transfer resistance of walling through the use of insulating materials. Since 2000, regulatory requirements for the design heat transfer resistance to walling in Russia have increased by an average of 3.5 times.

The main indicators characterizing physical-technical and operational properties of heat insulation materials are: density, heat conductivity and compressive strength. To be recommended for use as a heat insulating material it is necessary that the indicated parameters of the foam glass lie within the following limits: density – less than 300 kg / m^3; compressive strength - not less than 5 kgf / cm^2; heat conductivity coefficient – 0.04 ... 0.06 W / m • ° C.

The most important technical characteristic of thermal insulation materials is heat conductivity - the ability of material to transfer heat through its thickness, since the thermal resistance of the building envelope directly depends on it. In terms of quantity it is graded by the thermal conductivity coefficient λ, which shows the amount of heat passing through the sample of material 1 m thick and 1 m^2 size with temperature difference on opposing surfaces of 1 °C for 1 h. Density of material and its structure (type, size and location of pores) primarily affect thermal conductivity of heat insulation material. In production of insulating foam glass, it is necessary to strive to obtain material with minimum value of the volumetric density with the smallest possible diameter of closed pores.

Foam glass properties depend on the composition of the foaming mixture and foaming regimes.

For research, an unsorted cullet was used, consisting of sheet glass, dark bottle glass and light bottle glass in the ratio of ≈ 1: 1: 1. The blowing agent (calcium carbonate) was taken in the amount of 2% of cullet by weight. Moisture of the mixture was about 15%. The heat treatment process includes heating the foaming batch to the foaming temperature, holding at this temperature for complete foaming and cooling.

According to the results of laboratory studies, it was found out that the volumetric density of foam glass depends on the degree of dispersion of the foaming mixture consisting of crushed glass cullet and calcium carbonate. So the increase in the specific surface area from 3000 cm^2 / g to 6000 cm^2 / g resulted in the decrease in the volumetric density from 225 kg / m^3 to 50 kg / m^3.

Dependence of volumetric density of foam glass of maximum foaming temperature and the foaming duration (exposure to maximum temperature) are established. It was revealed that with the decrease in the foaming temperature, the process of structure formation proceeds a longer period of time. With a "steep" foaming, the same values of volumetric density of the foam glass can be achieved in a shorter period of time. As the results of the experiment show, minimum values of granules density are in the temperature range from 790 to 830 °C with a foaming time of 45–50 minutes.

The results of laboratory studies allowed us to establish the dependence of compressive strength of the samples on volumetric density. So, in the density range from 150 to 250 kg / cm^2 strength of the obtained foam glass samples was from 5 to 15 kg / cm^2.

Study of the relationship between the heat conductivity coefficient of foam glass and the volumetric density showed that in the density range from 150 to 250 kg / cm^2, the obtained samples of foam glass had heat conductivity range from 0.04 to 0.055 W / m •° C.

According to the results of research carried out as part of this work, obtained foam glass has the following (average) parameters: density – 200 kg / m^3; compressive strength – 10 kgf / cm^2; heat conductivity coefficient – 0.056 W / m • ° C.

Foam glass having these physical and mechanical properties meets the requirements for insulating materials.

To be recommended for use as structural heat insulating material, it is necessary that the specified parameters of the foam glass lie within the following limits: compressive strength – at least 10 kgf / cm^2; heat conductivity coefficient – 0.064 ... 0.10 W / m • ° C.

Wall blocks made of structural heat-insulating foam glass with strength of 10 kg / cm^2 can be used as a self-supporting structural heat-insulating material in high-rise residential and industrial construction and wall blocks with strength of about 30 kg / cm^2 – as bearing structural heat-insulating material in low-rise (up to 3 floors) house building. This makes it possible to dispence with additional structural layers, which significantly cuts the cost (materials and labor) of construction, reduces the wall thickness, while increasing the useful area of the building.

As a result of laboratory experiments foam glass samples were obtained, combining the properties of both structural and insulating materials.

It was found that foam glass samples with average density of 631.8 – 681.2 kg / m^3 and compressive strength of 4.6 – 5.6 MPa can be obtained from a batch containing 70% of cullet by way of changing the ratio of calcium carbonate (chalk) and fly ash.

Foam glass samples with average density of 598.5 – 645.2 kg / m^3 and compressive strength of 3.4 – 4.7 MPa can be obtained from a batch containing 80% of cullet by way of changing the ratio of calcium carbonate and fly ash. When the blend contains 90% of cullet foam glass samples were obtained with an average density of 576.6 – 603.1 kg / m^3 and a compressive strength of 3.3 – 3.6 MPa by changing the ratio of carbonate-calcium and fly ash.

The conditions for grinding the glass mixture components were determined, which made it possible to increase the physicomechanical properties of the foam glass of the formulations developed. It has been established that in order to increase the average density and strength of foamglass, it is advisable to grind cullet in a ball mill. At the same time, the optimal duration of mechanical activation of calcium carbonate was 2–5 minutes, fly ash – 10–15 minutes.

It was also proved the effectiveness of pre-heat treatment of foaming mixtures based on the developed compositions to improve the structural properties of foam glass. Increasing the strength of foamed glass is achieved at the temperature of heat treatment of these mixtures in the range of 550 – 700 °C.

As a result of preliminary heat treatment of a batch containing 70% of cullet the average density of the foam glass is 659.8 – 715.8 kg / m^3 and compressive strength of 6.4 – 8.0 MPa. When the content of cullet is 80% the average density of foam glass drops to 634.4 – 684.6 kg / m^3 and compressive strength to 4.9 – 7.2 MPa. When the content of cullet is 90% the average density of foam glass samples reaches 620.4 – 658.7 kg / m^3 and compressive strength of 4.7 – 5.3 MPa.

The decrease in the amount of calcium carbonate from 2% to 1.5% led to an increase in density and strength (1.2 – 1.4 times).

In all the samples obtained, the coefficient of heat conductivity met heat insulation requirements.

5. Conclusion

The authors carried out theoretical substantiation of the main directions of scientific work as well as the study of raw materials. Experimental studies of technological regimes were conducted.

The research work held at the Department of Building Structures of the Novgorod State University named after Yaroslav the Wise allowed to obtain samples of heat insulating and structural and heat insulating foam glass that meets the specified physical and mechanical parameters.

The results of the work on this topic (optimal compositions and technological parameters of production) can be handed over for implementation to enterprises engaged in the production of building materials and having the necessary equipment. Such enterprises are plants for the production of ceramic bricks or crockery.

In the future it is planned to study the physical and mechanical properties of foam glass samples. It is necessary to further examine the strength, frost resistance and water absorption of heat insulation and structural and heat insulation foam glass samples.

References

[1] Demidovich B K 1975 *Foam glass* (Minsk: Science and technology Publ) p 248 (In Russian)

[2] Pavlov V E 2006 Foam glass with increased structural properties based on effusive rocks and cullet: author. diss. cand. of tech.sciences (Ulan-Ude) p 30 (In Russian)

[3] Belokopytova A S 2006 *Development of recycling processes of cullet by creating composite materials:* author. diss. cand. of tech.sciences (Moscow) p 20 (In Russian)

[4] Gorodov R V 2009 *Mathematical modeling of the heat treatment cycle of foam glass mixture:* author. diss. candof . tech. sciences (Tomsk) p 20 (In Russian)

IOP Conf. Series: Materials Science and Engineering **656** (2019) 012054 doi:10.1088/1757-899X/656/1/012054

Wood moisture accounting in creep equations

K A Varenik[1,3], A S Varenik[1], R S Sanzharovsky[1] and B V Labudin[2]

[1]Yaroslav-the-Wise Novgorod State University, ul. B. St.-Peterburgskaya 41 173003 Veliky Novgorod Russia
[2] M.V. Lomonosov Northern (Arctic) Federal University, Severnaya Dvina Emb. 17, 163002 Arkhangelsk, Russia

[3]E-mail: vkirillv89@mail.ru

Abstract. Wood as a structural material differs from other materials used in construction and industry, high variability of its properties. This is due, among other things, to the action of various variable factors – moisture content, temperature, inclination of fibers and others. The influence of wood moisture is especially strong – as a result of moistening the strength of wood considerably decreases in comparison with durability in absolutely dry condition. In addition, wood is a significantly nonlinear structural material. Non-linearity is manifested both in short-term and prolonged loading. Research of wood in time can be divided into two directions: longterm durability of wood and creep of wood. Modern models of wood creep are based on linear theories. Analysis of experimental data of various researchers on wood creep, has shown that the area of linear creep is observed at rather low levels of loading. At medium and high levels of experimental loading the existing models of linear wood creep theory do not correspond to experimental data. All this points to the actuality of the development of methods of calculating wooden structures for creep (linear and nonlinear), taking into account the wood moisture.

1. Introduction

The experimental studies of wood during the short-term and long-term loading showed its considerable nonlinear behaviour when different schemes of loading are applied. Non-linearity is found even when the strain is little. Taking this into account it should be noted that reconsidering the creep of wood is one of the key trends in improving the standards of wooden structures design.

Creep strains are influenced by several factors. One of these is wood moisture. Moisture increase causes creep strain speed.

The issue of wood moisture accounting in rheological models as well as in linear creep equations has been paid attention to by both Russian and foreign scientists. But up until now significant gaps still remain in studying the non-linear creep of wood.

A series of important experimental results are presented in A.S. Sogoyan's thesis [1]. For instance, he has proved the dependence of the creep of wood and modulus of elasticity on wood moisture. Prokopovich I.E. and Zenderidze V.A. [2] noted that though the dependence between moisture and strength of wood has been thoroughly studied before, the effect of temporary moisture on the characteristics of long-term creep ability is less known. The existing experimental data are insufficient for a relatively complete description of the behaviour of wooden structure elements during different moisture regimes including such a natural one as drying. There being an appropriate external analogy, the authors found it quite reasonable to use an apparatus designed to account for the description of the ageing effect upon the stress-strain state of a solid for the needs of describing the drying effect. This allowed to put forward an idea to justify the approach suggested by A.S. Sogoyan who proposes to consider the wood moisture effect as a result of a certain ageing connected with drying. Such an approach has found no proper practical application so far.

In A.M. Borovikov's paper [3] it is shown that wood deformability and strength considerably depend on moisture w and temperature t. Meanwhile the wood ageing effect is denied. The notion of standard moisture, which helps to recalculate deformability and temperature, has been introduced to compare the results. Continuing the research in this direction, R.B. Orlovich in his paper [4] writes a creep of wood integral equation taking into consideration equilibrium and relative moisture:

$$\varepsilon(t) = \frac{\sigma(t)}{E(\varpi)} - \int_{\tau_1}^{t} \sigma(\tau) \frac{\partial S(t,\tau,\varpi)}{\partial\tau} d\tau + \varepsilon_w(t) + \varepsilon_0(t),$$

where $\varpi = F[\varphi(t), T(t)]$ - equilibrium moisture;

$\varphi(t)$ – relative moisture;

$T(t)$ – temperature; $S(t) = [1 + \psi(t)]\frac{1}{E}$;

$\psi(t,\tau,\varpi) = \frac{\psi_\infty(\omega)[1-e^{-\beta(\omega)(t-\tau)}]}{t-\tau} \int_0^t F[\omega(\tau+\rho)]\, d\rho$; when $\omega = 12$; $\psi_\infty = 0.15$.

In the papers devoted to wood rheology different simplified mathematical formulas are used to describe experienced simple creep curves received under the constant experimental efforts. Y.M. Ivanov [5] added the data which helped to integrate the moisture effect on the creep strain:

$$\varepsilon_\text{п} = \varepsilon(t_0) \frac{10^{-2}}{0.735-0.02086\omega} t^{0.21},$$

where moisture ω is measured in percent.

For the creep of wood ultimate characteristics Y.M. Ivanov's empirical suggestion can be transformed into the following formula:

$$\varphi_\infty(w) = \frac{\varphi_\infty}{0.0221} \cdot \frac{10^{-2}}{0.735-0.02086w},$$

where w – wood moisture is measured in percent.

Then this dependence can be reduced to a specific type taking into account the level of loading $\left(\frac{\sigma}{R}\right)$ for the case of non-linear creep, where R – wood strength:

$$\varphi_\infty(\sigma, w) = \frac{\varphi_\infty\left(\frac{\sigma}{R}\right)}{0.0221} \cdot \frac{10^{-2}}{0.735-0.02086w}.$$

Eventually, the wood creep characteristic would look as:

$$\varphi(t,\tau) = \varphi_\infty(\sigma, w) \cdot \varphi_1(t,\tau),$$

where $\varphi_\infty(\sigma, w)$ – the ultimate characteristic of the creep which does not depend on strain $\varphi_\infty(\sigma)$, if the creep of wood is linear but it is affected only by wood moisture.

V.N. Volynskij in his paper [6] also points out that wood moisture is the most influential factor among other ones which affect wood physicomechanical characteristics. He also mentions that though many papers have been devoted to the study of this factor and wide statistics have been gathered, a new summary is required. When dealing with this problem one should define the type of stress state of wood, its temperature, density and other parameters.

The scientist has analysed the available data about the research carried out in the studies of the wood moisture effect on its performance. It has been found out that there are certain differences in tensile strengths on moisture during stretching, compression and curving of pure wood. In standards and reference books the basis of calculating the moisture conversion indexes is the dependency that works in the moisture range from 5 to 30 %.

The introduced conversion indexes are differentiated according to the wood species and density. Though, the latter has not been verified experimentally. There have been worked out some algorithms of calculating this index which shows the strength fraction of raw wood in correlation to the strength of absolutely dry wood according to the known reference data. It is demonstrated that these indexes do not depend on wood density.

Moreover, the scientist has proved that moisture affects the strength of lumber less than that of pure wood.

2. Accounting for the effect of wood moisture on its creep

Analysing the results of the mentioned authors' research as well as the papers of the foreign scientists, one can say there exist three different principles which allow to account the wood moisture effect on the creep of wood.

The first principle of the moisture-time analogy introduces the notion of the given time suggested by H. Leaderman [7]. This principle was not meant to be applied to solving design calculation problems, but to gain the decrease in the period the creep and relaxation of wood indexes defining experiments lasted. This proposition found in A. Martensson's paper [8] implies that with varying moisture and temperature characteristics, the rheological behaviour of wood can satisfy an equation of exactly the same structure as at constant moisture and temperature, but with a different time scale.

The given time ξ is determined by the ratio:

$$d\xi = \frac{dt}{\alpha(u)},$$

where $\alpha(u)$ is the function decreasing to zero, the initial value of which is given at $\alpha = 68.5277$ found when $u = 0$, and the final value at moisture $u = 0.3159$.

The modulus of wood elasticity at such a ratio does not depend on moisture, which is a certain disadvantage of the theory.

It is known that the data obtained by this method testifies to some qualitative confirmation of such an approach but fails to testify to any reliable quantitative predictions [9].

According to the second principle, the physicochemical dependences of moisture are used to describe the process of deforming wood together with viscoelastic bodies. The construction of complex models of a large number of springs and viscous resistances greatly increases the order of the resolving differential creep equation of wood, creates major problems in the numerical implementation of the problems of calculating wooden structures. To make it simpler, all the major components of the model are treated separately in the research; their speeds are integrated автономно, which introduces an error in the results. The results of the model descriptions received analytically do not match the range of viscoelastic and humid elements, selected during the initial model compiling and are unable to describe the elastic properties of wood.

The third principle of wood moisture accounting is borrowed from the experimental works of A.M. Borovikov and Y.M. Ivanova who suggested taking it into account whenever the changes in wood moisture affect its mechanical characteristics such as its strength index, modulus of elasticity and the creep of wood index. For these mechanical characteristics of wood the authors mentioned above suggested their own variants of regression equations when indexes depend on the wood moisture. Under the well-known law of time changing the wood moisture introduced by certain known scientists it becomes possible to detest the process of changing long-term deformations of wood and take this process into account in concrete calculations of the stress-strain state of wooden structures based on appropriate methods of the theory of creep and plasticity.

These arguments indicate the relevance of improving the creep models of wood, taking into account the change in the effect of wood moisture, in the framework of the theory of creep developed by N. Arutyunyan, V. B. Kolmanovsky, Yu.N. Rabotnov, A.R. Rzhanitsyn.

3. Creep equations accounting for wood moisture

According to the results of research, the authors of the article obtained refined and new creep models of compressed wood using various creep theories, which allow to take into account the nonlinear creep of wood and its moisture.

After analyzing the linear creep equations of wood proposed by various scientists, the integral creep equation of wood was converted to a differential form. This allows writing a generalized equation in the form corresponding to the equation of A.R. Rzhanitsyn:

$$\dot{\sigma}(t)\frac{1}{\gamma[1+k_{def}(w)]}+\sigma(t)=E_0(w)\frac{1}{\gamma[1+k_{def}(w)]}\dot{\varepsilon}(t)+\frac{E_0(w)}{1+k_{def}(w)}\varepsilon(t), \quad (1)$$

where γ - the coefficient for wood in the equations of Sogoyan A.S., Prokopovich I.E., Zedgenidze V.A., Yatsenko V.F., Orlovich R.B., Roshchina S.I.; k_{def} - Eurocode 5 coefficient of deformation; w - wood moisture; E_0 - elasticity modulus. Experimental and analytical data on dependencies $E_0(w)$ and $k_{def}(w)$ is advisable to use from the works of A. Borovikov and Ivanova Yu.M.

Rejecting the last term in equation (1), we obtain the structure of the wood creep equation proposed by E.N. Kvasnikov. Using this equation, it is impossible to obtain the value of the long-term critical force of a compressed wooden column. The structure of equation (1) explains the features of changes in the relaxation time $\left(n=\frac{1}{\gamma[1+k_{def}(w)]}\right)$ and the long-term modulus of wood deformation$\left(H=\frac{E_0(w)}{1+k_{def}(w)}\right)$, which is important when processing experiments on the long-term loading of compressed wood samples.

To describe the non-linear creep of wood we use the method suggested in the paper [10]. The non-linear creep of wood equation is written as follows:

$$\varepsilon(t)=\frac{\sigma(t)}{E_0}-\int_{t_0}^{t}(\sigma+\beta_Д\sigma^2)\frac{1}{E_0}\cdot\frac{\partial}{\partial\tau}\varphi(t,\tau)d\tau, \quad (2)$$

where $\varphi(t,\tau)=\varphi_\infty^0(w)\cdot\left(1-e^{-\gamma(t-\tau)}\right)$.

Taking the value $\varphi_\infty^0(w)$ is not dependent on time (w – moisture in %), we have the basic equation of nonlinear creep of wood:

$$\varepsilon(t)=\frac{\sigma(t)}{E_0}-\int_{t_0}^{t}[\sigma(\tau)+\beta_Д\sigma^2(\tau)]\frac{1}{E_0}\varphi_\infty^0(w)\frac{\partial}{\partial\tau}\left(1-e^{-\gamma(t-\tau)}\right)d\tau. \quad (3)$$

In this equation, it is possible to take into account the dependence of the modulus of elasticity of wood on moisture$w$, substituting into it the appropriate empirical formula:

$$\varepsilon(t)=\frac{\sigma(t)}{E_0(w)}-\int_{t_0}^{t}[\sigma(\tau)+\beta_Д\sigma^2(\tau)]\frac{1}{E_0(w)}\varphi_\infty^0(w)\frac{\partial}{\partial\tau}\left(1-e^{-\gamma(t-\tau)}\right)d\tau. \quad (4)$$

Non-linear equations (3) и (4) can be applied to woodwork when compressed, stretched or curved. They will differ in the numerical values of the coefficients $\beta_Д$ and$\varphi_\infty^0(w)$. These equations do not consider wood ageing. To do so, one should use the ageing Functions.

It is known that the expression $e^{-\gamma(t-\tau)}$under the integral in equation (3) allows to convert it into a differential equation, more convenient in solving a number of practical problems of calculating wooden structures.

The equation (3) is differentiable in time t:

$$\dot{\varepsilon}(t)=\frac{\dot{\sigma}(t)}{E_0}-\gamma\left[\int_{t_0}^{t}\gamma[\sigma(\tau)+\beta_Д\sigma^2(\tau)]\frac{1}{E_0}\varphi_\infty^0(w)e^{-\gamma(t-\tau)}d\tau\right]+\gamma[\sigma(t)+\beta_Д\sigma^2(t)]\frac{1}{E_0}\varphi_\infty^0(w). \quad (5)$$

From the result obtained, we find the value of the integral in square brackets:

$$\gamma[I]=\frac{\dot{\sigma}(t)}{E_0}-\dot{\varepsilon}(t)+\gamma[\sigma(t)+\beta_Д\sigma^2(t)]\frac{1}{E_0}\varphi_\infty^0(w).$$

Substituting this value into the original equation (3), we get rid of the integral, and we get the basic equation of nonlinear creep of wood in the form of a differential equation:

$$\dot{\varepsilon}(t)+\gamma\varepsilon(t)=\frac{\dot{\sigma}(t)}{E_0}+\gamma[\sigma(t)+\beta_Д\sigma^2(t)]\frac{1}{E_0}\varphi_\infty^0(w)+\gamma\frac{\sigma(t)}{E_0}. \quad (6)$$

At low loading levels ($\frac{\sigma}{R}\le 0.38$ for compressed wood) let us write the equation of linear creep of wood, assuming: $\dot{\varepsilon}(t)+\gamma\varepsilon(t)=\frac{\dot{\sigma}(t)}{E_0}+\gamma\sigma(t)\frac{1}{E_0}\varphi_\infty^0(w)+\gamma\sigma(t)\frac{1}{E_0}$. (7)

To solve the simplest problems of bending under non-linear creep conditions, a number of simplifications can be made. The nonlinear creep equation for wood will correspond to the use of the Voigt model with nonlinear force:

$$\dot{\varepsilon}(t)+\gamma\varepsilon(t)=\gamma[\sigma(t)+\beta_Д\sigma^2(t)]\frac{1}{E_0}\varphi_\infty^0(w). \quad (8)$$

This equation is also convenient for solving problems of cyclic loading of wooden structures, including those of oscillations.

4. Conclusion

Thus, the design models proposed in the article account for the rheological features of wood more accurately compared to existing theories, and also allow taking into account the influence of moisture on the mechanical characteristics of wood.

As a result, universal equations of linear and nonlinear creep of wood were obtained, taking into account its moisture. Substituting into these equations the values of the coefficients obtained experimentally, we can describe the work of wood in compression, tension and curving.

On the basis of the obtained creep equations for wood, it is possible in the future to conduct a study of the stress-strain state of compressed wooden structures that are in buckling conditions, with subsequent suggestions for refining the methodology for regulatory calculation.

References

[1] Sogoyan A S 1954 *Experimental studies of the creep of wood and its effect on the operation of certain wooden structures:* abstract of dissertation of a candidate of technical sciences (Erevan) p 27

[2] Prokopovich I E and Zedgenidze V A 1980 *Applied creep theory* (Moscow) p 239

[3] Borovikov A M 1970 *The effect of temperature and humidity on the elasticity, viscosity and plasticity of wood*: dissertation of a candidate of technical sciences (Voronezh) p 310 (In Russian)

[4] Orlovich R B 1991 *Long-term strength and deformability of structures made of modern wood materials at the main operational impacts*: abstract of dissertation of a doctor of technical sciences (Leningtad) p 50 (In Russian)

[5] Ivanov A M 1959 Wood Chart Mechanical Test Study *Building and architecture* **4** 116–122 (In Russian)

[6] Volynskij V N 2000 *Relationship and variability of the physicomechanical properties of wood* (Archangelsk: AGTU Publ) p 196 (In Russian)

[7] Leaderman H 1944 *Elastic and creep properties of filamentous materials and other highpolymers* (Washington: The Textile foundation) p 278

[8] Martensson A 1992 *Mechanical behavior of wood exposed to humidity variations* (Lund: Report TVBK-1006. Institute of Technology Publ) p 189

[9] Ferri G 1963 *Viscoelastic polymer properties* (Moscow: Foreign Literature Publishing House) p 535 (In Russian)

[10] Varenik K A 2014 *Nonlinear creep of wood* (Moscow: Fundamental research) **9** (part 8) 1653–1657 (In Russian)

Reokinetics of urethane epoxy oligomers hardening and formation of cold curing adhesive compositions based on them

E R Volkova[1], A I Slobodinuk[1,2], A I Savchuk[1], E V Pogoreltsev[1] and V N Strelnikov[1]

[1]Federal State Budgetary Institution of Science Perm Federal Research Center, Ural Branch of the Russian Academy of Sciences, ul. Akademika Koroleva, 3 614013 Perm, Russia

[2]E-mail: slobodinyuk.aleksey.ktn@mail.ru

Abstract. In order to obtain polymeric binders for cold cure structural adhesives based on urethane epoxy oligomers, cycloaliphatic amines such as aminoethyl piperazine and isophorone diamine are used as hardeners. The reokinetic features of the curing process of polytetraurethanhydroxyl compositions at temperature of 25 ± 1 ° C were investigated, and influence regularities of the reactive oligomers composition on the dynamics of viscosity increase and the viability time of the compositions were established. The rheological parameters of the hardening process of the investigated compositions were determined. The assessment of adhesion strength for detachment of cold-curing adhesive joints formed on the basis of EUO is presented. The proposed compositions can be used for bonding ceramic linings operating under extreme conditions of the Arctic.

1. Introduction

When forming cold-resistant polymer composites and cold-curing structural adhesives, the choice of components used in the synthesis is of fundamental importance. Analysis of scientific literature showed that when forming composites that are operated in harsh climatic conditions, it is advisable to use epoxyurethane oligomers (EUO) [1–3]. An important difference in EUO is that, with the correct combination of epoxy and urethane fragments in the polymer structure, the final products of their curing -polytetraurethanhydroxyl compositions (PUHX) are high-strength elastic materials [4–8] that are out of brittle state and at low temperatures.

An important part in the development of polymer compositions based on reactive EUO is changes kinetics studying in the rheological properties of reaction mixtures (reokinetic) during their curing. The viscometric method is the most accessible for studying the rheokinetics of curing, which allows to register the change in viscosity at different stages of structuring [9]. According to the results of the reaction mixture viscosity measurements, one determines the kinetic characteristics of the curing process: reaction rate constants and the composition viability.

In this paper, the laws governing the reaction mixture viscosity changes based on EUO and the speed of their curing when interacting with amines are studied in order to form a cold-curing adhesive composition. For the first time, the adhesive strength of compositions based on EUO to ceramics was studied.

2. Experiment description

2.1. Research material

As subjects of study, polymer compositions based on EUO are considered, which are cured with amines at the temperature 25 ± 1 °C. The original EUO were synthesized in two stages [3, 10]. In the first stage, the interaction of oligotetramethylene oxydiol with MM~1400 (pF 1400) with isophorone diisocyanate (IPDI) by the reaction of urethane formation results in oligomeric urethane diisocyanate (OUDI) having active isocyanate groups. In the second stage, as a result of the reaction of OUDI with epoxypropanol (glycidol), the epoxy urethane oligomer EUO is formed.

It is shown in researches [3, 10] that with the ratio of the reactive groups NCO: OH> 2 (at the first stage of EUO synthesis), the final reaction product is a mixture of two types of oligomers: EUO(I) with long and EUO(II) with short oligomeric blocks, hereinafter – long and short oligomers.

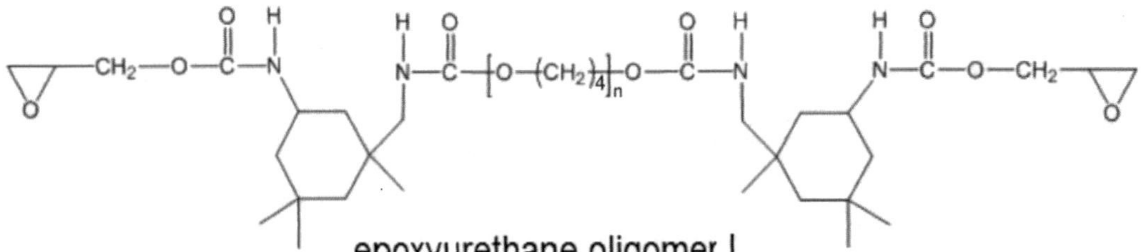

epoxyurethane oligomer I

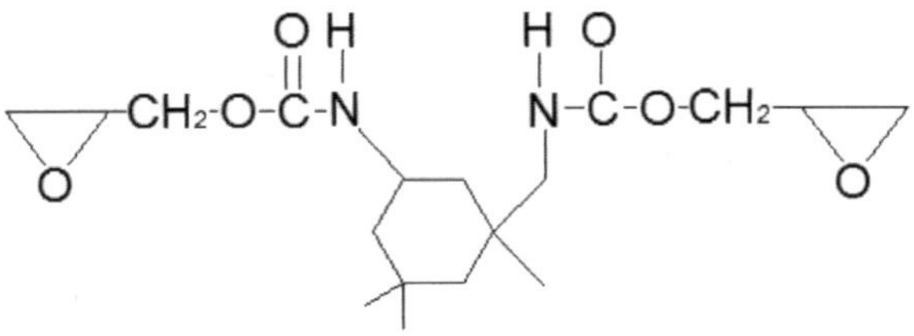

epoxyurethane oligomer II

The molar ratio of the NCO groups of the isocyanate and the OH groups of the oligodiol (NCO:OH) in the reaction mixture EUO was 2.0; 2.4; 2.7; 3.0 and 4.0.

As hardeners in the synthesis of PUHX, amine-type hardeners are selected:

• aminoethylpiperazine (AEP) with primary nitrogen atoms

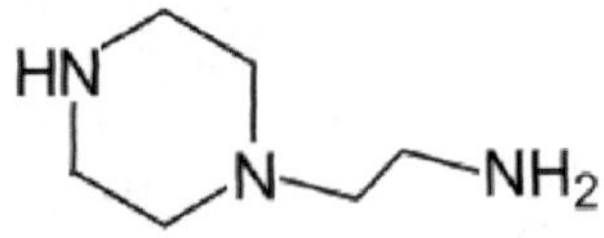

• isophorone diamine (IPDA), containing both primary, secondary and tertiary nitrogen atoms

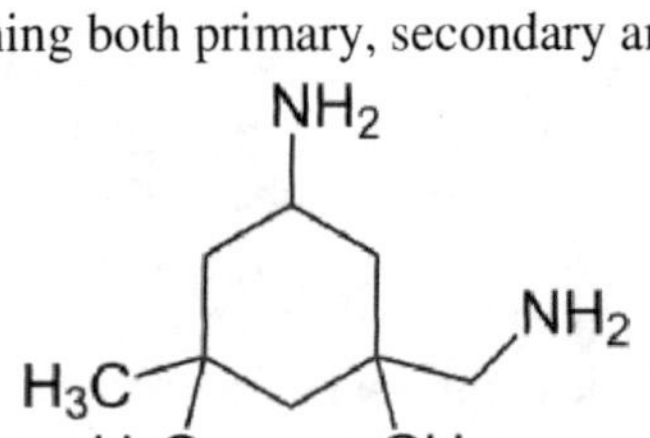

The molar ratio of EUO to AEP and IPDA was 1:0.87 and 1:0.75, respectively. Synthesis of PUHX was carried out under vacuum at a temperature 25 ± 1 °C for 5 minutes. Marking and composition of polymer samples are presented in table 1.

Table 1. Marking and composition of PUHX samples.

Oligodiol Designation	Diisocyanate Designation	Ratio NCO:OH	Hardener	Sample Code
		2.0:1		PF1400-2.0IPDI-3A[a]-0.87AEP
		2.4:1		PF1400-2.4IPDI-3A-0.87AEP
PF 1400	IPDI	2.7:1	AEP	PF1400-2.7IPDI-3A-0.87AEP
		3.0:1		PF1400-3.0IPDI-3A-0.87AEP
		4.0:1		PF1400-4.0IPDI-3A-0.87AEP
		2.0:1		PF1400-2.0IPDI-3A-0.75IPDA
		2.4:1		PF1400-2.4IPDI-3A-0.75IPDA
PF 1400	IPDI	2.7:1	IPDA	PF1400-2.7IPDI-3A-0.75IPDA
		3.0:1		PF1400-3.0IPDI-3A-0.75IPDA
		4.0:1		PF1400-4.0IPDI-3A-0.75IPDA

[a]3A is an epoxy fragment in the oligomeric block of EUO.

2.2. Research methods

Rheological studies were performed on a rotational viscometer RHEOTEST 2.1 (Germany) with a working cone-plate unit. The kinetics of curing polymer compositions was investigated at the temperature 25 ± 1 °C and shear rates of 180 s^{-1}. According to the test results, the kinetic curves η (*t*) were constructed, which characterize the change in the dynamic viscosity of the reaction mass over time, the constants of the viscosity growth rate k_η and the viability time were determined. During the time of viability of the *t** compositions during the curing process, a time to reach the viscosity of 100 Pa·s has been adopted.

Tensile strength (adhesion strength) $\sigma_{tensile}$ for adhesive joints of compositions based on EUO with ceramics based on aluminum oxide was determined at the temperature 23 ± 1 °C according to Russian National State Standard 14760-69 on samples "fungi" with a thickness of polymer layer (2 ± 0.1) mm and stretch speeds of 20 mm / min.

3. Results and discussion

Figure 1 shows the kinetic curves of compositions viscosity changes based on PF1400 *n*IPDI–3A during the curing of their AEP at the temperature 25 ± 1 °C.

It is seen that rheokinetic curves have 2 characteristic areas of viscosity change, corresponding to certain structural transformations occurring in the polymer mass during its curing [11]. At the initial stage, a rapid and significant increase in viscosity is observed, which is associated with an increase in the molecular weight of oligomeric chains and the formation of a fluctuation network as a result of intermolecular interactions leading to the formation of an internal spatial structure in the volume of the oligomer [12, 13]. At the same time, the intensity of the viscosity increase does not depend on the type of initial EUO, as evidenced by the same angle of inclination of the initial portion of the dependence $\eta(t)$ for all five compositions. At the second stage of curing, associated with the formation of branched and cross-linked structure [14], the viscosity increase rate slows down. When describing the kinetics of the process, the second, final stage of curing of the EUO, when the reaction mass loses its fluidity ($\eta \rightarrow \infty$) is decisive. The dependence $\eta(t)$ at the final stage of curing (figure 1) can be described by the exponential equation

$$\eta = \eta_0 \cdot \exp(k_\eta \cdot t), \qquad (1)$$

where η_0 is the initial viscosity, k_η is the viscosity increase constant, *t* is the curing time.

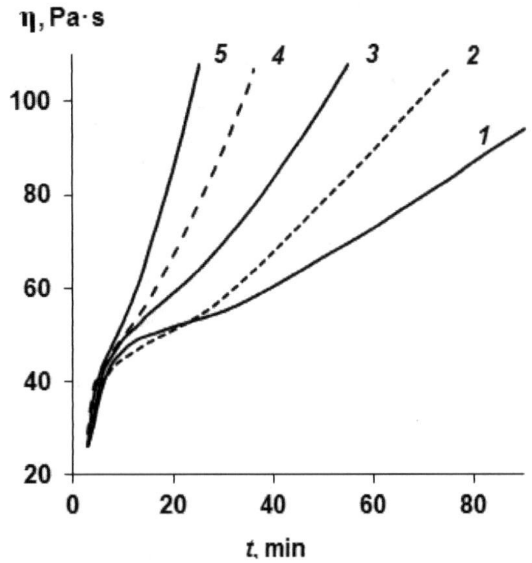

Figure 1. The dependence of the dynamic viscosity on the curing time at a temperature of 25 ± 1 °C for compositions:
1 – PF1400-2.0IPDA-3A-0.87AEP;
2 – PF1400-2.4IPDA-3A-0.87AEP;
3 – PF1400-2.7IPDA-3A-0.87AEP;
4 – PF1400-3.0IPDA-3A-0.87AEP;
5 – PF1400-4.0IPDA-3A-0.87AEP.

Equation (1) presented in semi-logarithmic coordinates makes it possible to determine the viscosity increase constants k_η graphically from the dependence slope ln $\eta(t)$ [9, 11] shown in figure 2.

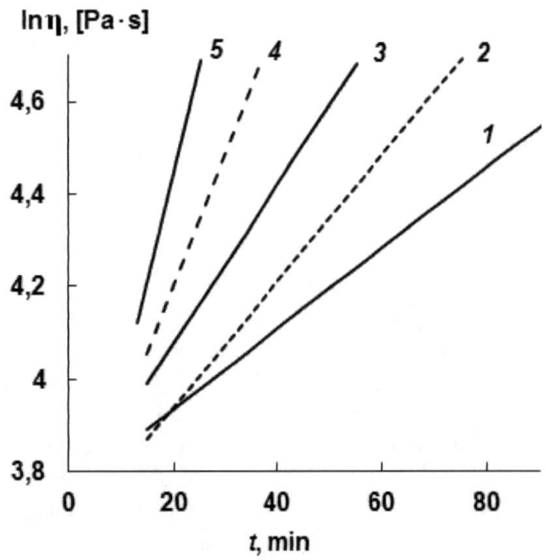

Figure 2. The dependence of ln η on the curing time at a temperature of 25 ± 1 °C for compositions:
1 – PF1400-2.0IPDA-3A-0.87AEP;
2 – PF1400-2.4IPDA-3A-0.87AEP;
3 – PF1400-2.7IPDA-3A-0.87AEP;
4 – PF1400-3.0IPDA-3A-0.87AEP;
5 – PF1400-4.0IPDA-3A-0.87AEP.

Similar rheokinetic studies were conducted for compositions based on oligomer PF1400–n IPDI–3A, curable with IFDA at a temperature of 25 ± 1 °C. The values of the curing parameters kη and t*, determined from the rheological curves (figures 1, 2) and the adhesive strength characteristics of the adhesive joint are presented in table 2.

Table 2 shows that the constant k_η monotonously increases with growth in the isocyanate component in the ratio of the reactive groups NCO:OH during the synthesis of EUO and reaches its maximum value at NCO:OH = 4.0:1. So replacing PF1400–2.0 IPDI–3A with PF1400–4.0 IPDI 3A when curing their AEP, allows to increase k_η by ~5.5 times, and reduce t^* composition by more than 4 times, which inevitably leads to a reduction in the curing time of the finished material.

Table 2. Rheokinetic parameters of compositions curing process at a temperature of 25 ± 1 °C and adhesive strength of adhesive bonding indicators.

No	Compositions Code	$k_\eta \times 10^3$ min^{-1}	t^* min	$\sigma_{tensile}$, MPa
1	PF1400-2.0IPDI-3A-0.87AEP	8.7	>90	6.2
2	PF1400-2.4IPDI-3A-0.87AEP	13.9	89	7.2
3	PF1400-2.7IPDI-3A-0.87AEP	17.3	51	8.1
4	PF1400-3.0IPDI-3A-0.87AEP	29.7	33	10.1
5	PF1400-4.0IPDI-3A-0.87AEP	47.8	23	12.4
6	PF1400-2.0IPDI-3A-0.75IPDA	3.2	>150	7.5
7	PF1400-2.4IPDI-3A-0.75IPDA	11.2	135	8.2
8	PF1400-2.7IPDI-3A-0.75IPDA	13.3	90	10.1
9	PF1400-3.0IPDI-3A-0.75IPDA	14.5	79	11.4
10	PF1400-4.0IPDI-3A-0.75IPDA	16.7	52	12.9

As mentioned above, the oligomer PF1400–*n*IFDI – 3A, is a mixture of EUO of two types: "long" and "short". The higher the concentration of the "short" EUO, the greater the content of oligomeric fragments with terminal epoxy groups in the reaction mixture, which are capable of entering into chemical interaction with the amine hardener. It is obvious that the change in the rheokinetic parameters of curing is associated with different mobility of the "oligomer-polymer" segments formed during the epoxyurethane polymer formation. The presence in the reaction system of short oligoester segments, leads to a restriction of the mobility of long macromolecular chains, and, as a result, to an increase in the viscosity of the composition and a decrease in its viability.

From table 2 it can be seen that the highest rate of the EUO curing is observed when using AEP as a hardener. This is probably due to the presence of a tertiary nitrogen atom in the AEP, and the tertiary amines, being Lewis bases, work as catalysts, initiating the polymerization process. Thus, AEP performs two functions simultaneously: a hardener and a catalyst, which affects the curing reaction kinetics.

The adhesion strength $\sigma_{tensile}$ of adhesive joints based on EUO and hardeners (aminoethylpiperazine and isophorone diisocyanate) was evaluated. The results are presented in Table 2. It can be seen that the composites synthesized on the basis of PF1400-4.0 IPDI-3A have higher adhesion strength values compared to other materials. The use of PF1400-4.0 IPDI-3A as an oligomeric base instead of PF1400-2.0 IPDI-3A contributes to an increase $\sigma_{tensile}$ by 1.72 times during the IFDA curing, and 2.0 times when the AEP is cured.

4. Conclusion

The study of the curing epoxyurethane oligomer kinetics with amino ethyl piperazine and isophorone diamine at a temperature of 25 ± 1 °C showed that with an increase in the isocyanate component during the synthesis of the initial EUO, a noticeable increase in the viscosity increase constant and a decrease in the pot life of the polymer composition are observed.

Comparative evaluation of adhesion strength of adhesive joints showed that an increase in the ratio of reactive NCO:OH groups during the EUO synthesis leads to an increase in the adhesion strength of polytetraurethanhydroxyl composites, which is probably due to an increase in the effective density of the mesh due to an increase in the proportion of hard hydroxyurethane segments and intermolecular hydrogen bonds.

It was found that the most promising when forming cold-curing adhesive compositions with high curing speed and level of adhesive strength is a model composition based on EUO synthesized at the ratio NCO:OH = 4:1 and containing aminoethylpiperazine as a hardener.

Acknowledgments
This work was supported by the Ministry of Education and Science of the Russian Federation project number 14.604.21.0192 (UI PNI RFMEFI60417X0192).

References

[1] Hepburn C 1992 *Polyurethane elastomers* (London New York: Elsevier appl. Science) p 441

[2] Cornille A, Serres J, Michaud G, Simon F, Fouquay S, Boutevin B and Caillol S 2016 Syntheses of epoxyurethane polymers from isocyanate free oligo-polyhydroxyurethane *European Polymer Journal* **75** 175–189

[3] Strel'nikov V N, Senichev V Yu, Slobodinyuk A I, Savchuk A V, Volkova E R, Makarova M A, Nechaev A I, Krasnosel'skikh S F and Ukhin K O 2018 Preparation and Properties of Frost-Resistant Room-Temperature-Curable Compounds Based on Oligoethertetraurethane Diepoxides of Various Chemical Structures *Russian Journal of Applied Chemistry* **91(3)** 463–468

[4] Jin H, Zhang Yu, Wang Ch, Sun Yi, Yuan Z, Pan Yo, Xie H and Cheng R 2014 Thermal, mechanical, and morphological properties of soybean oil-based polyurethane/epoxy resin interpenetrating polymer networks (IPNs) *Journal of Thermal Analysis and Calorimetry* **117(2)** 773–781

[5] Ghozali M, Triwulandari E and Haryono A 2015 Preparation and characterization of polyurethane-modified epoxy with various types of polyol *Macromolecular Symposia* **353(1)** 154–160

[6] Prokopchuk N R, Krutsko E T and Morev F V 2015 Modification of epoxy resins by polyisocyanates *Journal of Nature Science and Sustainable Technology* **9(1)** 121–127

[7] Nuraini L, Triwulandari E, Ghozali M, Hanafi M and Jumina J 2017 Synthesis of Polyurethane/Silica Modified Epoxy Polymer Based on 1,3-Propanediol for Coating Application *Indonesian Journal of Chemistry* **17(3)** 477–484

[8] Kausar A 2018 An investigation on epoxy/poly(urethane-amide)-based interpenetrating polymer network reinforced with an organic nanoparticle *Materials Research Innovations* **22(2)** 58–68

[9] Volkova E R, Strelnikov V N, Borisova I A, Slobodinyuk A I and Savchuk A V 2018 The Effect of the Isocyanate-Hydroxyl Ratio on the Structure and Properties of Hard Polyurethanes *Polymer Science Series D* **11(3)** 292–296

[10] Strel'nikov V N, Senichev V Yu, Slobodinyuk A I, Savchuk A V, Volkova E R, Makarova M A, Belov Yu L, Derzhavinskaya L F and Selivanova D G 2018 Preparation and Properties of Frost-Resistant Materials Based on Compounds of Oligoether Urethane Epoxides and Diglycidyl Urethane *Russian Journal of Applied Chemistry* **91(12)** 1937–44

[11] Malkin A Ya and Kulichikhin S G 2008 *Rheokinetics: Rheological Transformations in Synthesis and Reactions of Oligomers and Polymers* (John Wiley & Sons Inc) p 326

[12] Pierre-Gilles de Gennes 1979 *Scaling concepts in polymer physics* (London: Cornell University Press) p 368

[13] Ianniruberto G and Marrucci G 2014 Convective constraint release (CCR) revisited *Journal of Rheology* **58(1)** 89–102

[14] Hu Y Th 2014 Mechanisms of shear thickening in transient guar network *Journal of Rheology* **58(6)** 1789–1807

[15] Mezhikovskiy S M and Irzhak V I 2008 *Chemical Physics of Oligomers Curing* (Moscow: Science) p 269 (in Russ.)

Resonance magnetoelectric characteristics of Terfenol-D /Pb($Zr_{0.52}Ti_{0.48}$)O_3/Ni asymmetric three layered composites

Ying Liu[1,2], Chao Yang[1], Peng Zhou[1,2], Gopalan Srinivasan[2], Yajun Qi[1], Kun Liang[1], Tianjin Zhang[1] and D A Filippov [3,4]

[1]Department of Materials Science and Engineering, Hubei University, Wuhan 430062, P.R. China
[2]Physics Department, Oakland University, Rochester, Michigan 48309, USA
[3]Yaroslav-the-Wise Novgorod State University, ul. B. St. Peterburgskaya, 41 173003, Veliky Novgorod, Russia

[4]E-mail: Dmitry.Filippov@novsu.ru

Abstract. Three resonance modes, a first-order bending resonance mode and nonlinear and first-order longitudinal resonance modes, were observed in Terfenol-D/lead zirconate titanate(PZT)/Ni asymmetric magnetoelectric(ME) laminated composites. The ME outputs and resonance modes were tuned by the ratio of the thickness of the Ni negative magnetostrictive layer to the Terfenol-D positive magnetostrictive layer that of the(t_{Ni}/t_T). As (t_{Ni}/t_T) increased from 0.4 to 2.0, the strength of the first-order longitudinal resonance mode was decreased, while the first order bending resonance was enhanced. The asymmetric threelayered composites exhibited a large ME coefficient of 250mV/ (cm Oe) under a bias magnetic field of 400 Oe at quasi-static frequency (1 kHz). This work sheds some light on the design of ME composites with tunable ME resonance modes and multifrequency operation.

1. Introduction

Magnetoelectric (ME) composites are currently attracting increasing interest and have provoked a great number of research activities, driven by their potential applications in novel multifunctional devices such as sensors and transducers [1]. The direct ME effect, which describes the appearance of an electric polarization ***P*** upon applying a magnetic field ***H***, is usually evaluated by the magnetoelectric voltage coefficient (MEVC) α_E [1, 2]. It is well known that the ME effect of composites is extrinsic, strongly depending on the intrinsic piezoelectric or piezomagnetic properties of constituents and coupling interaction across magnetic/piezoelectric interfaces.

Various properties of ME laminated composites have been optimized for their use in potential applications, such as varying the magnetostrictive and piezoelectric materials; optimizing the composite structure, shape, and size; changing interfacial bonding conditions; and modulating the operation mode of ME coupling [3–7]. The symmetric threelayered composite structure formed by sandwiching a piezoelectric layer between two identical magnetostrictive layers is a classical ME composite structure [8, 9]. Compared with a symmetric threelayered ME composite, an asymmetric threelayered ME composite can operate in not only the longitudinal vibration mode, but also the bending vibration mode because of the flexural deformation caused by the asymmetric stress distribution in the ME composite [10–12]. When the ME composite operates in the bending vibration mode, its resonance frequency will be much lower than that of other modes, which can markedly decrease the eddy current loss for the magnetostrictive phase. This is especially relevant for rare earth alloys with large magnetostriction, such as Terfenol-D, because it can result in inefficient ME energy conversion. Meanwhile, the composite size could be kept quite small and ME coupling is strong. However, the bending vibration mode usually appears with the longitudinal vibration mode at comparable resonance intensity [10–12].

In this paper, we design a new asymmetric threelayered composite that can be operated in both bending and longitudinal modes. The resonance strength of each mode can be tuned by the thickness ratio of the positive and negative magnetostrictive layer. The results illustrate the possibility of implementing asymmetric threelayered composites in multifunctional devices with multifrequency and multimode operation.

2. Methods

Figure 1 depicts a schematic diagram of the asymmetric threelayered composite structure. The piezoelectric $Pb(Zr_{0.52}Ti_{0.48})O_3$ (PZT) layer with dimensions of 20×5×1 mm was polarized along the thickness direction. The magnetostriction orientation of the positive and negative magnetostrictive layers composed of $Tb_{1-x}Dy_xFe_{2-y}$ (Terfenol-D) and Ni flakes with dimensions of 15×5×0.5 mm, respectively, was along the longitudinal direction. The Terfenol-D, PZT, and Ni plates were bonded together by epoxy adhesive. The quick-drying epoxy adhesive could cure rapidly and ensure good mechanical coupling and strong bonding between layers at room temperature. Bottom Ni layers with thicknesses of 0.2, 0.5, and 1 mm were used to give composites with thickness ratios of the positive Terfenol-D to the negative Ni magnetostrictive layer t_{Ni}/t_T of 0.4, 1.0, and 2.0, respectively.

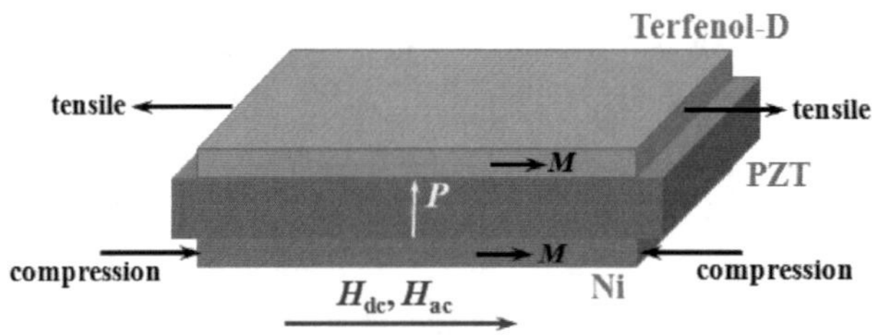

Figure 1. Schematic diagram of the asymmetric Terfenol-D/PZT/Ni threelayered composite.

The ME effects of the composites were measured by applying both bias constant (H_{dc}) and alternating (δH) magnetic fields parallel to the length direction. A shield cover was used to apply H_{dc} as well as shield external signals. Small Helmholtz coils were used to excite an ac magnetic field with amplitude of 3 Oe. The output voltage signal was detected using a lock-in amplifier (Stanford, SR830). All measurements were carried out at room temperature. The MEVC $\alpha_{E,31}$ was calculated using the formula of $\alpha_{E,31} = \delta V/(t_{PZT}\delta H)$, where δV is the voltage variation of PZT and t_{PZT} is the thickness of the PZT layer.

3. Results and discussion

Unlike a symmetric threelayered ME composite, an asymmetric threelayered ME composite can exhibit not only the longitudinal vibration mode, but also the bending vibration mode because of the flexural deformation caused by its asymmetric stress distribution. Accordingly, asymmetric threelayered laminated composites were designed with Terfenol-D with positive magnetostriction as the surface layer and Ni with negative magnetostriction as the bottom layer, as shown in figure 1. Composites with different Ni layer thicknesses of 0.2, 0.5, and 1 mm were fabricated to modulate the bending vibration, giving samples with t_{Ni}/t_T of 0.4, 1.0 and 2.0 respectively.

According to resonance theory, a bending resonance frequency is expected and can be calculated by the following formula [12, 13]:

$$^{n}f_{b1} = \frac{\pi t}{4\sqrt{3l^2}}\sqrt{\frac{\bar{Y}}{\bar{\rho}}}\beta_n^2 , \quad (1)$$

where t is the total thickness of ME composite materials, l is the length of ME composite materials, $\bar{\rho}$ is the average density of the sample, and $\beta_n = (n + 0.5)$, where n is the order of flexural vibration ($n \geq 1$). $\bar{Y}$is the average value of Young's modules, which can be calculated according to the following formula:

$$\bar{Y} = ({}^{T}Y\,{}^{T}t + {}^{Ni}Y\,{}^{Ni}t + {}^{P}Y\,{}^{P}t)/({}^{T}t + {}^{Ni}t + {}^{P}t) \quad (2)$$

in which ${}^{P}t$, ${}^{T}t$, and ${}^{Ni}t$ are the thickness of PZT, Terfenol-D, and Ni layers, respectively; ${}^{P}Y = 1/s_{11}^{P}$, ${}^{T}Y = 1/s_{11}^{T}$, ${}^{Ni}Y = 1/s_{11}^{Ni}$ are the Young's modules, and s_{11}^{P}, s_{11}^{T}, and s_{11}^{Ni} are the elastic compliances of PZT, Terfenol-D, and Ni, respectively.

For the longitudinal mode, we can use the calculations in our previous work [14].The constitutive equations for the magnetostrictive and piezoelectric phases are:

$${}^{P}S_1 = \frac{1}{{}^{P}Y}{}^{P}T_1 + {}^{P}d_{31}E_3 , \quad (3)$$

$${}^{T}S_1 = \frac{1}{{}^{T}Y}{}^{T}T_1 + {}^{T}\lambda_1(H) , \quad (4)$$

$${}^{Ni}S_1 = \frac{1}{{}^{n}Y}{}^{Ni}T_1 + {}^{Ni}\lambda_1(H) , \quad (5)$$

$${}^{P}D_3 = {}^{P}\varepsilon_{33}\,{}^{P}E_3 + {}^{P}d_{31}\,{}^{P}T_1 , \quad (6)$$

where ${}^{P}S_1$, ${}^{T}S_1$, ${}^{Ni}S_1$ are the strain tensor components of the PZT, Terfenol-D and Ni layers, ${}^{P}E_3$, ${}^{P}D_3$ are vector components of the electric field and electric displacement; ${}^{P}T_1$, ${}^{T}T_1$, ${}^{Ni}T_1$ are the stress tensor components of the piezoelectric and the magnetostrictive phases; ${}^{P}d_{31}$ is piezoelectric coefficients; ${}^{T}\lambda_1(H)$, ${}^{Ni}\lambda_1(H)$ are magnetostriction of Terfenol-D and Ni layers, respectively; and ${}^{P}\varepsilon_{33}$ is the tensor components of permittivity. The field dependent magnetostriction of Terfenol-D and Ni layers are shown in figure 2.

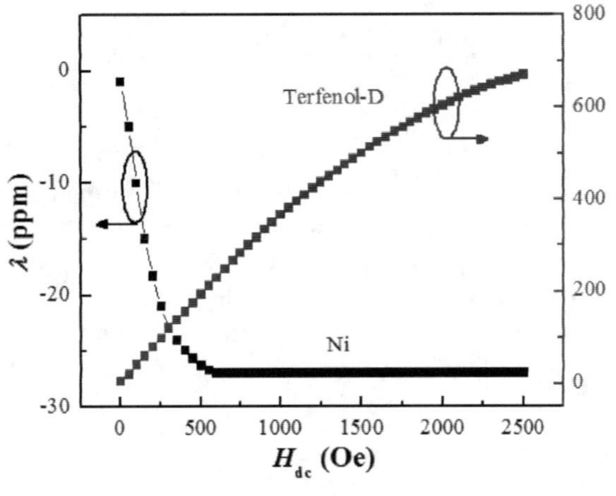

Figure 2. Dependence of the longitudinal magnetostriction λ_l of Ni and Terfenol-D on bias magnetic field H_{dc}.

IOP Conf. Series: Materials Science and Engineering **656** (2019) 012056 doi:10.1088/1757-899X/656/1/012056

Magnetostriction is nonlinear function of ac and bias magnetic fields. We can get the following function by taking into account the first and second terms of series:

$$\lambda_1(H\ ) = q_{11}H_1 + q_{11}^{(2)}H_1^2, \tag{7}$$

where $q_{11} = \left.\frac{\partial\lambda_1}{\partial H_1}\right|_{H_1=H_{bias}}$, $q_{11}^{(2)} = \left.\frac{\partial^2\lambda_1}{\partial H_1^2}\right|_{H_1=H_{bias}}$ are linear and nonlinear piezomagnetic coefficients, H_1 is ac magnetic field with angular frequency ω. The first term in Eq.(7) give the deformation of magnetic with frequency ω and second term give the deformation of magnetic with frequency 2ω [15-18]. These coefficients are also nonlinear functions vs magnetic field, the dependencies between field and piezomagnetic coefficients of Terfenol-D and Ni layers are shown in Fig. 3.

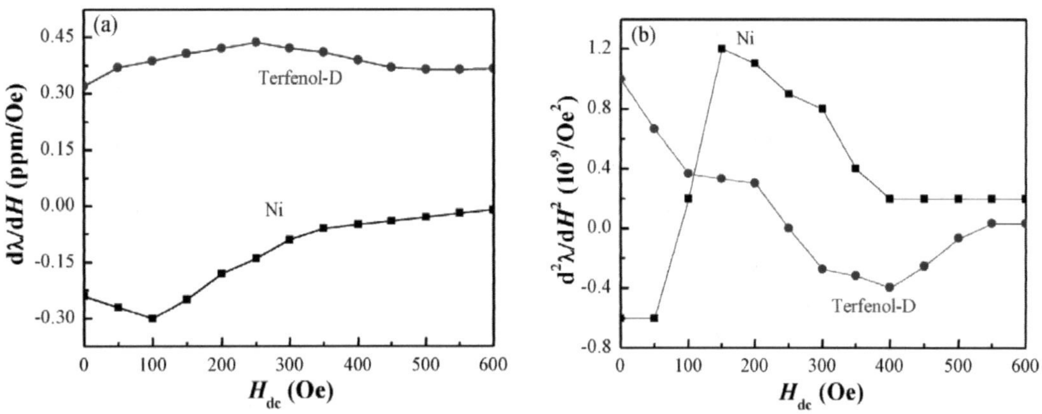

Figure 3. Dependence of the linear q_{11} and nonlinear $q_{11}^{(2)}$ piezomagnetic coefficients of Ni and Terfenol-D on bias magnetic field H_{dc} for thickness t_{Ni} = 0.5 mm and t_T=0.5 mm.

In figure 3, the linear and nonlinear piezomagnetic coefficients for Terfenol-D and Ni exhibit different signs. This leads to the opposite contribution to the value of ME effect for longitudinal mode.

Using approximation given by Equation (7), the Equation (4) and Equation (5) can be rewritten as:

$$^{T}S_1 = \frac{1}{^{T}Y}{}^{T}T_1 + {}^{T}q_{11}H_1 + {}^{T}q_{11}^{(2)}H_1^2, \tag{8}$$

$$^{Ni}S_1 = \frac{1}{^{Ni}Y}{}^{Ni}T_1 + {}^{Ni}q_{11}H_1 + {}^{Ni}q_{11}^{(2)}H_1^2, \tag{9}$$

The ac magnetic field with angular frequency ω produces the deformation with angular frequencies ω and 2ω in Terfenol-D and Ni layers. These deformations transmitted via mechanical interaction in PZT layer and electric field induced by piezoelectric effect. Since the layers are thin enough so that we could neglect the change in the displacements of the medium along the thickness of the layer, and assume that the displacements of the medium, the magnetic and the piezoelectric layer are equal, i.e. ${}^{T}u_l={}^{P}u_1={}^{Ni}u_l=u_1$. Therefore, in the first approximation the equation of motion for the medium has following form:

$$\bar{\rho}\frac{\partial^2 u_1}{\partial t^2} = \frac{\partial \bar{T}_{11}}{\partial x_1}, \tag{10}$$

where $\bar{\rho} = ({}^T\rho^T t + {}^{Ni}\rho^{Ni} t + {}^P\rho^P t)/({}^T t + {}^{Ni} t + {}^P t)$, $\bar{T}_1 = ({}^T T_1{}^T t + {}^{Ni}T_1{}^{Ni} t + {}^P T_1{}^P t)/({}^T t + {}^{Ni} t + {}^P t)$ are the average values of the density and stress tensor of the sample.

Using the method of calculations, which was presented in our previous work [14], we can obtain the MEVC as following expressions:

$$\alpha_E = \frac{{}^P Y\, {}^P d_{31}}{{}^P\varepsilon_{33}({}^T Y^T t + {}^{Ni}Y^{Ni} t + {}^P Y^P t)} \cdot$$

$$\left[({}^T Y^T t \cdot {}^T q_{11} + {}^{Ni}Y^{Ni} t \cdot {}^{Ni} q_{11}) \frac{tg(\kappa^{(1)})}{\Delta_A^{(1)} \kappa^{(1)}} + ({}^T Y^T t \cdot {}^T q_{11}^{(2)} + {}^{Ni}Y^{Ni} t \cdot {}^{Ni} q_{11}^{(2)}) H_1 \frac{tg(\kappa^{(2)})}{\Delta_A^{(2)}\, \kappa^{(2)}} \right]. \quad (11)$$

Here we introduce following denotes $\Delta_A^{(1)} = 1 - K_P^2 (1 - \frac{{}^P Y\, {}^P t}{({}^T Y^T t + {}^{Ni}Y^{Ni} t + {}^P Y^P t)} \frac{tg(\kappa^{(1)})}{\kappa^{(1)}}$ and $\Delta_A^{(2)} = 1 - K_P^2 (1 - \frac{{}^P Y\, {}^P t}{({}^T Y^T t + {}^{Ni}Y^{Ni} t + {}^P Y^P t)} \frac{tg(\kappa^{(2)})}{\kappa^{(2)}}$; $K_P^2 = \frac{{}^P Y ({}^P d_{31})^2}{{}^P\varepsilon_{33}}$ is the squared coefficient of electromechanical coupling; $\kappa^{(1)} = k^{(1)} l/2$ and $\kappa^{(2)} = k^{(2)} l/2$ are dimensionless parameters. The dispersions relation between wave number and angular frequency are as following expression:

$$k^{(1)} = \sqrt{\frac{\bar{\rho}}{\bar{Y}}}\omega\,, k^{(2)} = \sqrt{\frac{\bar{\rho}}{\bar{Y}}}\, 2\omega\,. \quad (12)$$

The first term in square brackets in Equation (11) describes the contribution to MEVC from the linear effect ME, and the second term in square brackets in Equation (11) describes the contribution from the nonlinear effect.

If $\Delta_A^{(1)} = 0$ or $\Delta_A^{(2)} = 0$, the parameter MEVC α_E has a resonance peak. The increase of resonance is observed near the value of dimensionless parameters $\kappa^{(1)} \approx (2n-1)\frac{\pi}{2}$ and $\kappa^{(2)} \approx (2n-1)\frac{\pi}{2}$, where n = 1, 2, ···. We can get the resonance frequencies from linear and nonlinear ME effect by using dispersion relation (12) and the following expressions:

$$f_l^{(1)} = \frac{1}{2l}\sqrt{\frac{\bar{Y}}{\bar{\rho}}},\; f_l^{(2)} = \frac{1}{4l}\sqrt{\frac{\bar{Y}}{\bar{\rho}}} = \frac{f_l^{(1)}}{2}\,. \quad (13)$$

Thus the value of resonance frequency for nonlinear (quadratic) ME interaction is half of the linear ME effect. For the calculation, we can use following parameters of the structure: the parameters of plate $Tb_{1-x}Dy_xFe_2$ (Terfenol-D) is 15×5×0.5 mm and thicknesses of Ni layers are 0.2, 0.5, and 1 mm; the piezoelectric PZT layer with dimensions of 20×5×1 mm; the material parameters of Terfenol-D, PZT, and Ni are: ρ_P= 7.5 ×10^3 kg/m^3,ρ_T= 8.5 ×10^3 kg/m^3,ρ_{Ni}= 8.9×10^3 kg/m^3, $s_{11}{}^P$ = 15 ×10^{-12} m^2/N,$s_{11}{}^T$ = 16 ×10^{-12} m^2/N,$s_{11}{}^{Ni}$ = 4.5 ×10^{-12} m^2/N; ${}^P d_{31}$=-175×10^{-12} m/V. The values of linear and nonlinear piezomagnetic coefficients for Terfenol-D at H_{bias} = 400 Oe were ${}^T q_{11}$ =0.35×10^{-6}1/Oe; ${}^T q_{11}^{(2)}$ = - 0.4×10^{-9}1/Oe^2. The values of linear and nonlinear piezomagnetic coefficients for Ni are dependent on the thickness of nickel layer. The parameters we used for calculations are: ${}^{Ni}t$=0.2 mm, ${}^{Ni}q_{11}$ =-0.05×10^{-6}1/Oe; ${}^{Ni}q_{11}^{(2)}$ =0.1×10^{-9}1/Oe^2;${}^{Ni}t$=0.5 mm, ${}^{Ni}q_{11}$ =-0.06×10^{-6}1/Oe; ${}^{Ni}q_{11}^{(2)}$ =0.2×10^{-9}1/Oe^2; ${}^{Ni}t$=1.0 mm, ${}^{Ni}q_{11}$ =-0.15×10^{-6}1/Oe; ${}^{Ni}q_{11}^{(2)}$ =0.25×10^{-9}1/Oe^2. Using these parameters we get the values of resonance frequencies, as shown in table 1.

Table 1. Calculated resonance frequencies of Terfenol-D/PZT/Ni composites.

^{P}t (mm)		1	
^{T}t (mm)		0.5	
^{Ni}t (mm)	0.2	0.5	1.0
Length vibration, linear effect $f_l^{(1)}$ (kHz)	83	86	93
Length vibration, nonlinear effect $f_l^{(2)}$ (kHz)	41.5	43	46.5
Flexural vibration frequency f_b (kHz)	13.5	16.5	20

These theoretical predictions were confirmed by experimental measurements. First, the linear magnetostriction of each composite was measured using a standard strain gauge technique in the directions parallel to H_{dc} at room temperature. The dependence of the in-plane parallel magnetostriction$\lambda_{//}$ on H_{dc} for Terfenol-D and Ni layers both with a thickness of 0.5 mm is shown in figure 2. These two materials exhibit different magnetostrictive properties. The Ni layer exhibits negative magnetostriction, whereas the Terfenol-D layer shows positive magnetostriction. When H_{dc}>600 Oe, the magnetostriction of Ni was saturated. Conversely, the magnetostriction of Terfenol-D was not saturated even when H_{dc} was as high as 2.5kOe. Although the magnetostriction of the Ni flake was small, it presented a relatively large linear piezomagnetic coefficient $d\lambda/dH$ of −0.3 ppm/Oe at 100 Oe, which then decreased to nearly zero as H_{dc} increased further, as shown in figure 3(a). The $d\lambda/dH$ value of Terfenol-D was positive, with a maximum of about 0.43 ppm/Oe at 300 Oe. With increasing H_{dc}, $d\lambda/dH$ decreased almost linearly. Figure 3 (b) shows the field dependency of nonlinear piezomagnetic coefficient $d^2\lambda/dH^2$ at bias magnetic field near 400 Oe. As shown in figure 3, the linear and nonlinear piezomagnetic coefficients for Terfenol-D and Ni have the opposite signs.

Figure 4 shows the $\alpha_{E,31}$values of the Terfenol-D/PZT/Ni composites as a function of frequency (f) at H_{dc} = 400 Oe. The thickness of Terfenol-D was 0.5 mm and that of the Ni layer was 0.2, 0.5, or 1 mm to give t_{Ni}/t_T of 0.4, 1.0, and 2.0, respectively. As shown in figure 4, three resonance peaks located around 20, 40, and 80 kHz were observed, which correspond to the first-order bending vibration, nonlinear and first-order longitudinal vibration, respectively, according to the above calculation results. The multiple resonance modes appear in the low-frequency region because of the asymmetric stress distribution in the Ni and Terfenol-D layers.

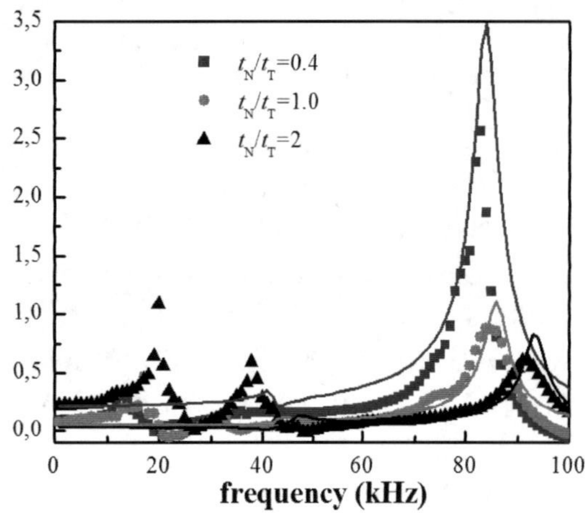

Figure 4. Frequency dependence of MEVC at H_{dc} = 400 Oe for composites with different t_{Ni}/t_T.Scattered data denote experimental results and solid line with same color denote theoretical calculations for longitudinal modes.

The magnetostriction of Ni is negative, whereas that of Terfenol-D is positive. For the bending modes it leads to increase the bending moment, because the Terfenol-D and Ni layers are located on opposite sides of the neutral layer. The magnitude of the induced voltage depends not only on the magnitude of the bending moment, but also on the location of the neutral layer relative to the axis of the piezoelectric. If the neutral layer is located on the axis of the piezoelectric, then the resulting voltage will be zero, since the tensile and compressed layers induced an opposite directions electric field. The bending moment is also increasing with increasing the thickness ratio t_{Ni}/t_T. In addition, it leads to increase the asymmetry of the neutral layer location of relative to the axis of the piezoelectric. All of these lead to an increase of the induced voltage and the MEVC for bending mode with increasing the t_{Ni}/t_T, which is in full accordance with figure 4.

For the longitudinal mode, when the magnetic field was applied, the Ni layer is compressed and the Terfenol-D layer is tensile. The value of MEVC, according to Equation (11), is proportional the sum of ${}^TY^T t \cdot {}^T q_{11} + {}^{Ni}Y^{Ni} t \cdot {}^{Ni} q_{11}$. When the Ni layer was thinner than the Terfenol-D layer, and as long as there is a relationship ${}^TY^T t \cdot {}^T q_{11} > {}^{Ni}Y^{Ni} t \cdot {}^{Ni} q_{11}$, the tensile deformation was dominant. As the t_{Ni}/t_T increased, the tensile and compression deformations began to compete and the value of MEVC is decreased with increase the thickness of Ni-layer. When the thickness of the Ni layer has increased so much that the relationship: ${}^{Ni}Y^{Ni} t \cdot {}^{Ni} q_{11} > {}^TY^T t \cdot {}^T q_{11}$ has been fulfilled, since then the coefficient value begins to increase with increasing nickel layer. The low-frequency operation of ME composites can lower the eddy current loss in the magnetostrictive phase, resulting in efficient ME energy conversion. Furthermore, ME composites that operate at a bending resonance mode with low frequency are also advantageous to miniaturize devices, because the normal transverse resonance frequency of layered ME composites decreases with increasing length [19].

Figure 5 shows the $\alpha_{E,31}$values as a function of H_{dc} at f = 1 kHz for the three Terfenol-D/PZT/Ni threelayered composites with various t_{Ni}/t_T. The dependence of $\alpha_{E,31}$ on H_{dc} was similar for the three samples. With increasing H_{dc} from zero, $\alpha_{E,31}$ increased to a maximum value at around H_{dc} = 400 Oe and then decreased gradually. The negative $\alpha_{E,31}$of the threelayered composite with t_{Ni}/t_T = 1.0 is mainly caused by the negative magnetostriction of the Ni layer at low H_{dc}. At low H_{dc}, $d\lambda/dH$ of the Ni layer is comparable with that of the Terfenol-D layer (figure 3(a)) when these two layers are of equal thickness, resulting in negative $\alpha_{E,31}$. With increasing H_{dc}, the magnetostriction of Ni became saturated and the $d\lambda/dH$ values approached zero. In contrast, the Terfenol-D layer did not show saturated magnetostriction even at 2.5 kOe.

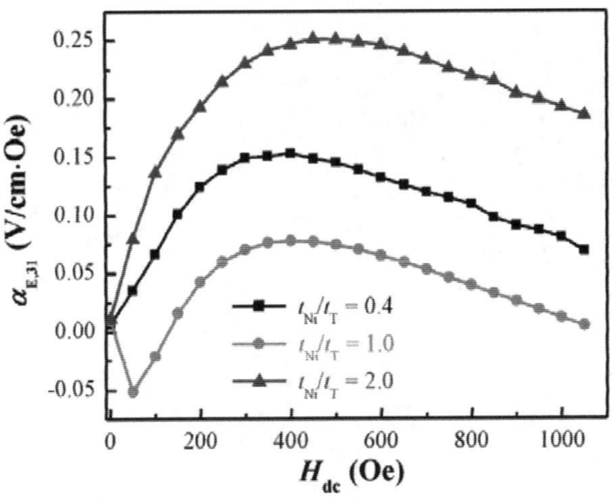

Figure 5. ME voltage coefficient $\alpha_{E,31}$ as a function of biasmagnetic field H_{dc} at f =1 kHz for composites with different t_{Ni}/t_T.

Moreover, the $\alpha_{E,31}$ values were proportional to dλ/dH and strongly depended on the volume fraction of the piezoelectric phase v [20, 21]. The composite with t_{Ni}/t_T = 2 exhibited the largest $\alpha_{E,31}$ values throughout the measured H_{dc} range, whereas that with t_{Ni}/t_T = 1 displayed the lowest $\alpha_{E,31}$values. This may result from the competition between the contributions ofdλ/dH of Terfenol-D and Ni layers and the volume fraction v.

The composite with t_{Ni}/t_T = 2has a thicker Ni layer than those in the other composites, which may result in large deformation and increase dλ/dH. However, the thicker Ni layer will decrease v, which will in turn decrease $\alpha_{E,31}$ [20, 21]. In this composite, the larger contribution from dλ/dH dominated its behavior. Although a thinner Ni layer can increase v, the effective vibrational deformation of both Terfenol-D and Ni layers results in the small $\alpha_{E,31}$ values of the composite with t_{Ni}/t_T = 1, as shown in figure 5. This also can be interpreted from the plots of $\alpha_{E,31}$ against v, which display a double maximum when the longitudinal strain and flexural strain in the asymmetric threelayered composite are both taken into consideration, as predicted in our previous work [22].

4. Conclusion

Terfenol-D/PZT/Ni asymmetric ME laminated composites were designed to generate a bending vibration mode with the aim of realizing multifrequency and multimode operation in ME devices. In these Terfenol-D/PZT/Ni asymmetric three layered composites, three resonance modes, first-order bending resonance and nonlinear and first-order longitudinal resonance modes were observed. The ME outputs and resonance mode were tuned by t_{Ni}/t_T of the composites. As t_{Ni}/t_T increased from 0.4 to 2.0, the strength of the first-order longitudinal resonance was suppressed considerably, whereas the first-order bending resonance and nonlinear longitudinal resonance became stronger. The asymmetric three layered composites exhibited a large ME coefficient of 250mV/cm·Oe under a magnetic field of 400 Oe at quasi-static frequency. This work provides an approach to design ME devices with tunable ME resonance modes and multi frequency operation.

5. Acknowledgment

This work was financially supported by the National Natural Science Foundation of China (51372074, 11574073, and 51472078) and Hubei Key Laboratory of Ferro & Piezoelectric Materials and Devices (201706) and the work was jointly supported by the Russian Foundation for Basic Research and Belarussian Foundation for Basic Research (Russian project no. 18-52-00021 and Belarussian project no. F18R-300).

References

[1] Ma J, Hu J, Li Z and Nan C W 2011 *Adv. Mater.* **23** 1062–1087
[2] Jayachandran K P, Guedes J M and Rodrigues H C 2018 *Sci. Rep.* **8** 4866
[3] Srinivasan G, Rasmussen E T, Gallegos J, Srinivasan R, Bokhan Y I and Laletin V M 2001 *Phys. Rev.* B **64** 214408
[4] Zhai J, Li J, Viehland D and Bichurin M I 2007 *J. Appl. Phys.* **101** 014102
[5] Lasheras A, Gutiérrez J and Barandiarán J M 2016 *Appl. Phys. Lett.* **108** 222903
[6] Nan C W, Liu G and Lin Y 2003 *Appl. Phys. Lett.* **83** 4366–4368
[7] Lu M C, Mei L, Jeong D Y, Xiang J, Xie H and Zhang Q M 2015 *Appl. Phys. Lett.* **106** 112905
[8] Bi K, Wang Y G, Wu W and Pan D A 2010 *J. Phys.* D: *Appl. Phys.* **43** 132002
[9] Dong S, Cheng J, Li J F and Viehland D 2003 *Appl. Phys. Lett.* **83** 4812–4814
[10] Chen L, Li P, Wen Y and Zhu Y 2015 *J. Alloys Compd.* **646** 1032–1035
[11] Bi K, Wang Y G, Pan D A and Wu W 2010 *Scr. Mater.* **63** 589–592
[12] Cheng J H, Wang Y G, and Xie D 2014 *Appl. Phys. Lett.* **104** 252411
[13] Wan J G, Li Z Y, Wang Y, Zeng M, Wang G H and Liu J M 2005 *Appl. Phys. Lett.* **86** 202504
[14] Filippov D A, Laletin V M and Galichyan T A 2014 *Appl. Phys.* A **115** 1087–1091
[15] Fetisov L Y, Fetisov Y K, Sreenivasulu G and Srinivasan G 2013 *J. Appl. Phys.* **113** 116101
[16] Burdin D A, Chashin D V, Ekonomov N A, Fetisov Y K, Fetisov L Y, Sreenivasulu G and

Srinivasan G 2013 *J. Appl. Phys.* **113** 033902
[17] Laletin V M, Filippov D A and Firsova T O 2014 *Tech. Phys. Lett.* **40** 237–240
[18] Filippov D A, Laletin V M and Firsova T O 2014 *Phys. Solid State* **56** 980–984
[19] Bi K, Wang Y G and Wu W 2011 *Sens. Actuators* A **166** 48–51
[20] Bichurin M I, Petrov V M and Srinivasan G 2003 *Phys. Rev.* B **68** 05440
[21] Filippov D A, Laletsin U and Srinivasan G 2007 *J. Appl. Phys.* **102** 093901
[22] Petrov V M and Srinivasan G 2008 *Phys. Rev.* B **78** 184421

IOP Conf. Series: Materials Science and Engineering **656** (2019) 012057 doi:10.1088/1757-899X/656/1/012057

Digital forms of describing Industry 4.0 objects

D A Zakoldaev[1], A G Korobeynikov[1], A V Shukalov[1] and I O Zharinov[1]

[1]Faculty of Information Security and Computer Technologies, Saint Petersburg National Research University of Information Technologies, Mechanics and Optics, 49, Kronverksky Av., Saint Petersburg, 197101, Russia

E-mail: mpbva@mail.ru

Abstract. The task is to project item digital model which is planned to be manufactured in cyber and physical production of the Industry 4.0. Item digital model can be classified as item digital replica, item digital shadow and item digital twin. Item digital model is a set of construction, program and technological documentation prepared electronically with automatic system of projection. Item digital shadow is a set of item digital replica documentation enhanced with mathematical equations with restrictions to describe physical and chemical properties of the item materials and its components. Details to describe the item digital shadow are defined with types of tests done with modelling and virtualization. Item digital twin is a finished form of item properties description done electronically (drawings, schemes, formulas and other) and successful physical tests of the whole cycle. The scheme is given for the item life cycle of the item designing item digital model. The scheme of production section and Industry 4.0 smart factory virtual environment components is described oriented to manufacture the items by its digital models.

1. Introduction

The development of information technologies implemented in some industries a digital approach [1, 2] in describing processes and items. The most popular today are aviation, maritime affairs, automobiles, rockets and space and some other industries to transfer the company business processes in the digital environment.

Company business processes [3, 4] as a digital object is a set of works being done in company divisions to develop or manufacture an item (technologies, materials and etc.) and then implement it. Business processes digitalizing is used to create [5, 6] in the company an informative and telecommunication environment which is used for electronic documents [7] of this company and has some means of automatic projecting (automatized) of for the product manufacturing.

Item designing or machine designing as a digitalizing object [8, 9] is a set of construction and technological, mechanical, optical, kinematics, mathematical and other description forms of material properties and components which interact. Item digital form permits to develop and create products electronically which has [10, 11] the necessary conditions to use in practice the modelling technologies and company business process virtualization.

Item life cycle in digitalizing includes three forms to describe processes or the items:

– digital replica;
– digital shadow;
– digital twin,

which may develop in time and support each other.

IOP Conf. Series: Materials Science and Engineering **656** (2019) 012057 doi:10.1088/1757-899X/656/1/012057

2. Digital replica of projection object

Item digital replica is a set of construction, program and technological documents in its electronic form to define the item appearance, properties, order of manufacturing and servicing of item in exploitation. Item digital replica has drawings, schemes and text documentation and a 3D-model (D-Dimension) of the item itself as an assembly unit.

The item digital replica documentation includes:

– technological appliances drawing which are necessary to create item in production;

– automatic work places scheme to check the quality and item functionality;

– software components with tests of integrated control and checking tests for the manufacturer plant and other.

Generally in radio and electronics item designing the item digital replica contains:

– item life cycle automatic control system for construction, program and technological documentation, including 3D-model and programs of numeric control machines prepared for the item;

– item life cycle automatic control system for construction, program and technological documentation prepared for automatic work places and technological appliances (documentation of auxiliary production);

– automatic control system of the item life cycle and control system of technological appliances life cycle;

– production, accounting and attributive data of the division name which is responsible for item development and manufacturing, the item itself and its assembly units, material consumption rates, sum indications of project and production works, name and contacts of component supplier and other.

3. Digital shadow and digital twin of projection object

Projection object digital shadow is a set of mathematical equations with a restriction system and a bunch of technical documentation in its electronic form to describe item properties relevant for the item expertise.

Item digital shadow is a more advanced form of the projection object appearance unlike a digital replica and can be formed after modelling procedures of item behavior in exploitation and the virtualization test process. Modelling and virtualization are key project procedures to automatize the activity of the designer in the Industry 4.0 digital factory.

The main goal of modelling and virtualization procedure is an equivalent of natural tests with a physical item which are being made in projection companies of the Industry 3.0. So digital shadow of the projection object describes item properties according to the tests made on its digital replica with modelling and virtualization. The first projection stages require details to be described as a digital replica and modelling cannot be done for each influence which is expected when the item is exploited. Item digital shadow is a description form being developed in time after some adjustment of mathematical models and when the documents are corrected.

The end digital form of the projection object is a digital twin. A digital twin is a set of mathematical equations with some restrictions and its electronic technical documents to describe the item properties which is good for any field of study known today. So item digital twin in its level of details is a reachable form of item description.

Figure 1 shows the scheme which in practice is used for informative development of the projection item description forms from a digital replica through a digital shadow to a digital twin.

Having analyzed figure 1 we can see that the main influence of the object description development in time as a digital shadow is for the data obtained after the item exploitation. The exploitation data (item exploitation history) can be aggregated in groups:

– data of item servicing and its repair;

– item incidents data (refusals) in its exploitation classified by the type of defects;

– current statistical data after the item exploitation (accuracy of functionality, time between failures, item tactical and technical characteristics and other) under different factors influence and other.

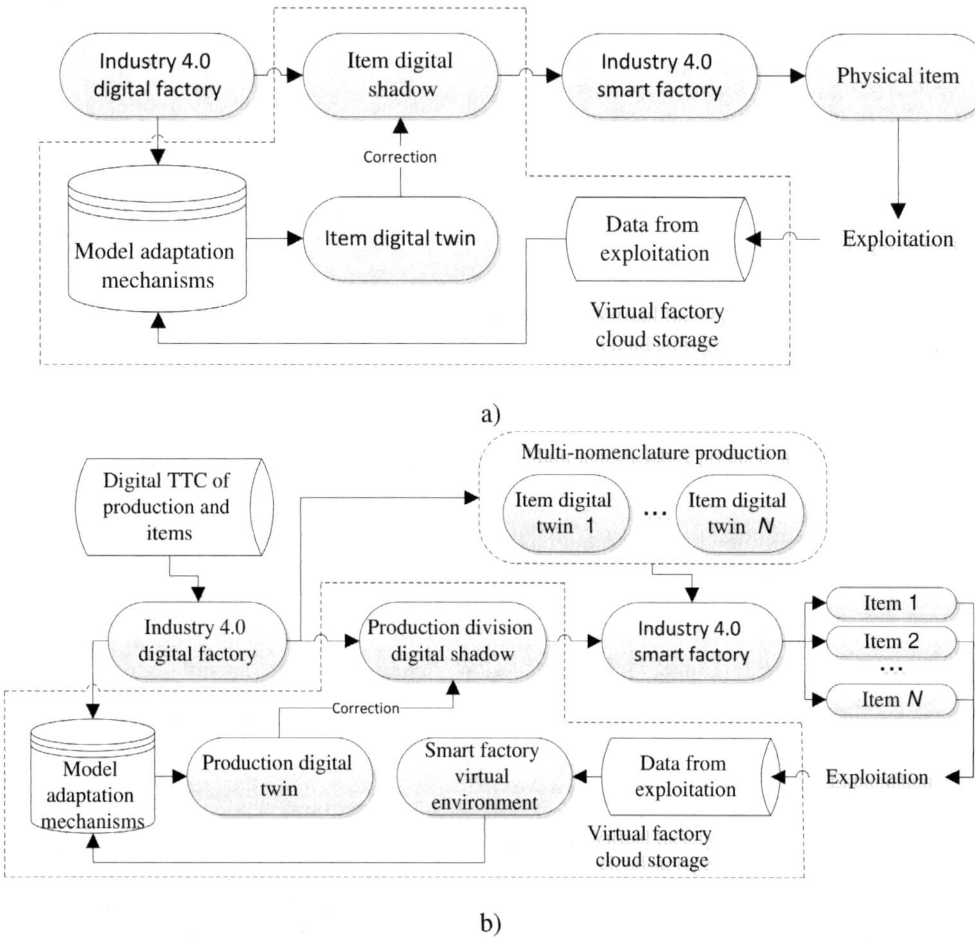

Figure 1. Scheme of informative development of the projection object description forms: a) item designing component, b) automatic mechanical assembly production (TTC – Tactical and Technical Characteristics).

4. Digital description of company business processes

There are several forms to describe company business processes by the processing model, conceptual model, reference model and other. To project Industry 4.0 company the reference model is the most suitable one in its properties.

Reference model of production company is a way of description of company business processes based on the quality criteria of technological operation completion and capable of modelling with computers the activity of a smart factory. Reference model of production is from practice of business activity in industry and is used the scheme method of IDEF (Integration Definition for Function Modeling) after company business processes audit to find out the cloning procedures and lacking procedures among company divisions. The main differences of reference model are:

– it is universal, a reference model can describe a type of companies of particular activity and not a single company;

– it is repeatable, a reference model can be used again when the new digital company is created if it was obtained after the existing company;

– it is optimal, a reference model includes organization and technical and economic solutions obtained after a real business;

– it is clear, a reference model is a graphical scheme with blocks (name of company divisions) and associated links (oriented arrows to joint blocks);

– it is scalable, a reference model can be detailed to the level of some company divisions interacting with each other to complete some technological operations;

– it is transparent, a reference model is based on samples of business processes which can be used for new forms of divisions interaction;

– it is adequate, a reference model has the company structure and connections after the technological operations being realized into the company, quality control after quantity criteria (expenses for processes completion, resource load level and other).

5. Application of digital models in Industry 4.0 production

Item technical documentation in digital form requires in the production division of smart factory cyber and physical technological equipment with numeric control. To unite cyber and physical systems engaged in item manufacturing in a single route map may create the technological section production which scheme is given in figure 2.

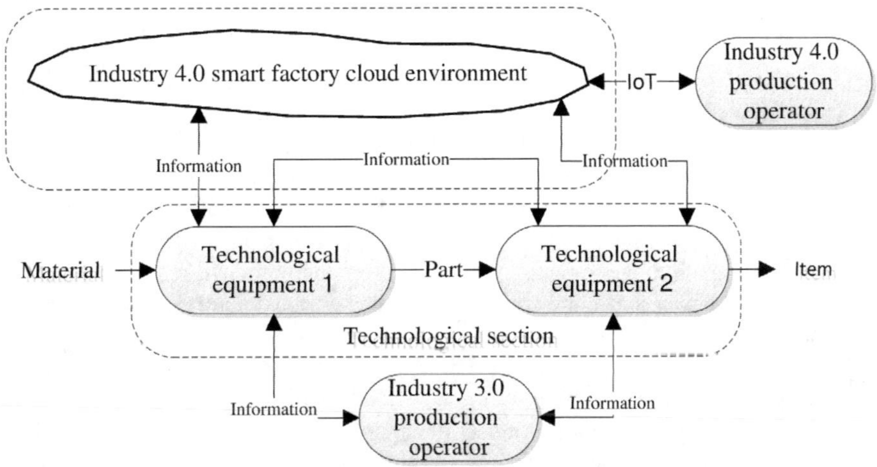

Figure 2. Production section scheme and smart factory components engaged in the item manufacturing.

Cyber and physical systems interaction is done automatically with information exchange protocols to transfer data among the machines about technological operation completion state and data transferring to the smart factory cloud. Production data operator access is done with channels of IoT united in a single informative and telecommunication system of company virtual environment and physical machines of cyber and physical systems.

Industry 4.0 smart factory virtual environment is a set of digital models to describe:

– the item being manufactured;

– item manufacturing technological process;

– cyber and physical system (control object) to manufacture the item.

Smart factory virtual environment components scheme is given in figure 3.

Interaction of digital models in virtual environment is done with program interface based on digital production semantics. Digital production semantics are formed in terms of model and processes ontology description which is the semantics dictionary of Industry 4.0.

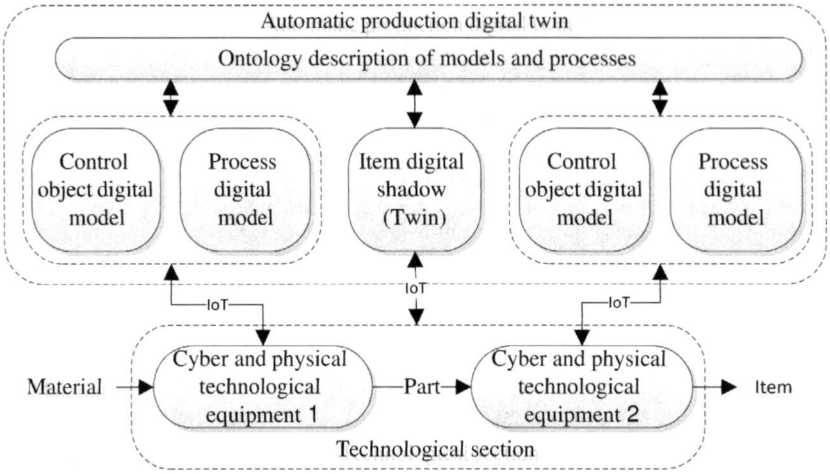

Figure 3. Smart factory virtual environment components scheme engaged in the item manufacturing.

6. Conclusion

To develop a digital replica, digital shadow and digital twin of the projection objects in the conditions of Industry 4.0 paradigm informative and telecommunication systems must be created to give the company the necessary conditions to develop digital platforms and technologies and the way of their effective interaction. Digitalizing of industrial companies sees the project and production processes as a whole new level which gives the perspectives of digital data development as a key factor of production modernization.

The transfer of project and production data in a digital format makes Industry 4.0 company functionality being independent and under the conditions of virtual factory informative infrastructure creation to unite some independent companies. Company independence is because they can transfer digital project and production data to different digital platforms without changing too much the data structure. This data acceptability is the base for digital economy projects.

Item designing and machine designing digital replicas can be viewed as a technological storage made in Industry 3.0 to transfer the project and production companies to function with the standards of Industry 4.0 where the item technical documentation is a digital shadow and digital twin.

To develop the item digital shadow and digital twin is based on informative and telecommunication system creation to aggregate and process, storage and give the designer (production operator) project and production data in its convenient form. Industry 4.0 informative and telecommunication system key properties are integrity, confidentiality, user authentication and availability of the information being transferred to all personnel of the digital company ecosystem.

References)

[1] Talkhestani B A, Jazdi N, Schloegl W and Weyrich M 2018 *Procedia CIRP* **72** 159–164
[2] Hu L, Nguyen N-T, Tao W, Leu M C, Liu X F, Shahriar M R and Nahian S M 2018 *Procedia manufacturing* **26** 1193–1203
[3] Qi Q, Tao F, Zuo Y and Zhao D 2018 *Procedia CIRP* **72** 237–242
[4] Kunath M and Winkler H 2018 *Procedia CIRP* **72** 225–231
[5] Gurjanov A V, Zakoldaev D A, Shukalov A V and Zharinov I O 2018 *IOP Conference Series: Materials Science and Engineering* **327** 022110
[6] Zakoldaev D A, Shukalov A V, Zharinov I O and Zharinov O O 2018 *Journal of Physics: Conference Series* **1015** 052033
[7] Cai Y, Starly B, Cohen P and Lee Y-Sh 2017 *Procedia manufacturing* **10** 1031–42

IOP Conf. Series: Materials Science and Engineering **656** (2019) 012057 doi:10.1088/1757-899X/656/1/012057

[8] Lohtander M, Ahonen N, Lanz M, Ratava J and Kaakkunen J 2018 *Procedia manufacturing* **25** 55–61
[9] Kousi N, Gkournelos C, Aivaliotis S, Giannoulis C, Michalos G and Makris S 2019 *Procedia manufacturing* **28** 121–126
[10] Bogatyrev V A, Parshutina S A, Poptcova N A and Bogatyrev A V 2016 *Communications in computer and information science* **678** 337–348
[11] Negri E, Fumagalli L, Cimino C and Macchi M 2019 *Procedia manufacturing* **28** 201–206

Information and telecommunication system of digital factory and smart factory Industry 4.0

D A Zakoldaev[1], A V Shukalov[1], I O Zharinov[1] and O O Zharinov[2]

[1]Faculty of Information Security and Computer Technologies, Saint Petersburg National Research University of Information Technologies, Mechanics and Optics, 49, Kronverksky Av., Saint Petersburg, 197101, Russia
[2]Department of Problem-Oriented Computing Complexes, Saint Petersburg State University of Aerospace Instrumentation, 67, Bolshaya Morskaia str., Saint Petersburg, 190000, Russia

E-mail: mpbva@mail.ru

Abstract. The task is to create information and telecommunication system of digital and smart factories of Industry4.0. Components are defined for information and telecommunication system, their properties and purpose are described. The scheme of digital factory information and telecommunication system is given. This scheme includes the physical equipment (machines) and virtual machines (cloud services and models) to support project activity of the designer. The scheme of smart factory information and telecommunication system is given. The scheme unites the cyber and physical production equipment and test equipment used for manufacturing on the basis of digital informative technologies. The given schemes are important in practice to organize the informative exchange among cyber and physical systems and designers of the project companies and operators of production companies to manufacture and transfer into exploitation the items of instrument-making, tool engineering, and mechanical engineering.

1. Introduction

Industry 4.0 digital and smart factories project and production activity require [1, 2] to create a company infrastructure oriented for electronic documents and implementation of digital production technologies. The base of such an infrastructure is information and telecommunication system [3, 4] to support the main stages of the item life cycle for its projection and manufacturing and document accompanying in the item exploitation.

Digital and smart factories information and telecommunication systems have [5, 6] two main components: technical means and program means to support net technologies of the company. Technical means of the information and telecommunication system could be [7, 8]:

– company server equipment for deployment of project and production data storage;

– routers, splitters, brandmauers and other communication means to organize the interaction of personnel in local and global nets of the company;

– personal computers (instrumental computer) and office equipment (plotters, printers and other) which is the part of the designer automatic work places for technical documentation;

– technical equipment (cyber and physical systems) oriented to function in automatic production sections, control programs and data which are transferred in the company net;

– individual communicators for production operators to access the remote net resources;

– physical lines of information transfer or fiber optical channels and a device which receives and sends for wireless data exchange among the company personnel and other.

Program means of the information and telecommunication system could be [9, 10]:

– automatic projection systems to develop and create construction (CD), program (PD) and technological (TD) documentation for the item;

– system of automatic control of the production technological preparation when they launch the new production lines;

– automatic control system of item engineering data to support the company production activity;

– budget system and electronic document creation in a company and other.

To make technical and program means function synchronically they have to develop and implement new solutions [11, 12] oriented to work with the digital economy standards.

2. Digital factory information and telecommunication system

Industry 4.0 digital factory is a new type of projection company oriented to prepare item digital technical documentation. The digital factory scheme of information and telecommunication system is given in figure 1.

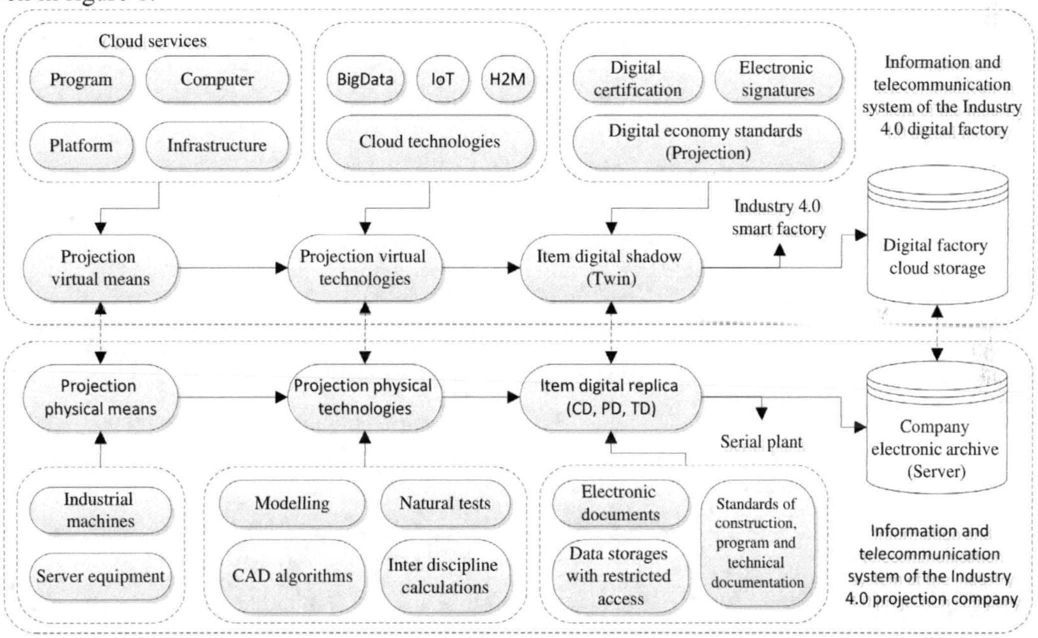

Figure 1. Industry 4.0 digital factory scheme of information and telecommunication system (CAD - Computer Aided Design).

Digital factory information and telecommunication system unites two components: virtual and physical. Physical component of information and telecommunication system could be [13, 14]:

– physical means of projection of items (industrial computers, server equipment and other) which is the part of device equipment designer automatic work place;

– physical technologies of projection which designers apply to create the item technical documentation. Physical technologies also include inter-disciplinary calculations, mathematical modelling, results and means of natural tests, algorithm of project activity automatizing and other;

– physical result of digital factory project activity is an item digital replica.

Item digital replica is prepared according to the requirements of:

– unified system of construction documentation;
– unified system of program documentation;
– unified system of technical documentation.

Virtual components of information and telecommunication system could be:

– virtual means of projection which are cloud services of program, virtual machine, platform, infrastructure types;

– virtual technologies of projection to support informatively the company project activity. The virtual technologies of projection include the technology of industrial Internet of Things (IoT), BigData, Humane-to-Machine (H2M) technology, and other;

– project activity virtual result is the item digital shadow (digital twin).

Physical and virtual results of the project activity is a general digital product which will be transferred to the smart factory to be manufactured and stored in archive of the digital factory. Digital factory archive as a physical machine is a server base and in the virtual level it is a data cloud storage.

3. Smart factory information and telecommunication system

Industry 4.0 smart factory is a new type of production company oriented to create item in automatic mode. The smart factory scheme of information and telecommunication system is given in figure 2.

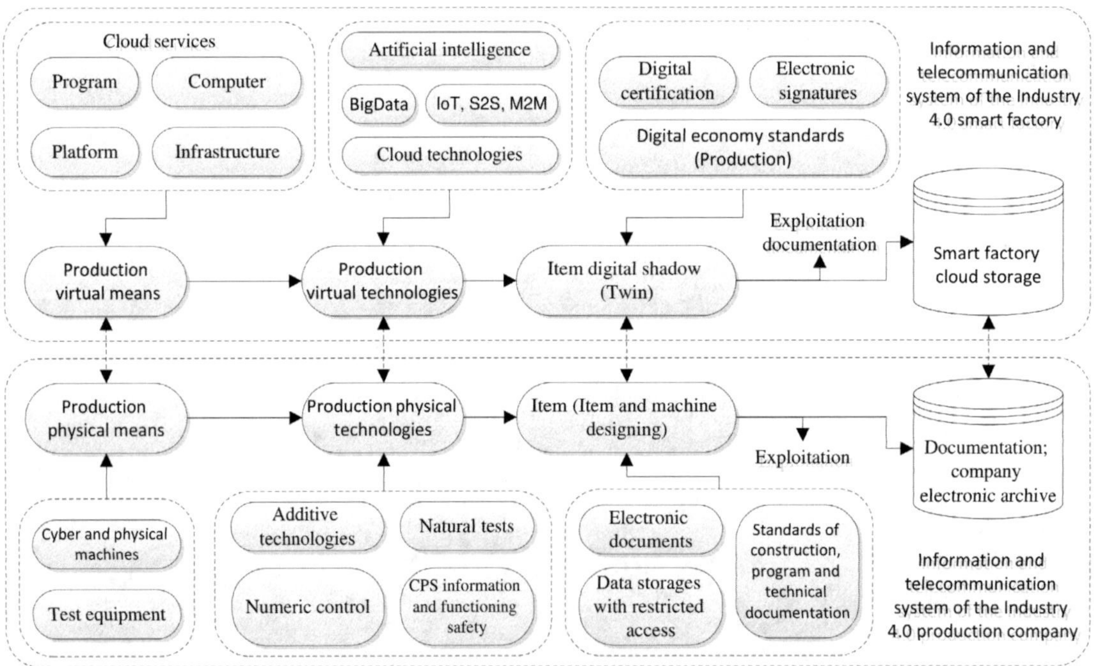

Figure 2. Industry 4.0 smart factory scheme of information and telecommunication system.

Smart factory information and telecommunication system unites two components: virtual and physical. Physical component of information and telecommunication system could be:

– physical means of production (cyber and physical equipment to manufacture the item, test equipment to check the item resistance to the external influencing factors and other);

– physical technologies of production to organize the functionality of technological sections in automatic mode to manufacture the item without humans. Physical technologies of production include additive technologies, technologies of cyber and physical systems (CPS) safety, technologies of numeric control and other;

– physical result of the smart factory production activity is a ready for use item (a group of items) with its electronic documents for exploitation.

Virtual component of information and telecommunication system could be:

– virtual means of production which are cloud services of program, computer, platform, infrastructure types;

– virtual technologies of production to manufacture items automatically. Virtual technologies of production include artificial intelligence, industrial Internet of Things, technologies of Systems-to-Systems (S2S), Machine-to-Machine (M2M), BigData and other;

– virtual result of production being developed in time to create an item digital shadow (digital twin). Item digital twin is done according to the standards of digital economy which is popular for the production activity with electronic signatures of technological operations executives and digital quality certifying system of the item being manufactured.

Physical and virtual results of Industry 4.0 smart factory production activity is a general digital which is the item and its informative model being developed in time. Item informative model is completely similar to the original item and is stored in the smart factory cloud. The development of informative model (digital twin) is explained with exploitation environment and its influence on the item and model adaptation mechanism (technical documentation) of the new factors influences which were never met before in a digital factory or a smart factory of Industry 4.0.

Smart factory cloud in physical level is a server which is the part of the company electronic archive.

4. Conclusion

Information and telecommunication system of the project and production activity of Industry 4.0 company is an infrastructure object to unite physical and virtual components. Physical and virtual components help each other and are informatively the same based on application of adequate physical devices of mathematical models. The division of objects and technologies to the components that match is done with dualism approach to create perspective companies oriented to function with digital economy standards.

Development of Industry 4.0 company information and telecommunication system is based on implementation to the infrastructure new protocols to transfer the data and new formats of data which are good for methods and means of the numeric control program for a cyber and physical production.

The application of information and telecommunication systems in item designing companies helps to organize the key technological operations completion without humans. Human personnel role in production is to control remotely with communication environment means.

References

[1] Pereira T, Barreto L and Amaral A 2017 *Procedia manufacturing* **13** 1253–60
[2] Ancans A, Petersons E, Ancans G, Stetjuha M, Ipatovs A and Stankevicius E 2019 *Procedia computer science* **149** 206–214
[3] Kurganov V, Gryaznov M and Dorofeev A 2018 *Transportation research procedia* **36** 392–397
[4] Asaul A, Malygin I and Komashinskiy V 2017 *Transportation research procedia* **20** 25–30
[5] Ordonez A, Eraso L, Ordonez H and Merchan L 2016 *Procedia computer science* **95** 353–360
[6] Gurjanov A V, Zakoldaev D A, Shukalov A V and Zharinov I O 2018 *IOP Conference Series: Materials Science and Engineering* **327** 022110
[7] Zakoldaev D A, Shukalov A V, Zharinov I O and Zharinov O O 2018 *Journal of Physics: Conference Series* **1015** 052033
[8] Kuhi K, Kaare K K and Koppel O 2015 *Procedia engineering* **100** 1348–53
[9] Sivri S D and Krallmann H 2015 *Procedia engineering* **100** 1032–39
[10] Panchenko S, Trubchaninova K and Korago I 2019 *Procedia computer science* **149** 177–184
[11] Lim Ch-H, Kim M-J, Heo J-Y and Kim K-J 2015 *Procedia CIRP* **30** 72–77
[12] Bogatyrev V A, Parshutina S A, Poptcova N A and Bogatyrev A V 2016 *Communications in*

computer and information science **678** 337–348
[13] Haug K C, Kretschmer T and Strobel T 2016 *Telecommunications policy* **40**(4) 291–306
[14] Fernandez-Miranda S S, Marcos M, Peralta M E and Aguayo F 2017 *Procedia manufacturing* **13** 1229–36

IOP Conf. Series: Materials Science and Engineering **656** (2019) 012059 doi:10.1088/1757-899X/656/1/012059

Client-server technologies at the enterprises of Industry 4.0

D A Zakoldaev[1], A V Gurjanov[2], A V Shukalov[1] and I O Zharinov[1]

[1]Faculty of Information Security and Computer Technologies, Saint Petersburg National Research University of Information Technologies, Mechanics and Optics, 49, Kronverksky Av., Saint Petersburg, 197101, Russia
[2]Director, Stock Company «Experimental Design Bureau «Electroavtomatika» named after P A Yefimov, 40, Marshala Govorova St., Saint Petersburg, 198095, Russia

E-mail: mpbva@mail.ru

Abstract. The task is to project the distributed architecture of calculation net for a digital and smart factory of Industry 4.0. The calculation net architecture defines the calculation resources components and their interaction to organize designer project activity or to manufacture a set of items in a serial plant. You may use in Industry 4.0 factory calculation net architecture client and server technologies. All factory technological equipment is classified by the type of the operation being completed: thin client, thick client, universal client and server equipment. Thin client of cyber and physical production is the simplest communication device with installed software which functions as a terminal and being used by the operator. Thick client of cyber and physical production is a complete production machine (test equipment) which is capable to work automatically by the programed algorithm. Universal client of cyber and physical production combines the properties of thin and thick clients. A cyber and physical system is a universal client example of the smart factory calculation net. Digital factory net thin client is a communication device with software which is used to control the modes of digital models virtual tests. Digital factory net thick client is a work station to conduct digital models virtual tests with your own calculation resources. Digital factory net universal client is a PC which is a part of the designer automatic work place and which is used to develop and create construction, program and technological documentation and also item digital models. Server equipment is a terminal server, data server and application server to exchange the information in multi-level hierarchy calculation net. Some schemes of digital and smart factories calculation net architecture are given.

1. Introduction

The implementation of digital informative technologies into item designing company projection activity and serial plants production activity led to [1, 2] the necessity to enhance the company calculation net. Traditional ways [3, 4] of distributed calculations and client exchange of project and production data were meant to create in a company a local calculation net with a central server for PC (Personal Computer) requests incoming to the designer automatic work places and requests from the technological equipment. This option of company calculation nets now is out of any possibility being developed further on the modern stage.

Perspective companies of Industry 4.0 (digital factory, smart factory) oriented [5, 6] to create distributed calculation nets the infrastructure elements of which are placed in the company cloud environment and as a part of production workshops technical support. The main client of calculation nets of Industry 4.0 project and production companies could be [7, 8]:

– cyber and physical systems for automatic completion of item manufacturing technological operations;

– automatic test equipment for physical tests of item being manufactured to prove the item quality;

– automatic work places PC with technical and software means to help the designer to prepare the item documentation;

– work stations (highly-productive industrial computers) which are used for digital models virtual tests with imitation modelling to prove the quality of item technical documentation;

– communication devices for project company designers and production company operators to control the item life cycle stages and other.

Different technical means and their software installed requires some new solutions [9, 10] form the calculation net architecture designers to optimize the net load among clients and information traffic of course circulated in the company. The best solution for project and production company calculation net architecture is to implement in the Industry 4.0 company infrastructure the client and server technologies [11, 12].

2. Smart factory distributed calculation net architecture

To unite different means of production with its integrated calculation resources into a single cyber and physical workshop with its own virtual environment they may use server and client technologies. Client and server technologies define the interaction order of clients into the company calculation net which are part of final devices category (clients) and also clients which are part of control devices category (servers) [13].

The scheme of distributed calculation net to unite the clients of the Industry 4.0 cyber and physical production is given in figure 1.

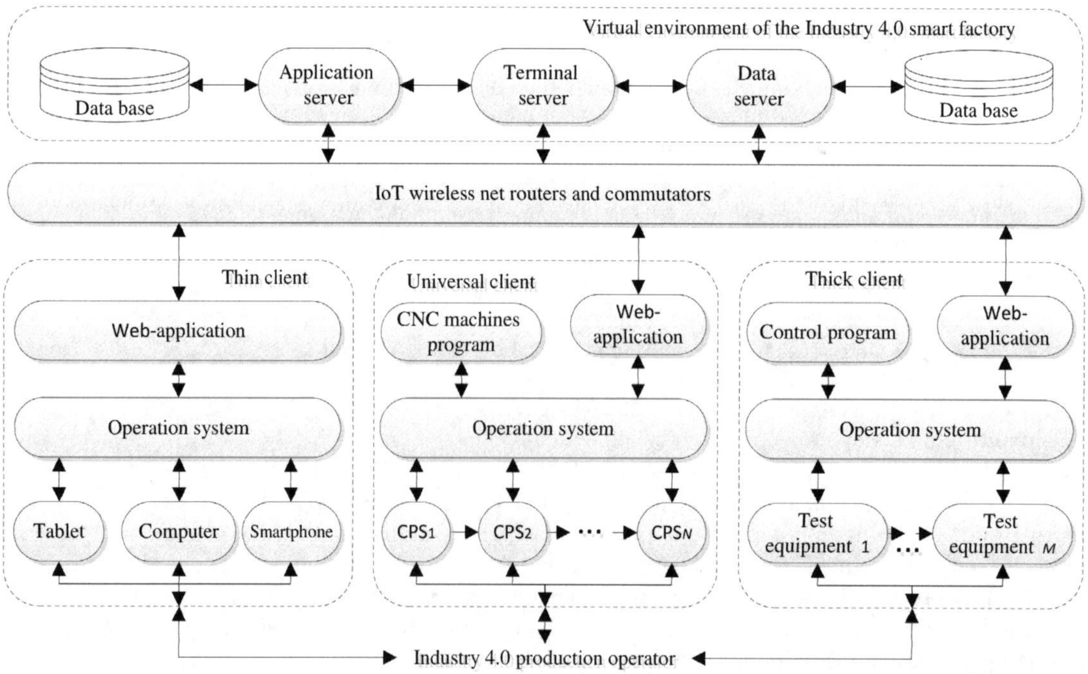

Figure 1. Calculation net (architecture) of Industry 4.0 smart factory cyber and physical production (CNC – Computer Numerical Control, CPS – Cyber-Physical System, IoT – Internet of Things).

Calculation net architecture in its physical level of cyber and physical production can be described with three main types of final devices:

– thin client;
– thick client;
– universal client

united in a single wireless net of IoT with routers, commutators and other net equipment.

Hardware and software classification of the smart factory calculation net to client and server components is necessary to divide the calculation load and functions among information exchange process participants in production.

Thin client is a final calculation device with installed communication application (web-application) which provides production operator with the possibility to work in the terminal mode. Web-application is controlled with local operation system to load up and storage into the memory in its solid devices. Thin client contains minimum configuration of calculation resources and interacts with remote terminal server in the client and server mode.

Communication software gives the operator in production the human and machine interface of interaction described graphically in the user environment. To show the operator visually the results of user requests for communication software and also they may use the elements of augmented reality and virtual reality. A thin client cannot do mathematical operation to process project and production data and the requests of smart factory operator it can only re-direct all requests to the terminal server directly where they will be completed. A thin client example in cyber and physical production is a communicator (tablet, smartphone, laptop and other) functioning in a session with the terminal server with IoT protocols and which the operator can use to monitor the conditions of technological operations completion and situation control of the smart factory production activity. So the production operator is the initiator of an information exchange and the company calculation net which are done with software and technical means of a thin client.

A thick client is an end calculation device with installed application software oriented for the information processing. The thick client interacts with IoT net with a server by the technology of client and server. For a thick client a server is a remote data storage (production data, technological process algorithms, programs of numeric control and other) and a thick client process all the data and show the results to the operator with its own calculation resources. A thick client example in the Industry 4.0 production is a piece of automatic test equipment which with commands of the control program downloaded from the factory cloud environment initiates a check of the manufactured item resistance to the external factors influence. Work mode of test equipment is programmed by the smart factory operator according to its technical documentation. Functionality of compressors, gears, refrigerator and other test equipment is controlled with modes and algorithms as some codes of the control program.

A universal client is an end calculation device which is capable to work with the operator as a terminal and is capable to produce items automatically with integrated calculation means. A universal client example in Industry 4.0 smart factory client and server architecture is a cyber and physical system. Thick client function of cyber and physical system is supported with integrated controller to complete technological operations in the work chamber with algorithms of numeric control. CNC program is placed in a remote application and is delivered to the cyber and physical systems through the net according to the technological operations to be completed in production to manufacture the item. Thin client function in cyber and physical system is a terminal interface placed in the CPS control which is used for CPS programming by the operator in production.

The server part of the smart factory net infrastructure has a terminal server, application server and project and production data server placed in the company cloud environment (virtual). Application server and data server gives the operator and cyber and physical and test equipment access which are active internet agents to the company data bases. If the server part is divided in some independent servers it increases significantly the productivity of cyber and physical production calculation net in general. The server part software supports the multi-user mode of information exchange so that why server resources of the smart factory cloud environment may connect a significant number of cyber and physical systems.

Client and server technologies to organize smart factory calculation net help to create environments of information exchange which can be scaled. Calculation net scaling is of module nature of the net constitution and it helps to increase the number of clients from different platforms and installed software without making the net topology being too complicated. In this case you need to make work together platforms and applications as a part of the same production company. To unite cyber and physical, test and communication equipment from different manufacturers and of different hardware platforms and operation systems (Windows, Linux, iOS and other) into a calculation net they need single communication protocols (for example TCP/IP protocols, Transmission Control Protocol/Internet Protocol). Universal communication protocols unify operator access methods and production equipment methods to the data with standard interface supported with hardware and software (communication software, operation systems and other) of cyber and physical systems.

3. Digital factory distributed calculation net architecture

A scheme of Industry 4.0 digital factory distributed calculation net architecture is given in figure 2. The net main components are: server equipment, thin, thick and universal clients connected to the wireless IoT net.

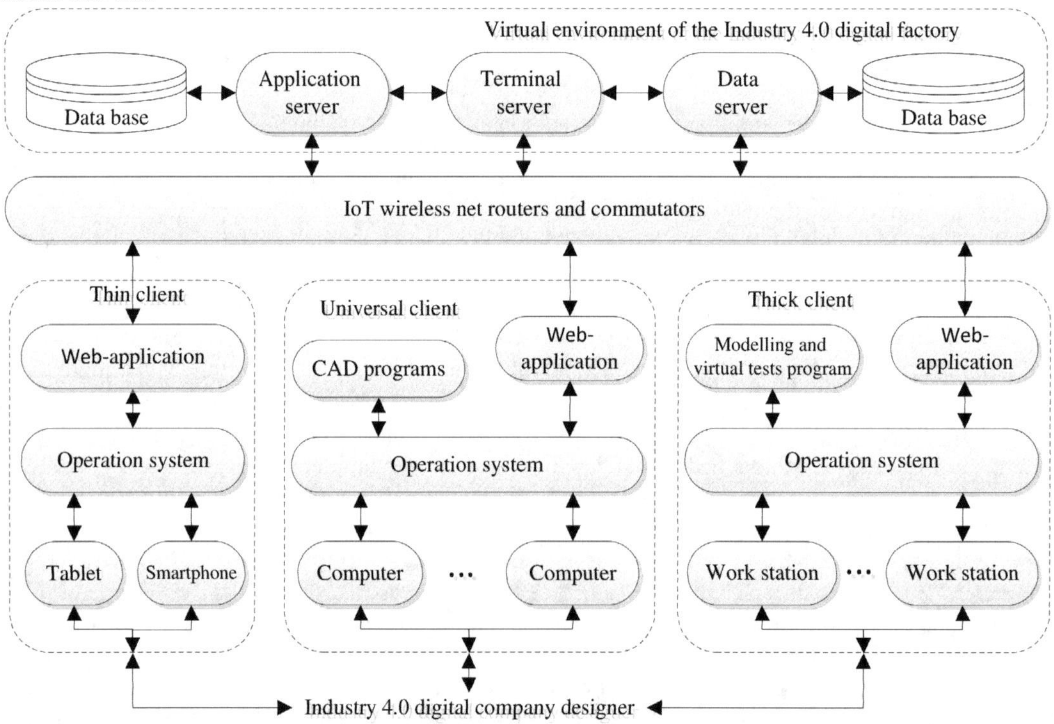

Figure 2. Industry 4.0 digital factory distributed calculation net architecture (CAD – Computer Aided Design).

Digital factory thin client is a communication device (tablet, smartphone) with installed web-application which is controlled by the operation system. Digital factory thin client function is to support designer project procedure oriented for digital replicas tests, item digital shadow and digital twin with modelling and virtualization. Web-application is for designer man-machine interface which supports Humane-to-Machine (H2M) technologies to control remotely modes and types of the item tests in the company.

Digital factory thick client is a work station (industrial computer) with its own calculation resources which is used for virtual tests of item digital replica, digital shadow and digital twins. Virtual tests are conducted with digital modelling placed in solid devices of universal clients. Virtual tests of digital models are done with tests and scenarios placed in server equipment. The virtual tests are controlled by the digital factory designer with technical and program means of a thin client.

Digital factory universal client is a computer (instrument PC) which is a part of the designer automatic work place with installed software for:

– automatic item designing and creation of construction, program and technological documentation;

– preparation of item 3D (D - Dimension)-model and its components and other.

Web-application of universal client provides the interaction of Industry 4.0 digital company designer and the server equipment to set up program components of automatic work place.

4. Conclusion

Industry 4.0 project and production company calculation net productivity is depended significantly on the pass-through ability of the wireless channels of IoT connection and the server equipment calculation resources. Also the productivity is highly depended on the number of net clients, transferring traffic volume and the number of hierarchy levels of calculation resources to divide company business logistics, project and production data, program applications and clients.

The increased network productivity in the net level must be done with implementation in the company not only wireless nets but also optical fiber nets to transfer the data. Optical fiber net can support data transferring industrial standards with Ethernet protocols and can be used to connect the stationary cyber and physical system and test equipment in production. An end communication device (tablet, smartphone and other) may preserve the information exchange with wireless channels of IoT connections.

Structural decisions how to increase calculation net productivity in the company is based on division of cloud environment server part into several independent components. Those components of high productivity are for user requests functioning in terminal mode and client serving who ask for company data base information can be done with application servers and data servers. This kind of approach will help to implement in companies highly-productive calculation nets which traffic among servers and clients is controlled with program and technical means (routers, commutators and other).

References

[1] Khan I, Rehman H, Al-khatib M H, Anwar Z and Alam M 2018 *Future generation computer systems* **89** 239–248

[2] Deboosere L, Simoens P, Wachter J D, Vankeirsbilck B, Turck F D, Dhoedt B and Demeester P 2011 *Future generation computer systems* **27(6)** 681–693

[3] Hocking M 2011 *Network security* **6** 17–19

[4] Tang W, Lee J-H, Song B, Islam M, Na S and Huh E-N 2011 *Procedia environmental sciences* **11** 499–504

[5] Gurjanov A V, Zakoldaev D A, Shukalov A V and Zharinov I O 2018 *IOP Conference Series: Materials Science and Engineering* **327** 022110

[6] Zakoldaev D A, Shukalov A V, Zharinov I O and Zharinov O O 2018 *Journal of Physics: Conference Series* **1015** 052033

[7] Chongdarakul W, Sophatsathit P and Lursinsap C 2018 *Future generation computer systems* **78** 901–919

[8] Freeman O and Pattinson H M 2010 *Futures* **42(4)** 304–312

[9] Ting D 2011 *Network security* **2** 16–17

[10] Browne D, Kempf R, Hansen A, O'Neal M and Yates W 2013 *Procedia computer science* **16** 373–382

[11] Varekova P, Varekova I and Cerna I 2010 *Electronic notes in theoretical computer science*

260 243–259
[12] Bogatyrev V A, Parshutina S A, Poptcova N A and Bogatyrev A V 2016 *Communications in computer and information science* **678** 337–348
[13] Wu M-E, Chang Sh-Y, Lu Ch-J and Sun H-M 2014 *Information sciences* **275** 348–359

Methods for processing structures based on solid solutions of $A^{III}B^{V}$ compounds

A V Zhelannov[1,3], B I Seleznev[2] and D G Fedorov[1]

[1]SC OKB-Planeta, Russia, 173004, Velikiy Novgorod, ul. Bolshaya Moskovskaya, 13a

[2]Yaroslav-the-Wise Novgorod State University, ul. B. St. Petersburgskaya, 41 173003, Velikiy Novgorod, Russia

[3]E-mail ZhelannovAV@okbplaneta.ru

Abstract. The technologies of plasma-chemical etching of $Al_xGa_{1-x}N$ epitaxial layers in chlorine-containing mixtures, which allow forming a microrelief on the surface of a semiconductor in the range from tens of nanometers to tens of micrometers, are presented. The modes of selective etching of GaN layers with respect to $Al_xGa_{1-x}N$ ($x = 0.25$) are determined.

1. Introduction

When developing device structures based on the $Al_xGa_{1-x}N$ solid solutions system, one of the key tasks is the development of a technology for local material removal. Such a technology is necessary for the formation of mesa-insulation of active regions of device structures, recesses for ohmic contacts [1] and for gate metallization.

Both chemical (“fluid”) and plasma (“dry”) etching methods can be used for etching of materials of the $A^{III}B^{V}$ group. "Liquid" methods of removing material involve the use of liquid solutions for chemical etching of masked areas of a semiconductor. These methods are simple to implement and do not require sophisticated technological equipment. But when working with AlGaN / GaN heterostructures, the “liquid” method did not become popular, because of the strong chemical bonding in nitride semiconductors, it is difficult to remove solutions that chemically etch the material. It was reported about photo-stimulated electrochemical etching of GaN, however, this method depends on many parameters and is poorly reproducible [2, 3].

A significant limitation of "liquid" etching methods for group III nitrides has caused considerable interest in the development of dry etching methods. Various methods of dry etching related to ion influence are used for the treatment of nitride semiconductors. Among them are ion etching, reactive ion etching, reactive ion etching using electron-cyclotron resonance and reactive ion etching in inductively coupled plasma, chemical ion-beam etching. As the most effective method of plasma etching, reactive ion etching with a source based on inductively coupled plasma can be distinguished, which allows to obtain a dense plasma with high homogeneity and controllability.

2. Experimental technique

The aim of this work is to obtain the technology of deep etching of gallium nitride using a mixture based on $Cl_2/BCl_3/Ar$ with rates of about 0.8–1.0 μm/min, and the technology of selective etching of GaN relative to AlGaN in a gas mixture based on $Cl_2/Ar/O_2$.

To establish the modes of gallium nitride etching, experimental studies of reactive ion etching were performed on a Sentech SI-500 unit (figure 1) equipped with a source of inductively coupled plasma. The testing of etching processes involved the selection of modes that ensure sufficiently high etching rates (up to 1 μm / min) to remove the semiconductor to a depth of 10 μm. As test samples, structures with an undoped layer of gallium nitride and a thickness of ~ 2.0 μm, grown by chemical vapor deposition using organometallic compounds on a sapphire substrate 2 inches in diameter, were used.

When choosing masking coatings, it is necessary that they have chemical resistance to aggressive plasma gaseous media, have high adhesion to the substrate, and are easily removed after etching without changing the semiconductor surface morphology [4].

In our experiments, a combined mask consisting of nickel (Ni) with a thickness of 0.25 μm and a sublayer of silicon dioxide (SiO_2) with a thickness of 0.3 μm was used. The sequence of technological operations for the formation of the mask is presented in figure 2.

Figure 1. Standard Sentech SI-500 etching unit.

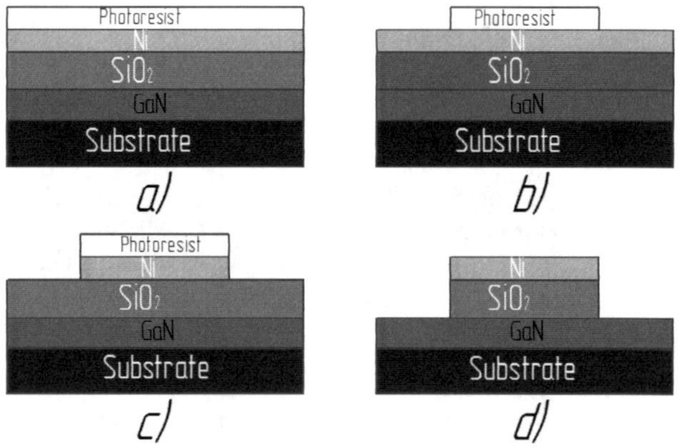

Figure 2. Formation of a SiO_2/Ni mask for etching gallium nitride: a – application of photoresist; b – creating a mask from photoresist; c – nickel ion etching; d – plasma-chemical etching of silicon dioxide and removing photoresist.

Mask formation begins with plasma-chemical deposition of a SiO_2 film. Silicon dioxide films were grown using monosilane with a flow rate of 155 cm^3/min and oxygen with a flow rate of 8.2 cm^3/min with an inductively coupled plasma source power of 150 W, high-frequency power 5 W, pressure in the chamber 5 Pa and temperature 250° C. The nickel film was sputtered by the magnetron method with the substrate heated to 250° C. After that, using optical photolithography, a photoresist mask was formed (figure 2a), which consists of alternating open and closed areas of the surface. To preserve the anisotropic profile, the removal of nickel and silicon dioxide layers was carried out using dry etching methods, namely: ion etching for nickel and plasma-chemical etching for SiO_2 (figure 2c). After etching the SiO_2/Ni layers, the photoresist was removed in dimethylformamide (figure 2d).

In the course of experiments on the etching of gallium nitride, the dependences of the etching rate on the parameters of the plasma process were established, including pressure in the reactor, gas flow rate, power of the source of inductively coupled plasma, high frequency power.

3. Measurement results

3.1. Effect of pressure in the chamber on the etching parameters

The pressure in the chamber affects the basic processes that are responsible for plasma etching. The mean free path is inversely proportional to pressure, so that the potential of the shell increases the energy of the ions bombarding the sample. The etching rate increases with increasing ion bombardment and therefore lower pressure may contribute to etching.

Increasing the pressure in the chamber leads to a decrease in the free path of the ions and an increase in the probability of electron and ion collisions; therefore, the plasma density decreases (figure 3).

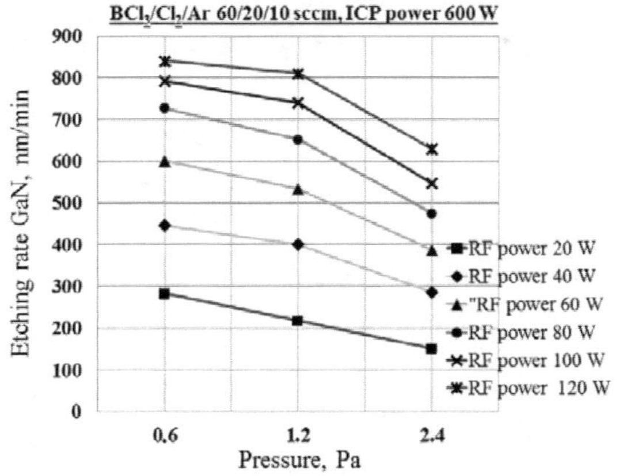

Figure 3. The dependence of etching rate of gallium nitride on the pressure in the chamber.

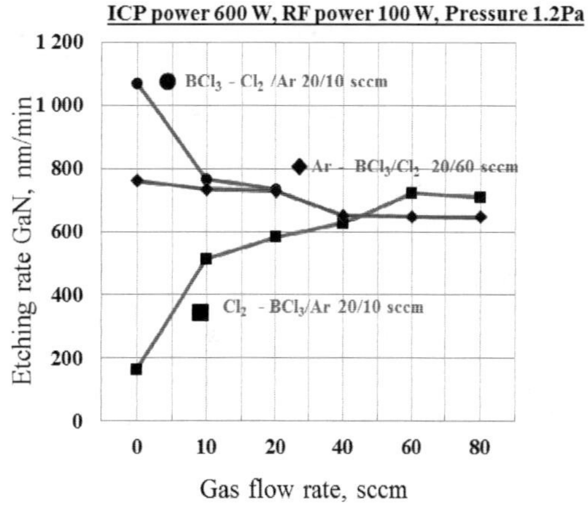

Figure 4. Dependence of etching rate of gallium nitride on the flow rate of the gas mixture components.

The etching rate of gallium nitride decreases monotonically with increasing pressure in the chamber from 0.6 Pa to 2.4 Pa, due to a decrease in the concentration of Cl_2 radicals in the plasma, or a decrease in the flux of ions coming to the surface of the substrate.

3.2. Effect of gas mixture flow on etching parameters

The correct choice of reactive gas or mixture of gases has a strong influence on the etching rate. However, the selection of the optimal gas environment is determined not only by the productivity of the process, but also by the achievement of high selectivity of etching. The etching rate increases rapidly with increasing feed rate or gas flow rate (figure 4).

It reaches a maximum, and then decreases with a further increase in gas consumption. Low etching rate at low gas consumption is determined by the insufficient number of chemically active particles formed in the discharge due to the lack of the original substance. The drop in rate at high flows can be explained by the fact that the active plasma particles do not have time to interact with the material being processed.

3.3. Effect of inductively coupled plasma source power on etching parameters

Figure 5 shows the change in the etching rate of gallium nitride with an increase in the power of an inductively coupled plasma source (ICP power). ICP power in the plasma process mainly regulates plasma density. As the plasma density increases with increasing ICP power, more reactive particles will be applied to the substrate, increasing the chemical aspect of etching. This is accompanied by the increase in etching rate. As can be seen in Figure 5, the etching rate of gallium nitride increases with~ 15 nm/min to ~ 545 nm/min with an increase in ICP power from 25 to 700 W at a high-frequency power value (RF power) of 20, 40 and 60 W.

IOP Conf. Series: Materials Science and Engineering **656** (2019) 012060 doi:10.1088/1757-899X/656/1/012060

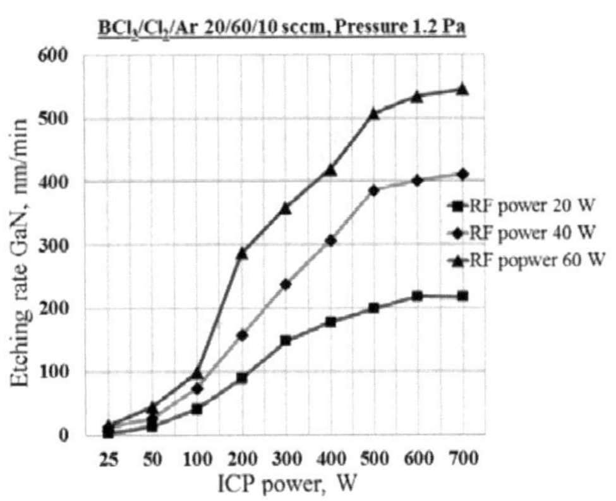

Figure 5. Dependence of the rate of gallium nitride etching on ICP power

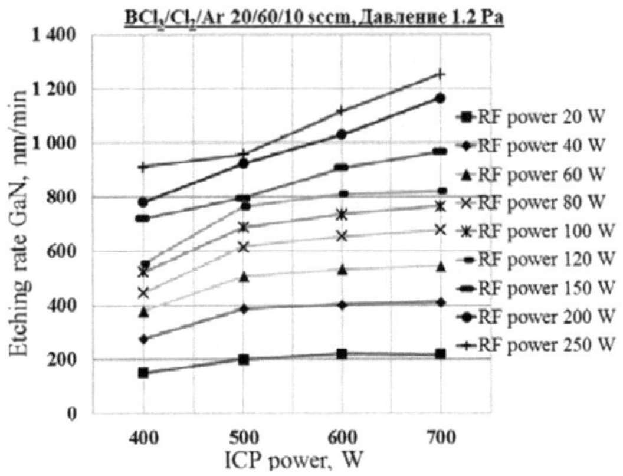

Figure 6. Dependence of etching rate of gallium nitride on ICP power

At the same time, as can be seen from Figure 6, for small values of high-frequency power (up to values of about 60 W), the etching rate is saturated when the ICP power of 600 W is reached. The saturation of the etching rate in this case can be caused either by saturation of the reactive particles on the surface, or by desorption of reactive substances on the surface before the occurrence of the reactions.

3.4. Effect of high-frequency power on etching parameters

The high-frequency power of the system increases the energy of the electrons and, therefore, increases the probability of ionization. As a result, the etching rate increases with increasing RF power (figure 7).

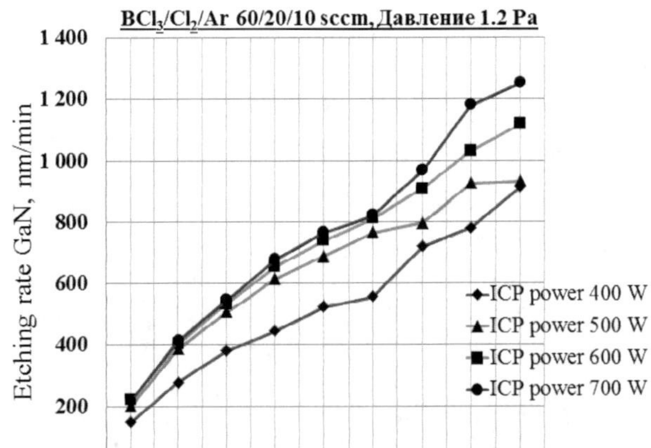

Figure 7. The dependence of etching rate of gallium nitride on RF-power.

Figure 7 presents the dependence of the etching rate of gallium nitride on RF power, where it is shown that the etching rate monotonously increases with increasing RF power. The etching rate is increased by increasing the contribution of the physical component of etching in combination with a strong component of chemical etching. The increase in the etching rate can be explained by the improved sputtering of the etching products, as well as by more efficient breaking of bonds in the semiconductor with an increase in the ion energy.

3.5. Selective etching of GaN / AlGaN for creating device structures

The etching modes discussed above do not provide for etching selectivity between gallium nitride and AlGaN solid solution. The development of such a process is necessary when using heterostructures with protective heavily doped layers during the formation of areas for barrier metallization [5]. The selectivity of etching of GaN/AlGaN heterostructures, as a rule, can be achieved by introducing oxygen into the gas mixture Cl_2/Ar. As soon as the AlGaN layer is exposed, the oxygen contained in the plasma reacts with aluminum to form an oxide barrier on the surface. This prevents further etching and also protects the surface from ion bombardment [6].

To implement this kind of process, structures with alternating GaN (80 nm)/$Al_{0.25}Ga_{0.75}N$ (80 nm)/GaN (2.2 μm) layers were used.

The samples were etched in Cl_2/Ar medium with the addition of various amounts of oxygen (from 0 to 10 cm^3/min) with an inductively coupled plasma source power of 75, 100 and 200 W and RF power 10, 30, 60 and 90 W. The chamber pressure is 1.2 Pa with a Cl_2/Ar flow of 60/10 cm^3/min. Figures 8 and 9 show the dependences of the etching selectivity on the oxygen consumption.

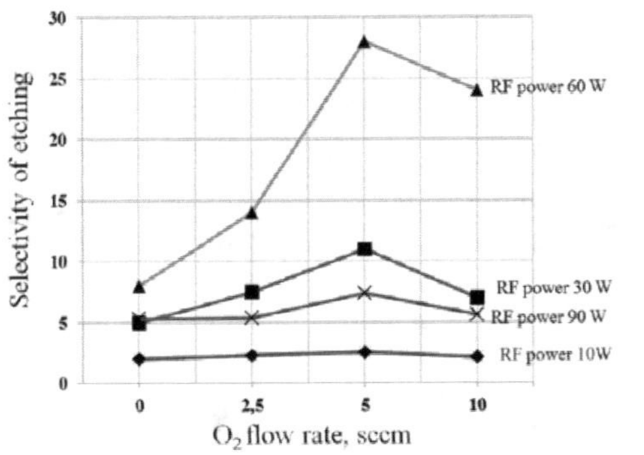

Figure 8. Dependence of GaN/AlGaN etching selectivity on oxygen consumption at various RF power values.

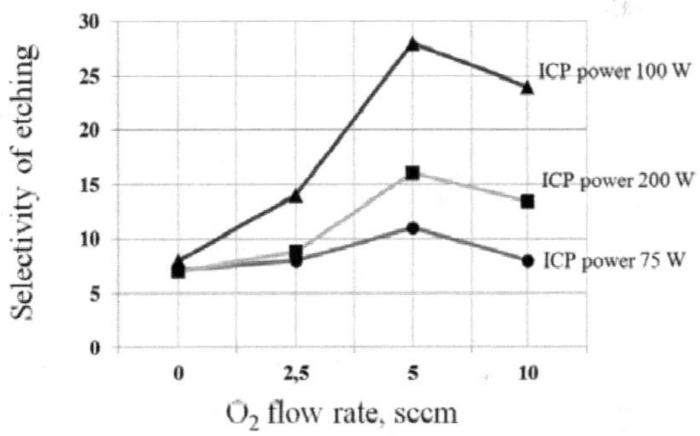

Figure 9. The dependence of the selectivity of etching of GaN/AlGaN from the consumption of oxygen at different values of ICP power.

As it can be seen from the plots, the addition of oxygen to the gas mixture reduces the etching rate of AlGaN. This is due to the oxidation of the AlGaN layer upon interaction with oxygen and the formation of a compound that is resistant to etching. In this case, the dependence of the etching selectivity on the oxygen consumption has a maximum at a flow rate of 5 cm^3/min and is about 30:1. A further increase in oxygen consumption leads to a decrease in selectivity due to a decrease in the etching rate of gallium nitride. In this case, the magnitude of the selectivity is also affected by the magnitude of the high-frequency power. Since the RF power determines the physical etching mechanism, then, at a certain its value, Al_2O_3 compound will be also etched. Thus, there is a maximum in the dependence of the etching selectivity on high-frequency power, as shown in figure 10.

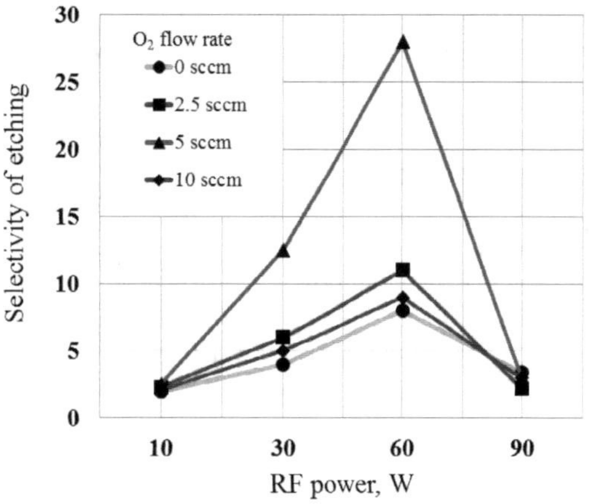

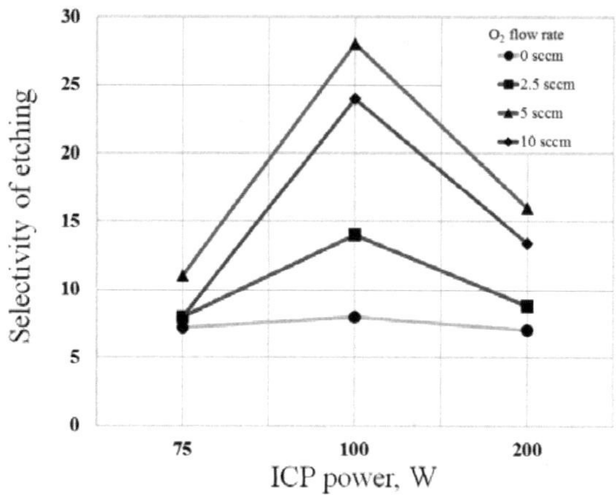

Figure 10. Dependence of GaN/AlGaN etching selectivity on RF power.

Figure 11. Dependence of GaN/AlGaN etching selectivity on ICP power.

The influence of the power of an inductively coupled plasma source on the rate and selectivity of etching of GaN and AlGaN was also investigated (figure 11). Other parameters of the etching mode remained constant: RF power of 30 W, a Cl_2/Ar flow of 60/10 cm^3/min, a chamber pressure of 1.2 Pa. The ICP power value in this case was varied from 75 to 200 W with different oxygen flow (0 and 5 cm^3/min). As one can see from the plots, the dependence of the rate and selectivity of etching has a maximum at an ICP value of 100 W. Smaller power values do not provide the necessary etching rates due to sufficiently low plasma density. When the power value is greater than 100 W, the AlGaN etching rate increases due to the denser plasma and, accordingly, the selectivity decreases.

4. Conclusion

Thus, in the course of the work, the processing modes of materials of the $A^{III}B^{V}$ group have been developed, allowing to obtain a surface microrelief in the range of up to ten micrometers. The influence of the main parameters of the plasma process on the etching rate of a semiconductor is considered. The modes of selective etching of GaN with respect to $Al_xGa_{1-x}N$ (x = 0.25) were studied and the etching selectivity was obtained on the order of 30:1.

Acknowledgments

The publication was prepared with the financial support of the Ministry of Education and Science of Russia within the framework of the project part of the state task, project No. 3.3572.2017.

References

[1] Gryglewicz J *at al* 2016 *Advances in Electrical and Electronic Engineering* **14(2)** 218-222
[2] Ma L *at al* 2006 *Proc. of the CSMANTECH Conf.* 105–108
[3] Usikov A S 2016 *Tech. Phys. Lett.* **42(9)** 80–87
[4] Protasov D Yu 2014 *Tech. Phys.* **84(9)** 96–99
[5] Wang W K 2005 *Electron Device letters* **26(1)** 5–7
[6] Han Y *at al* 2003 *Jpn. J. Appl. Phys.* **42(10A)** L1139–41

Neural network solution of the direct kinematics problem for a hexapod with ball-screw drives of legs

Yu A Zhukov[1], E B Korotkov[1], V V Zhukova[2] and A M Abramov[3,4]

[1]Baltic State Technical University "VOENMEH"
[2]Peter the Great St.Petersburg Polytechnic University
[3]Yaroslav-the-Wise Novgorod State University, ul. B. St. Petersburgskaya, 41 173003 Veliky Novgorod, Russia

[4]E-mail: Alexandr.Abramov@novsu.ru

Abstract. This research is a part of the work implemented by BSTU "VOENMEH" under the financial support of the Ministry of Education and Science of the Russian Federation for design and development of a precision mechanism with the parallel kinematics called "Hexapod". The object of research is the mechanism, in which the control of leg lengths is implemented by linear drives with ball-screw gears. The article proposes the solution of the extended inverse kinematics of the hexapod. The neural networks solution of the forward kinematics of hexapod is presented, i.e. the calculation of the position and orientation of the moving platform for given rotation angles of the ball-screw drives. The results of the synthesis of neural networks to achieve a given accuracy of solving the forward kinematics are presented. An estimate of the implementation of kinematic problems in the hexapod control system based on an ARM microprocessor is obtained.

1. Introduction

The researchers of BSTU "VOENMEH" and JSC "ISS" named after M.F. Reshetnev jointly created a multi-step mechanism with parallel kinematics [1, 2] (MPK) to ensure accurate positioning and orientation of on-board instruments and devices for space purposes [3].

The object of our research is the MPK ("Hexapod") based on six linear drives with screw gears, the design of which is shown in figure 1. The hexapod, designed according to the scheme of Stuart's platform [4], consists of a fixed base and a movable platform, controlled by six identical ball-and-screw drives – legs (bars, poles). Each leg is connected by universal joints with the base and platform. The leg of the hexapod consists of two half-bars, which are interconnected by a ball-screw drive. The kinematic diagram of the hexapod leg is shown in figure 2. Thus, the position and orientation of the platform is determined by the angles of rotation of the ball-screw gears α_{dr_i} of each leg of the hexapod.

The main task of the hexapod control system is to test the position of the platform set in Cartesian coordinates relative to the base with an accuracy of ± 10 μm and the orientation of the platform with an accuracy of ± 30 arc sec. Extreme working environment in conditions of open space defines strict requirements for the hardware of the hexapod. The control system is developed on the basis of a domestic radiation-resistant microcontroller [5]. Therefore, the tasks of assessing the feasibility and improving the quality indicators of the designed control algorithms are relevant and important.

When controlling a hexapod with screw gears, two kinematic problems, forward and inverse, are solved. In the majority of existing works on the subject of research, the solution of the inverse problem of kinematics [5, 6] is given as the calculation of the leg lengths of a hexapod for a given position and orientation of the platform. Thus, the inverse problem is used for planning trajectories with kinematic control of a hexapod and generating control signals with linear drives [7], for deriving the dynamic equations [6]. In [5] an algorithm for calculating control signals with stepper linear drives based on ball-screw drives is presented; an implementation of the inverse problem based on a domestic ARM-architecture controller is evaluated. The forward kinematics problem is defined [8] as the calculation of the position and orientation of the platform for given leg lengths. In [8, 9] the solution of the forward kinematics problem was obtained using iterative numerical methods. However, of particular interest is the approach based on artificial neural networks (ANN). In [10–14] approaches to the neural network solution of the forward kinematics problem of a hexapod and [13] methods of application in control problems of a hexapod are presented.

The kinematic tasks of the studied MPK with screw pairs differ from those discussed earlier: when controlling the MPK, the upper half bar ("upper leg", as shown in figure 2) rotates relative to the lower ("lower leg") at an angle γ, which affects the length of the leg. Therefore, for a hexapod with screw pairs, we introduce the following definitions of kinematic problems:

1) the extended inverse kinematics problem (EIKP) is to determine the rotation angles of the screw pairs α_{dr}, the relative rotation angles of the half-bar γ, and the leg lengths L from a given position and orientation of the moving platform;

2) forward kinematics problem (FKP) is to determine the position and orientation of the platform for given rotation angles of the screw pairs α_{dr}.

Of particular interest is to acquire the FKP of hexapod with screw pairs. There are no analytical solutions to the problem. In this paper, we propose to use the approximation capabilities of ANN to solve the FKP. Thus, the main purpose of this work is to obtain a neural network solution to the forward kinematics problem of a hexapod with screw transmissions and evaluate its feasibility in a control system based on the ARM microcontroller.

Figure 1. Hexapod design sample.

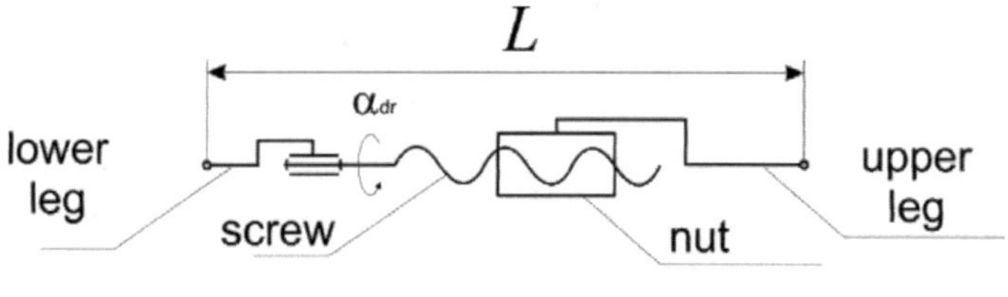

Figure 2. Kinematic diagram of hexapod leg with ball- screw transmission.

2. Hexapod kinematics

To solve the problems of kinematics, we introduce a coordinate system, as shown in figure 3. The fixed coordinate system OXYZ is connected to the base, the moving coordinate system O'X'Y'Z 'is connected to the platform. We define the initial "zero" position of a symmetrical hexapod, in which the legs have the same elongation. Thus, in this position, the O'X'Y'Z 'coordinate system with respect to the OXYZ coordinate system is shifted along the OZ axis by the h_0 parameter.

The length of the legs in the "zero" position will be defined as l_0 "zero length of the legs". We introduce the numbers of the legs, as shown in figure 3.

The position of the center O' of the platform relative to a fixed coordinate system is set using Cartesian coordinates *X, Y, Z*. Euler angles are used to determine the orientation of the platform [15] $\varphi,\ \theta,\ \psi$. Thus, the linear position and the angular orientation of the platform are given by the vector $\mathbf{q} = [\,X,\ Y,\ Z,\ \varphi,\ \theta,\ \psi]^T$.

The specified *working range of the hexapod* is: by coordinate X – ±100 mm, by coordinate Y – ±100 mm, by coordinate Z – ±25 мм, by angular coordinates ±7 deg.

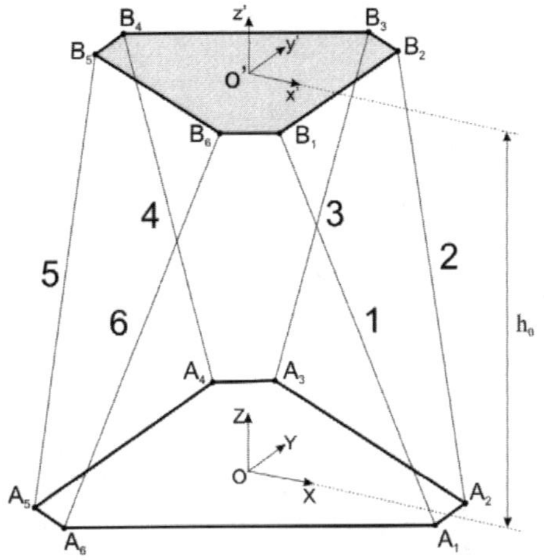

Figure 3. Kinematic diagram of hexapod.

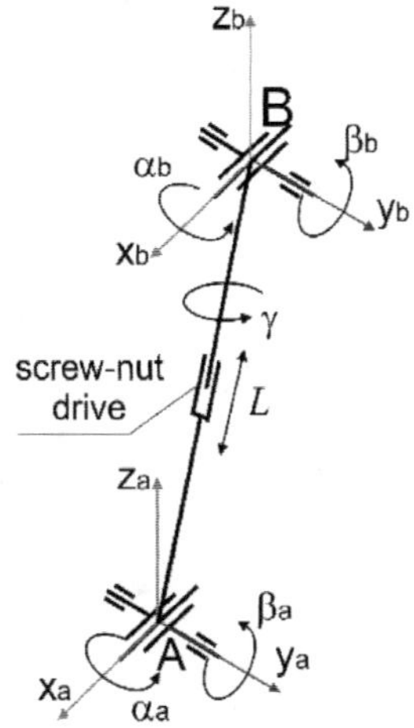

Figure 4. Leg coordinate systems with screw-nut transmission.

For the connection between the fixed and moving coordinate systems, we use the matrix transformation in homogeneous coordinates [15]:

$$\mathbf{R} = \mathbf{T} \times \mathbf{r'} \tag{1}$$

where $\mathbf{r} = [\,X,\ Y,\ Z,\ 1]^T$ is the vector of homogeneous coordinates of a point in the coordinate system OXYZ, T – index denoting vector transposition, $\mathbf{r}'$ is the vector of homogeneous coordinates of the same point relative to the moving coordinate system.

In the transformation (1), the expanded matrix **T** is defined by $\mathbf{p} = [\,X',\ Y',\ Z']^T$ – coordinates of the beginning of the moving system are relatively stationary - and the platform rotation matrix $\mathbf{R}_p$

$$\mathbf{R} = \begin{vmatrix} \boldsymbol{R_p} & & \boldsymbol{p} \\ \mathbf{0} & \mathbf{0} & \mathbf{0} & \mathbf{1} \end{vmatrix} \tag{2}$$

The rotation matrix for choosing Euler angles is defined by the expression

$$\mathbf{R}_p = \begin{vmatrix} \cos\varphi\cos\psi + \sin\varphi\sin\theta\sin\psi & \sin\varphi\sin\theta\cos\psi - \cos\varphi\sin\psi & \sin\varphi\cos\theta \\ \cos\theta\sin\psi & \cos\theta\cos\psi & -\sin\theta \\ \cos\varphi\sin\theta\sin\psi - \sin\varphi\cos\psi & \sin\varphi\sin\psi + \cos\varphi\sin\theta\cos\psi & \cos\varphi\cos\theta \end{vmatrix}, \tag{3}$$

Calculating from (1–3) the transformation matrix **T** for a given vector **q**, according to the coordinates of the points of attachment of the hinges of the legs to the base ($\mathbf{r}_{A_i}$ is the vector of homogeneous coordinates of a point A_i in a system OXYZ) and to the platform ($\mathbf{r}'_{B_i}$ is the vector of homogeneous coordinates of the point B_i relative to O'X'Y'Z'), we find the length of the *i*-th leg of the hexapod

$$L_i = \left\| \mathbf{r}_{A_i} - \mathbf{T} \times \mathbf{r}'_{B_i} \right\|. \tag{4}$$

For MPK with screw pairs, when implementing control tasks, it is necessary to calculate for each drive the angle of rotation of the screw-nut transmission α_{dr}, the value of which depends on the angle of rotation γ, as shown in figure 4.

Given the positions of the centers $\mathbf{p}_{A_i} = \left[x_{A_i}, y_{A_i}, z_{A_i} \right]^T$ and $\mathbf{p}_{B_i} = \left[x_{B_i}, y_{B_i}, z_{B_i} \right]^T$ of biaxial hinges A_i and B_i and the orientation of the biaxial hinges determined by the rotation matrices [15] $\mathbf{R}_{A_i}$ and $\mathbf{R}_{B_i}$, we formulate the matrix kinematics equation relating the transformations of turns of the coordinate systems of the i-th leg of the:

$$\mathbf{R}_{A_i} R_x\left(\alpha_{A_i}\right) R_y\left(\beta_{A_i}\right) R_z\left(\gamma_i\right) R_y\left(\beta_{B_i}\right) R_x\left(\alpha_{B_i}\right) = \mathbf{R}_{B_i}, \text{ where} \tag{5}$$

$R_x(\alpha) = \begin{vmatrix} 1 & 0 & 0 \\ 0 & \cos(\alpha) & -\sin(\alpha) \\ 0 & \sin(\alpha) & \cos(\alpha) \end{vmatrix}$ — matrix of rotation by angle α relative to the axis x,

$R_y(\beta) = \begin{vmatrix} \cos(\beta) & 0 & \sin(\beta) \\ 0 & 1 & 0 \\ -\sin(\beta) & 0 & \cos(\beta) \end{vmatrix}$ —matrix of rotation by angle β relative to the axis y,

$R_z(\gamma) = \begin{vmatrix} \cos(\gamma) & -\sin(\gamma) & 0 \\ \sin(\gamma) & \cos(\gamma) & 0 \\ 0 & 0 & 1 \end{vmatrix}$ —matrix of rotation by angle γ relative to the axis z.

We introduce the vector $S_i = (\mathbf{p}_{B_i} - \mathbf{p}_{A_i}) / L_i$, defining the ort of the z-axis of the coordinate systems of the lower and upper half-legs.

Knowing the orientation of the platform $\mathbf{R}_p$, base orientation $\mathbf{R}_o$ and initial orientations of hinges $R^0_{A_i}$, $R^0_{B_i}$ determined by design features, we calculate the orientations of the hinges of the base $\mathbf{R}_{0A_i}$ and the platform $\mathbf{R}_{0B_i}$

$$\mathbf{R}_{0A_i} = \mathbf{R}_o R^0_{A_i},\ \mathbf{R}_{0B_i} = \mathbf{R}_p R^0_{B_i}. \tag{6}$$

Kinematics equations for the lower half-legs

$$\mathbf{R}_{A_i} = R_x\left(\alpha_{A_i}\right) R_y\left(\beta_{B_i}\right), S_{A_i} = \mathbf{R}^T_{0A_i} \cdot S_i = \mathbf{R}^3_{A_i}. \tag{7}$$

where $\mathbf{R}^3_{A_i}$ is the third column of the matrix $\mathbf{R}_{A_i}$.

Similarly, for the upper half-leg –

$$\mathbf{R}_{B_i} = \left(R_y\left(\beta_{B_i}\right) R_x\left(\alpha_{B_i}\right) \right)^T, S_{B_i} = \mathbf{R}^T_{0B_i} \cdot S_i = \mathbf{R}^3_{B_i}. \tag{8}$$

where $\mathbf{R}^3_{B_i}$ is the third column of the matrix $\mathbf{R}_{B_i}$.

Using the definitions of the rotation matrices, it is easy to calculate the angles

$$\alpha_{A_i} = \operatorname{asin}\left(S^1_{A_i}\right), \tag{9}$$

where $S_{A_i}^1$ is the first element of the column S_{A_i},

$$\beta_{A_i} = \operatorname{atan2}\left(\frac{-S_{A_i}^2}{\cos(\alpha_{A_i})}, \frac{S_{A_i}^3}{\cos(\alpha_{A_i})}\right), \tag{10}$$

where $S_{A_i}^2, S_{A_i}^3$ are second and third elements of the column S_{A_i}, atan2 is the function of two-argument arctangent, similarly we obtain the angles

$$\beta_{B_i} = \operatorname{asin}\left(S_{B_i}^1\right), \tag{11}$$

where $S_{B_i}^1$ is the first element of the column S_{B_i},

$$\alpha_{B_i} = \operatorname{atan2}\left(\frac{-S_{B_i}^2}{\cos(\beta_{B_i})}, \frac{S_{B_i}^3}{\cos(\beta_{B_i})}\right), \tag{12}$$

where $S_{B_i}^2, S_{B_i}^3$ are second and third elements of the column S_{B_i}.

Finally, from equation (5) we get

$$R_z(\gamma_i) = \left(\mathbf{R}_{A_i} R_x(\alpha_{A_i}) R_y(\beta_{A_i})\right)^T \mathbf{R}_{B_i} \left(R_y(\beta_{B_i}) R_x(\alpha_{B_i})\right)^T \tag{13}$$

and define the angle

$$\gamma_i = \operatorname{atan2}\left(R_z^{2,1}, R_z^{1,1}\right), \tag{14}$$

where $R_z^{2,1}, R_z^{1,1}$ – the second and first elements of the first column of the matrix $R_z(\gamma_i)$.

Calculated from relations (4), (14) the length L_i and angle γ_i are applicable for calculating the total angle of rotation of the screw pair

$$\alpha_{dr_i} = 2\pi \frac{L_i - L_0}{h} \quad \gamma_i, i=1...6, \tag{15}$$

where L_i is the length of i-leg, corresponding to the desired position and orientation of the moving platform of the hexapod, L_0 is the leg length in the "zero" position, h is the pitch of the screw pair, γ_i is the angle of the axial rotation.

Thus, equations (1) – (15) give a solution to the extended inverse kinematics problem for each leg of a hexapod, which can be represented as a system of nonlinear algebraic equations

$$\boldsymbol{\alpha} = H(\mathbf{q}). \tag{16}$$

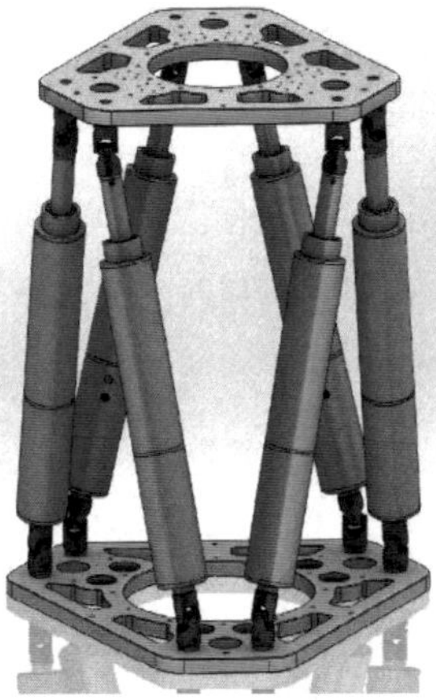

Figure 5. Solid state hexapod model.

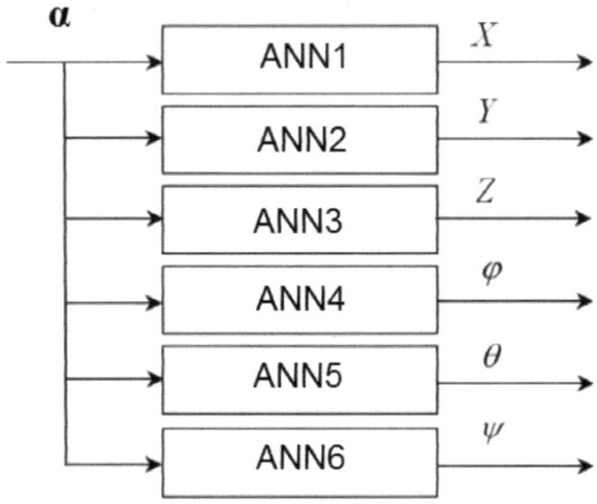

Figure 6. Neural network solution scheme of the direct kinematics problem.

The geometrical parameters needed to solve the extended inverse kinematics problem, the position of the hinges $\mathbf{p}_{A_i}, \mathbf{p}_{B_i}$ and the initial orientation of the hinges $R^0_{A_i}$, $R^0_{B_i}$ are found from the solid-state model of the hexapod shown in figure 5.

The solution of the forward kinematics problem can be represented by finding the inverse function of (16)

$$\mathbf{q} = H^{-1}(\boldsymbol{\alpha}) . \tag{17}$$

Since the analytical solution of problem (17) is difficult, it is proposed to use the ANN apparatus for approximating (17) the solution with a given accuracy.

3. Neural network solution of the forward kinematics problem

In this paper, for approximation (17), a separate neural network approximation of the components of the vector $\mathbf{q}$ is used, the scheme of which is shown in figure 6.

The use of this approach allows us to speed up and simplify the process of learning and synthesis of ANN for solving the FKP with a given accuracy, during which we determine: the algorithm for the formation of the training sample, the network architecture (network type, number of hidden layers, number of neurons in each layer), learning algorithm, matrix network implementation.

It is convenient to create and train a neural network suitable for the implementation of problems of approximation of multidimensional nonlinear functions using modern mathematical modeling environments, which include the Matlab Neural Network Toolbox [16] extension package. The toolkit of this package is used in the design of neural networks to solve the FKP.

The synthesis procedure was performed for six ANNs approximating the components of the vector $\mathbf{q}$; this paper demonstrates the results for the ANN1 network.

Three implementations of the structures of cascade neural networks of direct distribution for the FKP approximation were studied:

– a cascade network with one hidden layer CFNN1 (Cascade Feedforward Neural Network), which architecture is shown in figure 7,
– a cascade network with two hidden layers CFNN2, the architecture of which is shown in figure 7,
– a cascade network with three hidden layers CFNN3, the architecture of which is shown in figure 8.

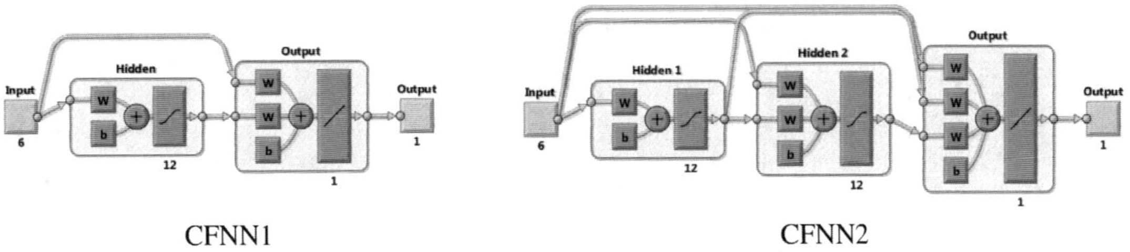

Figure 7. Structures of cascade neural networks of direct distribution CFNN1 and CFNN2.

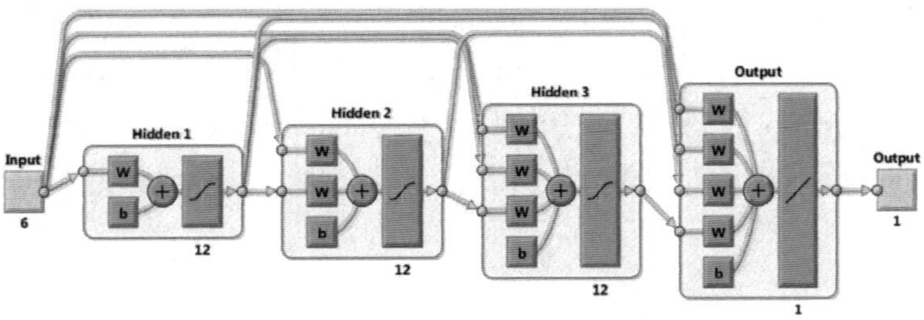

Figure 8. The structure of the cascade neural network of direct distribution CFNN3.

Based on the solution of the extended inverse kinematics problem (1) – (15), a training sample was formed as a linearly distributed array for the values of the hexapod working range. The array dimension for the training sample was $6 \times P^6$ (P=7). For these structures, ANN training was performed by the Levenberg-Marquardt method. The estimated accuracy of the approximation on the test sample is obtained Δ_{max} is the maximum absolute error of the approximation. The results of the accuracy estimation are summarized in table 1.

Table 1. Estimation of the approximation accuracy of cascade network architectures.

Network Type	Number of Neurons in the Hidden Layer	Δ_{max}, m
CFNN1	6	$6{,}86 \cdot 10^{-4}$
CFNN1	12	$1{,}31 \cdot 10^{-4}$
CFNN1	18	$1{,}04 \cdot 10^{-4}$
CFNN1	36	$1{,}81 \cdot 10^{-5}$
CFNN2	6	$7{,}76 \cdot 10^{-5}$
CFNN2	12	$1{,}08 \cdot 10^{-5}$
CFNN2	24	$1{,}4 \cdot 10^{-6}$
CFNN3	6	$9{,}15 \cdot 10^{-5}$
CFNN3	12	$3{,}45 \cdot 10^{-6}$
CFNN3	18	$1{,}25 \cdot 10^{-6}$

Analysis of the obtained results allows us to conclude that in order to achieve the required accuracy of the FKP solution for one coordinate in a given operating range, it is sufficient to use the CFNN2 network with 24 neurons in hidden layers or CFNN3 with 12 neurons in hidden layers.

For the implementation of the ANN, the CFNN2 structure was selected, for which the matrix expression system was obtained.

$$\begin{cases} \mathbf{y}_1 = \mathbf{K}_{inp}(\mathbf{x} - \mathbf{x}_0) + \mathbf{x}_{min} \\ \mathbf{y}_2 = F_{th}(\mathbf{W}_{11}\mathbf{y}_1 + \mathbf{b}_1) \\ \mathbf{y}_3 = F_{th}(\mathbf{W}_{21}\mathbf{y}_1 + \mathbf{W}_{22}\mathbf{y}_2 + \mathbf{b}_2) \\ \mathbf{y}_{out} = \mathbf{W}_{31}\mathbf{y}_1 + \mathbf{W}_{32}\mathbf{y}_2 + \mathbf{W}_{33}\mathbf{y}_3 + \mathbf{b}_3 \\ \mathbf{y} = \mathbf{K}_{out}(\mathbf{y}_{out} - \mathbf{y}_{min}) + \mathbf{y}_o \end{cases}, \quad (18)$$

Where matrices $\mathbf{W}_{11}, \mathbf{W}_{21}, \mathbf{W}_{31}, \mathbf{W}_{22}, \mathbf{W}_{32}, \mathbf{W}_{33}$ set the synaptic weights in the neural network layers, $\mathbf{b}_1, \mathbf{b}_2, \mathbf{b}_3$ are displacements in each layer, which values are obtained during the training of the ANN; $\mathbf{K}_{inp}, \mathbf{x}_0, \mathbf{x}_{min}$ are matrices of normalization coefficients of input $\mathbf{x}$ values (values of foot length vector L); $\mathbf{K}_{out}, \mathbf{y}_0, \mathbf{y}_{min}$ matrices of recovery coefficients for the output of the network $\mathbf{y}$ are determined by normalizing the training samples; F_{th} the activation function of the hyperbolic tangent is found from the expression

$$F_{th}(z) = \frac{2}{1+e^{-2z}} - 1. \quad (19)$$

4. Evaluation of the implementation of the FKP neural network solution in hexapod control systems

Currently, a digital hexapod control system based on the use of the 32-bit ARM RISC microcontroller K1986BE1T manufactured by JSC "PKK Milandr" [17] is implemented on the basis of the research laboratory of robotic and mechatronic systems of BSTU "VOENMECH". The microcontroller operates at a clock frequency of up to 144 MHz. It contains 128 Kbytes of one-time programmable ROM of programs with ECC and 48 Kbytes of ECC RAM.

The software for solving the hexapod FKP was made in the Keil μVision 5 environment. The ARM C / C ++ Compiler (armcc) compiler was used to assemble the microcontroller programs. Test results are presented for compilation using optimization (O2 level).

Figure 9. Hexapod drive control unit.

The implementation of kinematic problems in two versions of the representation of floating-point numbers: float (single precision) and double (double precision) is compared.
Based on the results of the synthesis of the cascade neural network (18) and (19), the algorithms for solving the FKP in the high-level C ++ language are obtained, performance and accuracy estimates (maximum positional Δ_{max} and angular approximation E_{max} errors) are presented in table 2.

Table 2. Evaluation of the implementation of the FKP neural network solution.

Implementation Type	T, ms	Δ_{max}, m	E_{max}, rad
float	18,8	$3{,}7{\cdot}10^{-6}$	$1{,}1{\cdot}10^{-5}$
double	28,3	$4{,}1{\cdot}10^{-6}$	$1{,}1{\cdot}10^{-5}$

Note that the implementation of the synthesized networks in the format of numbers with double precision double time of the algorithm execution increases, but the accuracy does not.

5. Conclusion

This paper presents the solution of the extended inverse problem of kinematics of a hexapod with screw pairs. Based on this, a neural network solution of the forward kinematics problem is obtained. We note the simplicity of the structure of the synthesized networks. For neural network approximation with a given accuracy, it is recommended to use cascade networks of direct distribution with two or three hidden layers based on the hyperbolic tangent as activation functions with a total number of neurons of no more than 48. In the space control hexapod control system under study, based on the domestic ARM microcontroller architecture, the execution time of the algorithms meets the requirements of positional control. For a given accuracy of positioning and orientation of the hexapod, the use of numerical algorithms with the implementation in the single-precision float format is recommended. Thus, the conducted studies have shown the effectiveness and expediency of using an artificial neural network apparatus to solve the forward kinematics problem of a hexapod with screw gears in leg drives.

Acknowledgments
This research work was carried out in accordance with the Resolution of the Russian Federation Government of April 9, 2010 No. 218 (PROJECT 218) within the framework of research and development activities carried out with the financial support of the Ministry of Education and Science of the Russian Federation (contract of December 1, 2015 No. 02.G25.31.0160). The research works are carried out by the by the Chief Executive Officer of the Research and Design - FSBI of HE BSTU "VOENMECH" named after D.F. Ustinov.

Reference

[1] Lung-Wen T 1999 *Robot Analysis, Mechanics of Serial and Parallel Manipulators* (New York: Wiley) p 520
[2] Merlet J P 2006 *Parallel Robots* (Springer) p 420
[3] Korotkov E B, Matveev S A and Jakovenko N G 2016 The ways of increasing quality rating of control system of machine with such strut-type structure as hexapod and tripod using Russian and global available electronic component *Questions of Radio Electronics* **8** 85–91
[4] Stewart D A 1965 Platform with Six Degrees of Freedom *Proceedings of the Institution of Mechanical Engineers* **180** 371–85
[5] Zhukov Yu A, Korotkov E B and Slobodzjan N S 2017 Radiation resistant microcontroller with Cortex-M4F core based control system of parallel kinematics mechanism designed for spacecraft's onboard devices movements *Questions of Radio Electronics* **7** 48–54
[6] Liu K, Fitzgerald J and Lewis F 1993 Kinematic Analysis of a Stewart platform manipulator *IEEE Transactions on Industrial Electronics* **40** 282–293
[7] Zhukov Yu A, Korotkov E B and Slobodzjan N S 2017 Simulation model of hexapod control systems with linear stepper drives *Questions of Radio Electronics* **7** 35–41
[8] Campa R, Bernal J and Soto I 2016 Kinematic modeling and control of the hexapod parallel robot *Proceedings of American Control Conference* 1203–1208
[9] Cardona M N 2015 A New Approach for the Forward Kinematics of General Stewart-Gough

Platforms *Proceedings of the 2015 IEEE 35th Central American and Panama Convention* 1–6

[10] Parikh P J and Lam S Y 2005 A Hybrid Strategy to Solve the Forward Kinematics Problem in Parallel Manipulators *IEEE Transactions on Robotics* **21** **(1)** 18–25

[11] Xinxin Guo, Guixi Ke, Fengwu Zheng and Lijie Zhang 2013 Forward Kinematics Analysis of the Stewart Parallel Platform Based on the Elman Recurrent Network *Proceedings of 5th International Conference on Intelligent Human-Machine Systems and Cybernetics* **2** 175 – 177

[12] Mohammed A. and Li S 2016 Dynamic Neural Networks for Kinematic Redundancy Resolution of Parallel Stewart Platforms *IEEE Transactions on Cybernetics* **46** (7) 1538–1550

[13] Zhukov Y A, Korotkov E B, Moroz A V, Zhukova V V and Abramov A M 2018 A control of hexapod based on the neural networks solution for forward kinematics *IOP Conf. Series: Mater. Science and Eng.* **441** 012002

[14] Zubizarreta A, Larrea M, Irigoyen E, Cabanes I and Portillo E 2018 Real time forward kinematic problem computation of the 3PRS robot using neural networks *Neurocomputing* **271** 104–114

[15] Fu K S, Gonzalez C S and Lee G 1987 *Robotics: Control, Sensing, Vision, and Intelligence* (Mcgraw-Hill Book Company) p 580

[16] Neural Network Toolbox Matlab [Official web site of MathWorks distributor in Russia and the CIS] 2019 Available at: https://matlab.ru/products/neural-network-toolbox/ (accessed: 15.05.2019) (In Russ.)

[17] Microprocessor 1986BE1T [Official web site of MILANDR GROUP] 2019. Available at: http://ic.milandr.ru/products/mikrokontrollery_i_protsessory/ (accessed: 15.05.2019). (In Russ.)

AUTHOR INDEX